The Biology of Aquatic and Wetland Plants

Aquatic plants play a critically important role in maintaining ecosystem health. They are natural biological filters in freshwater and estuarine wetlands; they contribute to the reproductive success of many organisms, some of which are harvested for food; they assist in flood control; and they are prominent elements in the aesthetics and recreational use of freshwater and estuarine habitats. Despite this globally recognized importance, wetlands have faced and continue to face threats from the encroachment of human activities. *The Biology of Aquatic and Wetland Plants* is a thorough and up-to-date textbook devoted to these plants and their interactions with the environment. The focus is on botanical diversity from the perspective of evolutionary relationships, emphasizing the role of evolution in shaping adaptations to the aquatic environment. By incorporating recent findings on the phylogeny of green plants, with special emphasis on the angiosperms, the text is broadly useful for courses in plant biology, physiology, and ecology. Additionally, a chapter on population biology and evolutionary ecology complements the evolutionary backdrop of hydrophyte biology by examining the details of speciation and applications of modern genetic approaches to aquatic plant conservation.

Key Features

- Synthesizes recent and seminal literature on aquatic and wetland plants
- Emphasizes evolutionary history as a factor influencing adaptations to the wetland environment
- Provides a global perspective on plant diversity and threats facing wetland ecosystems
- Highlights research needs in the field of aquatic and wetland plant biology
- Includes 280 figures, with more than 300 color photographs, and 41 tables to provide ease of access to important concepts and information

The Biology of Aquatic and Wetland Plants

Gary N. Ervin

CRC Press
Taylor & Francis Group
Boca Raton London New York

CRC Press is an imprint of the
Taylor & Francis Group, an **informa** business

Designed cover image: © Gary Ervin

First edition published 2023
by CRC Press
6000 Broken Sound Parkway NW, Suite 300, Boca Raton, FL 33487–2742

and by CRC Press
4 Park Square, Milton Park, Abingdon, Oxon, OX14 4RN

CRC Press is an imprint of Taylor & Francis Group, LLC

© 2023 Taylor & Francis Group, LLC

Library of Congress Cataloging-in-Publication Data
Names: Ervin, Gary N., author.
Title: The biology of aquatic and wetland plants / Gary N. Ervin.
Description: First edition. | Boca Raton, FL : CRC Press, 2023. | Includes bibliographical references and index.
Identifiers: LCCN 2022050213 (print) | LCCN 2022050214 (ebook) | ISBN 9781482232042 (hbk) | ISBN 9781032465395 (pbk) | ISBN 9781315156835 (ebk)
Subjects: LCSH: Aquatic plants. | Aquatic plants—Ecology. | Wetland plants. | Wetland plants—Ecology.
Classification: LCC QK102 .E78 2023 (print) | LCC QK102 (ebook) | DDC 581.7/6—dc23/eng/20230209
LC record available at https://lccn.loc.gov/2022050213
LC ebook record available at https://lccn.loc.gov/2022050214

ISBN: 978-1-482-23204-2 (hbk)
ISBN: 978-1-032-46539-5 (pbk)
ISBN: 978-1-315-15683-5 (ebk)

DOI: 10.1201/9781315156835

Typeset in Times
by Apex CoVantage, LLC

Access the Support Material: www.routledge.com/9781032465395

This work is dedicated to the memories of Robert and Carol Wetzel in gratitude for the support they provided to so many young aquatic scientists, the present author included.

Contents

Preface

I based this book on an aquatic botany class that I have taught eight times during my two decades at Mississippi State University. That class and the book have benefitted from my experiences as a faculty member at a land-grant university in a state heavily reliant on natural resources for the livelihoods of its people. Specifically, the text has benefitted from my need at times to teach such subjects as plant physiology, general botany, and ecology and from several years of teaching a class that I call "Living with Global Change." It also has benefitted from my involvement in research on invasive plant and insect species in both terrestrial and aquatic habitats. Perhaps most of all, though, this book has benefitted from timing. I'm writing this at a point in my career where I feel an urgent need to do work that has meaning for the soon-to-be ten billion people on our planet, at a time when my own research program is shifting to fill that need, and at a time when I have a bit of freedom to take a broad look at what we currently know and what we need to learn about aquatic and wetland plants.

When I first taught my aquatic botany class, I was fortunate to have the text that Julie Cronk and Siobhan Fennessy had just published, in the year 2001, *Wetland Plants: Biology and Ecology*. Theirs was an ambitious work covering much of the same content you'll find in the chapters that follow here, with the addition of chapters covering applied topics such as wastewater treatment, wetland restoration, and the use of wetland plants as biological indicators. Because some of those latter topics have been and continue to be covered in the excellent *Wetlands* text by William Mitsch and the late James Gosselink (recently in its fifth edition), I opted to only touch on them here when directly pertinent to other topics being covered.

In exchange for covering a narrower breadth of applied ecology of aquatic and wetland plants, I have attempted to provide the reader with a deeper understanding of the plants' biology. One of the key elements of the approach I have taken in this regard is to attempt to place the plants and their biology in an explicitly evolutionary context. Beginning in Chapter 2, with an overview of some of the major taxonomic groups of plants encountered in wetland ecosystems, I view botanical diversity in wetlands from the perspective of phylogenetic relationships of the plants, as well as timing of the divergence and diversification of major groups. The information here incorporates some of the most recently published findings on the phylogeny of green plants, with special emphasis on the angiosperms. In later chapters, the importance of higher taxonomic levels (orders and families) is highlighted when relevant, such as with the prevalence of fully aquatic plants in the order Alismatales or the pervasiveness of **monocot** taxa among examples of important wetland plant adaptations. A chapter on population biology and evolutionary ecology (Chapter 9) complements this evolutionary backdrop of hydrophyte biology by examining the details of speciation and application of modern genetic approaches to aquatic plant conservation.

I also have attempted to point out, throughout the text, areas that appear, based on my review of the literature, to need further study, as well as areas that have benefitted from relatively recent discoveries. For example, the phytoglobin-nitric oxide cycle is a very interesting adaptation of plants to dealing with hypoxia that has only recently been investigated in wetland plants. A mode of aerenchyma formation among wetland-adapted species (expansigeny) that has not received significant attention in the past is discussed, along with a potential source of information that could be used to explore the relative importance of this process among macrophytes. A couple of exciting findings in the area of pollination are the discovery that multiple species of aquatic invertebrates are capable of pollinating seagrasses and that some species of bladderwort (*Utricularia*) may be bird pollinated. The fact that there are so few known examples of these processes suggests a fertile area for basic natural history studies in pollination ecology. Another area where substantial discoveries could be made concern the role and ecological importance of pathogens and parasites of wetland plants, especially viruses. Although these agents have been investigated to a small degree as potential biological control agents, much more could be learned about the ecology of their interactions with wetland plants.

A final aspect of this text that I feel deserves mention here is my attempt to emphasize, throughout, the many threats that wetland ecosystems currently face. Wetlands are critically important ecosystems that provide humans invaluable ecological services such as flood mitigation, storm surge abatement, and purification of surface waters. Despite their importance, wetlands have faced threats from the encroachment of human activities throughout recorded human history. With our population sitting near eight billion currently and an additional two billion expected in the next three decades, those threats will only increase. Foremost among the threats that I address herein are alteration of natural hydrologic cycles, introduction of invasive species, **eutrophication**, and climate change. The threat posed by these activities for wetlands are unlikely to subside in the near term, heightening the need for a thorough understanding of the biology of aquatic and wetland plants as a means of optimally conserving wetlands and the services they provide. It is my hope that the connections I have attempted to make between biology of the plants and their responses to **anthropogenic** activities will help in this regard.

As with any work such as this, I am indebted to all those who have paved the way before me and who have provided assistance, encouragement, and inspiration during the writing process itself. I mentioned earlier the texts by Cronk and Fennessy and Mitsch and Gosselink that have been invaluable tools during my 25 or so years as a wetland biologist.

Similarly, Jessica Gurevitch, Sam Scheiner, and Gordon Fox produced a text (*The Ecology of Plants*) that I used for a decade in my own plant ecology course. *The Ecology of Plants* is one of my most-recommended resources for students with an interest in plant ecology, and those familiar with that text will certainly see portions of this book that have drawn on the presentation style of Gurevitch and colleagues.

Another critically important piece of work that has been a regular resource for writing this book and carrying out my own research is the series of limnology texts authored by my late mentor, Robert (Bob) Wetzel. Wetzel's 2001 *Limnology: Lake and River Ecosystems* has been within arm's reach during the entire process of writing this book, and you will see his work referenced throughout. This is largely due to his having advised approximately 70 graduate students and postdocs in essentially every aspect of aquatic biology. His greatest contribution to this book, however, was his willingness to encourage me to pursue doctoral studies under his guidance and to further ensure that I entered into academia as a profession. On an evaluation of one of my graduate seminar presentations, he wrote something along the lines of "It would be a shame if you did not go into teaching as a profession." He was right, and the longer I teach, the more I enjoy it. I would be grossly remiss here if I did not also mention Bob's late wife, Carol, who was a huge part of the lives of all of Bob's graduate students. Both Bob and Carol have been in my thoughts regularly throughout the time I have spent preparing for and then producing this book. I also am grateful to their son, Paul Wetzel, who has helped motivate me to complete this effort on more than one occasion.

Others from my graduate and undergraduate training who have inspired portions of this book include Art Benke, Bill Darden, Mark Dedmon, Laura Gough, Jim Grace, Bob Haynes, Eric Roden, Keller Suberkropp, and Amy and Milt Ward. The following colleagues and former students provided helpful reviews on portions of the book and/or engaged me with conversations about its content during the writing: Jason Bried, Matt Brown, Ryan Folk, Brook Herman, Heather Jordan, Tatiana Lobato de Magalhães, Lucas Majure, Carroll Mann, Nancy Reichert, Cory Shoemaker, Gray Turnage, Mark Welch, and Doug Wilcox. Ben Blassingame, Evelyn DiOrio, Jacob Hockensmith, Adrián Lázaro-Lobo, David Mason, Andy Sample, Sam Schmid, and Cory Shoemaker all were students during the development of this manuscript and all helped me in various ways to think through parts of the book during our conversations and research collaborations.

Guillermo Logarzo and Fernando McKay, colleagues from the Fundación para el Estudio de Especies Invasivas (FuEDEI; Buenos Aires, Argentina), provided photos for Chapters 2 and 3 and suggested the Argentine Iberá wetland area that was included in Chapter 3. Guillermo and another scientist from FuEDEI, Lora Varone, helped provide access to many wetlands across northern Argentina during 2008–2010, when I was fortunate enough to collaborate with

them on studies to better understand the ecology of their *Opuntia*–cactus moth system. Although not directly related to the research we were doing at the time, the opportunity to see those wetlands helped provide perspective on some of the invasive plant issues my collaborators and I study here in the United States today.

Pam Soltis provided helpful answers to questions I had about parts of Chapter 2, in addition to sending some useful reference material. Adrián Lázaro-Lobo suggested the inclusion of the Doñana wetlands of Spain, helped clarify some confusion about Spanish terminology during the writing of Chapter 3, and provided helpful, enthusiastic reviews of several chapters of the manuscript. Tatiana Lobato de Magalhães provided some amazing photos from her research systems in Querétaro, México, along with a review of Chapter 9. James Seago engaged me with discussions about the details of aerenchyma formation, provided some very helpful drawings of the early stages of aerenchyma formation, and donated photos of aerenchyma in *Hydrocotyle ranunculoides*. Others providing photos for inclusion here include Matthew Abbott—photos of *Sarracenia* from Elgin Air Force Base in Florida; Charles Bryson—photos of *Carex* species; John Gwaltney—photos of *Carex* from his southeasternflora.com website; Bartosz Jan Plachno—a micrograph from the interior of the floral tube of a bird-pollinated species of *Utricularia*; and Miguel Álvarez, of Sevilla, Spain—photos of waterbirds at Doñana National Park.

A wonderful friend and colleague, Travis Marsico, went above and beyond in his review of early versions of the first eight chapters of the book. He not only critiqued those chapters, but also used them to teach a course on wetland plants at Arkansas State University and required all the students to provide written critiques of their own. Connor Harris, Mathew Jones, Brendan Kosnik, Cameron Rhoden, Brody Ridge, Courtney Smith, Emily Sullivan, Chenoa Summers, Hayden Towner, and Amanda Trusty provided comments on those first eight chapters. The previous year, my own students provided critiques of the first seven chapters. Those students included Ben Blassingame, Kristen Cone, Quinn Cooley, Nicholas Engle-Wrye, Shelby Grice, Breanna Hasty, Damien Henderson, Conner Owens, Andy Sample, Nicholas Stewart, Cade Thornton, and Arielle Wynn.

Most importantly, however, my sweet wife, Lesia, who initially suggested I undertake this project, provided patient encouragement all along the way, even when the writing continued much later into the evenings than she would have liked. She also served as a soundboard for many of the areas I had difficulty articulating and helped with the finer details of interpreting many of the figures herein. My son Garrett, who was eight months old when I arrived at Mississippi State, spent many hours hanging out in the laboratory, greenhouse, and my office, and he has logged more time with me in the field than any of my graduate students. He is included here in a photo alongside *Eichhornia crassipes*, and he provided helpful critiques of many of my illustrations. His artistic eye is greatly appreciated. My stepsons Chandler and Lee, who each spent some of their summer vacations as student workers on aquatic plant management projects, likewise were encouraging and often discussed parts of the subject material, helping me refine the presentation of this material for a more general audience. I also owe a great deal of gratitude to my mother and sister, along with all the other wonderful women on both sides of my family, who encouraged me throughout my childhood and supported me during all the difficult times in my life. My mom and sister continue to be a regular source of love, encouragement, and inspiration.

Finally, John Oldshue, who is as near a brother as I could ever hope for, deserves enormous gratitude for his role in all of this. John encouraged me to enroll in the biology undergraduate program at the University of Alabama about two years after I dropped out of the forestry program at Mississippi State University. John then introduced me to my undergraduate advisor Bill Darden, who introduced me to Bob Wetzel, who turned me into a wetland ecologist. Many years later, John also introduced me to my wife Lesia, who, again, initially suggested producing this book.

As they say, all the gratitude is due to those mentioned earlier; all the fault lies with me. I undertook this work on my own, encouraged by Lesia, inspired by Wetzel's example, and with some measure of advisement from my colleague Ronn Altig. Ronn advised me never to write a book but added that if I couldn't follow that advice, to write the book myself. All errors and omissions thus are no one's fault but my own. Most will be unintentional, and for those, I ask your forgiveness. Where I have intentionally omitted things that readers felt important, I also ask forgiveness and hope that those readers will understand that this particular story, at this particular time, could not have been written differently.

About the Author

Gary N. Ervin received his BS (1996) and PhD (2000) in biological sciences from the University of Alabama, in Tuscaloosa. During his undergraduate and doctoral research, he studied the ecology of freshwater wetland plants, publishing several papers on the ecology of the rush, *Juncus effusus*. After completing his doctoral studies, Dr. Ervin held a postdoctoral research position in the Department of Entomology at the University of Arkansas, Fayetteville. While in Arkansas, he studied plant defense responses to insect herbivores, with an emphasis on oxidative biochemistry of plant–insect interactions. Dr. Ervin began his present faculty position in the Department of Biological Sciences at Mississippi State University in 2001. His research program at Mississippi State University has been focused on better understanding mechanisms influencing plant colonization and persistence, including research on species invasions.

Dr. Ervin and his collaborators have worked with invasive plants in forests, wetlands, and prairies, but also have studied interactions between plants and insect biocontrol agents. During his two decades on the faculty at Mississippi State, Dr. Ervin has taught courses on plant ecology, invasion ecology, plant biology, aquatic botany, and global change, and he has published dozens of scientific articles on both terrestrial and wetland plant ecology. Dr. Ervin is a long-time member of the Society of Wetland Scientists (SWS) and is certified as a professional wetland scientist by that organization. At the time of writing this text, he served as a member of the executive committee of the SWS South Central Chapter, on the mentoring committee of the SWS Latin American and Caribbean student mentoring program (HumMentor), and on the SWS education section editorial board for the online resource, *Foundations in Wetlands Science*.

1 Overview

Over the years, I have developed a deep connection to and a love for nature. Many of the experiences that have contributed to this connection have involved plants, the water, or both. As time passed and I learned more about how the natural world works, wetlands became a natural focus for my personal as well as professional pursuits. I eventually turned my love of wetlands and wetland plants into a career as a wetland scientist, and that ultimately resulted in my writing this book as a tool to teach others about the roles that plants play in the ecological functioning of wetlands and aquatic habitats.

As one begins to explore wetlands, it quickly becomes apparent that they are incredibly diverse, and they provide a wealth of services and benefits to us, both as individuals and as a society. We will see, as we progress through this book, that the diversity within and among wetlands comes about in part because of their diverse geography and geology, diverse connections with sources of water above, within, and below the soil, and, of course, the diversity of plants that inhabit aquatic and wetland ecosystems.

1.1 IMPORTANCE OF AQUATIC AND WETLAND PLANTS

Although aquatic and wetland plants share an affinity for growing in or near water, they are represented by an incredible diversity of life forms, even among the relatively small number of species that are fully adapted to aquatic life. Some curious examples are the large *Victoria* water lilies, shown in Figure 1.1 supporting a 34 kg child. In fact, the leaf of a *Victoria* lily growing in a London botanical park once held 194 kg of gravel before it took on water and sank (Seaman 1891). On the other end of the size spectrum are the incredibly tiny species of duckweeds (*Lemna*, *Wolffia*), some of which are barely one millimeter in length and are the smallest known flowering plants (Figure 1.1).

The simple fascination with diversity among aquatic plants has led many to develop an interest in water gardening, or the horticultural cultivation of aquatic plants for aesthetic purposes. Those plants also are of interest to aquarium owners for aesthetic purposes, but also for the benefits they provide to fish and other aquatic animals (e.g., oxygenation of water, habitat for insects and other invertebrates). In the United States alone, the horticultural trade of aquatic and wetland plants was a $1 billion industry in 2001 (Kay and Hoyle 2001). The sale of these plants to aquarium enthusiasts also constitutes a major international market, with $170 million of tropical fish and aquarium plants having been imported into Florida in 1993 (Maki and Galatowitsch 2004; Peres et al. 2018). One unfortunate consequence of this interest, however, is that imports and

trade in these plants result in numerous introductions of problematic plants, the control of which costs more than $100 million per year in the US (United States; Maki and Galatowitsch 2004). As impressive as these numbers are, perhaps the most important social and economic contribution of wetland plants can be attributed to a single species, rice (*Oryza sativa*; Figure 1.1). More than 400 million tons of rice are produced annually worldwide, serving as a principal source of calories for roughly half the world's population (Calpe 2006) and valued at approximately $160 billion (US) per year (www.fao.org/economic/RMM).

These plants play a vital role in the ecology of aquatic and wetland ecosystems, serving as the basal food resource for everything from bacteria and fungi to fish, waterfowl, and mammals such as beaver, nutria, and humans. Plants also serve as the structural component of habitat for many of these same groups of animals. For example, beavers (Figure 1.2) feed on a variety of wetland plants, from trees such as willow (*Salix* species), ash (*Fraxinus* species), and many others, to grasses, honeysuckle (*Lonicera* species), and water lilies (*Nymphaea* species; Roberts and Arner 1984). However, beavers also use trees and shrubs as structural reinforcement for the lodges where they live and the dams they use to create the ponds where they travel and forage (Figure 1.2).

An additional important contribution made by aquatic and wetland plants is the role they play in the purification of surface waters. This critical societal benefit has become increasingly important, on a per-acre or per-hectare basis, as the loss of wetland ecosystems worldwide continues concomitantly with global human population growth. This topic will be revisited within the appropriate sections of later chapters, particularly in Chapter 7, where the influence of plants on biogeochemical cycling is discussed.

1.2 WHAT THIS BOOK IS ABOUT

A comprehensive understanding of plant biology allows one to answer the question, "Why does *that plant* grow *there*?" This question appears quite simplistic, but a thorough answer involves an understanding of multiple, interacting factors that influence an individual plant's conception, development, dispersal, and interaction with its environment and neighboring species (not only of plants but other taxa as well). This understanding ultimately also requires knowledge of population biology and factors operating at multiple scales of space and time that enable individual populations to persist across generations. Taken together, all these factors will determine why a given plant may occur at its particular location, at the particular time one finds it there. My goal with this book is that the reader (you)

DOI: 10.1201/9781315156835-1

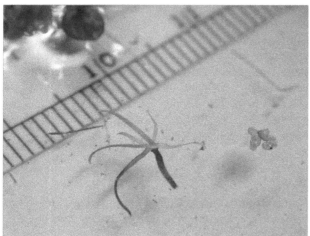

FIGURE 1.1   A Tiny Bit of Wetland Plant Diversity. Upper: Leaves of the *Victoria* Water Lily are Buoyant Enough to Support a Child, or Hundreds of Kilograms of Gravel! Lower Left: *Wolffiella* (Left) and *Wolffia* (Right, the Small Green Spheres) are the Tiniest of Flowering Plants. Lower Right: Rice (*Oryza Sativa*) Feeds Almost Half the Human Population. Photo of *Victoria* from Seaman (1891).

will be able to answer this question insightfully, drawing on multiple aspects of plant biology, with particular relevance to each individual's own reasons for asking, "Why does this plant grow here?"

John Harper (1977) summarized the aforementioned assertions about plant biology in an elegant illustration of plant "population behavior," which he used quite effectively to illustrate the individual and interactive roles of abiotic and biotic factors in determining the species of plants that coexist in a particular location (Figure 1.3). I have borrowed from Harper's idea and take a similar approach to build iteratively upon individual components

FIGURE 1.2 Beavers (*Castor* Species, Top Left) are Renowned Wetland Engineers. They Use Aquatic and Wetland Plants as Both Food and Building Material. Wood from Wetland or Riparian Trees is Used to Construct Lodges (Top Right) and as Structural Reinforcement for their Earthen Dams (Bottom). Photo of Beaver Taken by Neal Herbert of Yellowstone National Park (USA; Obtained from Flickr: flickr.com/photos/yellowstonenps).

of the environment (abiotic and biotic) to develop an integrated understanding of the biology of aquatic and wetland plants.

To set the stage for this process, some terminology must first be addressed, and the book's title provides an excellent starting point. This text is about the *biology* of *aquatic* and *wetland plants*. I hope the preceding paragraphs, along with Figure 1.3, have provided a general overview of what is meant here by "biology." Explanation of what qualifies as an aquatic plant or a wetland plant, however, requires a bit more background. We begin by examining the habitats themselves.

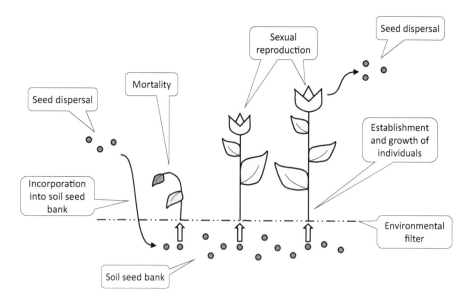

FIGURE 1.3   A Selection of Critical Aspects of Plant Biology that Influence the Occurrence of Plant Individuals and Species across Time and Space. Modified from Figures in Harper (1977).

1.2.1   WETLAND HABITATS DEFINED

Wetland habitats are complicated to define, because of their diversity and their dynamic nature. Owing to the legal importance of distinguishing wetland habitats from upland habitats in some parts of the world (in the US, for example), especially as it concerns urbanization or conversion of wetland habitats to other types of land use, there are fairly specific legal definitions for wetlands. Similarly, because wetland ecosystems have considerable ecological significance existing as ecotonal, or transitional, habitats between terrestrial and aquatic systems (US EPA 2016), there are some formalized scientific definitions for wetlands.

The 1987 US Army Corps of Engineers (US ACE) wetlands delineation manual (US Army Corps of Engineers 1987) defined wetlands as

> areas that are inundated or saturated by surface or ground water at a frequency and duration sufficient to support, and that under normal circumstances do support, a prevalence of vegetation typically adapted for life in saturated soil conditions.

The National Wetlands Working Group (1997) in Canada defined wetlands as areas

> saturated with water long enough to promote wetland or aquatic processes as indicated by poorly drained soils, hydrophytic vegetation and various kinds of biological activity which are adapted to a wet environment.

Environmental protection agencies in many other nations, as well as Wetlands International (Russia-based wetland organization) and the international Ramsar Convention, simply define wetlands as areas of interface between the water and the land or by using examples of different wetland habitat types. Noted wetland ecologist Paul Keddy defined wetlands as arising when "inundation by water produces soils dominated by anaerobic processes and forces the biota . . . to exhibit adaptations to tolerate flooding" (Keddy 2000). In contrast to the brevity of Keddy's definition, the US National Research Council used the following (López-Calderón and Riosmena-Rodríguez 2016):

> A wetland is an ecosystem that depends on constant or recurrent, shallow inundation or saturation at or near the surface of the substrate. The minimum essential characteristics of a wetland are recurrent, sustained inundation or saturation at or near the surface and the presence of physical, chemical, and biological features reflective of recurrent, sustained inundation or saturation. Common diagnostic features of wetlands are hydric soils and hydrophytic vegetation. These features will be present except where specific physicochemical, biotic, or anthropogenic factors have removed them or prevented their development.

Wetlands, then, are areas where shallow surface waters or saturated soils are present during a *biologically significant* portion of the year (see Tables 3.1 and 3.2 for specific examples and descriptions of types of wetlands). A shorter way of saying this is that wetlands are places where wetland plants grow. This seems a rather simplified (and completely circular) definition, but in practice, when an area is examined to determine whether it is a wetland, the presence of wetland plants is one of the primary criteria that is used. The area will be surveyed for the presence of water (or indicators that water has a substantial environmental impact on the site; the site's **hydrology**), soil that is indicative of submersed or saturated conditions (**hydric soil**), and plants that are characteristic of living in water or soil that is impacted by waterlogging or saturation for a biologically significant portion of the year (**hydrophytic vegetation**). We will talk extensively in later chapters about factors that determine this biological significance from the standpoint of the plants that are, or are not, adapted to living in wetlands and aquatic ecosystems. We also will examine hydrology and certain aspects of hydric soils in some detail in later chapters.

In this text, we will deal primarily with plants of freshwater ecosystems, and within that group, primarily with angiosperms (i.e., flowering plants). Plants of estuarine and marine systems will occasionally be mentioned, where especially relevant, but those who wish more detail in this area are encouraged to examine Bertness, Gaines, and Hay (2001) or Batzer and Sharitz (2006). As you will see later and in the next two chapters, this limitation still leaves us with thousands of plant species, spread across the entirety of the angiosperm lineages, so we have plenty of information to work with!

1.2.2 AQUATIC AND WETLAND PLANTS DEFINED

Given the previous definitions of the habitats concerned, a general description of the plants on which we will focus is

> species that are normally found growing in wetlands, i.e., in or on the water or where soils are flooded or saturated long enough for anaerobic conditions to develop in the root zone, and that have evolved some specialized adaptations to an anaerobic environment.
>
> (Cronk and Fennessy 2001)

A shorter way of saying this is that these plants live in areas where the soil is wet for a biologically significant portion of the year. To meet this biological significance criterion, the water must be present for such a duration that it will metabolically or physiologically impact plants that grow in those areas. The specific length of time will vary, depending on the species that are present, but the result is that plants living in wetland habitats will have specialized adaptations that allow them to avoid drowning (literally) while living in the water and/or soil that is saturated for some fraction of the year. This period of time usually is considered during the active growing season for the plant species of interest. As mentioned earlier, the adaptations and the environmental factors that make them so important will be discussed at great length in subsequent chapters.

Is there a difference, then, between aquatic plants and wetland plants? Some authors distinguish the two as follows: (1) **aquatic plants** are those plants that complete their life cycles with all vegetative (that is nonflowering) parts submersed in or supported by the water, and (2) **wetland plants** are all other plants adapted to and found living in wetlands or aquatic habitats. Various terms are used to describe these plants, including aquatic plants, wetland plants, hydrophytes, and macrophytes. You may see the latter three of these terms used somewhat interchangeably throughout the rest of this textbook. However, when the term "aquatic plants" is used in this book, this term will refer exclusively to those plants that spend the entirety of their lives with their nonreproductive parts submersed in or supported by the water.

Again, with respect to the identities of the aquatic and wetland plants treated here, this text will focus primarily on freshwater angiosperms, although numerous important examples will come from other of the vascular plants (e.g., ferns and allies, gymnosperms) and, to a lesser extent, nonvascular land plants (mosses) and some algae (e.g., the genus *Chara*). A more detailed examination of the taxonomy and systematics of the majority of the plants included in this text is provided in Chapters 2 and 3. An overview of a simplified morphological taxonomy is provided in the following section of this chapter.

A final topic related to defining wetland plants is the use, in the US, of a concept referred to as **Wetland Indicator Status**. Reed (1988) published the first list of this information as the result of a multi-agency collaboration among the US Fish and Wildlife Service (US FWS), the US Environmental Protection Agency (US EPA), the US ACE, and the US Department of Agriculture Soil Conservation Service (currently, the USDA Natural Resources Conservation Service; USDA NRCS). That document (Reed 1988) was produced to summarize information on wetland plant species of the US as part of a US FWS effort at identifying and classifying wetlands within the US. The list not only documented those plant species known to occur in wetlands of the US, but it also provided a "wetland fidelity rating system" that assigned an indicator rating to each species, based on the estimated likelihood that a species would occur in a wetland versus a non-wetland habitat. These indicator status ratings were assigned by a national panel comprising smaller regional panels made up of a total of 142 botanists and ecologists across the US, coordinated by an interagency review panel of "ecologists with a strong background in wetland botany" representing the four US agencies mentioned earlier.

The indicator status framework (below) consists of five indicator categories. The categories are:

1. **Obligate wetland species** (OBL): These species are expected to occur naturally at a frequency of > 99% under local conditions indicative of wetlands. That is, these species would virtually always be found directly exposed to standing water and/or saturated soil conditions during their active growing season, unless local conditions have been altered in some manner (most likely because of human activities).
2. **Facultative wetland species** (FACW): Occur under wetland conditions at 67%–99% frequency.
3. **Facultative species** (FAC): Occur under wetland conditions at 34%–66% frequency. In other words, these species are just as likely to occur under wetland conditions as they are non-wetland conditions.
4. **Facultative upland species** (FACU): Occur under wetland conditions at 1%–33% frequency.
5. **Obligate upland species** (UPL): Occur under wetland conditions less than 1% of the time.

Bear in mind that I have attempted to caveat the prior definitions and criteria to indicate, for example, that an OBL plant species has a virtual certainty (> 99% probability) of *a particular plant of that species* growing directly in soil and/or hydrologic conditions that are indicative of a wetland environment. For example, I have seen individuals of *Juncus effusus* and *Callitriche heterophylla* (both of which are OBL in my region of the US) growing in small pools of water situated in areas that absolutely would not be considered wetlands. However, the conditions experienced by the seeds and seedlings of those plants were entirely suitable for their survival, at least temporarily.

The point here that the plants that are the focus of this text are defined, as are all species characteristic of some broad **ecosystem** type, by their biological and ecological responses to conditions in the soil or water at places where individuals of a species can successfully establish. Thus, we arrive at the circular definition of wetlands that was presented in the previous section: wetlands are places where wetland plants grow. However, because of the nature of individual plants (their size, longevity, life history, etc.), the converse of this statement is not necessarily true. Species that are recognized as wetland plants will not always be found growing in wetlands; think again of a *Juncus effusus* individual growing in a small, isolated puddle. Similarly, because of the heterogeneity in wetlands, it is quite possible to find individuals of non-wetland species living in wetlands. Regardless of the species' identity and characterization, the local conditions where any individual plant occurs will always, by necessity, lie within some eco-evolutionary parameters that are characteristic of the species.

1.3 GROWTH FORMS AND ZONATION OF AQUATIC AND WETLAND PLANTS

The 2014 update of the US National Wetland Plant List (Lichvar et al. 2014) included more than 8,000 species of vascular plants recognized to occur in wetlands of the US. Among those 8,000 species is a considerable diversity of plant life, some of which will be discussed in the following two chapters. However, to simply this diversity and conceptually integrate the biology and morphological diversity of these plants—as a tool for teaching and for understanding—we will examine a simplified morphological taxonomy referred to here as aquatic and wetland plant growth forms.

Numerous life-form- or growth-form-based classifications have been used in describing the biology of these plants (Table 1.1; Figures 1.4 and 1.5). The one that is most used (e.g., Cronk and Fennessy 2001; Mitsch and Gosselink 2000; Wetzel 2001) and which will be used here is a four-category classification based on that presented by Sculthorpe (1967). As we can see from the information in Table 1.1, this four-category growth form classification is the simplest of those that have been presented in the literature. Hutchinson's (1975) "classification of life-forms and growth forms of higher aquatic plants" was, to me,

an incredibly complex system that accounted not only for those traits specified in Table 1.1 (where the plant is rooted, general plant morphology, and where the majority of leaves typically are found), but also such additional characteristics as leaf arrangement on the stem, leaf morphology, position of inflorescences (flowers and related floral structures) relative to the water surface, and whether the plants produce structures to aid in carnivory (i.e., "traps"). In fact, Hutchinson's system, which covered 7.5 pages of his text, is so complicated that the previous sentence summarizing its complexity was almost a paragraph itself!

Arber's (1920) "biological classification of hydrophytes" was based on a previous classification of water plants by German botanist Dr. Johann Heinrich Rudolf Schenck, whose *Biology of Water Plants* was published in 1886. Arber divided aquatic plants first into those rooted in the soil or otherwise, the latter of which were then subsequently divided based on whether they exist at, versus below, the water's surface, and then based upon whether they produce roots at all. The rooted plants, on the other hand, were divided based on the degree to which they occupy terrestrial habitats (essentially terrestrial, sometimes terrestrial, or primarily aquatic), and then by position and type of vegetative and reproductive structures.

Given the simplicity of the four-category classification presented by Sculthorpe, it may seem that Hutchinson's and Arber's systems were unnecessarily complicated. However, even Sculthorpe noted on multiple instances that there are many transitional morphologies among his four categories, as well as some individual species that fail to fit into a single classification themselves. The latter was illustrated with a few species that regularly demonstrate a phenomenon termed **heterophylly**, such as *Polygonum amphibium* (now named *Persicaria amphibia*) and many species in the genus *Callitriche*. Heterophylly is the tendency for individual plants to produce markedly differing leaf or stem morphologies in response to environmental cues and will be discussed briefly later in this chapter and again in more detail in Chapter 6.

1.3.1 OVERVIEW OF PLANT ANATOMY

Before moving too far into the details of the four growth forms, a brief refresher in plant anatomy will be helpful. In Figure 1.6, I have illustrated most of the more commonly used terms that you will see throughout the book. We can divide these terms into those that are related to the sexual and the vegetative (nonsexual) aspects of plant growth and structure. Among the terms in Figure 1.6 related to sexual functions, we can consider the flowers, florets, inflorescences, peduncles, scapes, seeds (which are illustrated but not named), and genet versus ramet. I suspect that everyone reading this book knows generally what a flower is, but the terms floret and inflorescence may be less familiar. **Florets** are essentially just individual small flowers that form part of a larger **inflorescence**, or cluster of flowers. The inflorescence usually will be supported atop or along

TABLE 1.1

Reconciliation of three disparate attempts at categorizing growth forms among aquatic and wetland vascular plants.

Sculthorpe (1967)	Hutchinson (1975)[1]	Arber (1920)	Aquatic and Wetland Genera, Examples
Plants rooted in sediments or on substrate			
Emergent	Rooted in sediment with part of vegetative structure above water[2]	Sometimes terrestrial but sometimes producing submersed-type leaves[2]	Carex, Juncus, Phragmites, Sagittaria, Typha, many others
	Rooted in sediment with part of vegetative structure above water[2]	Producing aerial, floating, and submersed leaves[3]	
Floating-Leafed[4]	Rooted in sediment with predominantly floating leaves[5]	Producing aerial, floating, and submersed leaves[3]	Brasenia, Hydrocotyle, Nuphar, Nymphaea
Submersed[6] (or submersed, in this text)	Leaves entirely submersed, stems elongate[7]	Plants normally submersed with no floating leaves produced, and having a long, highly branched axis[8]	Elodea, Myriophyllum, Najas, Potamogeton
	Leaves entirely submersed, forming a rosette[7]	Plants normally submersed and forming a rosette of linear leaves[9]	Isoetes, Lobelia, Sparganium, Vallisneria
	na[10]	Plants entirely submersed with thallus attached to the substratum	Podostemum, Tristicha
Plants not rooted in sediments			
Free-Floating	Free-floating, at the water surface[11]	Plants with aerial inflorescences and floating leaves or shoots: roots produced but not usually penetrating the soil	Eichhornia, Lemna, Limnobium, Spirodela
		Plants with aerial inflorescences and floating leaves or shoots: plants not producing roots	Wolffia
	Free-floating, within the water column[12]	Plants entirely or partially submersed: roots produced but not usually penetrating the soil	Lemna (trisulca)
		Plants entirely or partially submersed: plants not producing roots	Ceratophyllum, Utricularia, Wolffiella

[1] Each of the groups described here is divided by Hutchinson into three to seven subgroups, based primarily upon vegetative morphology.

[2] Hutchinson's treatment of these (divided into seven subgroups) is much more diverse, as Arber deals primarily with truly aquatic species in her "biological classification of hydrophytes."

[3] Arber divides these into two subgroups, dependent upon whether flowering is associated with dominance by aerial or floating leaves. The description of this group by Arber is difficult to reconcile with the Sculthorpe and Hutchinson systems because she deals only with truly aquatic species.

[4] Sculthorpe divides these into two subgroups, based on whether the plants produce above- versus belowground connections among modules.

[5] Hutchinson divides these into five subgroups, dependent upon morphology of the floating leaves.

[6] Sculthorpe divides these into three subgroups, consistent with those of Arber, based on whether the plants produce elongate stems, produce rosettes, or are of a thalloid (i.e., more or less flattened) form.

[7] Hutchinson divides these into three subgroups, dependent upon leaf morphology.

[8] Arber divides these into four subgroups, dependent upon whether the inflorescence is aerial or submersed and whether pollination takes place underwater.

[9] Arber divides these into two subgroups, dependent upon whether the inflorescence is aerial or submersed.

[10] Plants with these characteristics were not specifically addressed by this author's treatment.

[11] Hutchinson divides these into five subgroups, dependent upon shoot morphology (reduced, as in Lemna, etc., or various types of rosettes).

[12] Hutchinson divides these into three subgroups, dependent upon the morphology and organization of leaves.

NOTE: The footnotes here belong with, and are a continuation of, Table 1.1.

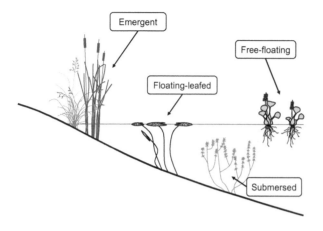

FIGURE 1.4   The Four General Wetland Plant Growth Forms, Based on the System Published by Sculthorpe (1967). See Text for Detailed Descriptions of Each.

some stemlike structure that may be called a **peduncle** or a **scape**, depending on the type of plant and the specific physical organization. Within the inflorescence, individual flowers sometimes may be supported at the ends of pedicels, which are more or less tiny peduncles. Pedicels are unlabeled in Figure 1.6, owing to the scale of the illustrations. Seeds will be produced within the ovaries, within the flowers, and these also are not labeled in this figure because of scale.

The terms genet and ramet are related to the modularity of plants. That is, plants exist as repeating modular structures, and in those that live for multiple growing seasons and undergo multiple rounds of sexual reproduction (i.e., **perennial** species), we often see these modules separate physically from the parent plant over time. Despite their physical separation, these modules remain genetically identical to one another, because they arise clonally through asexual growth. The collection of these genetically identical clones, whether connected or living separately from one another, is referred to as a **genet**, because of their genetic relatedness. Another way of thinking of this is that all of

FIGURE 1.5   Examples of Plants Exhibiting the Four General Wetland Plant Growth Forms in Figure 1.4. Clockwise, from Top Left: *Eriocaulon* Species (Emergent); *Nymphaea Odorata* (Floating-Leafed); *Hydrilla Verticillata* (Submersed); *Eichhornia Crassipes*, *Azolla Caroliniana*, and Mixed Duckweeds (Free-Floating).

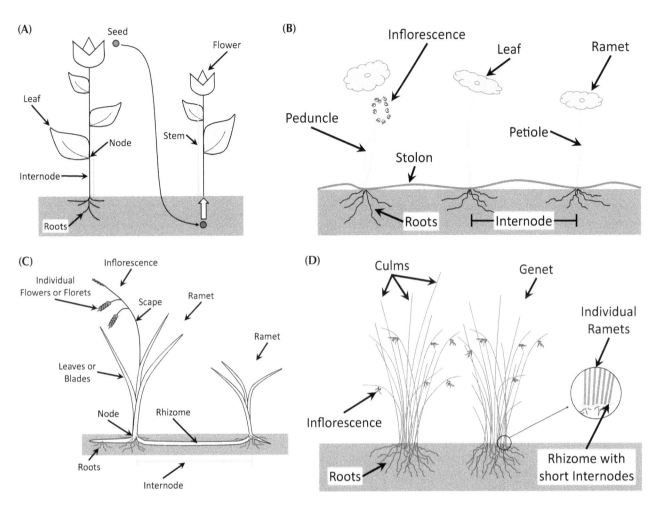

FIGURE 1.6 General Overview of Plant Anatomy. (A) Annual Plant Species Begin Each Generation from a Seed and thus have a much Simpler Structure than do Perennial Species. Perennial Species Exist as Repeating, Genetically Identical Modules, Each of which has the Potential to Live Separately from the Parent Plant. Perennial Species Develop Along Main Stems that often will be Either (B) Aboveground Stolons or (C–D) Belowground Rhizomes. The Length of Internodes also Varies Among Species (C–D), with this Influencing the Species that may Perform Best under Different Wetland Conditions, as we will see in Later Chapters.

the structures that developed from one single seed are collectively referred to as a genet. The individual modules, however, can be referred to as **ramets**, again, whether they remain connected or have separated for any reason.

In Figure 1.6, panel A represents an **annual** plant species, where we see each individual arising from a seed and then dying, with new individuals all arising from individual seeds. In this case, each unit that we might see in a population of this species would be both a genet and a ramet, with only one ramet per genet. If you think that this seems highly redundant and unnecessary, you are correct, and we typically do not use these terms when discussing annual species for this reason. The other three panels of Figure 1.6 represent perennial species, wherein we have collections of ramets forming the genets that are illustrated here.

The vegetative structures illustrated in this figure include leaves (also blades or culms), **petioles**, stems (including stolons and rhizomes), roots, nodes, and internodes. The distinction among leaves, blades, and culms varies relatively widely among different groups of plants. A leaflike structure

that might be called a blade in one group may be called simply a leaf, or perhaps some other term, in another group. Use of the term **culm**, similarly, varies somewhat among groups, but generally, this term refers to an unbranching stemlike structure, usually emerging from a node that branches at ground level. Sometimes culms are sexual structures, serving to support inflorescences (much like a form of peduncle), and we see this term used particularly frequently with grasses and sedges. Other times, culms may serve both vegetative and sexual roles. In species that have leaves supported by "stems" or stalks that are clearly differentiated from the main bladelike portion of the leaf, we refer to that supporting structure as a petiole. "Culms" and "blades" do not usually have petioles, which is perhaps another distinction of those structures from other leaflike structures. For the most part, these structures (leaves, blades, vegetative culms) serve as the primary sunlight gathering, photosynthetic organs in the plants that we will be studying in this book.

Although most people will tend to think of stems as being vertical structures that support the bodies of plants in an upright position, many stems (and perhaps most of those that would be produced by the plants in this book) have a horizontal orientation. The two most common forms of these are **stolons** (Figure 1.6B) and **rhizomes** (Figure 1.6C–D). The difference between stolons and rhizomes is whether they are produced aboveground (solons) or belowground (rhizomes). Both stolons and rhizomes comprise a series of repeating nodes and internodes. The **nodes** are the points along the stem where branching may occur and where leaves and/or flowers may be produced. **Internodes** are simply the sections of the stem between the nodes. Roots, the primary organ serving in uptake of nutrients and water, also may be produced at nodes, especially when nodes come in contact with the water or saturated soil. Note, however, that some plant species also produce aerial roots, so contact with water is not a requisite for root production.

1.3.2 Emergent Plants

Emergent plants are characterized by roots anchored in or on the soil or other substrate and aboveground parts (stems, leaves, and inflorescences) that extend above the surface of the water (Table 1.1; Figures 1.4 and 1.5). Note that these species, owing to their occupying the shallowest zones of wetlands, may occasionally (or always, depending on the species) be found in areas with no standing water at the base of the plant. Plants exhibiting the emergent growth form could lie almost anywhere along the OBL–FACU indicator category continuum. This growth form includes the widest diversity of life histories and body plans, among the four categories, with both herbaceous and woody plant species represented.

Trees and shrubs fit within this category, especially with the various caveats I have provided in the prior description. However, because Sculthorpe, Hutchinson, and Arber dealt almost exclusively with aquatic herbaceous plants, wetland trees received little mention in their works. Sculthorpe included mangroves (*Rhizophora*, *Bruguiera*) and cypress (*Taxodium* species) in the context of their specialized adaptations for root aeration, but otherwise, woody species were largely absent from his text. For many reasons, those interested in wetlands might have an interest in trees and shrubs (as well as some of the "woody" grasses, i.e., cane or bamboo), so I felt it important to emphasize their position within this growth form framework. They will continue to receive attention where appropriate in other parts of the book.

Emergent species rely on aerial reproduction (with primarily wind or insect pollination), soil or sediment as the sole source of nutrients (usually), and atmospheric CO_2 as the source of inorganic carbon. Collectively, emergent species occur in almost any type of wetland or riparian habitat, and some species exhibit the greatest **productivity**, or annual production of **biomass**, among aquatic and wetland plants (Table 1.2). Their increased productivity results from a combination of the vertical position of their leaves above the water surface, which gives better access to light, and the direct contact the leaves (and thus the stomatal openings) have with the atmosphere. An additional advantage afforded

to emergent species is the more rapid cycling of nutrients within the soil and sediments surrounding their roots than is the case for species growing in deeper, more permanent waters. This increased rate of nutrient cycling results from increased oxygenation of the sediments in areas supporting emergent species. This increased oxygenation is driven by the periodic drawdown of water at the wetland edge, by the ability of many of the plants to transport oxygen into the sediments surrounding the roots, and by increased gas exchange through the plants' stomata.

1.3.3 Floating-Leafed Plants

Floating-leafed species typically are also rooted in the sediments but with leaves that float at the water's surface. A large percentage of these species have oval, round, or cordate (heart-shaped) leaves with stomata on the upper surface, where they make direct contact with the air above. This allows exchange of oxygen, carbon dioxide, water vapor, and other gases between the plant and the atmosphere. Most of these species also have long, flexible petioles; this is a characteristic that we will see again in Chapter 6. These species typically have inflorescences that are elevated to some degree above the water's surface (albeit usually only a short distance) or may float at or on the water's surface. Because of the positioning of their leaves and inflorescences, these species can substantially shade the water column and may thus impact other plant species, including algae, within waters they occupy. This can sometimes lead to reduced oxygen availability within the water column, which may, itself, have cascading effects within wetlands dominated by floating-leafed species. Floating-leafed species tend to be at the lower end of the range of biomass production rates among aquatic and wetland species (Table 1.2), possibly because of their relatively inefficient three-dimensional usage of the areas they inhabit. Compare, for example, the floating-leafed versus submersed growth forms in Figure 1.4 with regard to the proportion of the water column occupied by each.

Some species that usually fall within this growth-form category may exhibit **heterophylly**, wherein they produce submersed leaves, floating leaves, and/or aerial leaves on the same plant (hetero- = different; -phyll = leaf). Typically, leaves of differing position relative to the water column will be produced at different times within the growing season and will also, because of the effects of one or more environmental variables on leaf physiology, have differing shapes. This latter feature is how the term heterophylly is derived to describe this phenomenon.

1.3.4 Submersed Plants

Submersed species produce most of their biomass and spend most of their lives beneath the water's surface. The exception to this is when some of these species produce flowers that emerge above the water's surface. Thus, these species, as with the floating-leafed species discussed earlier and the

TABLE 1.2

Annual biomass production among wetland plant growth forms. Values were obtained from multiple original sources by the two secondary sources cited in the far right column. Note that in some cases biomass is given as aboveground biomass only (e.g., stems, leaves, flowers, fruit), whereas other data represent a summation of above- and belowground biomass.

Group	Species	Location	Annual Productivity (kg per m² per year)			Salinity	Aboveground or All	Source
			Min	Mean	Max			
Emergent Growth Form								
Grasses	*Echinochloa polystachya*	Amazon basin		9.90		Freshwater	Aboveground only	Wetzel 2001
Grasses	*Panicum hemitomon*	Louisiana		1.70		Freshwater	Aboveground only	Mitsch and Gosselink 2000
Grasses	*Phragmites australis*	Western Australia		19.24		Freshwater	Aboveground only	Wetzel 2001
Grasses	*Zizania aquatica*	Delaware	0.63		1.45	Freshwater	Aboveground only	Wetzel 2001
Grasses	*Glyceria maxima*	Czech Republic	0.90		4.30	Freshwater	Summed above- and belowground	Mitsch and Gosselink 2000
Grasses	*Phragmites australis*	Czech Republic	1.00		6.00	Freshwater	Summed above- and belowground	Mitsch and Gosselink 2000
Grasses	*Phragmites australis*	Denmark		1.40		Freshwater	Summed above- and belowground	Mitsch and Gosselink 2000
Grasses	*Distichlis spicata*	Louisiana	1.16		1.29	Saltmarsh	Aboveground only	Mitsch and Gosselink 2000
Grasses	*Distichlis spicata*	Louisiana		1.97		Saltmarsh	Aboveground only	Mitsch and Gosselink 2000
Grasses	*Distichlis spicata*	Mississippi		1.07		Saltmarsh	Aboveground only	Mitsch and Gosselink 2000
Grasses	*Spartina alterniflora*	Louisiana	1.47		2.90	Saltmarsh	Aboveground only	Mitsch and Gosselink 2000
Grasses	*Spartina alterniflora*	Louisiana		1.38		Saltmarsh	Aboveground only	Mitsch and Gosselink 2000
Grasses	*Spartina alterniflora*	Mississippi		1.09		Saltmarsh	Aboveground only	Mitsch and Gosselink 2000
Grasses	*Spartina cynosuroides*	Louisiana		1.13		Saltmarsh	Aboveground only	Mitsch and Gosselink 2000
Grasses	*Spartina patens*	Louisiana	1.34		1.43	Saltmarsh	Aboveground only	Mitsch and Gosselink 2000
Grasses	*Spartina patens*	Louisiana		4.16		Saltmarsh	Aboveground only	Mitsch and Gosselink 2000
Grasses	*Spartina patens*	Mississippi		1.24		Saltmarsh	Aboveground only	Mitsch and Gosselink 2000
Grasses	*Spartina alterniflora*	Alabama		8.25		Saltmarsh	Summed above- and belowground	Mitsch and Gosselink 2000
Rushes	*Juncus militaris*	Rhode Island		0.44		Freshwater	Aboveground only	Wetzel 2001
Rushes	*Juncus effusus*	South Carolina		1.86		Freshwater	Summed above- and belowground	Mitsch and Gosselink 2000
Rushes	*Juncus effusus*	Alabama		9.83		Freshwater	Summed above- and belowground	Wetzel 2001
Rushes	*Juncus roemerianus*	Louisiana	1.81		1.96	Saltmarsh	Aboveground only	Mitsch and Gosselink 2000
Rushes	*Juncus roemerianus*	Louisiana		3.30		Saltmarsh	Aboveground only	Mitsch and Gosselink 2000
Rushes	*Juncus roemerianus*	Mississippi		1.30		Saltmarsh	Aboveground only	Mitsch and Gosselink 2000
Rushes	*Juncus roemerianus*	Alabama		10.66		Saltmarsh	Summed above- and belowground	Mitsch and Gosselink 2000
Sedges	*Carex atheroides*	Iowa		2.86		Freshwater	Aboveground only	Mitsch and Gosselink 2000
Sedges	*Carex* species	Minnesota		0.74		Freshwater	Aboveground only	Wetzel 2001
Sedges	*Carex lacustris*	New York	1.08		1.74	Freshwater	Summed above- and belowground	Mitsch and Gosselink 2000
Sedges	*Carex lacustris*	New York		1.17		Freshwater	Summed above- and belowground	Wetzel 2001
Sedges	*Carex rostrata*	Finland		1.42		Freshwater	Summed above- and belowground	Wetzel 2001

(Continued)

TABLE 1.2 (*Continued*)

Annual biomass production among wetland plant growth forms. Values were obtained from multiple original sources by the two secondary sources cited in the far right column. Note that in some cases biomass is given as aboveground biomass only (e.g., stems, leaves, flowers, fruit), whereas other data represent a summation of above- and belowground biomass.

Group	Species	Location	Annual Productivity (kg per m² per year)			Salinity	Aboveground or All	Source
			Min	Mean	Max			
Sedges	*Cyperus papyrus*	Kenya		7.34		Freshwater	Summed above- and belowground	Wetzel 2001
Sedges	*Schoenoplectus lacustris*	Czech Republic	1.60		5.50	Freshwater	Summed above- and belowground	Mitsch and Gosselink 2000
Sedges	*Scirpus fluviatilis*	Iowa		0.94		Freshwater	Summed above- and belowground	Mitsch and Gosselink 2000
Other monocots	*Acorus calamus*	Czech Republic	0.50		1.10	Freshwater	Summed above- and belowground	Mitsch and Gosselink 2000
Other monocots	*Sparganium eurycarpum*	Iowa		1.07		Freshwater	Aboveground only	Mitsch and Gosselink 2000
Other monocots	*Typha latifolia*	Oregon	2.04		2.21	Freshwater	Summed above- and belowground	Mitsch and Gosselink 2000
Other monocots	*Typha angustifolia*	Texas		5.14		Freshwater	Summed above- and belowground	Wetzel 2001
Other monocots	*Typha × glauca*	Iowa		2.30		Freshwater	Aboveground only	Mitsch and Gosselink 2000
Other monocots	*Typha latifolia*	Michigan		6.00		Freshwater	Aboveground only	Wetzel 2001
Other monocots	*Typha orientalis*	Australia		4.38		Freshwater	Aboveground only	Wetzel 2001
Other monocots	*Typha* species	Wisconsin		3.45		Freshwater	Summed above- and belowground	Mitsch and Gosselink 2000
Other monocots	*Typha* species	Minnesota		2.50		Freshwater	Aboveground only	Wetzel 2001
combined range			1.00		10.00		unspecified	Wetzel 2001
Floating-Leafed Growth Form								
Water lilies	*Brasenia schreberi*	Japan	0.16		0.24	Freshwater	Aboveground only	Wetzel 2001
Water lilies	*Nymphaea odorata*	Alabama		0.29		Freshwater	Aboveground only	Wetzel 2001
Water lilies	*Nymphaea tetragona*	Japan	0.52		0.56	Freshwater	Aboveground only	Wetzel 2001
Water lilies	*Nymphoides peltata*	Netherlands		0.56		Freshwater	Aboveground only	Wetzel 2001
Water lilies	*Nymphoides peltata*	Japan		0.18		Freshwater	Aboveground only	Wetzel 2001
combined range			0.10		0.56		unspecified	Wetzel 2001
Free-Floating Growth Form								
Other monocots	*Eichhornia* species	Louisiana	1.50		4.40	Freshwater	Aboveground only	Wetzel 2001
combined range			0.30		5.00		unspecified	Wetzel 2001
Creeping Emergent Growth Form								
Eudicot	*Nasturtium* species	New Zealand		2.20		Freshwater	Aboveground only	Wetzel 2001
Submersed Growth Form								
Alismatales	*Elodea canadensis*	Czech Republic		0.50		Freshwater	Aboveground only	Wetzel 2001
Alismatales	*Lagarosiphon ilicifolius*	Zimbabwe		0.60		Freshwater	Aboveground only	Wetzel 2001
Alismatales	*Ruppia* species	California		0.06		Saline Lake	Aboveground only	Wetzel 2001
Sedges	*Scirpus subterminalis*	Michigan		0.57		Freshwater	Aboveground only	Wetzel 2001
Other monocots	*Sparganium emersum*	Denmark		0.54		Freshwater	Aboveground only	Wetzel 2001
Eudicot	*Lobelia dortmanna*	New Hampshire		0.06		Freshwater	Aboveground only	Wetzel 2001
combined range			0.01		1.50		unspecified	Wetzel 2001

free-floating species that follow, would fall within the relatively small group of truly aquatic plant species. Many of these species (100–200 species) have so evolved to aquatic life that they even rely on the water to carry out pollination, rather than relying on wind or animals to transport pollen during sexual reproduction. For those species whose flowers never reach the surface of the water, pollination by water (or **hydrophily**) is a necessity.

Most submersed species are rooted in the sediments, as was the case with emergent and floating-leafed species.

However, there are some species that exist submersed within the water column that do not produce roots at all (in the genera *Ceratophyllum*, *Utricularia*, and *Wolffiella*, for example), and these must rely on the water column for mineral nutrition. One exception to this is the genus *Utricularia*, which produces submersed bladders evolved for capturing tiny aquatic animals and protozoa (or other life that may be in proximity of the bladders when they open; Figure 1.7). These captured organisms then decay within the bladders, and the resulting released nutrients can

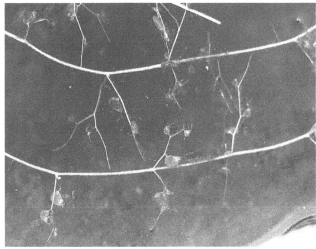

FIGURE 1.7 Top and Bottom Left: Bladders in *Utricularia Biflora* that are Used to Capture Aquatic Invertebrates to Aid in Access to Mineral Nutrition. Bottom Right: *Utricularia Biflora* Flowers.

be used by the plants, as with terrestrial carnivorous plants such as pitcher plants, sundews, and flytraps. This will be touched on again in Chapter 7.

For those that are rooted in the sediments, however, most of their mineral nutrition will be taken from the sediments, as with the previous two growth forms. However, because most of their aboveground parts are within the water column, and most have leaves and stems that are very thin (two to a few cells thick, in fact), they do have some opportunity for nutrient uptake from the water column. In nutrient-rich wetlands, however, these plants usually will be colonized by periphytic algae, bacteria, and aquatic fungi that will usually consume dissolved nutrients before they can reach the plants. In more oligotrophic systems (i.e., less nutrient-rich), nutrient availability within the sediments will suffi-ciently exceed that in the water column to make nutrient uptake by roots the more efficient strategy for these plants.

Because most photosynthetic tissues for submersed spe-cies are below the water's surface, they rely on dissolved carbon (CO_2 or HCO_3^-) as their carbon source for photo-synthesis (Wetzel 2001). This means that submersed spe-cies have substantially lower carbon availability, resulting in lower productivity than emergent species and some of the free-floating species that produce canopies above the water's surface (Table 1.2).

1.3.5 Free-Floating Plants

As you probably would guess, **free-floating plants** are those whose leaves and stems float freely on or beneath the water's surface. For those species within this group that produce roots, those roots typically will be suspended within the water column, where the plants will obtain mineral nutri-tion. Occasionally, free-floating species that produce roots will find themselves in sufficiently shallow water (or perhaps stranded during low water levels) that their roots may grow into the soil or sediment, and they will become rooted. In these cases, the plants will likely benefit greatly from having

access to stores of mineral nutrients within the decompos-ing organic sediments until such time as water levels rise, potentially causing the plants to break free and return to a free-floating habit. In spite of not having access to nutrients in soil or sediments, free-floating plants sometimes exhibit extensive vegetative growth, and some are widely recog-nized as highly productive, problematic aquatic weeds (e.g., *Eichhornia* [*Pontederia*] *crassipes*, *Pistia stratiotes*, *Trapa natans*; Table 1.2). As is the case with emergent and floating-leafed species, free-floating aquatic plants usually will be pollinated by wind or by insects.

1.3.6 A Few Exceptions to the Earlier Classification

As discussed earlier, the classifications used by Hutchinson and Arber relied on multiple ecological and morphologi-cal plant traits to produce a greater diversity of categories. Even with that finer resolution, exceptions were noted by each of those authors. For example, Hutchinson mentioned that some of what he referred to as submersed plants pro-duce stem apices above the water's surface, and Arber gave several examples of rooted species (i.e., species that pro-duce roots) that frequently occur unattached from the sedi-ments (or other substrate). It should be no surprise, then, that if we move to a growth form categorization having only four categories, we will see many exceptions. When I talk about this in the classroom, I analogize this to biologists' affinity for placing things into virtual "boxes" or "bins" as we attempt to make sense out of overwhelming biological diversity. Inevitably, no matter how many boxes we use, we will have some taxa that do not fit cleanly into only one box.

In the case of wetland plant growth forms, I have noted earlier a few species of submersed plants that lack roots (in the genera *Ceratophyllum*, *Utricularia*, and *Wolffiella*); thus, although they are submersed, they technically float freely within the water column (Figure 1.8), and some spe-cies of *Ceratophyllum* and *Utricularia* often can be found

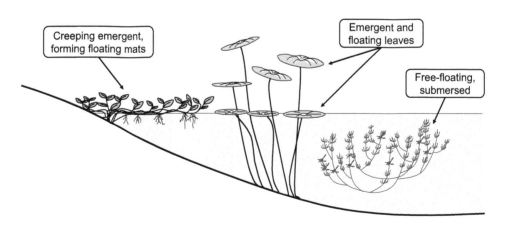

FIGURE 1.8 Examples of Plant Phenotypes that Fail to Fit Well into One of the Four Standard Growth Forms from Figure 1.4. Species that Would Fit these Examples Would be *Ludwigia Peploides* (Creeping Emergent, Mat Forming), *Nelumbo Lutea* (Emergent and Floating Leaves), and *Ceratophyllum Demersum* (Free-Floating Submersed).

with portions of their leaves and stems at or just below the water's surface. *Nelumbo lutea* likewise defies placement into a single growth-form category, as it often shifts the position of leaves based on density of neighbors (Figure 1.8). As the water's surface becomes more crowded across the growing season, or because of high density of neighbors early in the growing season, *N. lutea* will begin to elevate leaves above the water, producing a dense canopy of emergent leaves that may shade even its own floating leaves produced earlier in the year (Figure 1.9). A third grouping of plants are rooted species, such as *Ludwigia peploides, L. octovalvis, L. grandiflora, Rotala rotundifolia,* and *Alternanthera philoxeroides,* that produce a sprawling canopy of stems, leaves, and flowers in dense mats across the water's surface (Figure 1.8). Rejmánková (1992) referred to species exhibiting this growth habit as "creeping emergent" species, and I have adopted that terminology here. Creeping emergent species often occur around margins of lakes and ponds and throughout wetlands, sometimes presenting challenges to managers of those water bodies by outcompeting other plant species, obstructing boat traffic, or hampering other commercial or recreational uses of the water.

1.3.7 ZONATION

On the other hand, the tendency for many species to stay within their growth-form "boxes" results in a somewhat distinct pattern of zonation of vegetation in wetland ecosystems. These general patterns combine with other characteristics of the lake or wetland, such as substrate type (sand, silt, organic sediments), exposure to wind (and thus severity of sediment erosion), and trophic status of the water (i.e., degree of nutrient enrichment) to give an incredible diversity of zonation patterns in nature (Hutchinson 1975). Nevertheless, growth form is closely tied to adaptations of the individual species to dealing with environmental **stressors** and, thus, dictates the range of environmental conditions under which a species can survive and reproduce.

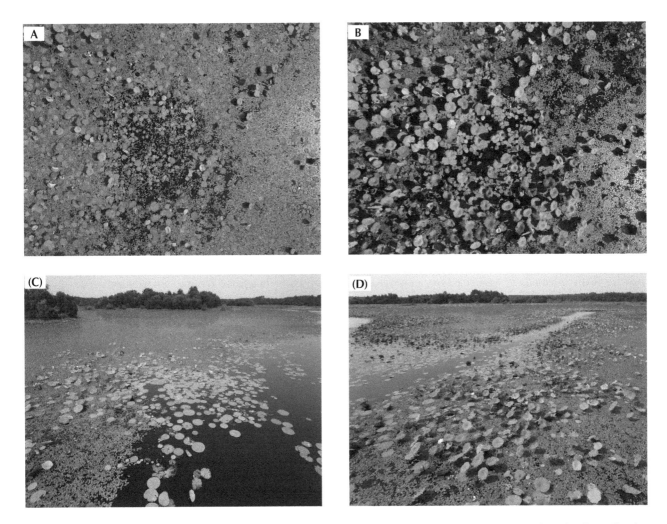

FIGURE 1.9 Vertical Position of Leaves of *Nelumbo Lutea* Can Shift Over Time, Illustrating the Difficulty of Placing Some Species into a Single Morphological Classification. Upper Photos are Aerial Images of the Same Experimental Plot, Both Taken from a Height of Eight Meters, on June 3, 2019 (A), and August 29, 2019 (B; Note White Stake in Upper Left of Both Photos). Lower Photos (C–D) Show a Closer View of these Plant Canopy Configurations. The Largest Leaves are *Nelumbo Lutea*, Growing Among *Nymphaea Odorata* and *Brasenia Schreberi* at the Noxubee National Wildlife Refuge (Mississippi, USA).

As depicted in Figures 1.4 and 1.8, rooted plants typically are restricted to the shallower depths of the lake or wetland. This is determined by the depth at which the plants can grow with their roots in the sediments while also maintaining access to sufficient light and inorganic carbon for photosynthesis. Even for an individual species, this depth will vary, depending on latitude (which dictates light intensity), trophic status of the water body (which will influence water clarity), underlying geology (which can influence water pH and thus the dominant form of inorganic carbon), and neighboring species, among other factors. Because of these factors, a particular species may grow in deeper or shallower water, depending on attributes of the particular site, but rooted species almost always will be restricted to the shallower depths at a site, in comparison with free-floating species.

Among the rooted growth-form categories, the emergent species (including creeping emergents, as in Figure 1.8) tend to be restricted to shallower depths than floating-leafed or submersed species. This enables the emergent species to maintain access to the atmosphere for gas exchange (i.e., intake of CO_2 and release of O_2), while rooted in the soil or sediments for water and nutrient uptake. Floating-leafed species tend to occur at slightly deeper maximum depths than emergent species, based on their abilities to move gases and other materials through long, flexible leaf petioles, between the leaf at the water's surface and the submersed stems, roots, and even to other leaves. Finally, the rooted submersed species typically will be able to establish at the deepest depths, with each species' limits for capturing light and carbon for photosynthesis playing a major role in determining depth limits within this growth form.

Free-floating species are not limited by water depth, *per se*. At shallower depths, they would likely be limited by the presence of other plant species where, for example, emergent species can produce sufficiently tall canopies that they overtop and outcompete the free-floating species for light. Likewise, floating-leafed and submersed species are at a disadvantage when competing with emergent species for light, and finally, the submersed species might be outcompeted by a dense canopy of floating leaves upon the water's surface. This all sets up conditions wherein a species may be limited in one direction by environmental constraints to growing at a particular depth and by biological constraints in the other direction from species that have found the appropriate conditions at a shallower depth. These interactions with the water depth and neighboring species results in the typical zonation of emergent species, followed by floating-leafed, then submersed species as one reaches deeper waters within the wetland.

1.4   IN SUMMARY

Aquatic and wetland plants are beautiful, interesting, and vital to the functioning of aquatic and wetland ecosystems. Because of this last attribute, they are also integral to defining whether a particular area is, or is not, a wetland. At the

same time, they are in some ways defined by their tendency to occur in wetland habitats.

Wetlands plants, then, are those species adapted to environments where, under normal conditions, soils or sediments are flooded or saturated for a biologically relevant period of time, during which oxygen-deficient conditions develop within the root zone, impacting plant growth, development, and/or reproduction. Aquatic plants are those species with adaptations that allow them to complete their lifecycles with all vegetative parts submersed in or supported by the water.

As a matter of convenience, we tend to group plants into a small set of growth forms, based primarily on the position of the vegetative parts relative to the water's surface, in combination with whether they are rooted in the soil or sediments. Thus, we often refer to wetland plants as belonging to an emergent, floating-leafed, free-floating, or submersed growth form. Each of these growth forms may be characterized by a general suite of adaptations to the position they typically inhabit within wetland or aquatic habitats and to general patterns of annual biomass production among growth forms. These plants' adaptations, along with the tendency for plants of a particular growth form to occur under conditions appropriate to those adaptations, lead to the formation of recognizable zones within aquatic and wetland ecosystems.

1.5   FOR REVIEW

1. What three elements of an area define whether it is a wetland?
2. What is a wetland plant? How does that differ from an aquatic plant?
3. What is meant by the phrase "a biologically significant period of time" when defining wetland plants?
4. Why do plants need to have special adaptations for living in aquatic and wetland habitats?
5. What is Wetland Indicator Status?
6. Aside from Duncan Sculthorpe, what other authors have devised aquatic plant growth form (or life form) classifications?
7. What are the four standard growth forms used in this book?
8. Which growth form tends to have the greatest levels of annual biomass production? Why?
9. Which growth form tends to have the lowest levels of annual biomass production? Why?
10. What is the/a general order in which we see the four growth forms under "typical" wetland plant zonation?
11. Why do plant zonation patterns exist at all?

1.6   REFERENCES

Arber, Agnes R. 1920. *Water Plants: A Study of Aquatic Angiosperms*. Cambridge, UK: Cambridge University Press.

Batzer, Darold P., and Rebecca R. Sharitz. 2006. *Ecology of Freshwater and Estuarine Wetlands*. Berkeley, CA: University of California Press.

Bertness, Mark D., Steven D. Gaines, and Mark E. Hay. 2001. *Marine Community Ecology*. Sunderland, MA: Sinauer Associates, Inc.

Calpe, Concepcion. 2006. "Rice International Commodity Profile Prepared by Concepción Calpe Markets and Trade Division." *Food and Agriculture Organization of the United Nations*, December, 23pp.

Cronk, Julie K., and M. S. Fennessy. 2001. *Wetland Plants: Biology and Ecology*. Boca Raton, FL: CRC Press.

Harper, John L. 1977. *Population Biology of Plants*. New York: Academic Press, Inc.

Hutchinson, G. Evelyn. 1975. *A Treatise on Limnology: Volume III, Limnological Botany*. New York: John Wiley & Sons, Inc.

Kay, Stratford H., and Steve T. Hoyle. 2001. "Mail Order, the Internet, and Invasive Aquatic Weeds." *Journal of Aquatic Plant Management* 39 (1): 88–91.

Keddy, Paul A. 2000. *Wetland Ecology: Principles and Conservation*. Cambridge, UK: Cambridge University Press.

Lichvar, R. W., M. Butterwick, N. C. Melvin, and W. N. Kirchner. 2014. "The National Wetland Plant List: 2014 Update of Wetland Ratings. U.S. Army Corps of Engineers." *Phytoneuron* 41 (April): 1–42. https://pdfs.semanticscholar.org/857f/5145355622a681577690758251ce61178103.pdf.

López-Calderón, Jorge Manuel, and Rafael Riosmena-Rodríguez. 2016. *Wetlands. Encyclopedia of Earth Sciences Series*. National Academies Press. https://doi.org/10.1007/978-94-017-8801-4_399.

Maki, Kristine, and Susan Galatowitsch. 2004. "Movement of Invasive Aquatic Plants into Minnesota (USA) through Horticultural Trade." *Biological Conservation* 118 (3): 389–96. https://doi.org/10.1016/j.biocon.2003.09.015.

Mitsch, William J., and James G. Gosselink. 2000. *Wetlands*. 3rd ed. New York: John Wiley & Sons, Inc.

National Wetlands Working Group. 1997. *The Canadian Wetland Classification System*. Waterloo: Wetlands Research Branch, University of Waterloo.

Peres, Cleto Kaveski, Richard Wilander Lambrecht, Diego Alberto Tavares, and Wagner Antonio Chiba de Castro. 2018. "Alien Express: The Threat of Aquarium e-Commerce Introducing Invasive Aquatic Plants in Brazil." *Perspectives in Ecology and Conservation* 16 (4): 221–27. https://doi.org/10.1016/j.pecon.2018.10.001.

Reed, Porter B. 1988. "National List of Plant Species That Occur in Wetlands." Report 88(24), US Fish and Wildlife Service National Wetlands Inventory, Washington, DC.

Rejmánková, Eliška. 1992. "Ecology of Creeping Macrophytes with Special Reference to *Ludwigia peploides* (H.B.K.) Raven." *Aquatic Botany* 43 (3): 283–99. https://doi.org/10.1016/0304-3770(92)90073-R.

Roberts, Thomas H., and Dale H. Arner. 1984. "Food Habits of Beaver in East-Central Mississippi." *The Journal of Wildlife Management* 48 (4): 1414–19.

Sculthorpe, C. D. 1967. *Biology of Aquatic Vascular Plants*. New York: St. Martin's Press.

Seaman, William H. 1891. "The *Victoria regia*." *Proceedings of the American Society of Microscopists* 13: 163–70.

US Army Corps of Engineers. 1987. *Corps of Engineers Wetlands Delineation Manual*: 92 pages et annexes. Vicksburg, MS: US Army Corps of Engineers Environmental Laboratory. http://el.erdc.usace.army.mil/wetlands/pdfs/wlman87.pdf.

US Environmental Protection Agency. 2016. "National Wetland Condition Assessment 2011: A Collaborative Survey of the Nation's Wetlands." US Environmental Protection Agency Report EPA-843-R-15–005.

Wetzel, Robert G. 2001. *Limnology: Lake and River Ecosystems*. San Diego, CA: Academic Press.

2 Evolutionary Relationships among Aquatic and Wetland Plants

2.1 WHY EVOLUTIONARY HISTORY IS IMPORTANT

As we saw in Chapter 1, much of this text deals with adaptations of plants for life in environments strongly influenced by water. To best understand the importance of those adaptations, I believe it is also important to understand the evolutionary history of the plants that live in these areas. There sometimes is such an incredible diversity of phylogenetic, or evolutionary, lineages in certain ecosystems (mangrove forests, for example) that evolutionary history might be thought to be entirely unimportant. Despite the breadth of groups of trees and shrubs that eventually came to inhabit mangrove ecosystems, however, we find that most of the families of plants containing mangrove species also include species found in other types of wetlands. Understanding similarities among those evolutionary diverse lineages can yield insight into features of wetland ecosystems that drive evolutionary change or adaptations that provide species the greatest chances for success there. In other cases, there is striking consistency in the lineages that dominate certain habitat types. For example, all truly aquatic marine plants come from just a few families in a single order, the Alismatales. This may suggest that rare or potentially difficult to achieve adaptations were critical for survival in marine habitats. In all cases, the evolutionary origins of the dominant plant species will be a focus of the overview of aquatic and wetland plant diversity provided in this and the following chapter.

2.1.1 THE DISORDERLINESS OF TAXONOMIC ORGANIZATION

When I was an undergraduate college student, we were taught a very clean taxonomic system that included seven major levels of organization: kingdom, phylum, class, order, family, genus, and species, substituting division for phylum when studying plants. Many discoveries have been made in the ensuing decades about evolutionary relationships among branches on the tree of life, and now, one school of formal taxonomic classification currently begins at the level of domains, rather than kingdoms, and kingdoms now fall within domains. Another approach uses what are referred to as unranked taxonomic groups (Adl et al. 2019). I lean more towards the latter in this book, largely because the higher-order groupings are not terribly important in this context, regardless of their hierarchical relation to one another, and we will be working primarily at the level of order and below.

The operationally important parts of the taxonomic hierarchy, for our purposes in this book, are given in Table 2.1. You may first notice an additional category has been added (subclass) and half of the previous list of organizational levels has been omitted. An ironic thing about adding subclasses and dropping classes is that the level of class was dropped because one of the two classes of angiosperms is no longer valid (recall that we will emphasize angiosperms in this book). These two classes were the monocots (Liliopsida) and **dicots** (Magnoliopsida). It turns out that the "dicots" are an invalid **phylogenetic grouping** (a grouping based on evolutionary relationships), in that the "dicots" violate the requirement that named groups, or clades, be **monophyletic** (Figure 2.1). A monophyletic group is one that contains the most recent common ancestor of members in the group, as well as *all* descendants of that ancestor. The "dicots," which we place in quotes because it is not a valid clade, violate this requirement because they are split phylogenetically into two parts by the monocots, which, it turns out, are a valid clade. That is, there are no non-monocot branches emerging *from within* the monocots.

So, we will not use classes of plants in this text, but we may occasionally recognize subclasses. Subclasses are useful in that they allow us to group things at a somewhat intermediate level, just for sake of some semblance of organization (remember the virtual bins I mentioned in Chapter 1), but some orders are not nested within a subclass. Nymphaeales, the order that contains the water lilies, for example, does not sit within a recognized subclass. The next level of organization, the order, is by far much more biologically useful, in part because the taxa within orders have much more in common with one another than do the various orders within a subclass.

That all having been said, it is at the level of order where we can really start to see some obvious biological similarities among the taxa of interest. For example, the grasses (family Poaceae), are in an order (Poales) with many other families of plants that superficially look like grasses, with long, linear, blade-like leaves. These species often are collectively called **graminoids**, owing to their grasslike appearance. These families include the grasses (Poaceae), rushes (Juncaceae), sedges (Cyperaceae), and others sharing this graminoid morphology, or shape, although the term graminoid often is restricted to the grass, sedge, and rush families. The shared morphology and monophyletic evolutionary history of these species give them various other similarities in resource acquisition, pollination, and the ways in which they interact with their surroundings. The Poales, however, are in a subclass (Commelinidae) with a

DOI: 10.1201/9781315156835-2

TABLE 2.1

Taxonomic ranks commonly used in this text. *Oryza sativa* **is the most common species of cultivated rice, and** *Sagittaria latifolia* **is a species of "arrowhead."**

Category/Rank	Naming Convention	Examples	
Subclass	-idae suffix	Commelinidae	Alismatidae
Order	-ales suffix	Poales	Alismatales
Family	-aceae suffix	Poaceae	Alismataceae
Genus	Italicized and capitalized	*Oryza*	*Sagittaria*
(Genus) species	Italicized, a binomial including generic name	*Oryza sativa*	*Sagittaria latifolia*

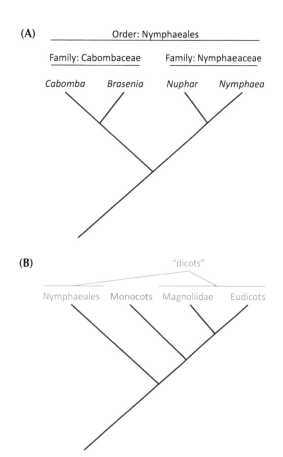

(A)

(B)

FIGURE 2.1 Examples of a Monophyletic Clade (A) Versus a Paraphyletic Clade (B). In the First Tree, all of the Named Groups of Branch Tips (Cabombaceae, Nymphaeaceae, and Nymphaeales) are Valid Clades Because they Each Include the Most Recent Common Ancestor and all Descendants of those Groups (Without Including Taxa that are not Members of the Clade). In the Paraphyletic Clade (B), the Taxa Included in the Group Given in Quotes ("Dicots") Include a Most Recent Common Ancestor, but the "Dicots" do not Include all Descendants of that Ancestor. The Monocots Descend from that most Recent Common Ancestor but are not Included Among those Taxa Historically Referred to as Dicots. Furthermore, We No Longer Refer to Nymphaeales or the Magnoliids as Dicots, and we Use the Term Eudicot when we Refer to the Actual Dicot Species. see Judd et al. (2016) and Soltis et al. (2018) for More Detailed Discussion on Monophyly.

fairly diverse suite of taxa that includes palms (Arecales: Arecaceae), dayflowers (Commelinales: Commelinaceae), and pickerelweed and water hyacinth (Commelinales: Pontederiaceae), among others. An interesting feature of this subclass, however, is that many of the orders, including those just mentioned, have a very high proportion of species adapted to aquatic and wetland habitats.

Within the level of order, we see progressively greater degrees of similarity among the **genera** (plural of genus) within a family, and finally among species within a genus (Figure 2.2). Nevertheless, a somewhat surprising truth about organizing taxa into these groups is that, despite our efforts, the groups are not as equivalent as we might like to think (Soltis et al. 2018; Stevens 2017). Generally, the taxa in each genus or family or order are analyzed using similar approaches and the results interpreted in the same manner, but the resulting groups may not be equivalent. That is, what constitutes a family, for example, in one order may be very different in terms of genetic and/or morphological diversity compared with a family in another, distantly

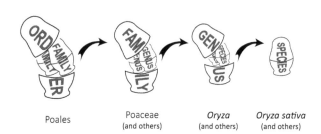

FIGURE 2.2 Taxonomic Organization Conceptually Resembles a Series of Matryoshka Dolls ("Russian" Nested Dolls). At Each Level, we can "Open Up" a Group and Find One or More Subordinate Groups Nested Inside, with Each of those also Containing One or more Subordinate Groups, and so on. The Numbers of Subordinate Groups at Each Level will be Determined by the Best Available Evolutionary Evidence Describing How Closely the Taxa within that Level are Related to One Another. The Numbers of Subordinate Groups and their Placement within the Higher-Level Groups can Change Over Time, as New Evidence is Gathered. Rice (*Oryza Sativa*) is Used here as an Example, as Discussed in the Text.

related order. The one thing each of these levels, and the named clades within them share, however, is that they must be conclusively monophyletic (Figure 2.1). So, for example, the order Poales contains an ancient, and probably extinct, most recent common ancestor to all the grasses, sedges, cattails, etc., along with all the members of those groups. The family Poaceae (the grasses), in turn, contains the most recent common ancestor of all the grasses, along with all genera of grasses, and the genus *Oryza* would include the most recent common ancestor of all species of *Oryza*, along with all species descended from that ancestor. This includes *Oryza sativa*, which is the name of the most commonly cultivated species of rice. This organizational scheme is essentially the taxonomic equivalent of a nesting doll (Figure 2.2), except that each time you open a layer, you might find multiple sets of nested dolls inside, instead of only one. In each of these examples, the most recent common ancestor would be represented by the branching point that gave rise to all known lineages within a clade. In Figure 2.1, for example, the branching point between the Cabombaceae and the Nymphaeaceae represents the most recent common ancestor to both families. We very rarely know the precise identity of this ancestor, but the approximate age of its existence can be inferred through statistical analyses of genetic data.

An interesting development related to some of these taxonomic details is that the group of scientists who are collaborating to disentangle the evolutionary relationships among the angiosperms (a **community** of scientists under the name "Angiosperm Phylogeny Group," or APG) is gradually moving away from a focus on the higher-level categories. That is, the APG focuses primarily at the level of order, family, genus, and species, and uses unranked, or rank-free, groupings above the level of order as organizational tools only (Soltis et al. 2018; Stevens 2017). This same approach has been taken by an international collaborative group that studies the eukaryotic tree of life, for many of the same reasons (Adl et al. 2019; Keeling and Burki 2019; Burki et al. 2019). Aside from the problem noted earlier regarding the validity of monocots versus "dicots," a very practical issue is that if we are constrained to a ranked system, the entire system might need to be rebuilt periodically as a result of the rapidity with which we are learning about evolutionary relationships among known species, while at the same time discovering new species (and genera, families, etc.) in some broad corners of biology (among the "protists," for example; Burki et al. 2019).

In this book, I use the APG classifications as a referenceable standard against which other existing classifications can be compared for a given taxon. Because the APG system is relatively new (initiated in 1998; Soltis et al. 2018) and is a living classification system, with new developments constantly being published, there is widespread disagreement between the APG classification of many plant species and classifications that are or have been in use up to now. I will point out some of the more important disagreements as we run across them in the following sections of this chapter. However, after this chapter, I primarily will use names of species as reported in scientific articles that I cite.

2.1.2 RECONSTRUCTING EVOLUTIONARY TREES

Before moving on to look at evolutionary relationships among specific groups of plants, we should briefly examine how those relationships are determined. I have borrowed some data from a paper by Donald Les and colleagues (Les et al. 1999) to help illustrate this process. Les et al. (1999) used a set of 68 **phenotypic characters** (visible or measurable attributes of the plants), along with three distinct regions of DNA to evaluate the phylogenetic relationships of eight genera in the order Nymphaeales. I borrowed a subset of the phenotypic character data for four of those genera (*Cabomba*, *Brasenia*, *Nuphar*, and *Nymphaea*; Figure 2.3) to show, at a very basic level, how one would go about determining a phylogeny for these four taxa, given the characters and character states shown in Table 2.2.

In Figure 2.4, I have mapped changes in character states for the 17 characters given in Table 2.2 across a tree suggesting the relationships of these four genera of plants. Four of these characters do not vary among the four genera, and we might assume that these would be characteristics shared with the most recent common ancestor of these four genera. Given that we lack an outgroup for comparison (**outgroup** = group that is not among our focal taxa), we can only assume those four character states to represent the ancestral, or **plesiomorphic**, state for these four traits. Without an outgroup, we also can only investigate these taxa to determine their relative relationships to one another, without regard to which branch may have diverged first. I cheated a bit in Figure 2.4 by referencing the **topology** (that is, the general branching pattern) of this tree against the full tree published by Les and colleagues.

One important note about these **bifurcating** trees (trees with two-way splits at each node) is that, at any bifurcation in the tree, we can imagine ourselves to freely rotate the branch at that node. So, in this case, I could have just as correctly placed *Cabomba* on the second branch tip, with *Brasenia* on the first, and similarly for the *Nuphar–Nymphaea* split, or the first main branch itself, placing the *Nuphar-Nymphaea* split on the left of the diagram. Another way to say this would be that we should read along the branches to determine relationships, rather than across the tips. As long as the relative placement of the nodes along the path from the base of the tree to the branch tips remains constant, as do the lengths of the branches (which is unimportant in this example), the information being conveyed by the tree remains the same.

Another important consideration here is that all the extant taxa at the branch tips have been evolving for the same amount of time following the split from their most recent common ancestor. The joint condition that the clades, or groups, can be rotated on a node and that the branch tip taxa have had the same amount of time to accumulate modifications since a given split means that we must

FIGURE 2.3   Examples of Species in the Four Genera Used in the Example from Table 2.2 and Figure 2.3. (A) *Cabomba Caroliniana*, (B) *Brasenia Schreberi*, (C) *Nuphar Advena*, and (D) *Nymphaea Odorata*.

be careful when imagining the appearance of the shared ancestor. It is easy to be tied into the notion that the taxon on the tip positioned closest to apex of the tree "evolved from" something that looked like the taxon on the tip of its sister branch. It is equally possible that both tips diverged from a shared common ancestor that looked nothing like either of the extant taxa.

When a character changes states, we refer to that as an apomorphy. **Apomorphies** shared among taxa at the tips of a branch are referred to as **synapomorphies**, while **autapomorphies** are those that are present in only one of the taxa in a clade, where a clade is a node along with all the tips that share that node. Note that this description of a clade is essentially the same that we used to describe a monophyletic group in the previous section. A node represents the most recent common ancestor, and the tips represent descendants from that ancestor.

In the tree in Figure 2.4, we see that there were five (of 17) characters that changed at some point after the *Nuphar–Nymphaea* clade split from the *Cabomba–Brasenia* clade, but before *Nuphar* and *Nymphaea* diverged from their most recent common ancestor. On the other branch, three (of 17) characters changed in the *Cabomba–Brasenia* clade after

that same early split from the common ancestor of all four groups, but before *Cabomba* and *Brasenia* split from their most recent common ancestor. Further changes in eight of the characters occurred later (with one of them, number 14, changing independently in two taxa), leading to the eventual appearance of these four genera as we now know them.

As we saw in Figure 2.1, this specific tree yields three monophyletic groups, or **clades**, above the level of genus. The node that subtends *Cabomba* and *Brasenia* is the base of what is known as the family Cabombaceae, a family that according to the APG classifications given by Stevens (2017) contains only these two genera and something like six species (Table 2.3). The right half of this tree is known as the Nymphaeaceae, which contains three to six genera, depending on the source that one consults. Some authors consider *Nymphaea* to be broadly defined to include not only those species of *Nymphaea*, strictly speaking, but also species that are sometimes assigned to the genera of *Euryale*, *Ondinea*, and *Victoria*. The latter is used by Stevens (2017) and was the basis for the data given in Table 2.3. These families account for two monophyletic clades in Figure 2.4; the third named clade in this tree is the order Nymphaeales, which contains all four of these genera and, thus, the

TABLE 2.2
Selected morphological characters used by Les et al. (1999) in phylogenetic analysis of the Nymphaeaceae. The specifics of these plant characters will make much more sense as we progress further in the book, but for now, just know that they represent various physical features of species in these genera. The form that each character (or trait or attribute) exhibits in each of these species is coded by a 0, 1, or 2, and a brief description is given in the rightmost column. For example, *Brasenia* and *Cabomba* both have flowers with five or fewer petals (coded as 0 for character #7), whereas *Nuphar* and *Nymphaea* both have flowers with more than five petals (coded as 1).

Character	Character states	Brasenia	Cabomba	Nuphar	Nymphaea
1	duration: 0 = perennial	0	0	0	0
2	floating leaf margins: 0 = not strongly upturned	0	0	0	0
3	shed pollen: 0 = individual grains	0	0	0	0
4	male gametophyte: 0 = two-celled pollen	0	0	0	0
5	floating leaf shape: 0 = cordate, 1 = peltate	1	1	0	0
6	winter buds: 0 = absent, 1 = present	1	1	0	0
7	petals: 0 = 5 or fewer, 1 = more than 5	0	0	1	1
8	stamen insertion: 0 = cyclic, 1 = spiral	0	0	1	1
9	stamens: 0 = 50 or fewer, 1 = more than 50	0	0	1	1
10	submersed leaves: 0 = present at maturity, 1 = absent at maturity	1	0	0	0
11	laticifers: 0 = absent, 1 = present	0	1	1	1
12	vessel distribution: 0 = roots/rhizomes, 1 = confined to roots, 2 = confined to stems	0	2	1	1
13	pollination: 0 = insects, 1 = wind	1	0	0	0
14	petal nectaries: 0 = absent, 1 = adaxial, 2 = **abaxial**	0	1	2	0
15	floral habit: 0 = aerial, 1 = aerial and floating	0	0	0	1
16	sepals: 0 = more than 4, 1 = 4, 2 = less than 4	2	2	0	1
17	fruit maturation: 0 = above water, 1 = under water	0	0	0	1

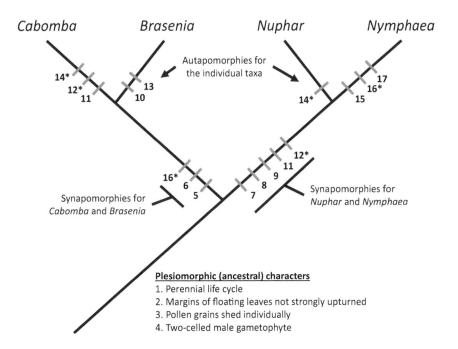

FIGURE 2.4 Reconstruction of the Relationships Among Our Four Nymphaeales Genera, Based on the Characters Given in Table 2.2. Traits Indicated by an Asterisk Evolved Independently on Different Branches of the Tree in this Reconstruction, Sometimes with Different Phenotypes. For Example, Trait #12 (Distribution of Vessels within the Vascular Tissues) Began, in this Tree, with Vessels Present in Roots and Rhizomes, but are Confined Only to Roots in *Nuphar* and *Nymphaea* and are Present Only in Stems in *Cabomba*. Data from Les et al. (1999).

TABLE 2.3

A summarization of commonly encountered families of aquatic and wetland plants, modified and updated from Sculthorpe (1967). Approximate numbers of genera and species per family obtained from Stevens (2017). Approximate molecular ages (million years before present) obtained from Li et al. (2019). Abbreviations: E = emergent, FL = floating-leafed, S = submersed, FF = free-floating, H = heterophyllous.

Order	Family	Noteworthy Genera	Approx. genera	Approx. species	Growth Forms	Age (Mya)
Nymphaeales	Hydatellaceae	*Hydatella*	1	10	E, S	159
	Cabombaceae	*Brasenia, Cabomba*	2	6	FL, H	114
	Nymphaeaceae	*Barclaya, Nuphar, Nymphaea*	3	60	FL	114
Alismatales	Alismataceae	*Alisma, Caldesia, Echinodorus, Hydrocleys, Limnocharis, Sagittaria*	15	88	E, FL, S, H	102
	Butomaceae	*Butomus*	1	1	H	97
	Hydrocharitaceae	*Blyxa, Egeria, Elodea, Halophila, Hydrilla, Limnobium, Najas, Ottelia, Thalassia, Vallisneria*	18	136	FF, S	97
	Aponogetonaceae	*Aponogeton*	1	50	FL, S	93
	Scheuchzeriaceae	*Scheuchzeria*	1	1	E	87
	Juncaginaceae	*Cycnogeton, Tetroncium, Triglochin*	3	30	E	74
	Araceae	*Lemna, Pistia, Spirodela, Wolffia, Wolffiella*	144	4,000–6,500	FF	73
	Potamogetonaceae	*Groenlandia, Potamogeton, Zannichellia*	4	102	FL, S, H	52
	Zosteraceae	*Phyllospadix, Zostera*	2	14	S	52
	Posidoniaceae	*Posidonia*	1	9	S	40
	Cymodoceaceae	*Cymodocea, Halodule, Syringodium, Thalassodendron*	5	16	S	36
	Ruppiaceae	*Ruppia*	1	10	S	36
Poales	Typhaceae	*Sparganium, Typha*	2	25	E, FL, H	121
	Mayacaceae	*Mayaca*	1	10	E, S	117
	Eriocaulaceae	*Eriocaulon, Syngonanthus*	7	1,160	E	115
	Xyridaceae	*Xyris*	5	325	E	112
	Restionaceae	*Empodisma, Sporadanthus*	64	545	E	106
	Cyperaceae	*Carex, Cladium, Cyperus, Eleocharis, Eriophorum, Rhynchospora, Schoenoplectus, Scirpus*	98	5,695	E	79
	Juncaceae	*Juncus, Luzula*	8	442	E, S	79
	Poaceae	*Andropogon, Echinochloa, Glyceria, Leersia, Oryza, Panicum, Paspalum, Phragmites, Saccharum, Spartina, Zizania, Zizaniopsis*	707	11,337	E	76
Commelinales	Pontederiaceae	*Heteranthera, Pontederia*	2	33	E, FL	82
Piperales	Saururaceae	*Anemopsis, Gymnotheca, Houttuynia, Saururus*	4	6	E	68
Ceratophyllales	Ceratophyllaceae	*Ceratophyllum*	1	6	S	159
Ranunculales	Ranunculaceae	*Aconitum, Anemone, Caltha, Clematis, Ranunculus*	62	2,525	E, FL, S, H	87
Proteales	Nelumbonaceae	*Nelumbo*	1	2	H	83
Saxifragales	Haloragaceae	*Myriophyllum, Proserpinaca*	9	145	E, S, H	53
Myrtales	Lythraceae	*Ammannia, Cuphea, Decodon, Lythrum, Rotala, Sonneratia, Trapa*	31	650	E, FF, Trees	70
Myrtales	Onagraceae	*Epilobum, Ludwigia*	22	656	E, FL, H	70
Malpighiales	Podostemaceae	*Podostemum, Tristicha, Weddellina, Zeylanidium*	54	300	S	75
Malpighiales	Elatinaceae	*Bergia, Elatine*	2	35	E, S	64
Caryophyllales	Droseraceae	*Aldrovanda, Dionaea, Drosera*	3	205	E	87
Caryophyllales	Polygonaceae	*Brunnichia, Persicaria, Polygonum, Rumex*	55	1,110	E	78
Cornales	Hydrostachyaceae	*Hydrostachys*	1	20	S	50–90
Ericales	Sarraceniaceae	*Darlingtonia, Heliamphora, Sarracenia*	3	32	E	87
Asterales	Menyanthaceae	*Menyanthes, Nymphoides*	5	58	E, FL	80
Asterales	Asteraceae	*Ambrosia, Aster, Baccharis, Bidens, Eupatorium, Mikania, Pluchea, Senecio, Solidago, Xanthium*	1,620	25,000	E	75
Lamiales	Plantaginaceae	*Bacopa, Callitriche, Hippuris, Limnophila, Plantago*	90	1,900	E, S	49
Lamiales	Lentibulariaceae	*Genlisea, Pinguicula, Utricularia*	3	350	E, FF	30

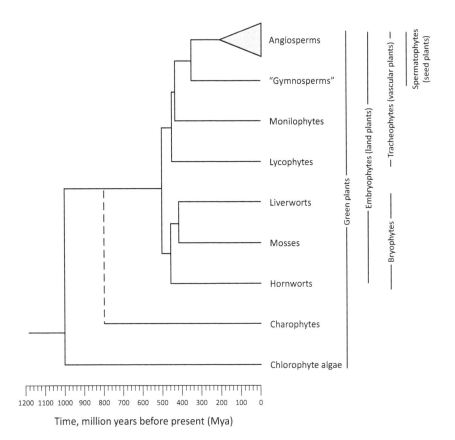

Time, million years before present (Mya)

FIGURE 2.5 A Chronogram Displaying Approximate Divergence Times for Major Groups within the Green Plant Phylogeny. Data Obtained from Bell, Soltis, and Soltis (2010), Clarke, Warnock, and Donoghue (2011), Guo et al. (2021), Heckman et al. (2001), Judd et al. (2016), Karol et al. (2001), Li et al. (2019), McCourt et al. (2016), Sanderson (2003), and Soltis et al. (2018). Divergence of the Charophytes is Represented by a Dashed Line Because the Timing Still Appears to be Somewhat Uncertain. Divergence Times within the Bryophytes were Estimated from Phylogenies in Harris et al. (2020) and Li et al. (2020).

Cabombaceae and the Nymphaeaceae, along with the third family in the Nymphaeales, the Hydatellaceae, that was not included in my example here but is listed in Table 2.3.

With this very general overview of how evolutionary relationships are reconstructed and why some groups are named while others are not, we can now consider some of the more important aquatic and wetland plant-containing clades. We will begin with a very broad perspective, looking at what some authors refer to as the green plants (Judd et al. 2016), including green algae, charophytes, and others, then transitioning through the nonvascular land plants, followed by a few groups of early diverging vascular plants before arriving at the angiosperms (Figures 2.5 and 2.6).

Incidentally, the taxa included in Table 2.3 are all angiosperms and were included based on either their inclusion by Sculthorpe (1967) in his table of families of aquatic vascular plants, by their frequent appearance in the plant surveys discussed in Chapter 3, or my own knowledge of their reputation as important components of other wetlands not included in Chapter 3. Table 2.3 is not meant to be an all-inclusive list of plants; it is meant to give a general idea of some of the more common angiosperms and of the broad diversity one may encounter across wetlands globally.

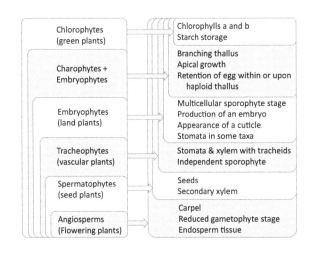

FIGURE 2.6 Hierarchical Taxonomy and Sharing of Characters within Selected Groups of Green Plants. On Both Sides of the Diagram, Outermost Boxes Include all Taxa (Left) or all Characters (Right) Contained in all the Boxes Inside them. For Example, the Embryophytes Include the Tracheophytes, which Include the Spermatophytes, Which Include the Angiosperms. the Embryophytes Share all the Characters from Chlorophylls a and b through Stomata (In Some Taxa). Stomata are the only Exception here, in that Stomata were Lost in some Moss Taxa. Based on Information from Judd et al. (2008).

2.2 THE GREEN PLANTS AS A STARTING POINT

The green plants, also known as Viridophytes (Judd et al. 2016) or Viridiplantae (Adl et al. 2019), but recently termed Chloroplastida (Adl et al. 2019), include the chlorophyte algae ("green algae") and the streptophytes. The streptophytes comprise the land plants along with some plants that have commonly been referred to as "green algae," which include a group of algae in the order Charales, known as the charophytes. This latter group is important in the context of understanding aquatic and wetland plants not only because they are evolutionarily the sister group to the land plants (Judd et al. 2008; Karol et al. 2001), but also because of their morphological and ecological similarities to a number of species of aquatic macrophytes.

As depicted in Figure 2.5, and as we currently understand, there was a major split roughly one billion years ago (1,000 million years ago [Mya]) between the chlorophyte algae and the streptophytes, which include the angiosperms. The branch within the streptophytes that led to the Charales diversified sometime prior to about 470 Mya to give the current members of that order (Karol et al. 2001).

2.2.1 CHAROPHYTES

According to Heckman et al. (2001), land plants may have appeared around 700 Mya, but this timing was disputed by Sanderson (2003), who placed the appearance of land plants at approximately 450–500 Mya (somewhere around the timing of the split between bryophytes and vascular plants). Discrepancies such as these, which are not the subject of this text, arise from differences in DNA sequences used in reconstructing evolutionary trees (both the identities and the numbers of sequences used), the number of taxa included in the reconstructions, the methods used to analyze those sequence data, dates of fossils used for calibration of trees, and the methods by which those calibrations are employed (Sanderson 2003; Soltis et al. 2018).

The timing of that split is somewhat important in this section because the Charales have been determined by some to be the **sister group** to land plants, meaning the split between Charales and the groups that went on to give rise to land plants had to have occurred prior to diversification, or species differentiation, within either of those groups (i.e., the Charales and the land plants). More specific to the Charales themselves, Karol et al. (2001) and McCourt et al. (2016) estimate that early diversification within this group began somewhere around 450–500 Mya, similar to the timing of early diversification in the land plants. McCourt et al. (2004) suggest that the divergence between ancestors of Charales and ancestors of the land plants must have occurred prior to about 500 Mya (as suggested by the prior data), but that the timing was unknown. However, others have suggested that the Charales are not a sister group to the land plants, but, instead, are sister to a clade that contains the land plants and one or more other algal lineages (Wickett et al. 2014). These elements of uncertainty about the timing of divergence and the exact position of Charales relative to land plants are the rationale for the dotted line connecting Charales to the land plant lineage in Figure 2.5.

Regardless of discrepancies on the timing of the major split between Charales and the land plants, it is widely accepted that land plants arose out of an "algal" lineage that also produced the macroalgae of the order Charales. In other words, there is strong support for a shared ancestor of the Charales and the land plants. Thus, perhaps surprisingly, the aquatic and wetland plants that are the major focus of this book are derived from lineages that, some 500-plus million years ago, left the water and diversified as the dominant **autotrophs** of terrestrial ecosystems. Recent evidence further suggests that the earliest plants may have been assisted in their colonization of land by fungal symbionts, with evidence of this provided through widespread presence of fungal-derived genes among all land plant lineages (Wang et al. 2021).

There is one extant family within the Charales (the Characeae), and it comprises six genera: *Chara*, *Lamprothamnium*, *Lychnothamnus*, *Nitellopsis*, *Nitella*, and *Tolypella* (McCourt et al. 2016). There are hundreds of species among these six genera, with *Chara* and *Nitella* being the most common genera; the remaining genera include 20 or fewer species each (McCourt et al. 2016). As mentioned earlier, charophyte algae are ecologically and morphologically similar to some aquatic angiosperm taxa (Figures 2.7 and 2.8). Under the right conditions, charophytes are the dominant plants in aquatic habitats, serving as a trophic base and structural habitat for numerous consumer groups (McCourt et al. 2016). Typical habitats for charophyte algae are slow moving, hard (alkaline) waters with abundant light and oxygen and relatively low concentrations of nutrients such as nitrogen and phosphorus (McCourt et al. 2016).

2.2.2 EARLY DIVERGING EMBRYOPHYTES— THE BRYOPHYTA

The group that I refer to here as early diverging embryophytes includes the hornworts, mosses, and liverworts and is termed the Bryophyta, or the bryophytes. As discussed in the previous section, the branch leading to this split between the bryophytes and tracheophytes (the vascular plants) began to diversify approximately 450–500 Mya, and the bryophyte lineages were among the first of the embryophytes to dominate terrestrial systems, even though they still relied strongly on water for reproduction and lacked vascular tissues (xylem and phloem) for the transport of water and dissolved nutrients. We do, however, see the development of stomata (openings in the plant epidermis for gas exchange) during the diversification of these groups (Judd et al. 2008), with mosses and hornworts possessing stomata, although in the mosses they are not always

FIGURE 2.7 *Chara* Species from a Pond in East-Central Mississippi. Characteristics to Note are the Whorled Branching Pattern (Multiple Lateral Branches Arising from Individual Nodes), the Unbranched Lateral Branches, and the Small Spiny Nodes Along the Lateral Branches, from Which Spores are Produced.

functional (e.g., in *Sphagnum* species) and have been lost in multiple moss lineages (Kenrick and Crane 1997). The appearances of stomata and a waxy cuticle in the evolution of early land plants (Figure 2.6) were important innovations for migration onto land because they allowed, to some degree, plants to control the rate at which water escapes plant tissues.

In terms of life cycle, in the embryophytes we see the development of a multicellular sporophyte stage. A key distinction between the bryophytes and the vascular plants is that the gametophyte stage (which already was represented by a multicellular form) dominates the life cycle in the bryophytes, whereas the sporophyte is the dominant generation in the vascular plants (Figure 2.9). In the bryophytes, the sporophyte simply takes the form of the stalked sporangium that grows from the apices of gametophyte branches (Figure 2.10).

From a wetland perspective, perhaps the most influential lineage within the bryophytes are the *Sphagnum* mosses in the order Sphagnales. *Sphagnum* species have a broad range of habitat affinities, ranging from entirely aquatic forms to those that inhabit "moderately damp" environments (Hutchinson 1975), but they are perhaps best known as foundation species in acidic bogs of northern boreal ecosystems (Keddy 2000; Mitsch and Gosselink 2000). Relatively high rates of biomass production (among the mosses), slow decomposition rates of dead *Sphagnum* tissues, chemical adsorption of important plant nutrients, and acidification of the wetland soil environment all make *Sphagnum* species ecologically important in peat accumulation and vegetation dynamics of boreal wetlands (Keddy 2000; Mitsch and Gosselink 2000; Turetsky et al. 2010). However, there are common occurrences of other bryophyte taxa in wetlands and aquatic habitats, such as *Ricciocarpos natans*, a species of aquatic liverwort, that occur in many wetlands of the southeastern US (Figure 2.11).

2.2.3 NON-ANGIOSPERM TRACHEOPHYTES

Within the tracheophytes (Figure 2.6), we see the appearance of xylem tissues composed of tracheids, which are dead, tube-like, water-conducting cells connected by a wall with openings called pits (Judd et al. 2008). Although they are dead, passage of water (and dissolved ions) between the individual tracheids is controlled by a thickened, lignified primary cell wall called a pit membrane (Judd et al. 2008; Kenrick and Crane 1997). These water-conducting vascular tissues allowed for an increase in plant size, owing to the ability to maintain water supply to tissues more distant from the soil or other water source. We also see in the tracheophytes the emergence of an independent sporophyte generation (Figures 2.6 and 2.9), accompanied by a progressive reduction in the size of the gametophyte generation. This progression eventually gives us the pollen grain (male gametophyte) and the embryo sac (female gametophyte, contained within the ovule) that we see in the angiosperms. These will be covered in much greater detail in Chapter 8.

Noteworthy clades within these non-angiosperm tracheophytes include the lycophytes (clubmosses, spike mosses, and isoetids; Figure 2.12), the monilophytes ("ferns," horsetails, whisk ferns; Figure 2.13), and the "gymnosperms" (Judd et al. 2008; Figures 2.14 and 2.15). The gymnosperms receive quotes here because if we consider all **extant** (currently living) and extinct species of naked seed plants together, they form a non-monophyletic group containing a monophyletic group of the extant cycads, ginkgo, gnetophytes, and conifers. There are many aquatic and wetland taxa within the lycophytes, monilophytes, and extant gymnosperms, a few of which will be specifically mentioned here; these and others will appear in later chapters, when relevant to a particular topic. Among the lycophytes, the isoetids (e.g., the genus *Isoetes*) are of interest because they are relatively widespread

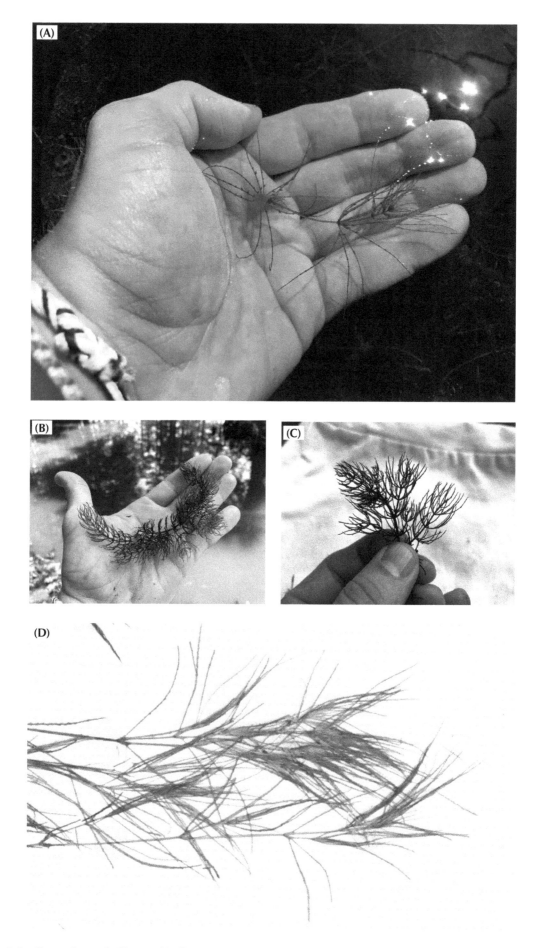

FIGURE 2.8 Comparison of *Chara* with Two Morphologically Similar Aquatic Angiosperms. (A) *Chara Species*, (B–C) *Ceratophyllum Demersum*, (D) *Najas Minor*. Note with C. *Demersum* that the Leaves are (B) Much Denser and (C) Dichotomously Branched (Each Leaf Divides into Two Parts). (D) Leaves in *Najas* are Opposite (Two per Node), but Dense Branching can Make them Appear Whorled. Older Leaves of *N. Minor* also may Become Stiff, with Spiny Teeth. Photo of *N. Minor* from the University of Alabama Herbarium (with Permission); Specimen Collected by Robert R. Haynes.

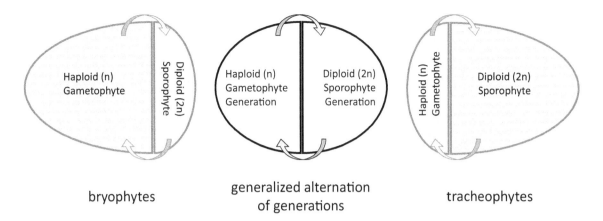

bryophytes

generalized alternation
of generations

tracheophytes

FIGURE 2.9 Alternation of Generations in Plants. Through Alternating Cycles of Sexual and Asexual Reproduction, Plants Will Produce Diploid, then Haploid Generations (Center Diagram). The Early Diverging Embryophytes (i.e., the Bryophytes, Left) are Characterized by having a much Smaller Diploid Sporophyte Generation that Relies on the Gametophyte for Nutrition (Illustrated by the Contrasting Sizes of the Haploid and Diploid Portions of the Figure). In Contrast, Tracheophytes have Evolved a Larger, Independent Sporophyte (Cycle Illustrated on the Right), Which is the Form of the Plant with which we are Most Familiar. The Gametophytes in Angiosperms are Represented by Pollen Grains and the Embryo Sac within the Developing Ovule.

FIGURE 2.10 In Wetter Climates or Habitats, Mosses Frequently Blanket the Forest Understory and Decaying Wood or Serve as a Major Foundation Layer for Wetland Habitats, as in the Boreal Peatlands. Upper Image: Blanket of Mosses Along a Small Glacial Lake Near Mendenhall Glacier, Juneau, Alaska. Lower Images: Mosses Growing on Decaying Tree in a Mississippi Forest Understory; the Stalked Structures Growing Out of the Gametophyte Thalli are the Sporophyte Generation, Each Topped by a Spore-Producing Sporangium (White Arrows in Bottom Right).

FIGURE 2.11   *Ricciocarpos Natans* ("Fringed Heartwort") is a Relatively Common Liverwort Species in Wetlands of the Eastern US. I have Seen it in many Wetlands I've Visited, Where it Looks a Bit Like Shredded Leaves Floating on the Water.

(Hutchinson 1975) and have some interesting photosynthetic pathways that will be explored in Chapter 6.

The monilophyte horsetails (Figure 2.13) often occupy littoral wetland margins or other **ecotonal** (transitional) habitats (Hutchinson 1975). The leptosporangiate ferns include 12,000 or so species that are commonly referred to as ferns (note, however, that there are some "ferns" in two other clades within the monilophytes, the Marattiales and the Ophioglossales; Judd et al. 2008). There are many wetland-associated species in this group, as well as some fully aquatic genera (e.g., *Azolla*, *Salvinia*, and *Marsilea*; Figure 2.16). *Azolla* species are noteworthy because of their symbiosis with nitrogen-fixing cyanobacteria (*Anabaena azollae*). Because of this, *Azolla* species sometimes are used an organic fertilizer in rice cultivation (Watanabe, Berja, and Del Rosario 1980).

As we move from the lycophytes and monilophytes into the spermatophyte (seed plant) clades, we see not only the

FIGURE 2.12   Clubmosses (Lycophytes), Growing in a Forest Understory with Miscellaneous Mosses, near Mendenhall Glacier, Juneau, Alaska. Upper: *Lycopodium Clavatum*. Lower: *Spinulum Annotinum*.

FIGURE 2.13   A few Examples of Wetland Monilophytes. (A) *Equisetum* Species Along Margins of a Glacial Lake Near Mendenhall Glacier. (B–C) Closer View of *Equisetum* from the Birmingham Botanical Gardens (Alabama, USA). (D) Bracken Fern Growing at the Birmingham Botanical Gardens. (E) "Fiddlehead" Structure of a Newly Emerging Fern Leaf. (F) Sori, or Clusters of Sporangia, on the Underside of a Leptosporangiate Fern Leaf.

FIGURE 2.14 *Ginkgo Biloba*, the Only Remaining Species in its Order (Ginkgoales, Ginkgoaceae). Upper Two Photos from a Plant at the Birmingham (Alabama) Botanical Gardens (Seeds, Not Fruit, Shown Upper Right). Lower Three Photos are of a Plant on the Mississippi State University Campus Trees Tour. This Species is Commonly Used as a Landscaping Tree because of its Attractive Foliage, which Turns Bright Yellow in Autumn.

FIGURE 2.15 (A–B) *Taxodium Ascendens* Cones and Branchlets/leaves in Everglades National Park, Florida. (C–D) *Taxodium Distichum* Swamp and Closer View (D) of the "Knees" or Pneumatophores Protruding from an Exposed *Taxodium Distichum* Root.

production of secondary xylem (i.e., wood) but also the emergence of the seed as a protective structure for the developing embryo (Figure 2.6). The seed is formed after fusion of sperm with an egg cell within the female gametophyte. The female gametophyte is produced within a spore, which is produced within the megasporangium, contained within the ovule (Judd et al. 2008). This entire complex, encased within the ovule, which serves as the outer **integument** (protective layer), then becomes the seed, within which the embryo will be nourished by female gametophyte tissue during germination and early establishment. The seed is

an important protective structure that further aided domination of terrestrial ecosystems by the spermatophytes because of the protection it provided against desiccation, among other threats. We will see this information again in Chapter 8, when we discuss sexual reproduction.

The only clade within the extant gymnosperm lineages that we encounter in aquatic and wetland habitats is that of the conifers, of which there are approximately 630 extant species (Judd et al. 2008). Among these are *Taxodium* species (bald cypress, pond cypress; Figure 2.15) of littoral, palustrine, and forested wetlands and many species in the

FIGURE 2.16   Aquatic Ferns from the Genus *Azolla* (A–B) and *Salvinia* (C–D).

Pinaceae that may be found in peatlands and forested wetlands (see, for example, Table 3.5). Other important wetland conifers are those of the boreal peatland ecosystems in North America and Eurasia. In North America, we see spruce (*Picea glauca*, *P. mariana*), fir (*Abies balsamea*, *A. lasiocarpa*), tamarack (*Larix laricina*), and pine species (*Pinus banksiana*, *P. contorta*) as important components of boreal peatlands (Pojar 1996). In boreal peatlands of Siberia, Cajander larch (*Larix cajanderi*), Siberian spruce (*Picea obovata*), and Scots pine (*Pinus sylvestris*) are often the dominant tree species (Lloyd, Bunn, and Berner 2011; Ludwig et al. 2018). There also is a monotypic genus of conifers in eastern Asia, *Glyptostrobus*, containing the species *G. pensilis*, that is morphologically and ecologically similar to *Taxodium* (Earle 2021).

2.3   EVOLUTION AND DIVERSIFICATION OF THE ANGIOSPERMS

In case you have not been counting, so far, we would have encountered maybe 2,000 species in the Charales, approximately 18,000 bryophyte species, 1,200 species in the lycophytes, something more than 12,000 species in the monilophytes, and roughly 1,000 species of extant gymnosperms (Judd et al. 2016). This gives us approximately 34,000 species of plants in the streptophyte clade, excluding the flowering plants (angiosperms). In contrast, depending on the sources being consulted, the angiosperms number somewhere from 257,000 to 400,000 species (Judd et al. 2016; Soltis et al. 2018).

Current estimates are that the angiosperms began to appear and diversify somewhere around 200–210 Mya (Guo et al. 2021; Li et al. 2019; Figures 2.5 and 2.17). Analyses conducted by Li et al. (2019) incorporated chloroplast DNA sequences from 80 genes obtained from 2,351 species representing 1,390 genera across 353 (of 416) angiosperm families, with at least one member from each of the 64 orders recognized by the Angiosperm Phylogeny Group. Additionally, Li et al. (2019) made use of DNA sequence data from 163 species (77 genera, 12 families) across the cycads, ginkgo, gnetophytes, and conifers (the extant gymnosperms) as a reference group for detailing the angiosperm tree. This was the first example to use such a massive amount of data to define evolutionary relationships among the angiosperms. Divergence times for angiosperm groups determined by Li et al. (2019) are used extensively throughout the remainder

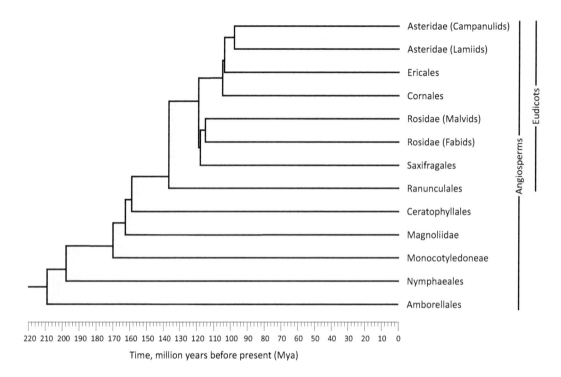

FIGURE 2.17 Chronogram Displaying Approximate Divergence Times for Major Groups within the Angiosperms (Flowering Plants). Data Obtained from Li et al. (2019; Using Chloroplast Genes) and Guo et al. (2021; Using Nuclear Genes). Green Shaded Areas Represent the Range of Divergence Times Given in those Two References. Note also that those Two References Disagree about the Order of Divergence of the Magnoliids and the Monocotyledoneae. See Text for Further Information on this.

of this chapter (including in Tables 2.3, 2.4, and 2.5) to help place important groups of aquatic and wetland plants within a time-specific evolutionary context.

It should be noted, however, that plants possess three distinct sources of DNA that could be used for conducting such analyses: the nucleus, the chloroplast, and the mitochondrion (Doyle 1993). As noted by Li et al. (2019), chloroplast genes have historically been the source of DNA upon which these studies have relied; however, as whole-genome sequencing has become more accessible, we have seen more use of nuclear genes for these analyses. Increased access to sequencing of the nuclear genome is allowing greater insight into the evolutionary relationships of plants (and other organisms), in part because of the greater number of genes and greater genetic diversity represented in the nuclear genome (Doyle 1993). In Figure 2.17, we can see some examples of the degree to which analyses of the nuclear genome agree, or disagree, with similar analyses using only chloroplast genes. Generally, the divergence times for major groups of plants show close agreement (within ten million years or less) in the analyses of chloroplast genes by Li et al. (2019) and the similar analyses of nuclear genes by Guo et al. (2021). One major difference, however, between those two analyses was the order in which the Magnoliids and the monocots (Monocotyledoneae) diverged from their sister groups. The analyses of chloroplast genes by Li et al. (2019) suggested the Magnoliids diverged earlier, but within one million years of the monocots. Other analyses using nuclear genomes agree with Guo et al. (2021) that the monocots

were the earlier diverging of these two groups (e.g., Yang et al. 2020). It has been suggested that some of the disagreement between chloroplast-based analyses and those using nuclear genomes may be the result of potential hybridization that occurred deep in angiosperm history between ancient monocots and dicots, or that another, similar, evolutionary mechanism has blurred genetic distinctions among these groups (Guo et al. 2021; Yang et al. 2020).

2.3.1 Nymphaeales

The branch that produced the Nymphaeales diverged from the remaining angiosperms approximately 190–200 Mya, and the order itself has an early diversification time of about 114 Mya (Figure 2.17, Table 2.3). Thus, the Nymphaeales represent one of the earliest diverging lines of angiosperms, nestled between the Amborellaceae, represented by only one extant forest understory tree species, and the order Austrobaileyales, which comprises three families (~100 species) of trees, shrubs, and vines (Soltis et al. 2018; Stevens 2017). For those who are more interested in relationships among these groups, Soltis et al. (2018) devote a full chapter to discussing the various similarities and differences among these three basal orders of angiosperms.

The Nymphaeales now comprises three families, which are well represented among the aquatic and wetland flora (Table 2.3) and two of which we saw earlier in the chapter (Figure 2.3). The Nymphaeaceae, as mentioned previously, includes either three or six genera, depending on the

TABLE 2.4

Core mangrove plant species. These are species commonly found among the dominant vegetation within core mangrove habitat. Original list from Spalding, Kainuma, and Collins (2010). Numbers of genera per family and approximate ages (million years before present) of genera and subfamily groups obtained from Stevens (2017). Geography: IWP = Indian Ocean–West Pacific; AEP = Atlantic Ocean (and Caribbean)–East Pacific.

Species	Geography	Genera in Family	Family Age, Mya (Li et al. 2019)	Family Age, Mya (Stevens 2017)	Genus or Subfamily Age, Mya
		Malpighiales—Rhizophoraceae			
Bruguiera cylindrica	IWP	16	66	42	–
Bruguiera exaristata	IWP	16	66	42	–
Bruguiera gymnorhiza	IWP	16	66	42	–
Bruguiera hainesii	IWP	16	66	42	–
Bruguiera parviflora	IWP	16	66	42	–
Bruguiera sexangula	IWP	16	66	42	–
Ceriops australis	IWP	16	66	42	–
Ceriops decandra	IWP	16	66	42	–
Ceriops tagal	IWP	16	66	42	–
Kandelia candel	IWP	16	66	42	–
Kandelia obovata	IWP	16	66	42	–
Rhizophora apiculata	IWP	16	66	42	40–50
Rhizophora mangle	AEP	16	66	42	40–50
Rhizophora mucronata	IWP	16	66	42	40–50
Rhizophora racemosa	AEP	16	66	42	40–50
Rhizophora samoensis	IWP	16	66	42	40–50
Rhizophora stylosa	IWP	16	66	42	40–50
		Myrtales—Combretaceae			
Conocarpus erectus	AEP	14	89	46–102	–
Laguncularia racemosa	AEP	14	89	46–102	23
Lumnitzera littorea	IWP	14	89	46–102	–
Lumnitzera racemosa	IWP	14	89	46–102	–
		Myrtales—Lythraceae			
Sonneratia alba	IWP	31	70	46–95	–
Sonneratia apetala	IWP	31	70	46–95	–
Sonneratia caseolaris	IWP	31	70	46–95	–
Sonneratia griffithii	IWP	31	70	46–95	–
Sonneratia lanceolata	IWP	31	70	46–95	–
Sonneratia ovata	IWP	31	70	46–95	–
		Sapindales—Meliaceae			
Aglaia cucullata	IWP	50	69	36–96	54–78
Xylocarpus granatum	IWP	50	69	36–96	39–75
Xylocarpus moluccensis	IWP	50	69	36–96	39–75
Ericales—Tetrameristaceae					
Pelliciera rhizophorae	AEP	3	46	25–45	–
		Lamiales—Acanthaceae			
Avicennia alba	IWP	220	30	50–90	40
Avicennia bicolor	AEP	220	30	50–90	40
Avicennia germinans	AEP	220	30	50–90	40
Avicennia integra	IWP	220	30	50–90	40
Avicennia marina	IWP	220	30	50–90	40
Avicennia officinalis	IWP	220	30	50–90	40
Avicennia rumphiana	IWP	220	30	50–90	40
Avicennia schaueriana	AEP	220	30	50–90	40

TABLE 2.5

Fringe mangrove plant species. These are species that may be found as subordinate taxa among the vegetation within core mangrove habitat or in transitional, fringe habitat peripheral to mangrove swamps. Original list from Spalding, Kainuma, and Collins (2010). Numbers of genera per family and approximate ages (million years before present) of genera and subfamily groups obtained from Stevens (2017). Geography: IWP = Indian Ocean–West Pacific; AEP = Atlantic Ocean (and Caribbean)–East Pacific.

Species	Geography	Genera in Family	Family Age, Mya (Li et al. 2019)	Family Age, Mya (Stevens 2017)	Genus or Subfamily Age, Mya
Polypodiales—Pteridaceae					
Acrostichum aureum	AEP	53	–	87–106	–
Acrostichum aureum	IWP	53	–	87–106	–
Acrostichum danaeifolium	IWP	53	–	87–106	–
Acrostichum speciosum	IWP	53	–	87–106	–
Arecales—Arecaceae					
Nypa fruticans	IWP	188	96	60–110	50–85
Malpighiales—Euphorbiaceae					
Excoecaria agallocha	IWP	218	60	90–102	–
Excoecaria indica	IWP	218	60	90–102	–
Fabales—Fabaceae					
Cynometra iripa	IWP	766	84	68–92	58
Mora oleifera	AEP	766	84	68–92	58
Myrtales—Lythraceae					
Pemphis acidula	IWP	31	70	46–95	–
Malvales—Malvaceae					
Camptostemon philippinense	IWP	243	77	40–71	26
Camptostemon schultzii	IWP	243	77	40–71	26
Heritiera fomes	IWP	243	77	40–71	–
Heritiera globosa	IWP	243	77	40–71	–
Heritiera littoralis	IWP	243	77	40–71	–
Osbornia octodonta	IWP	131	84	85–95	55
Caryophyllales—Plumbaginaceae					
Aegialitis annulata	IWP	29	78	17–43	–
Aegialitis rotundifolia	IWP	29	78	17–43	–
Ericales—Ebenaceae					
Diospyros littorea	IWP	4	73	54–57	38–42
Ericales—Primulaceae					
Aegiceras corniculatum	IWP	58	71	46–80	53
Aegiceras floridum	IWP	58	71	46–80	53
Gentianales—Rubiaceae					
Scyphiphora hydrophylacea	IWP	614	70	55–87	–
Lamiales—Acanthaceae					
Acanthus ebracteatus	IWP	220	30	50–90	70–80
Acanthus ilicifolius	IWP	220	30	50–90	70–80
Lamiales—Bignoniaceae					
Dolichandrone spathacea	IWP	110	29	52	–
Tabebuia palustris	AEP	110	29	52	–

source being consulted. *Nymphaea* commonly is divided into the genera *Nymphaea*, *Ondinea*, *Euryale*, and *Victoria* (Figure 2.18), although some recent genetic studies suggest these might be combined into the genus *Nymphaea* (Stevens 2017). The other two genera in this family are *Nuphar* and *Barclaya* (Table 2.3).

The other two families in the Nymphaeales are the Hydatellaceae and the Cabombaceae. Cabombaceae

FIGURE 2.18 *Euryale Ferox* (Top) and *Victoria Cruziana* (Bottom) at the Fairchild Botanical Garden in Miami, Florida. Species Identifications Courtesy of Dr. Brett Jestrow, Fairchild Botanical Garden Director of Collections.

includes only two genera (*Cabomba* and *Brasenia*), while the Hydatellaceae includes one or two genera. Again, depending on the source consulted, one may find the Hydatellaceae split into *Hydatella* and *Trithuria*, or one may find only the genus *Hydatella* or only the genus *Trithuria* (Soltis et al. 2018; Stevens 2017).

Sculthorpe (1967) provides a detailed table (his Table 1.2) with information on then-recognized families of aquatic vascular plants (recall that these would include everything from the lycophytes onward). At that time, the genus *Nelumbo* (Figures 1.9 and 2.19) was included in the Nymphaeaceae because both groups produce embryos with two seed leaves, or **cotyledons**, and produce flowers that are somewhat similar superficially. We now know that species in *Nelumbo* (there are only two) belong to their own family (Nelumbonaceae), in the order Proteales, among what are referred to as the **basal eudicots** (i.e., the true, or core, dicots). The species of *Nelumbo* are: *N. lutea* and *N. nucifera* (Flora of North America Editorial Committee 1993; Stevens 2017).

FIGURE 2.19   *Nelumbo Lutea*, One of Two Species in the Nelumbonaceae, Showing this Species' Flowers (Upper Photos) and Round, Peltate, Water-Repellent Leaves (Lower Photos). In Chapter 1, the Variable Positioning of Leaves in this Species was Mentioned; Some Leaves may Float on the Water's Surface (Lower Left), while Others are Elevated Well above the Water, and often above Neighboring Vegetation (Lower Right).

2.3.2 MONOCOTS

The branch giving rise to the monocots diverged from the remaining angiosperms (the Magnoliids + Ceratophyllales + eudicots) around 162–170 Mya (Figure 2.17). There are approximately 52,000 species of monocots, with around half of these occurring in either the Poaceae (~9,000 grass species) or the Orchidaceae (~18,000 orchid species; Soltis et al. 2018). Much of the order-level diversification within the monocots occurred around 126 to 138 Mya (Figure 2.20), giving rise to lineages of six orders, including those containing the Poaceae (Poales) and the Orchidaceae (Asparagales).

The monocots are of considerable importance in aquatic and wetland habitats, as suggested by their prevalence in Table 2.3 (Alismatales, Poales, Commelinales) and their frequency in wetlands from around the world represented in Table 3.3. Of particular importance are the Poaceae (grasses) and Cyperaceae (sedges), along with Juncaceae (rushes), Typhaceae (cattails), and Restionaceae (a southern hemisphere family morphologically resembling a cross between grasses and sedges) of the Poales (Figures 2.21–2.25) and all the families in the Alismatales, as will be discussed in a later section. Because of the prevalence of monocots among the wetland and truly aquatic taxa, many of the examples used throughout this book will make use of monocot species.

Quite a few changes in the taxonomic classification of monocots have taken place since Sculthorpe (1967) produced his table of families of aquatic vascular plants. Some of the more important changes are described here, based on information provided by Stevens (2017). Several genera that previously were included in the Butomaceae are now known to belong to the Alismataceae. *Butomus* is now the only genus in Butomaceae, and it has only one described species, *Butomus umbellatus* (Figure 2.26). The Zannichelliaceae, as described in Sculthorpe (1967), has now been split, with the genera *Althenia* and *Zannichellia* falling into Potamogetonaceae, while *Amphibolis*, *Cymodocea*, *Halodule*, *Syringodium*, and *Thalassodendron* have been kept together under the new family name Cymodoceaceae. The genus *Najas* (Najadaceae; Figure 2.8) has now been grouped into the Hydrocharitaceae, and all genera formerly in the Pontederiaceae (e.g., *Eichhornia*; Figure 2.27) have now been collapsed into either the genus *Heteranthera* or *Pontederia* (Pellegrini et al. 2018). The Lemnaceae (commonly referred to as the duckweeds; Figure 1.1) now are considered part of the Araceae, along with more recognizable Araceae species, such as *Pistia stratiotes* (Figure 2.28). Finally, Sparganiaceae (*Sparganium*; Figure 2.29) has been collapsed into the Typhaceae, with those species keeping the generic name *Sparganium*.

Another item of note in the monocots is that the relative positions of some groups within the clade remain ambiguous, owing to discordance among results obtained with differing methodologies. For example, some studies place the order Acorales as basal, or sister, to all other monocots, while others place them sister to the Alismatales deeper within the monocots (Les 2020). Along those lines, the Alismatales sometimes are considered as a subclass that contains four orders rather than as a single order, as has been adopted by the APG and illustrated in Figure 2.20. Finally, assignment of species to the Liliales and Asparagales also appears to have been debated as a result of conflicting results among published studies on these orders as well (Les 2020).

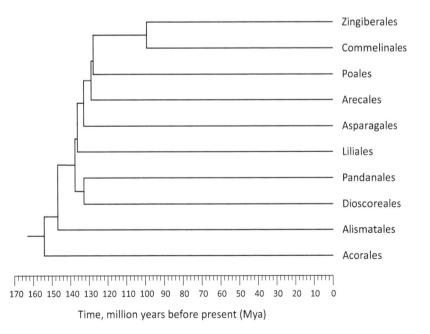

FIGURE 2.20 Chronogram of Monocot Orders, Displaying Approximate Divergence Times for Major Orders within the Clade. Data Obtained from Li et al. (2019).

(A)

(B)

(C)

(D)

FIGURE 2.21   A few Wetland Species from the Poaceae. (A) *Panicum Hemitomon* from the MacArthur Agro-Ecology Research Center, Lake Placid, Florida. (B) *Leersia Oryzoides* from Noxubee National Wildlife Refuge, Mississippi. (C) *Panicum Repens* from South Mississippi and (D) Inflorescence of *P. Repens*.

2.3.3   MAGNOLIIDAE

The Magnoliid clade split from the remaining angiosperms around 164 Mya (30 million or so years after the split of the Nymphaeales; Figure 2.17). This group includes four fairly diverse orders (Magnoliales, Laurales, Canellales, and Piperales) with around 10,000 species total (Soltis et al. 2018; Stevens 2017). Some of the more common families representing the Magnoliids in wetland habitats are the Magnoliaceae (magnolias), the Lauraceae (bays, laurels, spicebush), Aristolochiaceae (birthworts), and Saururaceae (lizard tails; Figure 2.30). These include trees, shrubs, vines, and herbaceous plants across multiple wetland types, from understories

and canopies of forested wetlands to permanently inundated margins of lakes and other deep-water wetlands.

2.3.4   CERATOPHYLLALES

The Ceratophyllales has been difficult to place in phylogenetic analyses using morphological data and genetic data and in studies using both types of data (Soltis et al. 2018). At various points in time, the Ceratophyllales were thought to have arisen out of the Nymphaeales or from among the basal-most groups of angiosperms, prior to divergence of the monocots (Les 2018). Recent studies making use of massive amounts of nuclear and/or chloroplast genetic

FIGURE 2.22 Examples from One of the more Characteristic Genera of Wetland Sedges, Carex. (A) *Carex Aureolensis*, (B) *Carex Comosa*, (C) *Carex Hyalinolepis*, (D) *Carex Lupulina*, and (E) *Carex Decomposita*, Known Commonly as Cypress Knee Sedge. The Latter is Shown here Growing with Several other Wetland Species around the Base of a Cypress, *Taxodium Distichum*. Photos A–C Provided Courtesy of John R. Gwaltney at southeasternflora.com, and Photos D–E Provided Courtesy of Charles T. Bryson.

data have largely agreed on the general placement given in Figure 2.17 (Guo et al. 2021; Leebens-Mack et al. 2019; Li et al. 2019; Yang et al. 2020). The Ceratophyllales consists of a single family of aquatic plants (Ceratophyllaceae) with a single genus (*Ceratophyllum*; Figure 2.31) and five to 30 species (Les 2018; Soltis et al. 2018; Stevens 2017). The time of divergence for the Ceratophyllaceae from the other angiosperms is 155–160 Mya (Guo et al. 2021; Li et al. 2019), between that of the Magnoliids and the clade referred to as the core dicots, or eudicots (Figure 2.17).

The genus *Ceratophyllum* is very broadly distributed, occurring on all continents, except Antarctica; thus, it is one of the more commonly encountered genera among the fully aquatic species in freshwater habitats. As indicated in Chapter 1, *Ceratophyllum* is a genus of free-floating submersed plants that lack roots, making them a bit distinctive among the submersed species. Most angiosperm genera that bear a morphological resemblance to *Ceratophyllum* (e.g., *Najas*; Figure 2.8) will occur rooted in the soil or sediment, unless disturbed.

FIGURE 2.23   Additional Wetland Species from the Cyperaceae. (A) *Rhynchospora Corniculata* from Noxubee National Wildlife Refuge, Mississippi. (B) *Scirpus Cyperinus* from North Mississippi. (C–D) *Dichromena* [*Rhynchospora*] *Colorata* from Everglades National Park (Florida, USA).

2.3.5   EUDICOTS

The final stop in this overview of evolutionary relationships among the aquatic and wetland groups of green plants is the eudicot clade, those groups that diversified after divergence from the Ceratophyllales (Figure 2.17). There was considerable diversification within the eudicots during about 95–120 Mya (Figure 2.17; Li et al. 2019; Soltis et al. 2018). There are 160,000–300,000 species in the eudicots, depending on the source of species-level classifications, as mentioned previously. Some of the larger families in the eudicots are the Asteraceae (asters: ~25,000 species), Fabaceae (legumes: 20,000), Lamiaceae (mints: 7,300), Brassicaceae (mustards: 4,000), Rosaceae (roses: 3,000), Solanaceae (nightshades and tomatoes: 2,500), and the Ranunculaceae (buttercups: 2,500). Perhaps unsurprisingly, with this level of species diversity, all these larger families have representatives among the aquatic and wetland plants (Figure 2.32; Table 3.4).

There are a few smaller families of eudicots that contain some relatively common truly aquatic species.

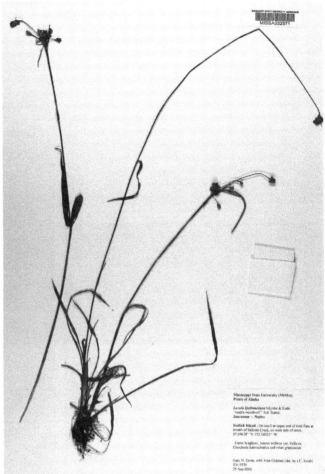

FIGURE 2.24   Representatives of the Juncaceae. Left: *Juncus Effusus*; Upper Left is an Individual Tussock; Lower Left is a Beaver Pond Wetland Whose Vegetation is Dominated Largely by *J. Effusus*. Right: a Specimen of *Luzula Kjellmaniana* that I Collected on Kodiak Island, Alaska (Identification Determined by J. C. Swab).

We have seen one of these in Chapter 1 and in section 2.3.1, when we addressed the similarity of *Nelumbo* to the Nymphaeaceae. The genus *Nelumbo* (Proteales, Nelumbonaceae; Figure 2.19) is among the basal eudicot groups, along with *Ranunculus*, of the Ranunculales. A few others worth noting are the Haloragaceae, Lythraceae, Onagraceae, and Plantaginaceae (Table 2.3, Figure 2.33). The first three of these include some species that are recognized as invasive weeds in multiple parts of the world: *Myriophyllum* species in the Haloragaceae, *Lythrum* and *Rotala* species in the Lythraceae, and *Ludwigia* species in the Onagraceae. Another interesting genus that we will see again in multiple chapters is the genus *Callitriche*, in the Plantaginaceae whose age is estimated at just under 50 Mya. *Callitriche* species display some interesting forms of **self-pollination** and commonly exhibit heterophyllous leaf morphologies. As with the monocots, we will see many examples from this group throughout the remaining chapters in this text.

2.4 TRULY AQUATIC PLANTS

As defined in Chapter 1, *aquatic plants* are those that complete their lifecycles with all nonflowering parts submersed in or supported by the water. We see examples of these throughout the green plants (except in the extant gymnosperms): charophytes, mosses, isoetids, ferns, and angiosperms from the Nymphaeales to the Lentibulariaceae, a family that appeared only around 30 Mya and includes the aquatic "carnivores" in the genus *Utricularia* (Figure 2.34, Table 2.3).

However, the monocot order Alismatales includes a disproportionately high diversity of aquatic plant species (Figure 2.35), and all of the marine-adapted aquatic angiosperms are found in five of the 14 families in the Alismatales: Hydrocharitaceae (~116 species total in this family), Zosteraceae (14), Cymodoceaceae (16), Ruppiaceae (10), Posidoniaceae (9) (Soltis et al. 2018; Stevens 2017). A few additional aquatic species are adapted to brackish waters and are scattered among the Aponogetonaceae, Juncaginaceae, and Potamogetonaceae (Les et al. 1997).

FIGURE 2.25 *Typha Latifolia*, Clockwise, from Top Left: General Growth Habit, Flowering/Fruiting Plant Growing in an Experimental Tank, and Showing Belowground Rhizome Connections Among Three Excavated Cattail Shoots in the Bottom Left.

Beyond simply receiving (or requiring) physical support by the water during growth and development, some species of aquatic plants have further adapted to the aquatic environment by using water to convey pollen from anther to stigma during reproduction (recall from the previous chapter that pollen transport by water is called *hydrophily*). This will be explored in much more detail in Chapter 8, but it is important to note here regarding aquatic plant origins and diversity. Hydrophilous species are present in all genera of Cymodoceaceae, four genera of Hydrocharitaceae, the Posidoniaceae, the Ruppiaceae, both genera of Zosteraceae, and all species that previously were classified as Zannichelliaceae, which now are classified under the Potamogetonaceae (Les et al. 1997).

2.5 TAXONOMIC DIVERSITY OF MANGROVE SPECIES

A final group of plants that we will examine before moving on to look at wetland types and their associated plant

diversity is the mangroves. These species are the dominant plants of mangrove wetlands, or mangals, which can be found along coastlines between about 30° N and 30° S latitude (Figures 2.36 and 2.37). Mangrove wetlands are dominated by a broad diversity of trees or shrubs (or occasionally ferns) from one family of leptosporangiate ferns (the Pteridaceae), one family of monocots (Arecaceae), and 14 families from across nine orders of eudicots (Tables 2.4 and 2.5). Even among those species commonly found to dominate core areas of mangrove wetlands, the families of these species arose over a span of time from about 30 to 90 Mya, and the ages of the genera themselves are estimated at 23–80 million years old (Table 2.4). Thus, adaptations to tolerate the salinity, wave action, tidal forces, and other stresses associated with living in marine coastal habitats appeared multiple times in mangrove species over a considerable period of evolutionary time. Among the adaptations that we see among mangrove trees are prop roots for stability (e.g., *Rhizophora mangle*, Figure 2.37), pneumatophores for root aeration (e.g., *Avicennia germinans*, Figure 2.37),

FIGURE 2.26 *Butomus Umbellatus* (Alismatales: Butomaceae). (A) Plants Exhibiting Emergent Growth Form at Low Water Level, (B) Plants at Almost Two Meters Depth, with Submersed Growth Form, (C) Vegetative Rhizome Buds, and (D) Pseudoviviparous Buds Produced within an Inflorescence.

and some interesting reproductive strategies in which seeds germinate before leaving the parent plant (see Chapter 8).

The broad evolutionary diversity among mangrove species contrasts somewhat with what we saw in the previous section, where we find all truly aquatic marine angiosperms within only one order (five families), although family ages for those species span a similar period as we see among mangrove trees (36–97 million years; Table 2.3). It also is interesting to note that all families represented among the "fringe" mangrove species (Table 2.5) are commonly found in other types of wetlands, often as shrub or tree species, suggesting potential exploitation of adaptations to general wetland conditions as part of expansion into the fringes of mangal ecosystems.

Mangrove ecosystems are incredibly important because of their high productivity, their function as nurseries for many fish and shellfish species, and their potential to stabilize shorelines and buffer inland ecosystems, as well as human population centers, from wind and storm surge associated with tropical storms. Unfortunately, their position along marine coastlines places these habitats at risk from ongoing sea level rise caused by coastal subsidence and climate change. We will visit mangals and mangrove species again in multiple chapters, when we learn about plant diversity in different types of wetlands (Chapter 3), adaptations for dealing with sediment anoxia (oxygen deficit, Chapter 6), and plant reproductive strategies (Chapter 8) and when we look at potential impacts of climate change (Chapter 11).

2.6 IN SUMMARY

An understanding of evolutionary history helps provide context for the suite of adaptations that we see among the

FIGURE 2.27   Common Species from the Pontederiaceae. (A–B) *Eichhornia* [*Pontederia*] *Crassipes* near Columbus, Mississippi; (C–D) *Pontederia Cordata* at the Birmingham (Alabama) Botanical Gardens.

diverse plant species that inhabit wetlands and aquatic habitats. Knowing the evolutionary origins of species of interest also can help us gain an increasingly sophisticated perspective on the question "Why does this plant live here?" and allow us to dig much deeper into ecological and evolutionary mechanisms that drive patterns we see in nature.

One outcome of gaining this increased insight into plants' evolutionary origins through increased efficiency in genetic sequencing and analyses of these complex datasets

has been the rearrangement of taxonomic organization at all levels. In the past two decades, we have seen the higher-level taxonomy reorganized to the point that many systematists have embraced the idea of rank-free taxonomies above the level of orders (even though many of the older names have been retained). At lower taxonomic levels (order and below), there has been considerable reorganization as we have discovered the more detailed subtleties of evolutionary lineages, with many families being split, new

FIGURE 2.28   Four Genera that Previously were Assigned to the Lemnaceae are now Classified Based on Genetic Evidence as Belonging to the Araceae. (A) *Lemna*, (B) *Spirodela*, (C) *Wolffiella*, and (D) *Wolffia*. *Wolffia* are the Tiny Somewhat Spherical Plants on the Surface of the Water and Stranded atop a Leaf of Brasenia Schreberi in Panel D. They also can be Seen off to the Right of the Star-Shaped *Wolffiella* in Panel C. (E–F) another Common Aquatic Member of the Araceae, *Pistia Stratiotes*, Showing General Growth Habit and a Close-Up of the Inflorescence.

FIGURE 2.29 (A–C) *Sparganium Americanum* from a Beaver Wetland in East-Central Mississippi. Note the Erect, Emergent Growth in Slow-Moving Waters of a and B, but Submersed-Floating Growth Nearer the Stream Channel in Panel C. (D) *Sparganium Angustifolium* from a Glacial Lake Near Mendenhall Glacier, Juneau, Alaska.

family names arising, and other families remaining intact but being placed within other pre-existing families. For the field biologist, these changes often seem no more than a complicating annoyance, but findings such as these can be incredibly informative in answering questions about the origins and mechanisms of biological diversification.

In this chapter, we traced the origins of aquatic and wetland angiosperms from the green algae and charophyte macroalgae through the early land plants, vascular plants, and finally the seed plants. Along the way, we also saw the

series of adaptations that paved the way for the ancestors of each of these lineages to progress from an aquatic existence to successful colonization, then dominance of terrestrial habitats, only to end up back in the water through the aid of such adaptations as hydrophilous pollination. Among the more important adaptations were the appearance of a waxy cuticle that helped to retain moisture, the appearance of stomata to allow gas exchange between the plant and the atmosphere, and vascular tissues that enabled more efficient movement of water throughout the plant.

FIGURE 2.30 *Saururus Cernuus*, in the Saururaceae (Order Piperales) of the Magnoliid Clade of Basal Angiosperms.

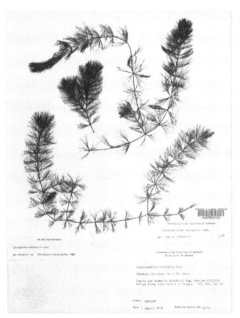
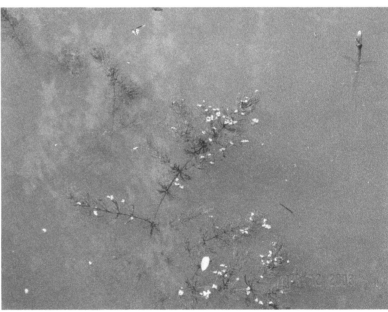

FIGURE 2.31 *Ceratophyllum Echinatum* from Huntsville, Alabama (Left) and *Ceratophyllum Demersum* from Holly Springs, Mississippi (Right). Photo of *C. Echinatum* from the University of Alabama Herbarium (With Permission); Specimen Collected by Robert R. Haynes and Identification Determined by Donald H. Les.

Finally, we paid special attention to a few groups of plants that commonly appear as important components of aquatic and wetland habitats. We saw a very early diverging group of angiosperms (the Nymphaeales) that have been quite successful among the aquatic plants, appearing commonly in wetlands across the world (we will see more evidence of this in the next chapter). Among the monocots, whose earliest ancestors split from the other angiosperms around 160 Mya (only about 50 million years after the appearance of the first angiosperms), we see the Poales and Alismatales as two additional lineages that are among the most important of aquatic and wetland plants. Within the Alismatales, we see several families of plants that have evolved adaptations to allow them to live a fully aquatic existence, in addition to finding here the only marine-adapted aquatic angiosperms. In contrast to this concentration of aquatic and wetland taxa within the

FIGURE 2.32 A Selection of Representatives from Six of the Largest Eudicot Families: (A) *Bidens* Species (Asteraceae), (B) *Apios Americana* (Fabaceae), (C) *Lycopus Virginiana* (Lamiaceae), (D) *Neobeckia Aquatica* (Brassicaceae), (E) *Rosa Species* (Rosaceae), and (F) *Ranunculus Aquatilis* (Ranunculaceae).

FIGURE 2.33 Common Wetland and Aquatic Genera from some of the Smaller Eudicot Families: (A–B) *Myriophyllum Aquaticum* and *Proserpinaca Palustris* (Haloragaceae), (C) *Rotala Rotundifolia* (Lythraceae), (D) *Ludwigia Decurrens* (Onagraceae), and (E) *Callitriche Heterophylla* (Plantaginaceae).

FIGURE 2.34   Two Species of *Utricularia* (Lentibulariaceae) Encountered in the Southeastern United States. (A–B) *Utricularia Biflora* (Mississippi) and (C–D) *Utricularia Foliosa* (Everglades National Park).

monocots, we see hydrophytes scattered across many orders of the eudicots, including 14 families among nine orders that have produced species adapted to life in mangrove swamps.

2.7   FOR REVIEW

1. What are major taxonomic groups within the green plants?
2. What are some major evolutionary adaptations that allowed plants to succeed as terrestrial organisms?
3. How do the aforementioned adaptations aid in survival on land?
4. What is monophyly, and how does it affect taxonomic organization?
5. What is alternation of generations? How does it differ between bryophytes and the angiosperms?
6. Why is it inappropriate to separate angiosperms into monocots and dicots?
7. What are some major groups of monocots in which we find aquatic and wetland macrophytes?

FIGURE 2.35 Representative Aquatic and Wetland Species from the Alismatales. (A) *Potamogeton Nodosus* (Potamogetonaceae), (B) *Hydrilla Verticillata* (Hydrocharitaceae), (C) *Alisma Subcordatum* (Alismataceae), and (D) *Sagittaria Latifolia* (Alismataceae).

8. What are some major groups of eudicots in which we find aquatic and wetland macrophytes?

9. What is significant about the Alismatales, from an aquatic plant perspective?

10. What are mangroves, phylogenetically and ecologically speaking?

2.8 REFERENCES

Adl, Sina M., David Bass, Christopher E. Lane, Julius Lukeš, Conrad L. Schoch, Alexey Smirnov, Sabine Agatha, et al. 2019. "Revisions to the Classification, Nomenclature, and Diversity of Eukaryotes." *Journal of Eukaryotic Microbiology* 66 (1): 4–119. https://doi.org/10.1111/jeu.12691.

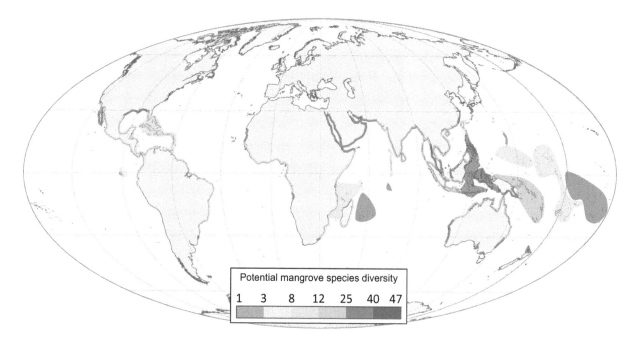

FIGURE 2.36 Approximate Global Distribution of Mangrove Tree Species. Color Scale Approximates the Maximum Potential Number of Mangrove Tree Species in a Local Area, Based on Data Provided by Spalding, Kainuma, and Collins (2010).

FIGURE 2.37 Common Species of Western Hemisphere Mangrove Ecosystems. (A) *Rhizophora Mangle* (Red Mangrove) in Puerto Rico, (B) *Rhizophora Mangle* Near Miami, Florida, (C) *Laguncularia Racemosa* (White Mangrove) in the Town of Antonina, Paraná, Brazil, (D) *Laguncularia Racemosa* Near Miami, (E) *Conocarpus Erectus* Near Miami, and (F–G) *Avicennia Germinans* Near Miami. Note in the Characteristic Prop Roots of *R. Mangle* ((A–B) and the Extensive Pneumatophore Systems Protruding from the Sediments Around *A. Germinans* (F–G). Photo of Puerto Rican *R. Mangle* Courtesy of Guillermo Logarzo, Fundación Para El Estudio De Especies Invasivas (FuEDEI), Buenos Aires, Argentina.

(F)

(G)

FIGURE 2.37 (Continued)

Bell, Charles D., Douglas E. Soltis, and Pamela S. Soltis. 2010. "The Age and Diversification of the Angiosperms Re-visited." *American Journal of Botany* 97: 1296–303.

Burki, Fabien, Andrew Roger, Matthew W. Brown, and Alastair G. B. Simpson. 2019. "The New Tree of Eukaryotes." *Trends in Ecology & Evolution2* In press.

Clarke, John T., Rachel C. M. Warnock, and Philip C. J. Donoghue. 2011. "Establishing a Time-Scale for Plant Evolution." *New Phytologist* 192: 266–301.

Doyle, Jeff J. 1993. "DNA, Phylogeny, and the Flowering of Plant Systematics." *BioScience* 43 (6): 380–89.

Earle, Charles J. 2021. "Gymnosperm Database." *Gymnosperm Database.* www.conifers.org/index.php.

Flora of North America Editorial Committee, eds. 1993+. 2021. "Flora of North America North of Mexico [Online]. 22+ Vols." http://floranorthamerica.org/.

Guo, Xing, Dongming Fang, Sunil Kumar Sahu, Shuai Yang, Xuanmin Guang, Ryan Folk, Stephen A. Smith, et al. 2021. "*Chloranthus* Genome Provides Insights into the Early Diversification of Angiosperms." *Nature Communications* 12 (1): 1–14. https://doi.org/10.1038/s41467-021-26922-4.

Harris, Brogan J., C. Jill Harrison, Alistair M. Hetherington, and Tom A. Williams. 2020. "Phylogenomic Evidence for the Monophyly of Bryophytes and the Reductive Evolution of Stomata." *Current Biology* 30: 2001–12.

Heckman, Daniel S. et al. 2001. "Molecular Evidence for the Early Colonization of Land by Fungi and Plants." *Science* 293: 1129–33.

Hutchinson, G. Evelyn. 1975. *A Treatise on Limnology: Volume III, Limnological Botany.* New York: John Wiley & Sons, Inc.

Judd, Walter S., Christopher S. Campbell, Elizabeth A. Kellogg, Peter F. Stevens, and Michael J. Donoghue. 2008. *Plant Systematics: A Phylogenetic Perspective.* 3rd ed. Sunderland, MA: Sinauer Associates, Inc.

———. 2016. *Plant Systematics: A Phylogenetic Approach.* 4th ed. Sunderland, MA: Sinauer Associates, Inc.

Karol, Kenneth G., Richard M. McCourt, Matthew T. Cimino, and Charles F. Delwiche. 2001. "The Closest Living Relatives of Land Plants." *Science* 294: 2351–53.

Keddy, Paul A. 2000. *Wetland Ecology: Principles and Conservation.* Cambridge, UK: Cambridge University Press.

Keeling, Patrick J., and Fabien Burki. 2019. "Progress towards the Tree of Eukaryotes." *Current Biology* 29 (16): R808–17. https://doi.org/10.1016/j.cub.2019.07.031.

Kenrick, Paul, and Peter R. Crane. 1997. *The Origin and Early Diversification of Land Plants: A Cladistic Study.* Washington, DC: Smithsonian Institution Press.

Leebens-Mack, James H., Michael S. Barker, Eric J. Carpenter, Michael K. Deyholos, Matthew A. Gitzendanner, Sean W. Graham, Ivo Grosse, et al. 2019. "One Thousand Plant Transcriptomes and the Phylogenomics of Green Plants." *Nature* 574 (7780): 679–85. https://doi.org/10.1038/s41586-019-1693-2.

Les, Donald H. 2018. *Aquatic Dicotyledons of North America: Ecology, Life History, and Systematics.* Boca Raton, FL: CRC Press.

———. 2020. *Aquatic Monocotyledons of North America: Ecology, Life History, and Systematics.* Boca Raton, FL: CRC Press.

Les, Donald H., Maryke Cleland, and Michelle Waycott. 1997. "Phylogenetic Studies in Alismatidae, II: Evolution of Marine Angiosperms (Seagrasses) and Hydrophily." *Systematic Botany* 22 (3): 443–63.

Les, Donald H., Edward L. Schneider, Donald J. Padgett, Pamela S. Soltis, Douglas E. Soltis, and Michael Zanis. 1999. "Phylogeny, Classification and Floral Evolution of Water Lilies (Nymphaeaceae; Nymphaeales): A Synthesis of Non-Molecular, rbcL, matK, and 18S rDNA Data." *Systematic Botany* 24: 28–46.

Li, Fay-Wei, Tomoaki Nishiyama, Manuel Waller, et al. 2020. "*Anthoceros* Genomes Illuminate the Origin of Land Plants and the Unique Biology of Hornworts." *Nature Plants* 6: 259–72.

Li, Hong Tao, Ting Shuang Yi, Lian Ming Gao, Peng Fei Ma, Ting Zhang, Jun Bo Yang, Matthew A. Gitzendanner, et al. 2019. "Origin of Angiosperms and the Puzzle of the Jurassic Gap." *Nature Plants* 5 (5): 461–70. https://doi.org/10.1038/s41477-019-0421-0.

Lloyd, Andrea H., Andrew G. Bunn, and Logan Berner. 2011. "A Latitudinal Gradient in Tree Growth Response to Climate Warming in the Siberian Taiga." *Global Change Biology* 17 (5): 1935–45. https://doi.org/10.1111/j.1365-2486.2010.02360.x.

Ludwig, Sarah M., Heather D. Alexander, Knut Kielland, Paul J. Mann, Susan M. Natali, and Roger W. Ruess. 2018. "Fire Severity Effects on Soil Carbon and Nutrients and Microbial Processes in a Siberian Larch Forest." *Global Change Biology* 24 (12): 5841–52. https://doi.org/10.1111/gcb.14455.

McCourt, Richard M., Charles F. Delwiche, and Kenneth G. Karol. 2004. "Charophyte Algae and Land Plant Origins." *Trends in Ecology and Evolution* 19 (12): 661–66. https://doi.org/10.1016/j.tree.2004.09.013.

McCourt, Richard M., Kenneth G. Karol, John D. Hall, Michelle T. Casanova, and Michael C. Grant. 2016. "Charophyceae (Charales)." In *Handbook of the Protists*, edited by J. M. Archibald et al. Cham, Switzerland: Springer International Publishing.

Mitsch, William J., and James G. Gosselink. 2000. *Wetlands*. 3rd ed. New York: John Wiley & Sons, Inc.

Pellegrini, Marco O. O., Charles N. Horn, and Rafael F. Almeida. 2018. "Total Evidence Phylogeny of Pontederiaceae (Commelinales) Sheds Light on the Necessity of Its Recircumscription and Synopsis of PontederiaL." *PhytoKeys* 83 (108): 25–83. https://doi.org/10.3897/phytokeys.108.27652.

Pojar, Jim. 1996. "Environment and Biogeography of the Western Boreal Forest." *Forestry Chronicle* 72 (1): 51–58. https://doi.org/10.5558/tfc72051-1.

Sanderson, Michael J. 2003. "Molecular Data from 27 Proteins Do Not Support a Precambrian Origin of Land Plants." *American Journal of Botany* 90 (6): 954–56. https://doi.org/10.3732/ajb.90.6.954.

Sculthorpe, C. D. 1967. *Biology of Aquatic Vascular Plants*. New York: St. Martin's Press.

Soltis, Douglas, Pamela Soltis, Peter Endress, Mark W. Chase, Steven Manchester, Walter Judd, Lucas Majure, and Evgeny Mavrodiev. 2018. *Phylogeny and Evolution of the Angiosperms*. Chicago: University of Chicago Press.

Spalding, Mark, Mami Kainuma, and Lorna Collins. 2010. *World Atlas of Mangroves*. New York: Taylor and Francis.

Stevens, P. F. 2017. "Angiosperm Phylogeny Website." www.mobot.org/MOBOT/research/APweb/.

Turetsky, Merritt R., Michelle C. Mack, Teresa N. Hollingsworth, and Jennifer W. Harden. 2010. "The Role of Mosses in Ecosystem Succession and Function in Alaska's Boreal Forest." *Canadian Journal of Forest Research* 40 (7): 1237–64. https://doi.org/10.1139/X10-072.

Wang, Qin, Yan Wang, Jianhua Wang, Zhen Gong, and Guan Zhu Han. 2021. "Plants Acquired a Major Retrotransposon Horizontally from Fungi during the Conquest of Land." *New Phytologist*. https://doi.org/10.1111/nph.17568.

Watanabe, Iwao, Nilda S. Berja, and Diana C. Del Rosario. 1980. "Growth of *Azolla* in Paddy Field as Affected by Phosphorus Fertilizer." *Soil Science and Plant Nutrition* 26 (2): 301–7. https://doi.org/10.1080/00380768.1980.10431212.

Wickett, Norman J., Siavash Mirarab, Nam Nguyen, Tandy Warnow, Eric Carpenter, Naim Matasci, Saravanaraj Ayyampalayam, et al. 2014. "Phylotranscriptomic Analysis of the Origin and Early Diversification of Land Plants." *Proceedings of the National Academy of Sciences* 111 (45). https://doi.org/10.1073/pnas.1323926111.

Yang, Yongzhi, Pengchuan Sun, Leke Lv, Donglei Wang, Dafu Ru, Ying Li, Tao Ma, et al. 2020. "Prickly Waterlily and Rigid Hornwort Genomes Shed Light on Early Angiosperm Evolution." *Nature Plants* 6 (3): 215–22. https://doi.org/10.1038/s41477-020-0594-6.

3 Wetland Ecosystems and Plant Diversity

3.1 MAJOR WETLAND TYPES

As we saw in attempts at naming and organizing plant species, there sometimes are challenges in attempting to summarize the diversity of aquatic and wetland ecosystems into a classification scheme with only a few categories. Among North American wetland scientists, we can see a few key pieces of work that have attempted to summarize this diversity, with some welcome similarities among the classifications used (Table 3.1). Nevertheless, we also see that sometimes when one author wishes to simplify a category, another may find that multiple categories are needed to represent fully even the coarse-scale diversity within that group. For example, Mitsch and Gosselink's (2000) "peatlands" are separated into "bogs" and "fens" by Keddy (2000), while Keddy's "swamps" are subdivided into freshwater swamps, riparian ecosystems, and mangroves by Mitsch and Gosselink (2000).

[1] Cowardin et al. (1979) also used a "systems" tier to wetland classification, meant to convey overarching hydrogeomorphic or biogeochemical factors that may influence the ecological character of a given wetland. These included marine, estuarine, riverine, lacustrine, and palustrine settings for a given wetland. As explained in the accompanying text, I use "riparian," instead of "riverine," to refer to wetlands that are strongly influenced by stream or rivers inputs.

Keddy (2000) spent a few pages discussing this very issue in a section he named "The six basic types," but which started with a list of four types that he suggested should be a minimal starting point for wetland type standardization. In my own attempt at summarizing classification schemes for this book, I struggled with trying to optimize efficiency while representing a breadth of wetland types (note that I say "a breadth" rather than "the breadth"). For instance, I initially used Cowardin's "Riverine" systems (Cowardin et al. 1979) as a systems settings group, but I was uncomfortable calling oxbows "riverine," given their typically stagnant water flows. In the end, I opted for Mitsch and Gosselink's "Riparian" as a broader category of stream-associated wetlands to deal with that problem.

I have attempted to summarize in Tables 3.1 and 3.2 the interaction of dominant vegetation with **hydrogeomorphic setting** (the "shape" of the wetland and its connections with underlying geology and with inflows and outflows of water; Brinson 1993), and then to give examples of named wetland types within ten of the resulting vegetation × setting combinations (Table 3.2). I have also given a small number of alternate names to these specific wetland types, based on differences in regional vocabulary or language. There will

be many specific wetland types that are not listed, or perhaps even represented, in Table 3.2. For example, in Latin America there is a term (ciénega) for permanently inundated freshwater wetlands (Berlanga-Robles et al. 2008); however, depending on the region of the Spanish-speaking world, a ciénega may be dominated by herbaceous or by woody vegetation and may be fed by springs or streams or rivers. Thus, although this is a commonly used term in parts of the world, it was unclear to me which box(es) of Table 3.2 would include this term. Brinson (1993) further subdivided some of the wetland groups that I have included in Table 3.2, based on a functional incorporation of the role that hydrogeomorphic setting plays in influencing ecological properties of wetlands, including the vegetation. For example, under Brinson's hydrogeomorphic approach, my riparian wetlands might be segregated into headwater stream wetlands, and then high-, middle-, or low-gradient riparian or alluvial wetlands, depending on their position along the stream and local topographic relief.

Reflected in Tables 3.1 and 3.2 is the fact that, at the coarsest scale, most wetlands can be classified as some form of swamp, marsh, or peatland. This is exactly the point Keddy was making when he suggested that any classification of wetlands should minimally divide wetlands into these groups, although he split peatlands into bogs and fens. The most basic reason for starting with three broad wetland types, as I have done in Table 3.1, is that either wetlands accumulate peat to an appreciable degree or they do not. Those that do so share some strongly influential climatic characteristics that reduce decomposition of dead plant material to such a degree that the **detritus** from plant production and senescence accumulates and forms substantial peat deposits. Principal among those climatic factors are cold temperatures and relative low rates of evapotranspiration. In boreal regions, for example, temperatures tend to be low and growing seasons short; as a result, peat layers average a depth of 2.5 m in these colder regions ("Northern Peatlands: Role in the Carbon Cycle and Probable Responses to Climatic Warming" 1991). Wetlands that do not accumulate peat to an appreciable degree then can be separated into those dominated by herbaceous plants or by woody plants. This also is driven to a large degree by environmental factors, although those often are at a smaller geographic scope than are factors that determine rates of peat accumulation, generally speaking. Certainly, there are other ways that wetlands could be categorized, but this system seems, to me, to best capture a plant-focused system of wetland types.

Once these factors are taken into consideration, the local hydrogeomorphic setting of the wetland will determine

TABLE 3.1

Major wetland types, compared across sources.

This text	Mitsch and Gosselink (2015)	Keddy (2000)	Cowardin et al. (1979)[1]
Peatland	Peatland	Bogs	Moss-lichen wetlands
		Fens	
Herbaceous non-peatland	Freshwater marshes	Marshes	Emergent wetland
	Tidal marshes	Wet meadows	Aquatic bed
		Shallow water wetlands	
Woody non-peatland	Freshwater swamps	Swamps	Scrub-shrub wetlands
	Mangrove swamps		Forested wetlands
	Riparian ecosystems		

TABLE 3.2

Descriptions of and examples of names given to some of the more common types of wetlands and aquatic plant habitats. References: Berlanga-Robles et al. (2008), Bridgham et al. (1996), Cowardin et al. (1979), Keddy (2000), Lazaro-Lobo (personal comm.), Mitsch and Gosselink (2015).

Wetland Type and Setting	Names as Used in US/Canada, Descriptions, and Synonyms
	Peatland
Palustrine	**Bogs** are typically defined as peat-accumulating wetlands that are hydrologically isolated from groundwater or surface flows (both inputs and outputs), with pH usually below 6 because of this isolation and the vegetation that establishes in these wetlands. Vegetation usually dominated by *Sphagnum* mosses, sometimes intermixed with conifers or ericaceous shrubs (i.e., from the Ericaceae).
	Fens, in contrast to bogs, are connected to groundwater and/or surface inflows and outflows and thus have less acidic soils (usually in the range of 6 to 8). Vegetation usually dominated by graminoids (Poaceae, Cyperaceae) and a broad variety of shrub and tree species.
	Muskeg is a term used in Canada and Alaska (US) to refer to a peatland landscape, along with its wetlands.
	Pocosins are temperate peat-accumulating wetlands in the southeastern US, dominated by woody evergreen plant species.
	Europe: mire refers to any peat-accumulating wetland, while moor may refer to the wetlands themselves or to the peatland ecosystem containing the wetlands; *Denmark and Sweden*: mose is used to refer to peat-accumulating wetlands; *Spanish*: turbera is the Spanish term for bog ecosystems, but turbera alta or turbera ácida can be used to specify a bog, in contrast to a fen, which is referred to as a turbera baja; *New Zealand*: pakihi are infertile, acidic peat wetlands, usually on mineral soils of glacial plains or erosional outwash deposits.
Riparian	Scrub-shrub wetlands or herbaceous marshes (see later for more information on these wetland types) may form along floodplains of streams or rivers in peatland landscapes.
	Herbaceous non-peatland
Estuarine	**Salt marshes** are dominated by salt-tolerant grass species and typically occur in tidal zones of coastal ecosystems. Freshwater tidal marshes may also occur farther inland along river-supplying estuaries, where stream flows are impacted by coastal tidal actions, but where freshwater inputs are sufficient to maintain relatively low levels of salinity. *Spanish*: marisma de agua salada.
Marine	**Seagrass beds** are near-shore aquatic habitats in coastal marine systems dominated by herbaceous aquatic angiosperms, all of which belong to families in the order Alismatales (Cymodoceaceae, Hydrocharitaceae, Posidoniaceae, Ruppiaceae, and Zosteraceae). *Spanish*: pradera marina.
Lacustrine	**Littoral zone wetlands** may occupy the margins of lake ecosystems. Many authors restrict the "wetland" portions of lake littoral zones to waters no deeper than 2 m, although aquatic plants may grow at much deeper depths, given sufficient light availability.
Palustrine	**Marshes** are a broad category of shallow freshwater wetlands dominated by herbaceous vegetation. *Australia*: cumbungi swamps are wetlands dominated specifically by cattail (*Typha*) species; *Europe*: marshes dominated by *Typha* are known as reedmace swamps, while reedswamps are marshes dominated by reed (*Phragmites*) species; *Spanish*: marisma; *New Zealand*: raupo swamps are marshes dominated by *Typha* species.
	Potholes are depressional wetlands of glacial origin, dominated by a variety of herbaceous species, depending on degree of connectivity to surface and groundwater inflows and outflows. Potholes are highly prevalent in the northern plains of the US and in central Canada.
	Wet meadows or *wet prairies* occur where a shallower depth or shorter duration of flooding allows establishment of a greater diversity of plant species, including many species of grasses, sedges, and rushes.

TABLE 3.2 (*Continued*)

Descriptions of and examples of names given to some of the more common types of wetlands and aquatic plant habitats. References: Berlanga-Robles et al. (2008), Bridgham et al. (1996), Cowardin et al. (1979), Keddy (2000), Lazaro-Lobo (personal comm.), Mitsch and Gosselink (2015).

Wetland Type and Setting	Names as Used in US/Canada, Descriptions, and Synonyms
Riparian	*Oxbow* wetlands are former river channels, cut off from the main channel through erosional or depositional processes. *Australia*: billabongs.
	Woody non-peatland
Marine	**Mangrove swamps**, or *mangals*, are marine coastal wetlands (saline or brackish) dominated by species of mangrove trees or shrubs; they occur in tropical and subtropical regions. *Latin America*: manglares.
Palustrine	**Swamps** are wetlands dominated by trees, or sometimes shrubs. Note that the term "swamp" may refer to herbaceous-dominated wetlands in some regions (e.g., European reedswamps mentioned earlier). Swamps may occur in or around oxbow wetlands where sufficient time has passed for development of a forest canopy. *Spanish*: humedal (or pantano) arbolado, humedal (or pantano) boscoso.
Riparian	*Bottomland hardwood forests* are formed on alluvial plains, typically along larger streams or rivers, often remaining connected with the river via seasonal flooding. Also termed floodplain forests.
	Scrub-shrub wetlands, in contrast to floodplain forests, are dominated by woody shrubs such as willows (*Salix*), alder (*Alnus*), birch, (*Betula*), etc. *Europe*: carr.

which species are to be most successful at establishing and persisting within the wetland. This is largely due to the influence of hydrology on the suite of plant species that arrive and attempt to establish at the site, but also to the influence of geology on the quantity and quality of nutrients available within the water of the wetland (Alahuhta et al. 2021). As it pertains to the hydrogeomorphic setting, hydrology is influenced by a combination of the source(s) of inflowing water, the type(s) of outflows of water from the wetland (when present), and the shape of the basin within which the wetland forms (Brinson 1993).

Riparian wetlands, for example, are associated in some way with a stream or river, and as such, they will be strongly influenced either by the more or less unidirectional flow of water in the stream, if they are located near the stream channel itself, or by the periodic arrival of floodwaters and sediments contained within them, if they are situated farther from the channel. These wetlands often will be connected to both surface and groundwater inflows and outflows, because of the nature of **alluvial deposits** found in proximity to the stream or river. Sand-dominated alluvium, for instance, that is deposited nearer the stream channel tends to be highly porous and conducive to subsurface water movement. Riparian wetlands also tend to be somewhat linear in general spatial configuration, also a result of the linear nature of the associated stream or river. Another major factor impacting riparian wetlands is the frequency and intensity of disturbance resulting from the flow of water and/or scouring and debris deposition during floods.

Palustrine wetlands, on the other hand, are a diverse collection of inland, nontidal wetlands that are generally isolated from streams or lakes. If they are located in close proximity to open water, they are characterized by prevalence of emergent plant species, including trees, shrubs, and sometimes mosses and lichen (Cowardin et al. 1979; Mitsch

and Gosselink 2007). These wetlands will lack the general influence of directional water flow that riparian wetlands experience, although they may occasionally experience sediment inputs associated with flooding, if situated within a stream or river floodplain. All of these factors have implications for the plant species that occupy wetlands, as will be discussed in later chapters, along with similar constraints that exist in all wetland type × hydrogeomorphic setting combinations.

3.2 GLOBAL TOUR OF WETLANDS AND PLANT DIVERSITY

To illustrate some of the plant diversity among global wetlands, I have somewhat arbitrarily selected representative wetland areas from around the world (Figures 3.1 and 3.2). The areas on this tour that represent distinct wetland complexes ranging in size from about 70 km² to around 150,000 km² in area. The boreal peatlands area, which is really **circumboreal** (i.e., throughout the cold northern latitudes), is estimated at 2.4 to 4 million km², the high-altitude region containing the Qinghai-Tibetan wetlands similarly is about 2.5 million km² in size, and forested wetlands in the United States (the last of the wetlands areas discussed) collectively cover something in the neighborhood of 200,000 km² across the US. For the larger areas, I chose some specific examples for plant species inventories, for simplicity.

For each wetland area, I provide some general information about the area, some of the more important threats faced by each, and a listing of some of the representative aquatic and wetland plant species, based on floristic surveys in the area. The intent of the plant lists is twofold. First, I want to compare/contrast the plant diversity we find in wetlands around the world, looking within and across wetland types. Second,

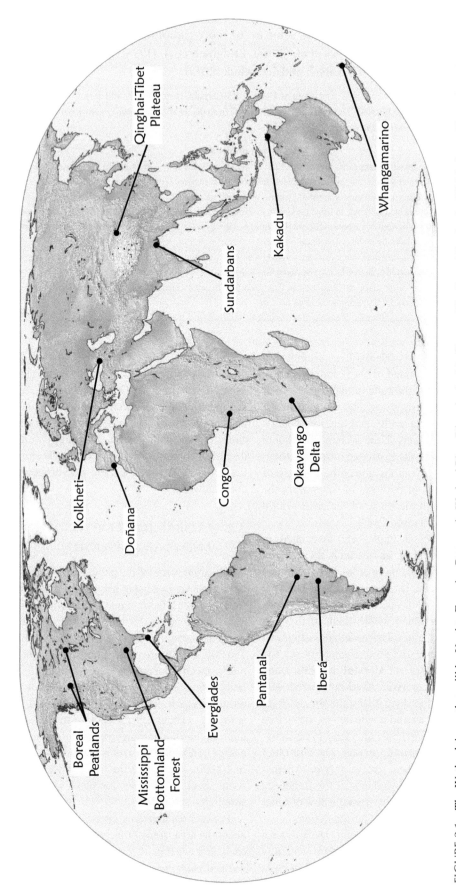

FIGURE 3.1   The Wetland Areas that will be Used to Examine Patterns in Wetland Plant Diversity in this Chapter. The Boreal Peatlands, Which Occur Throughout the Boreal Regions of the Northern Hemisphere, are Represented here by Two Canadian Peatland Areas.

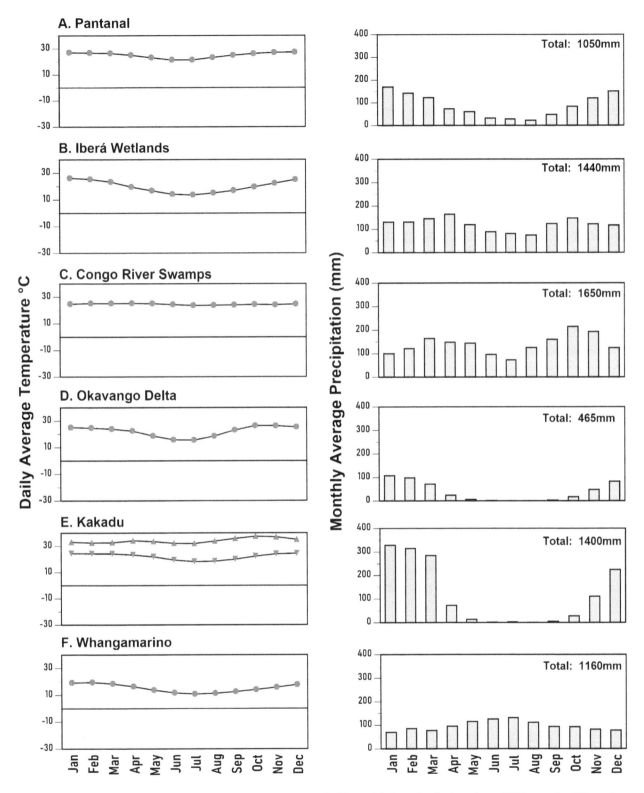

FIGURE 3.2 Climate Data for all but One of the Wetland Areas in Figure 3.1. Data for the Last Area (US Bottomland Forests) are Shown in Figure 3.24. Data were Obtained from the Worldclimate Database (www.worldclimate.com/), for the Nearest Weather Station for which Both Temperature and Precipitation Data were Available. Distances of the Nearest Station all were 100 Km or Less, with Nine Lying within 50 Km of the Pertinent Wetland. In Most Cases, Climate Data were Averages Covering the Latter Half of the 20th Century. For Kakadu Park, Australia, there were No Mean Daily Temperatures for a Weather Station within 100 Km of the Park; Maximum and Minimum Temperatures are Given for this Wetland Area (E).

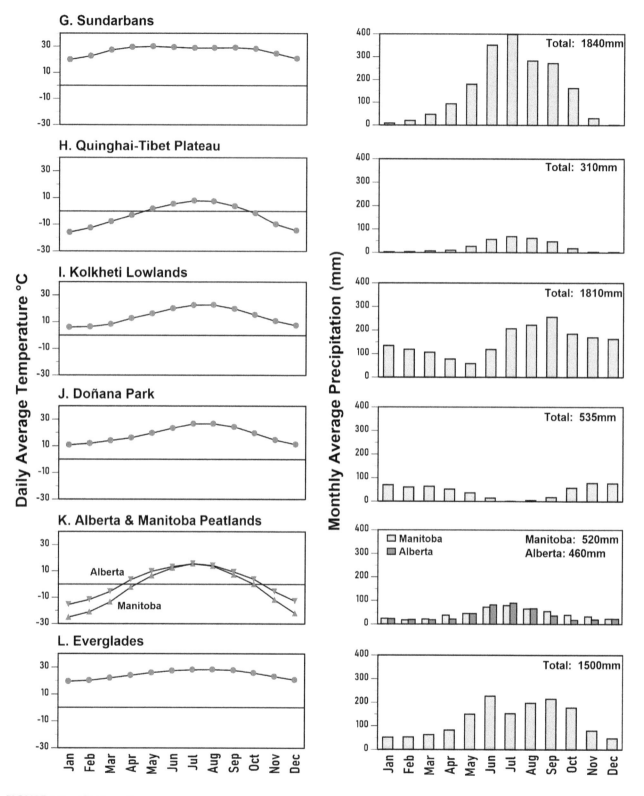

FIGURE 3.2   (Continued)

I want us to reflect a bit on the diversity of angiosperms that we saw in Chapter 2 and the degree to which those are represented in this "Wetlands World Tour." The lists are not meant to be exhaustive by any means and represent only a fraction of the wetland plant diversity in each region.

3.2.1   THE PANTANAL, BRAZIL

The Pantanal (a Portuguese word for "wetland") is an enormous complex of seasonally and permanently flooded wetlands, rivers, lakes, and intermingled upland habitats, including forests, savannas, and grasslands (Figures 3.3

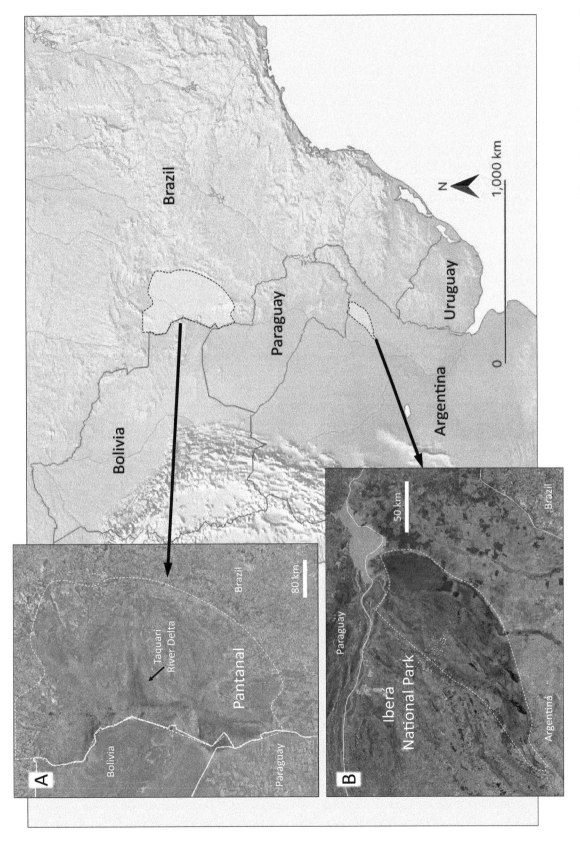

FIGURE 3.3 Aerial View of the Pantanal (A) and Iberá Wetlands (B). The Extents of the Wetland Areas are Indicated by the Dashed and Shaded Boundaries. Aerial Imagery from Esri (2022) "World Imagery" www.arcgis.com/home/item.html?id=10df2279f9684e4a9f6a7f08febac2a9.

and 3.4). Size estimates for the Pantanal wetland landscape range from 100,000 km^2 to almost 150,000 km^2 (Alho 2009; Prance and Schaller 2006), which make it comparable in size to many of the US states east of the Mississippi River (e.g., Louisiana, Mississippi, Alabama, etc.) or countries such as Cuba, Honduras, Iceland, and Nepal.

The hydrology of the Pantanal is driven in part by the complex of eight medium-sized rivers that all eventually flow into the Paraguay River as it snakes its way along the Paraguay-Brazil border (Alho 2009). The Paraguay River then continues southward across Paraguay and into Argentina, where it flows into the Paraná River just west of the Iberá wetlands that we will see shortly. In the Pantanal, one of the more noticeable hydrographic features is the expansive deltaic fan of the Taquari River, which flows east to west across the middle of the Pantanal, towards the Paraguay River (Figure 3.3).

The many large river systems in the Pantanal basin have carved out a complex topography over the past two million years or so (Alho 2009). Old and new river channels are now intermixed with one another, and the landscape possesses heterogeneous soils and geology, providing an incredible diversity of abiotic conditions. In addition to the diverse physiography, this area receives dynamically seasonal precipitation (Figure 3.2), and all these factors combine to foster the wealth of biological diversity for which the Pantanal is known.

The Pantanal is home to more than 260 species of fish, 80 species of amphibians, 113 reptile species, 132 mammals, and more than 460 species of birds (Alho 2009). These include such threatened or endangered species as the jaguar, marsh deer, giant anteater, giant armadillo, and maned wolf. There are a number of protected areas across the Pantanal, but the vast majority of the landscape lacks protected status. A portion of the Pantanal (a bit more than 223,000 ha) has been listed as Ramsar wetlands of international importance (referred to as "Ramsar sites"; ramsar.org), and 188,000 ha of this area is listed as a United Nations Educations, Scientific, and Cultural Organization (UNESCO) World Heritage site because of the biodiversity and ecological significance in the region (unesco.org).

3.2.1.1 Noteworthy Threats

As with many natural systems worldwide, the Pantanal has been subjected to a variety of human impacts. Unsustainable use of the natural resources provided by the Pantanal ecosystem is perhaps the biggest threat to this wetland landscape. Pollution of waters within the wetlands and rivers has resulted from urbanization within the watershed feeding the system. Much of the landscape in the highlands around the Pantanal has been converted to agricultural use to produce row crops, largely soybean, or for cattle ranching (Alho 2009). Conversion to row crop or cattle farming contributes to water degradation via runoff of nutrients, pesticides, or eroded soil and is usually preceded by removal of existing forest cover, which itself can result in significant erosion and runoff into adjacent streams and other aquatic ecosystems.

3.2.1.2 Representative Plant Species

The Pantanal has been visited by botanists and other natural historians for quite some time, although the first expeditions beyond the fringes of its main river tributaries did not occur until the early 1900s (Pott et al. 2011). Considerable work then was conducted in the late 1900s, including the publication of more detailed floristic studies of the Pantanal vegetation (e.g., Prance and Schaller 2006). Broad-scale descriptions of the dominant vegetation, including the commonly observed dominant plant species, were obtained from Prance and Schaller (2006) and Pott et al. (2011). The family- and genus-level diversity among the Pantanal vegetation types, along with that of the other wetlands we will visit, is summarized in Tables 3.3–3.5.

3.2.2 The Iberá Wetlands, Argentina

The Iberá wetlands in northeastern Argentina (known in Spanish as "Esteros del Iberá") are situated between the Paraná and Uruguay rivers, in the Argentine province of Corrientes, near the border between Argentina and Paraguay (Figures 3.3 and 3.5). This wetland complex covers an area of 12,000 to 13,000 km^2, making it one of the largest such complexes in the world (Pacella, Garralla, and Anzótegu 2011; Neiff et al. 2011; Úbeda et al. 2013). The Esteros del Iberá system comprises numerous wetland types, in addition to many open water lakes, streams, and floating wetland islands (locally called "embalsados"), which can accumulate organic sediment, largely formed from dead plant material, sometimes equal to the depth of the water column itself (Neiff et al. 2011).

The Iberá wetlands were added as a Ramsar site in 2002 (ramsar.org), but much of the wetland has been under some form of protection since 1983 (Úbeda et al. 2013). Justification for the Ramsar listing was based, in part, on the importance of these wetlands for some noteworthy threatened animal species, such as the yellow anaconda, knob-billed duck, Neotropical river otter, marsh deer, and at least two sensitive species of caiman (ramsar.org). The wetlands also are home to some 200 species of birds, at least 20 species of amphibians, and 80 or so fish species. The biodiversity of this area makes it a popular ecotourism destination in Argentina.

3.2.2.1 Noteworthy Threats

Recent climatic patterns have resulted in more frequent droughts in the Iberá region of Argentina (among other places), accompanied by warmer than usual temperatures, although some of these anomalies have been related to La Niña events (Úbeda et al. 2013). Reduced water levels, coupled with increased nutrient loadings from agriculture in the Iberá basin, were thought to contribute to high mortality rates of marsh deer and capybara during 2007–2008

TABLE 3.3

Numbers of genera from families of monocots found as common or dominant plant species in floristic surveys or other published natural history accounts of wetlands discussed in this chapter. Mississippi Bottomland Forests were excluded from this list because the studies selected to represent that area focused solely on tree and shrub species.

Monocot Families	Pantanal	Iberá	Congo	Okavango	Kakadu	Whanga-marino	Sundar-bans	Qinghai-Tibet	Kolkheti	Doñana	Peat-lands	Everg-lades
Alismataceae	3	1			1					1		1
Amaryllidaceae												2
Araceae	2	1							1	2	2	1
Arecaceae			2				2					
Asparagaceae											1	
Bromeliaceae												1
Commelinaceae	1											
Cyperaceae	5	7	1	4	4	4		4	2	3	3	4
Dioscoreaceae									1			
Eriocaulaceae		2										
Flagellariaceae							1					
Hydrocharitaceae	1	2	1	2	1					1		1
Iridaceae									1			
Juncaceae						1			1	1		
Juncaginaceae					1							
Marantaceae	1	1										
Orchidaceae						1						
Pandanaceae					1							
Poaceae	3	8	5	5	6		4	4		5	2	2
Pontederiaceae	2	1	1									1
Potamogetonaceae		1		1	1					2		
Restionaceae						2						
Ruppiaceae												1
Scheuchzeriaceae											1	
Smilacaceae									1			
Typhaceae	1	1		1							1	1
Xyridaceae		1										
Genera per site	19	26	10	13	14	9	7	8	7	15	10	15

(Úbeda et al. 2013). Nutrient-induced cyanobacterial blooms, composed largely of a cyanotoxin-producing species (*Cylindrospermopsis raciborskii*), were correlated directly with symptoms exhibited by some of the dead animals. Úbeda et al. (2013) also noted numerous species of introduced plants (sawgrass and cultivated pines) and animals (wild hogs and water buffalo) that had established in the Iberá area, potentially driven by human modification of the wetland basin. The potential thus exists for future climatic shifts to drive increases in the prevalence of invasive species along with a decrease in abundance of flora and fauna adapted to the historic conditions within the Iberá wetlands ecosystem.

3.2.2.2 Representative Plant Species

Plant species data for the Iberá wetlands were obtained from Neiff et al. (2011), who were interested in comparing the dry period of 2007–2008 with a "reference" time period from three decades earlier (1976–1977). They conducted plant surveys in permanent wetland areas adjacent to the Corrientes River and several of the larger water bodies of the Iberá wetlands basin, some of which are visible in Figure 3.3, along the southern border of the wetland area. Reference data had been collected from 36 plots across the study sites, with 137 plots sampled at the same locations in the spring and summer of 2007–2008. These surveys yielded occurrences for 120 plant species (including ferns and liverworts) in 1976–1977 and 117 species in 2007–2008. The most common families across both survey time periods were Asteraceae, Cyperaceae, Fabaceae, Onagraceae, Poaceae, and Polygonaceae (Tables 3.3–3.5). A similar suite of aquatic and wetland plant species were found in a separate survey of the Apipé Grande Island Nature Reserve, which lies immediately north of the Iberá

TABLE 3.4

Numbers of genera from families of non-monocot angiosperms found as common or dominant plant species in floristic surveys or other published natural history accounts of wetlands discussed in this chapter. Mississippi Bottomland Forests were excluded from this list because the studies selected to represent that area focused solely on tree and shrub species.

"Other" Families	Pan-tanal	Iberá	Congo	Oka-vango	Kakadu	Whanga-marino	Sundar-bans	Qinghai-Tibet	Kolk-heti	Doñana	Peat-lands	Everg-lades
Acanthaceae							1					
Amaranthaceae		1								3		
Apiaceae		1								3		
Apocynaceae							2					1
Araliaceae		1							1	1		
Asteraceae	2	4				1		4		2	1	2
Begoniaceae		1										
Boraginaceae									1			
Brassicaceae										1		
Cabombaceae		1										
Caprifoliaceae								1				
Caryophyllaceae										2	1	
Celastraceae									1		1	
Ceratophyllaceae				1								
Cistaceae										1		
Convolvulaceae	1			1			1					
Droseraceae						1					1	
Elatinaceae										1		
Fabaceae	2	2	1				3			2		
Haloragaceae		1				1				1		1
Hypericaceae									1	1		
Lamiaceae							1					
Lentibulariaceae				1	1	1					1	1
Lythraceae										1		
Melastomataceae		1										
Menyanthaceae				1							1	1
Nelumbonaceae				1								
Nymphaeaceae	1	1	1	1	1				1			1
Onagraceae		1		1	1					1	1	1
Orobanchaceae								1	1			
Plantaginaceae				1						1	1	
Polygonaceae	1	1	1	1				2	1	1		
Primulaceae								2	1	1		
Ranunculaceae									1	1	1	
Rosaceae								2	1		2	
Rubiaceae										1	1	
Scrophulariaceae												1
Solanaceae	1								1			
Verbenaceae					1							
Vitaceae	1								1			
Genera per site	9	16	3	6	7	4	8	12	12	25	12	9

TABLE 3.5

Numbers of genera from families of trees and shrubs found as common or dominant plant species in floristic surveys or other published natural history accounts of wetlands discussed in this chapter.

Tree/Shrub Families	Pantanal	Iberá	Congo	Okavango	Kakadu	Whangamarino	Sundarbans	Qinghai-Tibet	Kolkheti	Doñana	Peatlands	Everglades	Bottomlands
Acanthaceae							1					1	
Altingiaceae													1
Annonaceae													1
Aquifoliaceae													1
Asteraceae												1	
Betulaceae									2	1			2
Bignoniaceae	1												
Cannabaceae													1
Combretaceae	1											2	
Cornaceae									1				
Cupressaceae												1	1
Ebenaceae									1				1
Ericaceae						2				2	5		
Euphorbiaceae	1	1	1				1						
Fagaceae									1				2
Frankeniaceae										1			
Juglandaceae									1				1
Lecythidaceae				1									
Lythraceae							1						
Magnoliaceae													1
Malvaceae	1	1	1				3						
Meliaceae							1						
Moraceae			1						2				1
Myrtaceae					1	1	1			1			
Nyssaceae													1
Oleaceae									1	1			2
Pinaceae											2		1
Platanaceae													1
Rhamnaceae									1				
Rhizophoraceae							2					1	
Rosaceae									1				1
Rubiaceae		1											1
Salicaceae											1	1	2
Sapindaceae							1		1				1
Symplocaceae													1
Thymelaeaceae										1			
Ulmaceae									1				2
Viburnaceae									1				
Vochysiaceae	1												
Genera per site	5	3	3	0	2	3	11	0	14	6	9	7	26

Wetlands National Park, along the Paraná River (Fontana 2008).

3.2.3 CONGO RIVER SWAMPS, REPUBLIC OF THE CONGO AND DEMOCRATIC REPUBLIC OF THE CONGO

The Congo River ranks second globally in drainage area and annual **discharge** among all major river systems (Wetzel 2001) and, because of its tropical location, it hosts incredibly high biodiversity, as well as a substantial number of threatened and endangered species of plants and animals (Shumway et al. 2003). The area drained by the Congo River is roughly two-thirds the size of the area drained by the Amazon, and its annual water discharge is more than double the annual flow of the Mississippi River (Wetzel 2001). It also recently was discovered that the forested wetland floodplains of the Congo

FIGURE 3.4   Aerial View of the Brazilian Pantanal. © Filipe Frazao, Used under Creative Commons Attribution-Sharealike 3.0 (CC BY-SA 3.0) License (creativecommons.org/licenses/by-sa/3.0/deed.en).

FIGURE 3.5   Water-Level View of the Iberá Wetlands. Note the *Salvinia* (Aquatic Ferns) and *Eichhornia* [Pontederia] in the Foreground. Courtesy of Fernando Mckay, Fundación Para El Estudio De Especies Invasivas (Fuedei), Buenos Aires, Argentina.

River (and tributaries) in eastern Republic of the Congo and western Democratic Republic of the Congo contain some of the largest tropical peat reserves on the planet (Dargie et al. 2017, 2019), equivalent in carbon storage to all of the aboveground forests of the Congo River basin (Dargie et al. 2017).

An important characteristic of the region (Figure 3.6), relative to the wetlands themselves, is the low topographic relief along these major river systems. The very flat terrain (approximately a 7 cm drop in elevation per km of river) results in slow flows of the rivers and almost continuous inundation of portions of the floodplain (Dargie et al. 2017). This continuous saturation of the soil is responsible for the extensive peat deposits in the basin (~145,000 km² of tropical peatland forests; Dargie et al. 2019). The waterlogging results in slowed decomposition of dead organic matter from plants (and other organisms) and will be discussed extensively in later chapters.

The swamp forests that develop on these flooded peat soils include hardwood forests dominated by species from the genera *Uapaca* (Phyllanthaceae), *Carapa* (Meliaceae), and *Xylopia* (Annonaceae) and palm-dominated swamps, where *Raphia* species (Arecaceae) are the dominant trees (Dargie et al. 2017). Other similar swamps further downstream from this reach of the Congo River have been documented to support forests of *Mitragyna*, *Nauclea*, and *Stipularia* species from the Rubiaceae, as well as *Symphonia* (Clusiaceae), *Alstonia* (Apocynaceae), and *Xylopia* species (Annonaceae) (Pangou et al. 2003).

There are three Ramsar wetland sites in this portion of the Congo River Basin: Lac Télé and Grands Affluents in the Republic of the Congo and Lac Tumba in western Democratic Republic of the Congo (ramsar.org). These three Ramsar sites comprise ~130,000 km² along the western Congo River, roughly the size of Brazil's Pantanal wetland (ramsar.org).

3.2.3.1 Noteworthy Threats

Dargie et al. (2019) discuss multiple potential threats to the Congo River swamp ecosystems, with special focus given to the peat storage beneath those forests, owing to the implications of that peat storage for climate change. Aside from potential impacts of forest alteration on climate change, they also listed deforestation, oil and gas extraction, dam construction (for hydroelectric generation), and road construction as threats to these forested wetlands. They went on to point out, using an example from Indonesia, the potential compounding effects of having multiple of these factors simultaneously impacting the forests. In the Indonesian example, 10,000 km² of tropical forested peatland was to be converted to rice cultivation. Canals were dug to drain the wetlands and portions of the forest were cleared, and these activities coincided with an El Niño climatic event in 1997, resulting in half of those peatlands catching fire. Two years later, the government abandoned the program, and the wetlands continued to degrade, because of continued drainage via the canals that were constructed and drought-induced wildfires (Dargie et al. 2019).

3.2.3.2 Representative Plant Species

Some vertebrate populations in the Congo River basin have been studied somewhat extensively (reptiles, large mammals), but information still is lacking for fish, invertebrates, and plants (Dargie et al. 2019). The Ramsar web site (https://rsis.ramsar.org/ris/950) provides a list of plant species published in 2000 for the Lac Télé Community Reserve, a Ramsar site just west of the Ubangi River (which itself lies just west of the Congo), near the border between the Republic of the Congo and the Democratic Republic of the Congo. This list, however, gave no indication as to the habitat types in which the plants were observed (i.e., which would be considered wetland vs. upland species). Nevertheless, the list includes a number of genera that are readily recognizable as having numerous aquatic or wetland representatives, including *Pycreus* (Cyperaceae), *Hydrocharis* (Hydrocharitaceae), *Hibiscus* (Malvaceae), *Nymphaea* (Nymphaeaceae), *Panicum*, *Paspalum*, and *Vossia* of the Poaceae, and *Polygonum* or *Persicaria* (Polygonaceae).

A qualitative biodiversity survey was carried out by Shumway et al. (2003) as part of a Congo River Environment and Development Project (CREDP) funded by the United States Agency for International Development (USAID). That survey collected species-level occurrence data (i.e., lists of species that were present) on fish, birds, reptiles, amphibians, mammals, aquatic invertebrates, and, fortunately, plants, in three western provinces in the Democratic Republic of the Congo, one of which includes the portion of the Congo River that is discussed here.

The plant data for the Congo River Swamps in Tables 3.3–3.5 (aside from those species included in the Ramsar list mentioned earlier) were collected from nine sites along the Congo River, near its confluence with the Ubangi River, on the eastern edge of the Republic of the Congo (Shumway et al. 2003). Aquatic and wetland habitats sampled at these locations included flooded swamp forests, floating islands, and flooded grasslands. Limited sampling time spent at these sites was attributed to "immigration and logistical travel difficulties," which resulted in the observation of only 13 plant species, including three invasive plants (*Eichhornia crassipes*, *Panicum repens*, and the terrestrial invasive grass *Imperata cylindrica*). Additional aquatic plant genera observed at the other two study areas included *Echinochloa* (Poaceae), *Cyperus* (Cyperaceae), *Pistia* (Araceae), *Ipomoea* (Convolvulaceae), and *Vossia* (Poaceae), all of which are included in Tables 3.3 and 3.4 for other wetlands.

3.2.4 Okavango Delta, Botswana

The Okavango River Delta lies in northern Botswana (Figure 3.6), where the Okavango River flows into the Makgadikgadi saltpans of the Kalahari desert (Ramberg et al. 2006; Ashton, Nordin, and Alonso 2003). The river itself has its headwaters in Angola, where it is known as the Cubango River. The river runs along the border between

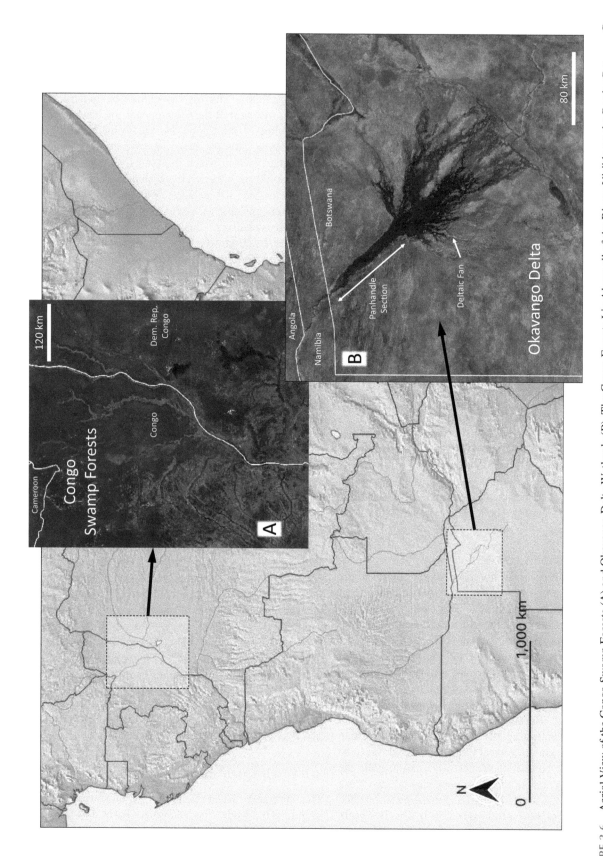

FIGURE 3.6   Aerial View of the Congo Swamp Forests (A) and Okavango Delta Wetlands (B). The Congo Forests Lie Along all of the Rivers Visible on the Border Between Congo and the Democratic Republic of the Congo and Southward. Aerial Imagery from Esri (2022) "World Imagery." www.arcgis.com/home/item.html?id=10df2279f9684e4a9f6a7f08febac2a9.

Angola and Namibia for approximately 350 km, then flows roughly 50 km across a narrow strip of eastern Namibia before terminating in the Botswanan Okavango Delta. The Okavango Delta is somewhat unusual in that, unlike most major river deltas, it essentially disappears into a desert at the river's mouth (Figure 3.6).

The Delta itself ranges in area from ~6,000 km^2 during the dry season (May–October, Figure 3.2D) to more than 15,000 km^2 in the wet season, with approximately 9,000 km^2 of wetlands during the wettest years and just under 5,000 km^2 of permanently flooded wetlands (Ellery and Tacheba 2003; Ramberg et al. 2006). The Delta begins in what is termed the "panhandle" section along the river, shortly after crossing into Botswana (Figure 3.6B), after which it spreads laterally into a deltaic fan comprising a complex of interconnected river channels and distributaries (i.e., channels that flow back out of a larger river, forming a functionally separate stream or river; Ramberg et al. 2006). The delta proper (that is, the deltaic fan at the river mouth) is constrained on its northern and southern limits by fault lines of the East African rift system that run from southwest to northeast across southern Africa. This system of geologic

faults later turns north, into the area where the African rift lakes Malawi, Tanganyika, and so on were formed approximately 2.5 Mya (Ramberg et al. 2006).

The Okavango Delta wetlands were added as a Ramsar site in 1996 and became the 100th UNESCO World Heritage site in 2014 (ramsar.org). The Delta is home to ~1,200 plant species, 32 species of large mammals (including elephants, lions, hyenas, cheetahs, and hippopotamus; Figure 3.7), 84 species of dragonflies/damselflies, more than 650 resident or transient bird species, and more than 70 species of fish (ramsar.org).

3.2.4.1 Noteworthy Threats

One major threat to the Okavango Delta is that the rivers that supply it lie predominantly in two other countries. Upstream plans for the installation of dams for hydroelectric generation and agricultural water extraction potentially threaten the water supply for this desert wetland system (Alonso and Nordin 2003; Ashton et al. 2003; Junk et al. 2006). Other threats include pollution, introduction of non-native species (including the aquatic fern *Salvinia molesta*), and increased human encroachment, because of ongoing

FIGURE 3.7 Elephants Grazing in the Okavango Delta Wetlands. © Thapthim, Used under CC BY-SA 3.0 (commons.wikimedia. org/w/index.php?curid=113509).

development of local settlements and ecotourism (Alonso and Nordin 2003).

3.2.4.2 Representative Plant Species

Conservation International partnered with the Field Museum in Chicago, IL (USA), in 1996 to develop a biodiversity assessment program called the Aquatic Rapid Assessment Program (AquaRAP). The focus of this program is to inventory biodiversity in important conservation sites around the world, such as the Okavango Delta (Alonso and Nordin 2003). An AquaRAP expedition was conducted in the Okavango Delta in June 2000. One focus of that expedition was documentation of the vegetation in four areas of the Delta: the upper panhandle, the junction of the panhandle and the fan, an area of permanent wetlands in the northeastern portion of the deltaic fan, and an area of seasonal wetlands near the center of the fan (Ellery and Tacheba 2003; Figure 3.6B).

This expedition sampled vegetation at 130 plots across the four areas, with 106 of those plots being classified as "wetland plots" and the remaining 24 plots as riparian woodland plots. A total of 233 plant species were recorded in this census, representing roughly 20% of the total documented floristic diversity of the Okavango Delta.

Plant taxa that were most common across the four areas examined by Ellery and Tacheba (2003) are indicated in Tables 3.3 and 3.4. For this particular study, taxa given in these tables are those that were present at no less than 10% of the total study plots (≥ 13 of the 130 plots), at least one-third of the plots within a study area, or in more than one of the four study areas. However, there were a few additional taxa that were present in 20% of the study plots within only one site. That is, they were relatively abundant within one of the four areas but absent from the other three. These included species from *Phragmites* (Poaceae), *Trapa* (Lythraceae), *Ottelia* (Hydrocharitaceae), *Fuirena* (Cyperaceae), and *Brasenia* (Nymphaeaceae).

Ramberg et al. (2006) found very similar suites of species in their review of extant biodiversity surveys of the Okavango Delta, and Ellery and Ellery (1997) likewise describe similar assemblages of dominant plants among the permanently flooded and seasonally flooded areas of the delta. Additional important wetland genera they identified that are not represented in the tables include *Ottelia* (Hydrocharitaceae), *Nesaea* (Lythraceae), *Trapa* (Lythraceae), *Fimbristylis* (Cyperaceae), and *Phragmites* (Poaceae).

3.2.5 Kakadu Wetlands, Australia

Kakadu National Park, in northern Australia (Figures 3.8 and 3.9), is listed in its entirety as a Ramsar site. The park includes rivers, coastal wetlands and near-shore marine habitats (including mangrove swamps), springs, riparian and floodplain ecosystems, *billabongs* (disconnected river channels, also called oxbows; Table 3.2), and the intervening terrestrial habitats in the ~20,000 km² park (ramsar.

org). Most of the park also is listed as a UNESCO World Heritage site because of archaeological and anthropological sites in the park, some dating back as far as 40,000 years ago (unesco.org).

The Kakadu region is dissected by five river systems whose watersheds range from about 1,400 km² to 12,000 km² in area, the largest of which are the South Alligator and East Alligator rivers, with a total area of ~22,000 km² between them (Finlayson et al. 2006). Similar to the Pantanal and the Okavango Delta systems, the Kakadu region of northern Australia has distinct wet and dry seasons, with the dry season falling during the southern hemisphere winter (May–September/October; Figure 3.2). During the dry period, there is considerable drying of the river floodplains throughout the region, which impacts vegetation dynamics, as well as seasonal patterns of other wildlife (Finlayson 2005; Finlayson et al. 2006).

3.2.5.1 Noteworthy Threats

The Kakadu rivers and wetlands face similar threats to other areas discussed in this chapter. A number of climate-related threats loom for the national park. Because of the park's proximity to the coast, saltwater intrusion into near-coastal freshwater environments is accompanying climate change–induced sea level rise, and changes in the timing and duration of fire seasons puts native species at risk, while favoring some introduced species (unesco.org). Additionally, introduced species of plants (*Mimosa pigra* [Fabaceae] and *Salvinia molesta*) and animals (Asian water buffalo—the same species introduced to the Iberá wetlands in Argentina) threaten the natural integrity of Kakadu's aquatic and wetland ecosystems (unesco.org).

3.2.5.2 Representative Plant Species

Finlayson (2005) conducted a review of the published work in freshwater wetlands of northern Australia (particularly those wetlands in what is known as the "wet-dry tropics"), and in that review, he provides detailed summaries of the dominant vascular plant species of a variety of wetland types in that region. The data from the summaries in that paper came from a variety of sources, including some of Finlayson's own work over a period of several years of wet seasons on the Magela River (a tributary of the East Alligator River, in the eastern edge of Kakadu National Park; Finlayson 2005). Data for the Kakadu wetlands in Tables 3.3–3.5 were obtained from this paper by Finlayson.

3.2.6 Whangamarino Wetland, New Zealand

The Whangamarino wetland complex is an area of peat bogs, fens, swamps (dominated by introduced *Salix* and *Alnus* species), and marshes covering approximately 7,000 ha (~70 km²) on the north island of New Zealand (Figure 3.8B). Almost 6,000 ha of the Whangamarino wetland area is listed as a Ramsar site because of its value as one of New Zealand's largest peatland complexes and as

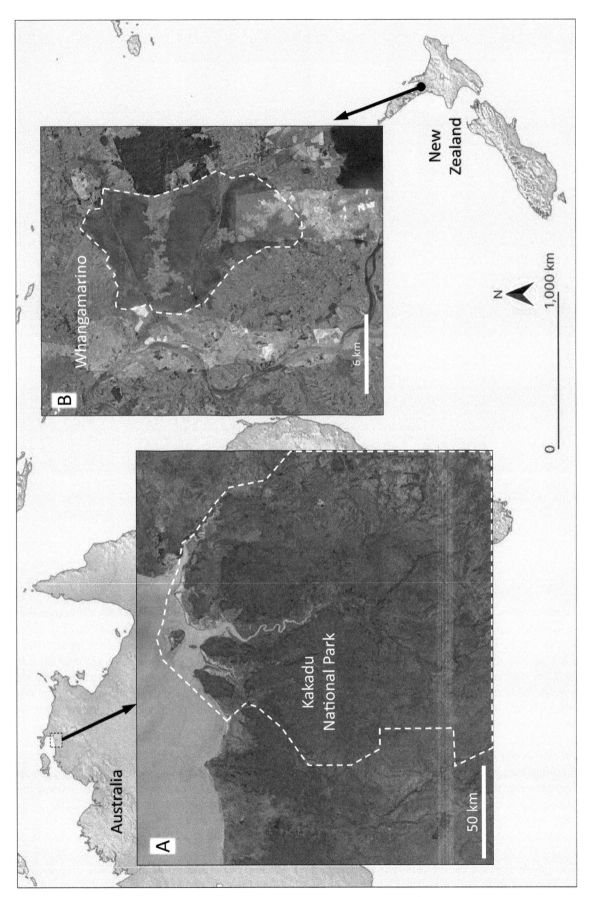

FIGURE 3.8   Aerial View of Kakadu National Park (A) and the Whangamarino Wetlands (B). The Approximate Extents of the Wetland Areas are Indicated by the Dashed White Boundaries. Aerial Imagery from Esri (2022) "World Imagery." www.arcgis.com/home/item.html?id=10df2279f9684e4a9f6a7f08febac2a9.

FIGURE 3.9   A View across Wetlands in Eastern Kakadu National Park. The Photo was Taken from Atop Ubirr Rock, a Site of Aboriginal Rock Art. © Christina Storey, Used under CC BY-SA 4.0, (commons.wikimedia.org/w/index.php?curid=79093667).

habitat for many rare and threatened species of birds and wetland plants (Duggan et al. 2013; ramsar.org).

This wetland complex was important for the Maori people, who inhabited the islands of New Zealand prior to European colonization. It was used by the Maori for hunting and fishing, and there were numerous Maori forts erected around the wetland area at the time of European colonization (mid-1800s; ramsar.org). Following arrival of Europeans, the wetlands were used as boat transportation corridors and waterfowl hunting, and some areas were used agriculturally. It was also at this time (~1885) that the willows now dominating portions of the wetlands were introduced into the Whangamarino area.

Fire is important in the ecology of these wetlands, facilitating the spread of introduced plant species in some areas but being important in habitat management for an endangered orchid (*Corybas carsei*; Figure 3.10) in others (Duggan et al. 2013; Norton and De Lange 2003). Paleobotanical studies in these and neighboring wetlands indicated that fire may have been important in structuring vegetation in these systems as far back as 12,000 years

ago (Norton and De Lange 2003). It is useful to note here that evidence for human influence on the vegetation of New Zealand is dated at only about 700–750 years before present in some northern wetlands (Newnham et al. 1995). Prior to human settlement of the island, fires appear to have occurred at roughly 100–244-year intervals, influenced by certain groups of plants (rushes, sedges, and grasslike restiad species) that provide sufficient aboveground fuel during drought periods to carry periodic wildfires (Norton and De Lange 2003).

3.2.6.1   Noteworthy Threats

Increasing agricultural land use within the Whangamarino watershed has led to increasing inflows of nutrient and sediments into these wetlands (Duggan et al. 2013). These activities also have been accompanied by introduction of numerous non-native plant species, many of which have spread throughout the area. In biodiversity surveys for the New Zealand Department of Conservation, Duggan et al. (2013) found that 72 of the 104 plant species recorded in the Whangamarino wetlands were non-native to the area;

FIGURE 3.10   The Endangered Orchid, *Corybas Carsei*, in Whangamarino Wetland, New Zealand. © Catherine Beard, Used under CC BY 4.0, (commons.wikimedia.org/w/index.php?curid=66044312).

this ratio more or less held across all groups of wildlife surveyed at Whangamarino (~70% of all species were non-native, across fish, lizards, mammals, birds, invertebrates, and plants). Finally, as mentioned earlier, some species of rare plants are also at risk because of modified fire regimes that reduce the competitive ability of these rare plants, to the advantage of aggressive introduced species.

3.2.6.2   Representative Plant Species

Dugganetal. (2013) mapped vegetation in the Whangamarino wetlands and classified the map into 16 classes of vegetation, based on dominant species. Approximately 55% of the wetland area was dominated by native plants species, with bogs being dominated entirely by native plants species, but only 6% of marsh habitats was dominated by native species. Fens were intermediate, with 45% of the area of fens being native species dominated.

Studies on the ecology of New Zealand's peatland ecosystems (Clarkson et al. 2004; Norton and De Lange 2003) and the Whangamarino wetland Ramsar information sheet (ramsar.org) served as the basis for the plant lists represented in Tables 3.3–3.5 for the Whangamarino wetlands. The plant species included in Clarkson et al. (2004)

represented Whangamarino and eight other similar wetland complexes in a 24,000 km² region of New Zealand's north island. These wetlands are classified as raised "restiad" bogs, based on the abundance of plants in the family Restionaceae, the most important of which in this region of New Zealand are in the genera *Empodisma* and *Sporadanthus* (Clarkson et al. 2004).

3.2.7   The Sundarbans, India and Bangladesh

The Sundarbans is one of the world's two largest continuous mangrove forests (~6,500 km²; Spalding, Kainuma, and Collins 2010) and is located in the combined delta of the Ganges, Brahmaputra, and Meghna rivers along the border between India and Bangladesh (Figures 3.11 and 3.12). This wetland is listed both as a Ramsar site of international importance and as a UNESCO World Heritage site (ramsar.org; unesco.org). The wetlands are home to the Sundarban Tiger Reserve and are considered a globally critical area for conservation of tigers. The mangrove forests of this ecosystem are important not only as habitat for many species of rare and threatened wildlife species, but also for their role in buffering the forces of tropical storms

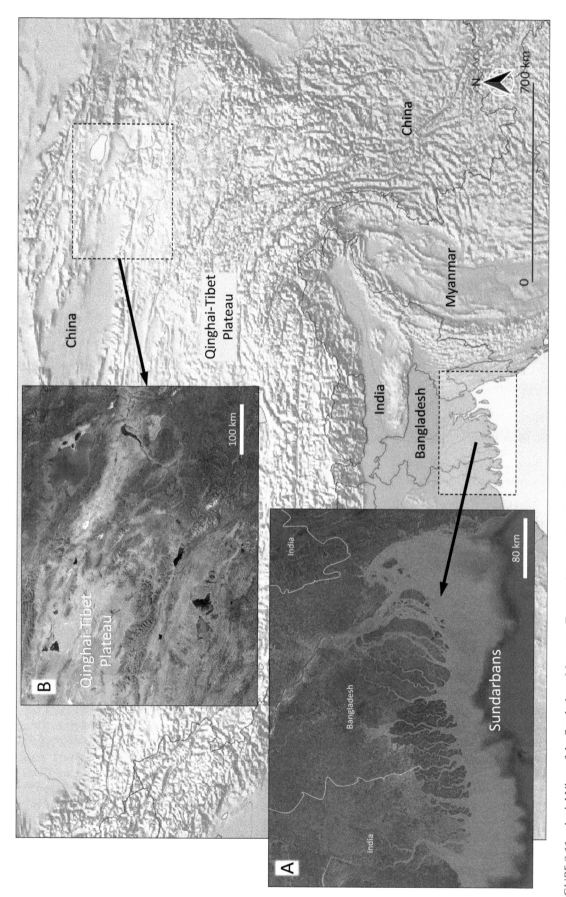

FIGURE 3.11   Aerial View of the Sundarbans Mangrove Forests (A) and a few of the Qinghai-Tibetan Plateau Lakes and Wetlands (B). Aerial Imagery from Esri. 2022. "World Imagery." www.arcgis.com/home/item.html?id=10df2279f9684e4a9f6a7f08febac2a9.

FIGURE 3.12   Shoreline of the Sundarban Mangrove Forest at West Bengal, India. Pneumatophores (Anatomical Structures Providing Root Aeration) are Visible Emerging from the Soil Near the Center of the Photo. © Pinakpani, Used under CC BY-SA 4.0 (commons. wikimedia.org/w/index.php?curid=65468981).

as they move inland into the delta and in serving as nursery ground for many species of fish and shellfish (Rahman and Asaduzzaman 2013).

Of the ~65 species of mangroves worldwide (Spalding, Kainuma, and Collins 2010), it has been estimated that there are 30–35 species in the Sundarban mangal wetlands (Gopal and Chauhan 2006). This high diversity is thought to derive, in part, from southeast Asia serving as a potential seat of evolutionary diversification for mangroves (Spalding, Kainuma, and Collins 2010), but also because of the diverse abiotic conditions resulting from tidal cycles, tropical climate, and a gradient of increasing salinity as one moves from east to west across the Sundarbans landscape (Gopal and Chauhan 2006; Iftekhar and Saenger 2008).

3.2.7.1   Noteworthy Threats

Construction of the Farraka Barrage (dam) in the 1970s resulted in as much as 40% of the dry season flow in the Ganges River being diverted for other uses (Rahman and Asaduzzaman 2013). This reduces freshwater flushing of the Sundarbans ecosystem, while simultaneously increasing

saltwater intrusion, leading to dieback of some of the more susceptible mangrove species and replacement with other species (Rahman and Asaduzzaman 2013). The Sundarban mangroves also are at risk to timber harvests, fisheries exploitation, agricultural encroachment, and other factors associated with the growing human populations in and around the delta (Rahman and Asaduzzaman 2013). Some 2.5 million or so people live in and immediately around the Sundarbans (Gopal and Chauhan 2006), and clearing of the mangrove forests for agriculture and other uses has taken place for at least 800 years (Iftekhar and Islam 2004).

3.2.7.2   Representative Plant Species

Because of their prevalence in this ecosystem, many published plant surveys of the Sundarbans focus principally on the mangroves themselves. Data represented in Table 3.3 for this wetland complex were obtained from the work of Iftekhar and Saenger (2008) and Harun-or-Rashid et al. (2009). The latter of these was a study aimed at evaluating the similarities between aboveground vegetation and soil seedbanks in the Sundarbans, in the context of potential response of these ecosystems to intense natural disturbances, such as floods, tidal surges, and hurricanes/cyclones (Harun-or-Rashid et al. 2009). That study examined not only the mangrove forest habitats, but also nearby grassland habitats that receive periodic tidally driven flooding, whereas the work by Iftekhar and Saenger (2008) focused solely on the mangrove tree species themselves.

3.2.8 QINGHAI-TIBET PLATEAU WETLANDS, CHINA

The Qinghai-Tibet Plateau (QTP) covers approximately 2.5 million km² in western China, just north of the Himalayan mountain range, at elevations from 3,000 to 5,000 m above sea level (Gao et al. 2013; Sun et al. 2014; Figure 3.11). Despite being a very cold and very dry region (Figure 3.2), the QTP hosts a wide diversity of wetland types, primarily dominated by herbaceous plant species, as we will see later. Lakes and wet meadow wetlands dominate the wetland coverage on the QTP, accounting for approximately 94% of the wetland surface area (~100,000 km² combined area of these wetland types), with freshwater and saline marshes accounting for most of the remaining QTP wetland area (Xue et al. 2018). The growing season for plants tends to be very short, at about three to four months in duration, when average temperatures briefly exceed freezing (mean annual temperature is approximately −4 °C; Gao et al. 2013).

The high altitude of the QTP and its location relative to large mountain chains (Himalayas, Kunlun, Pamir, and Hengduan mountain ranges; Sun et al. 2014) presents numerous physiological obstacles for plant species on the QTP. The mountain ranges that surround the QTP on all sides present a barrier to significant precipitation through their rain shadowing effect, and the high altitude results in very cold temperatures as well as very high UV radiation exposure (Li et al. 2016; Sun et al. 2014). Some examples of morphological adaptations present in these species are modified bracts (leaf-like structures) or layers of dense hairs that cover the flowers to absorb UV radiation and protect reproductive organs from cold damage and possessing a very compact growth habit that serves to accumulate organic matter for moisture and heat retention, referred to as a "cushion plant" morphology (Sun et al. 2014).

3.2.8.1 Noteworthy Threats

Since the 1970s, a substantial fraction of the QTP wetlands has disappeared. Losses averaged 6% to 8% area loss over this 40-year period; however, some wetland types experienced much higher losses, with freshwater marshes and saline marshes declining by roughly 50% during this time (Xue et al. 2018; Zhao et al. 2015). These losses were attributed largely to changes in climate, as lake area increased at the same time that other wetland habitats were declining. Increasing temperatures have led to increased rates of evaporation from shallow wetlands along with increased melting of accumulated snow and ice. Air temperatures on the QTP have increased over the past 50–60 years at about twice the average global rate, resulting not only in faster

evaporation of surface waters but also increased rates of glacier loss, which also contributes to increases in lake area and depth (Zhao et al. 2015). Along with the impacts of climate change in this region, the growing human population has placed increasing demands on the landscape in the form of higher densities of livestock grazing and intensification of land use change, all of which negatively impact all natural ecosystems, wetlands included (Zhao et al. 2015).

3.2.8.2 Representative Plant Species

It seems that recent recognition of the rates of loss of wetlands on the QTP, and the awareness that some of these losses are driven by the rapid rates of climate change in the region, have generated heightened interest in studying these wetlands, their plants, and potential responses of the plants to environmental stresses and change. This has resulted in recent publication of somewhat detailed listings of plant species from QTP wetlands. Xu et al. (2015) conducted an experimental study to examine plant species composition changes to wetland hydrology, using inventories of dominant plant species as the baseline for evaluating impacts of altered hydrology. Li et al. (2016) conducted more of an observational analysis of soil and other environmental characteristics that influence plant species composition in QTP wetlands, and Gao et al. (2013) were attempting to develop a locally relevant wetland classification approach to aid in optimizing wetland monitoring and management efforts in the region. Plant data from these three papers were used to extract the information that represents the QTP wetlands in Tables 3.3–3.5.

3.2.9 KOLKHETI LOWLANDS, GEORGIA

The Kolkheti Lowlands is a complex of wetlands and other habitats at the southeastern edge of the Black Sea, in present-day Georgia (Figure 3.13). The Kolkheti region, as a whole, comprises something in the range of 40,000 to 60,000 km² lying between the Black Sea to the west, the Caucasus Mountains to the north, and on the south by the Anticaucasus Mountains (Connor et al. 2007). Because of the shelter provided by the adjacent mountain ranges and the buffering provided by the Black Sea, this region served as a floristic and faunal refuge during the last glaciation, resulting in it hosting a mixture of species with previous Mediterranean, Siberian, and east Asian distributions (Connor et al. 2007; Denk et al. 2001). The Kolkheti region wetlands include bogs, fens, marshes, and forested wetlands (swamps) along the Rioni River and numerous other streams and rivers draining from the region's bordering mountain ranges (Figure 3.14; Connor et al. 2007; Denk et al. 2001).

The region also has importance for archaeology and anthropology. The Greeks had occupied the area by about 2,600 years ago, and they used its diverse timber and marsh plant species to build homes in the Rioni River delta, along which the present-day Kolkheti National Park was established (Connor et al. 2007). The combination of ecological importance as glacial refuge and archaeological

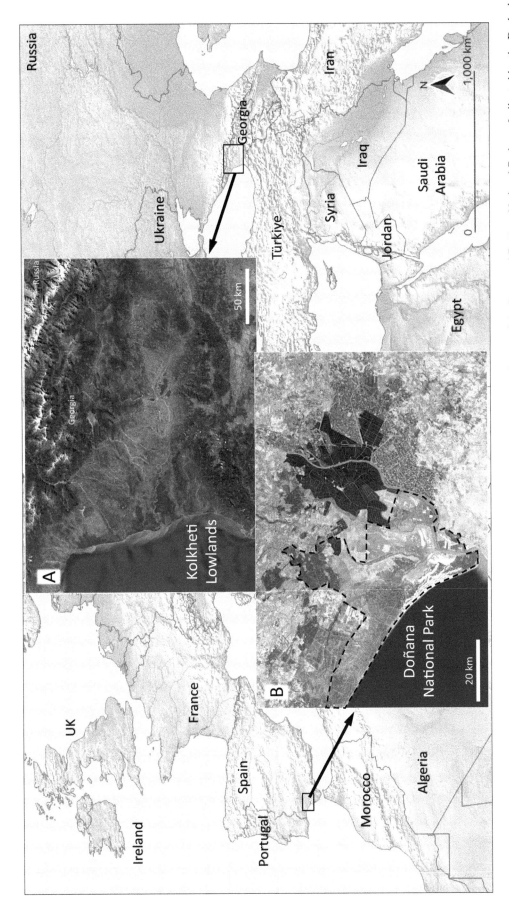

FIGURE 3.13 Aerial View of the Kolkheti Lowlands (A) and Doñana National Park (B). The Approximate Extent of the Protected Doñana National Park are Indicated by the Dashed Black Boundaries. Note the Dense Aggregation of Rice Agriculture Just East of Doñana National Park. Aerial Imagery from Esri (2022) "World Imagery" www.arcgis.com/home/item.htm l?id=10df2279f9684e4a9f6a7f08febac2a9.

FIGURE 3.14 One of the Protected Wetland Areas in the Western Edge of Kolkheti National Park, Approximately 1 Km from the Black Sea Coast. © Paata Vardanashvili, Used under CC BY-SA 4.0 (commons.wikimedia.org/w/index.php?curid=70267489).

importance of the region were instrumental in the Kolkheti wetlands being listed as a Ramsar site in 1997 as nominated as a UNESCO World Heritage site in 2007 (ramsar.org; unesco.org).

3.2.9.1 Noteworthy Threats

During a period of about 4,000 years ago until 2,000 years ago, there was a notable accumulation of charcoal in sediment cores obtained for paleoecological studies of the Kolkheti Lowlands (Connor et al. 2007). It is thought this was associated with either slash and burn agricultural practices or burning of the deltaic marshes. However, natural ecosystems of the Kolkheti region appear to have remained largely undisturbed following abandonment of early settlements at about 1,900 years ago (when the charcoal accumulation ceased) until the early 1900s, when forest clearing began to allow planting of tea, citrus, and other plantations (Connor, Thomas, and Kvavadze 2007). Current threats to the natural integrity of the relict habitats occupying this region include draining of wetlands for peat mining and agriculture (row crops, fruit trees, and cattle) and deforestation of the region's remaining old-growth forests (ramsar. org).

3.2.9.2 Representative Plant Species

Denk et al. (2001) conducted a detailed floristic inventory of the Kolkheti region, in an effort to document the diversity of plant species and habitats and to characterize the environmental conditions of the glacial refugee species in

this area. Over a period of three years, floristic data were collected along a west-to-east transect from dunes along the Black Sea coast to the forests of eastern Georgia, up to an elevation of 2,000 m above sea level. Denk and colleagues sampled more than 200 vegetation plots along this transect, documenting plant species (and some reptiles and amphibians) in dune habitats, bogs, fens, marshes, river floodplains, backwater swamps, and montane forests. Plant taxa representing the Kolkheti Lowlands in Tables 3.3–3.5 were those specifically mentioned as being dominant or abundant in areas sampled by Denk et al. (2001).

3.2.10 Doñana National Park, Spain

Doñana National Park, on the Atlantic coast of southwest Spain (Figures 3.13 and 3.15), encompasses around 112,000 ha (1,120 km^2) of wetlands and associated habitats in the delta of the Guadalquivir River, approximately 55 km southwest of the city of Seville. Wetland types on the park include salt marshes, interdune ponds, lagoons (large brackish or saline ponds or sheltered bays), canals, river floodplains, and more than 3,000 seasonal ponds (Díaz-Paniagua et al. 2010; García et al. 1993; Green et al. 2016). The park is listed as a Ramsar site and a UNESCO World Biosphere Reserve and World Heritage site because of the diversity of wetland habitats and the animal biodiversity supported by them, as well as the cultural importance of the park (ramsar. org; unesco.org). It is estimated that some six million or so birds pass through the park each year (Figure 3.15), with

FIGURE 3.15 In Addition to Wetlands, Doñana National Park is Home to Important Cultural Sites and Biodiverse Terrestrial Ecosystems. El Rocío, adjacent to the Park, (A) is an Annual Pilgrimage Destination for more than One Million Spanish Citizens. The Park itself is largely Forested with Native Pine Woodlands (B), Populated with Deer ((C–D) and the Elusive Iberian Lynx. Wetlands in Doñana Support Hundreds of Species of Resident and Migratory Birds, such as Geese (E) and Flamingos (F). Photo of Flamingos (F) Provided by Biologist Miguel Álvarez, of Sevilla, Spain.

more than 300 species of resident or migratory birds having been recorded here, including the endangered Spanish imperial eagle (Green et al. 2016; unesco.org). More than 100 species of invertebrates are known to occupy the park, along with 27 inland fish species (Green et al. 2016).

3.2.10.1 Noteworthy Threats

It is estimated that the delta once comprised around 1,800 km² of natural wetlands, 1,500 km² of which were

drained during the 1900s for conversion to agriculture, aquaculture, or salt ponds (Green et al. 2016). The Guadalquivir River was channelized, dredged, and diked off from the surrounding floodplain, both isolating the floodplain wetlands from riverine inputs and altering the natural hydrology of the Doñana marshes (Green et al. 2016). Modification of this river was, in part, intended to maintain and enhance access of large ships to the port of Seville, roughly 80 km upstream from the where the river meets the Gulf of Cádiz.

This shipping traffic alone poses certain risks to the delta ecosystems, such as introduction of non-native aquatic species and pollution. Other threats to the Doñana wetlands are those that are characteristic of human population growth and urbanization, including runoff from agriculture and urban environments and overexploitation of aquifers (which lowers the **water table** underlying the wetlands).

3.2.10.2   Representative Plant Species

García et al. (1993) studied the salt marsh vegetation of Doñana National Park to better understand how plant productivity and salt marsh plant species diversity are influenced by environmental variables such as soil salinity and hydrology. Díaz Barradas et al. (1999) analyzed plant species across a habitat transect in Doñana National Park to evaluate whether there were recognizable plant functional traits associated with habitats in which the species were most common. Their work included study plots in some of the park's interdune ponds. Finally, Díaz-Paniagua et al. (2010) conducted a relatively large-scale study of temporary ponds across a central 45 km² area of the park, surveying plants, macroinvertebrates, and amphibians in 90 ponds over a two-year period, then following this study with a more detailed monthly examination of 21 of those 90 ponds. These three studies were used to develop a list of common wetland plant taxa for the Doñana park wetlands for Tables 3.3–3.5.

3.2.11   Boreal Peatlands, Canada

Collectively, boreal peatlands cover somewhere in the range of 2.5 to 4 million km², out of an estimated total 7 to 9 million km² of inland wetlands across the globe (Mitsch and Gosselink 2000; Mitsch et al. 2013; Reis et al. 2017). As with the tropical forested peatlands of the Congo River basin, discussed earlier in section 3.2.3, the accumulation of peat in these ecosystems results from low evaporation, relative to precipitation inputs, and the reduced rates of decomposition that are caused by the ensuing oxygen deficits in the saturated wetland sediments. In contrast to the Congo, however, where the evaporation:precipitation ratio is dominated by high precipitation inputs in the African rainforests, this ratio in boreal peatlands is driven by low evaporation, resulting from low temperatures (Figure 3.2). Annual precipitation in the Canadian peatlands averages roughly one-third of that in the Congo, with approximately a third of that precipitation falling when temperatures are below freezing (Figure 3.2).

The accumulation of peat in these wetlands is important, not only for driving the ecology of these ecosystems through its influence on hydrology, soil/sediment biogeochemistry, and vegetation, but also globally, in its influence on the global carbon cycle. Boreal peatlands store approximately one-third of the world's terrestrial soil carbon (i.e., organic carbon that is contained within the soil vs. in vegetation, the atmosphere, or the oceans; Dargie et al. 2017; Gorham 1991). This puts them front and center as an area

of concern for driving feedback loops in the face of ongoing climate change, as we will see in the next section.

As indicated in Table 3.2, peatlands typically are segregated into two categories, bogs and fens, based on a combination of soil biogeochemistry and vegetation (which, as discussed in Chapter 1 and in more detail in Chapters 4–7, are quite intimately interconnected). Bridgham et al. (1996) and Gorham and Janssens (1992) examined patterns of plant species (especially mosses in the Sphagnaceae and Amblystegiaceae), pH, and other environmental factors. Both sets of investigators found very close association between surface water pH and moss species, with a marked distinction between peatlands dominated by bog species (pH below 5) and those dominated by fen species (pH above 6; Figure 3.16). Data included in the work of Gorham and Janssens (1992) were taken not only from their own work in Alaska, Minnesota, New York, Maine, Nova Scotia, and Newfoundland, but also from the work of other scientists in North America, Sweden, and Norway; thus, these patterns are not only consistent but widespread across boreal peatland ecosystems. Factors driving this relationship will be discussed further in the next chapter, when we specifically address the interaction between wetland hydrology and plant species.

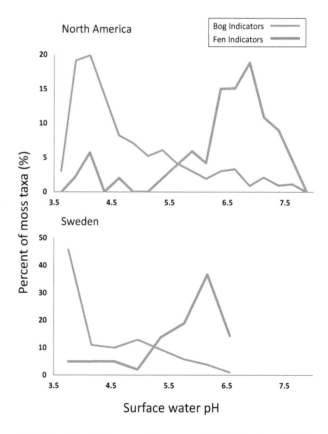

FIGURE 3.16   Gorham and Janssens (1992) Found a Striking Contrast in the Relative Occurrence of Moss Species Indicative of Bog Vs. Fen Habitats Along a Ph Gradient in North American and Swedish Boreal Peatlands. Bog Species Tended to be more Common in Wetlands with Surface Water Ph below 5, Whereas Fen Species Tended to Occur in Wetlands where the Ph was 6 or Higher.

3.2.11.1 Noteworthy Threats

Because the distribution of Boreal Peatlands overlaps considerably with regions that have been occupied by humans for millennia, these wetlands have experienced, and continue to experience, numerous forms of anthropogenic disturbance. Among the more important impacts humans have on peatlands are grazing of sheep and other livestock, fire (to increase forage production for livestock), drainage, agriculture, and peat extraction for use as fuel (Turetsky and St. Louis 2006).

Human-initiated fires to improve livestock grazing are not the only fires impacting peatlands, however. Recent estimates are that between 50,000 to 120,000 km^2 per year of boreal forest and **tundra** are burned because of wildfires (Turetsky and St. Louis 2006; Abbott et al. 2016), and the frequency and intensity of those fires have been increasing since the 1980s (Turetsky et al. 2011). This presents a complex environmental problem, as the increase in fires adds to the stocks of atmospheric carbon that are exacerbating climate change at a global scale, while the burned ground left behind absorbs solar radiation, resulting in greater depth of thawing of the permafrost, which adds to the flammable fuel loads of peatland ecosystems, further increasing future fire risk. This will be discussed more in Chapter 11.

3.2.11.2 Representative Plant Species

Thormann et al. (1999) surveyed dominant plant species in ten peatland wetlands of Alberta, Canada (54.2°–54.7°

N latitude; Figure 3.17), with pH ranging from 3.8 to 8.2, to assess the mycorrhizal status of dominant peatland vegetation (more about the mycorrhizal aspect of this work in Chapter 7). Camill (1999) surveyed vegetation in 30 sites with pH ranging from 4 to 6.5, spanning ~400 km across peatlands of central Manitoba, Canada (Figure 3.17). Their work was aimed at understanding the responses of vegetation to permafrost thaw and collapse resulting from ongoing climate change. Plant species lists from this work by Camill (1999) and Thormann et al. (1999, Thormann, Bayley, and Currah 2001) were used to develop a list of representative taxa for wetlands of the boreal peatland landscape (Tables 3.3–3.5), realizing that species at individual wetlands across some three million or so square kilometers of peatlands will show incredible variation.

3.2.12 The Everglades, Florida, United States

The Florida Everglades ecosystem covers (or historically covered) more than 11,000 km^2 of southern Florida (Loveless 1959; Harvey et al. 2017; Figure 3.18). Generally speaking, the Everglades landscape is a flat, former sea floor plain, dominated by freshwater marsh and wet prairie. However, at a finer scale, one finds not only marshlands but also subtropical forested hammocks (tree islands), freshwater sloughs, saltwater marshes, seagrass beds, and one of the largest mangrove forests in the western hemisphere (Figure 3.19; ramsar.org; unesco.org). The Everglades

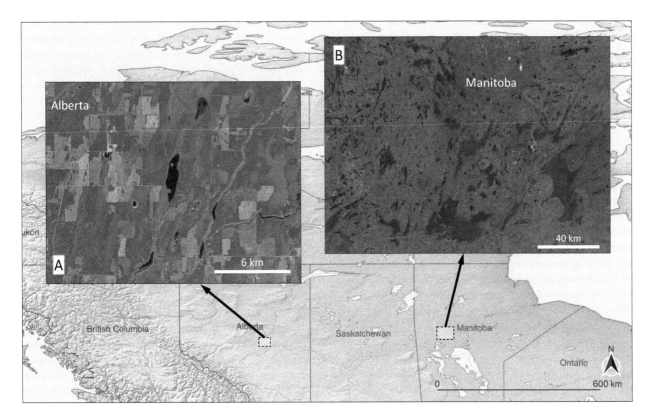

FIGURE 3.17   Aerial View of Landscapes in Alberta (A) and Manitoba (B), where the Peatlands Floristic Surveys were Conducted. Aerial Imagery from Esri (2022) "World Imagery" www.arcgis.com/home/item.html?id=10df2279f9684e4a9f6a7f08febac2a9.

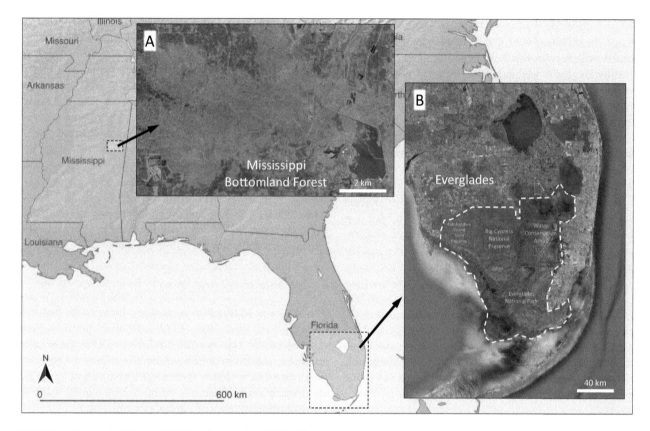

FIGURE 3.18 Aerial View of (A) Noxubee National Wildlife Refuge Bottomland Forests, with Rivers Illustrated, and (B) Southern Florida, Highlighting the Everglades Region (B). The Approximate Extent of Protected Areas Comprising the Everglades Wetland Complex is Indicated by the Dashed White Boundaries. Aerial Imagery from Esri (2022) "World Imagery" www.arcgis.com/home/item.html?id=10df2279f9684e4a9f6a7f08febac2a9.

National Park (~6,000 km²) is listed as a Ramsar site, UNESCO Biosphere Reserve, and UNESCO World Heritage site (ramsar.org, unesco.org).

The location of the Everglades, at the junction of freshwater and saltwater, subtropical and temperate climates, and urban areas and wilderness, has fostered a broad diversity of both native and non-native biota. The Everglades is home to more than 20 rare, threatened, or endangered animal species, including Florida panther, alligators (Figure 3.19), crocodiles, and the manatee (unesco.org), but also is home to more than 130 species of non-native animals, and the South Florida Water Management District mentions a priority list of 75 species of non-native plants (Rodgers et al. 2018).

The Everglades ecosystem has been the subject of a massive, US$15 billion restoration effort, beginning in the year 2000 (Borkhataria et al. 2017). The restoration centers on restoring flows of freshwater to the Everglades, following large-scale drainage and compartmentalization for agricultural and urban use that reduced water flows to about half that of the pre-drainage Everglades (Harvey et al. 2017). That drainage, and the accompanying rerouting of remaining flows, resulted in significant alterations to the ecological dynamics of the Everglades. Reduced water levels allowed oxidation of organic sediments within the wetlands, and decreased inputs of freshwater (along with increased inputs

of nutrients from upstream agriculture) greatly altered nutrient dynamics. Restoration of flows is intended to return the ecosystem to some semblance of its pre-drainage condition, albeit with a much different set of threats that could impact the end results of these efforts.

3.2.12.1 Noteworthy Threats

In addition to agricultural land use to the north and dense urbanization to the east (Mitsch and Hernandez 2013), non-native species pose a substantial threat to the ecological integrity of the Everglades ecosystem. Noteworthy among the more than 130 non-native, invasive animal species are Burmese pythons, green iguanas, cane toads, Nile monitors, spectacled caiman, boa constrictors, ball pythons, Northern African pythons, feral hogs, and giant African land snails (Rodgers et al. 2018). The invasive animals pose ecological threats to native animals as well as native plants of the Everglades, and considerable effort goes into attempting to minimize those impacts. The list of important invasive non-native plants includes many of the most widely known aquatic weeds from tropical and subtropical areas around the world. These include *Melaleuca quinquenervia*, *Lygodium microphyllum* (Old World climbing fern), *Eichhornia crassipes* (water hyacinth), *Panicum repens* (torpedograss), *Hydrilla verticillata*, *Pistia stratiotes* (waterlettuce), and

FIGURE 3.19 Florida Everglades National Park. (A) View across the Everglades National Park, West of Miami. (B) Closer View of a Cluster of Trees and Shrubs Jutting above the Surrounding Wet Meadow Vegetation. (C) Juvenile Alligator (*Alligator Mississipiensis*).

Schinus terebinthifolius (Brazilian pepper). Control efforts for *Melaleuca quinquenervia* alone exceed US$40 million over the 20 years that resources managers have worked to control that species in south Florida (Rodgers et al. 2018). In 2016 and 2017, more than US$7.5 million was spent on control efforts for five of these plant species (*M. quinquenervia*, *L. microphyllum*, *S. terebinthifolius*, *E. crassipes*, and *P. stratiotes*; Rodgers et al. 2018).

3.2.12.2 Representative Plant Species

Loveless (1959) conducted vegetation surveys in two lev-eed water conservation areas in the northern reaches of the Everglades, for the purposes of augmenting efforts at managing Everglades wildlife. The survey areas (now referred to as water conservation areas, or WCAs) remain under management by the Florida Game and Freshwater Fish Commission, and the levees that control flow of water through the Everglades are potential targets for removal as part of restoration efforts (sfwmd.gov). Almost 40 years later, David (1996) published vegetation surveys in one of the WCAs (WCA3) included in the work by Loveless in the 1950s, to evaluate the degree to which vegetation had changed because of hydrological modifications to the Everglades system.

Further south, work has been conducted to describe the vegetation of the south Everglades mangrove forests, which cover approximately 1,700 km² in the four southern Everglades counties (Odum et al. 1982). Davis et al. (2005) discussed the development of a conceptual model describing hydrology and salinity gradients that impact southern Everglades mangrove forests. Ross et al. (2000) conducted a comparative study of mangrove forest vegetation between the 1950s and 1990s to document changes in species composition following 50 years of hydrological change. Davis et al. (2005) gave an overview of only the dominant mangrove species, while the work of Ross et al. (2000) included not only the mangroves themselves, but also the associated plant species. The plant lists from these four publications are summarized in Tables 3.3–3.5.

3.2.13 BOTTOMLAND FORESTS, MISSISSIPPI, UNITED STATES

Our last stop on this wetlands tour brings us to my home, in the southeastern US (Figures 3.20 and 3.21). Bottomland forests historically formed in the dynamic floodplains of rivers and larger streams, where the erosional and depositional processes of untamed rivers created a mosaic of microhabitats suitable to a broad diversity of tree species. It is estimated that as much as 90% of these forested wetlands have been lost to land clearing and hydrologic alterations since the mid-1800s (Ervin et al. 2006; Mitsch and Gosselink 2000). The US Fish and Wildlife Service report on "Status and Trends" of US wetlands during 2004–2009 (Dahl 2011) estimated that there were roughly 200,000 km² of forested wetlands in the US in 2009, with a loss of just over 1% of these wetlands during that five-year timeframe.

The largest extent of bottomland forests in the US occurs within the Atlantic, East Gulf, and West Gulf Coastal Plains and Mississippi River Alluvial Valley geographic provinces in the southeastern US (Figure 3.22; Hodges 1997). Because of their occurrence in these river floodplain landscapes, bottomland forests experience a diversity of topographic, soil, and hydrological heterogeneity. Natural processes of erosion, deposition, and overbank flooding of unchannelized rivers results in migration of the main channel and formation of topographic features such as sandy levees along the channel, followed by flats and sloughs with

FIGURE 3.20   The Sam D. Hamilton Noxubee National Wildlife Refuge has been Home to much of the Research My Students and I Have Conducted During My Time at Mississippi State University.

FIGURE 3.21   Bottomland Forest Sites at Noxubee National Wildlife Refuge in March (Top) and May (Bottom).

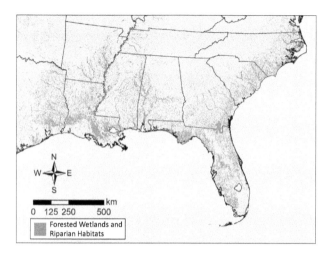

FIGURE 3.22 Approximate Current Extent of Forested Wetlands and Associated Riparian Habitats across the Southeastern United States. Land Cover Data Provided By Natureserve, in Collaboration with the International Union for Conservation of Nature.

deposition of finer-grained sediment particles, and eventually additional ridges and swamps, with occasional oxbows where old channels have been cut off during channel migration (Hodges 1997). Each of these different river floodplain habitat types will come to be dominated by different suites of tree species, adapted to local soil and hydrology, given sufficient time and propagule supplies.

Bottomland forests are important in helping to buffer floodwaters (where the river remains connected to the floodplain), in providing a diversity of habitat types for various animals (reptiles, amphibians, overwintering migratory waterfowl in the southeastern US), and in nutrient transformation and storage (because of the heterogeneity in soil type and hydrology). They also are important in storing carbon (Bernal and Mitsch 2012), which is expected to become an increasingly critical contribution to the global carbon cycle as boreal wetlands increasingly are exposed to warmer, drier conditions because of climate change.

3.2.13.1 Noteworthy Threats

Dahl (2011) estimated that forested wetlands in the US declined in area by almost 2,600 km² during 2004–2009. About 16% of these losses were in conversion to urban land use, and another 24% were associated with silvicultural activities, including drainage of land for forest harvests and subsequent conversion to commercial monoculture forestry. Drainage and conversion of forested wetlands has taken a toll on these ecosystems for more than 200 years, with the loss of old-growth stands of cypress-gum (dominated by *Taxodium* and *Nyssa* species) and Atlantic white cedar (*Chamaecyparis thyoides*) swamps in North America (Mitsch and Hernandez 2013). Unfortunately, potentially because of the length of time necessary for restoration of forests, forested wetlands tend not to be regarded as suitable targets for wetland restoration activities, resulting in more persistent losses of this wetland type, in comparison

to herbaceous species-dominated wetlands (Dahl 2011). Although land use change is the major driver of losses of bottomland forests, other factors impacting these ecosystems include climate change–induced drought, sea level rise, and saltwater intrusion (Dahl 2011; Mitsch and Hernandez 2013).

3.2.13.2 Representative Plant Species

Hodges (1997) gave an excellent and highly detailed overview of the general patterns of forest development in bottomland forests. In this paper, he described geomorphological processes involved in bottomland forest development and successional dynamics among species, and he places the ecology of these ecosystems within a forest management context. In 2003, my students and I conducted our own work in bottomland forests of the US Fish and Wildlife Service's Sam D. Hamilton Noxubee National Wildlife Refuge, which is situated along the floodplain of the Noxubee River in eastern Mississippi (Ervin et al. 2006; Figure 3.18). Our work was aimed at evaluating the impact of long-term seasonal inundation on bottomland forest tree assemblages, as a management approach to enhance habitat for overwintering waterfowl, and we collected data on tree species composition in the main canopy and understory across a roughly 60 km² area of the Noxubee River floodplain. These two papers were used to develop a list of representative plant taxa for US Bottomland Forests (Table 3.5).

3.3 GLOBAL PATTERNS OF WETLAND PLANT DIVERSITY

The plant surveys used to obtain species lists for the wetlands we have just seen are summarized in Tables 3.3–3.5. The plant diversity among these wetlands is presented at the level of families, with numbers of genera per family given for each wetland area. I chose this approach, rather than listing plant diversity at the level of individual species, for two reasons. First, the total number of species present across the plant lists of these wetlands would have been too long to represent with any practicality in a single table (or set of three tables). There were 256 genera represented among the numerous studies used to develop these lists, and often there were multiple species within a given genus even within an individual wetland area. For example, the sedge genera *Cyperus*, *Carex*, and *Eleocharis*, the grass genera *Panicum*, *Paspalidium*, among others, as well as several non-monocot genera (*Polygonum* [*Persicaria*], *Bacopa*, *Ludwigia*) all were represented by multiple species, even within one or more individual wetland areas. This resulted in there being several hundred individual species in the overall list of wetland plants from these sites.

The second, and more important, reason that I chose to present this plant diversity at the level of numbers of genera per family was to allow us to look for general patterns in the data. We have wetland areas from across the globe, representing tropical, subtropical, temperate, and boreal ecosystems. We have wetlands from very wet climates

(Congo River Swamp Forests, Kakadu National Park, the Sundarbans) and very dry climates (Okavango Delta, Doñana Park), as well as quite warm (the Pantanal, Kakadu National Park) and frigid climates (Qinghai-Tibet Plateau, Boreal Peatlands). Looking at the level of plant families across this diversity of environmental conditions, we should be able to tell quite easily whether there are any general patterns worth discussion in the wetland vegetation represented here.

Recalling numbers of species among the high-level groups represented here (i.e., monocots vs. non-monocot families), there are approximately 416 families of angiosperms. Only 76 of these are monocot families, accounting for 18% of the angiosperm families; this gives a monocot:non-monocot family ratio of about 1 to 4.5. If we were looking at extant species diversity, the ratio would be similar, with monocots representing about 25% of angiosperm diversity (McInnes et al. 2013). A comparison of Tables 3.3 and 3.4 shows we have 27 families of monocots and 40 non-monocot families represented among the wetland areas included in those tables; this is a ratio of 1 to 1.5, a much more even ratio than we would have expected by chance. As discussed in Chapter 2 and as we will see in later chapters, monocot lineages appear generally better adapted to aquatic and wetland conditions, so perhaps it unsurprising that we see these families represented at this higher rate.

Looking more closely at the data in these two tables, there appears to be a general trend for the number of monocot genera to decrease and for the number of non-monocot genera to increase from left to right in these tables. It turns out that the wetlands in the left half of these tables all were in the southern hemisphere, most within 30° latitude of the equator. Those in the right half all were in the northern hemisphere, most farther than 30° latitude from the equator. When we look at the ratio of monocot:non-monocot genera across these sites as a function of latitudinal distance from the equator, there is a general trend for an increase in the prevalence of monocot genera the closer the wetlands are to the equator (Figure 3.23). This is in line with general patterns that have been observed among monocots, with higher diversity of grasses, palms, and orchids in the tropics and temperate regions of the southern hemisphere than elsewhere (McInnes et al. 2013). Among strictly aquatic monocots, however, diversity tends to be slightly higher at higher latitudes than in the tropics, and aquatic or wetland species of Cyperaceae in North America outnumber those in Central America more than two to one (Crow 1993). Thus, although we see a general trend in wetlands for monocot taxa to outnumber non-monocot taxa nearer the equator, some individual taxonomic groups may show patterns that appear to contradict those general trends. Alahuhta et al. (2021) found similarly complex spatial patterns in freshwater plant diversity and concluded that overall species diversity of aquatic and wetlands plants appeared to peak in the mid-latitudes. A potential reconciliation of all these patterns is that there may be a transition in taxa along a

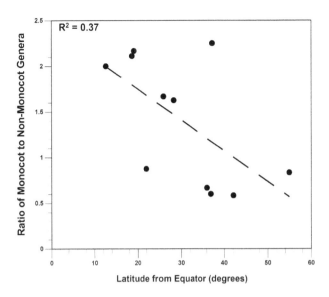

FIGURE 3.23 Monocot Genera were more Common than Dicot Genera in Plant Surveys of Wetland Areas Nearer the Equator, Among the 11 Wetland Areas for which Published Floristic Surveys of Herbaceous Species were Obtained.

latitudinal gradient from the equator to the poles, with mid-latitude wetlands hosting species from each end of this gradient and a shift in the subsets of species that dominate at each end.

Regardless of the patterns of change in overall plant diversity as we progress away from the equator, we do see consistent representation among key families. Across the sample of wetland areas in this chapter, the most common monocot families were the sedges (Cyperaceae), grasses (Poaceae), and the Hydrocharitaceae, followed by the Alismataceae, Araceae, and Typhaceae (cattails). Recall that Alismataceae, Araceae, and Hydrocharitaceae all belong to the Alismatales, which is the order containing the most fully aquatic species and the only marine-adapted aquatic angiosperms. Among the non-monocot families, Asteraceae (asters), Nymphaeaceae (water lilies), and Polygonaceae (smartweeds) were the most common, with slightly higher occurrence than the Onagraceae (primroses), Lentibulariaceae (represented by the bladderworts, *Utricularia* species), and Fabaceae (legumes). Among these families, you may remember that the Asteraceae and Fabaceae are the two most speciose families of the eudicots; this accounts in part for their frequency among these wetland examples. Nymphaeaceae and Lentibulariaceae have a large proportion of aquatic or wetland species among their members, and there are some very common wetland genera in the Onagraceae and Polygonaceae (e.g., *Ludwigia* and *Polygonum* [*Persicaria*]).

Unfortunately, few of the floristic studies in the wetland areas here focused specifically on tree species. Even in the Congo River Swamp Forests, which by definition should be dominated by trees, we find few plant species listed in the published literature at all (Dargie et al. 2019). Three of these wetland areas, however, were dominated by trees: the

Sundarbans, the Kolkheti Lowlands, and the Bottomland Forests of the southeastern US.

The Sundarbans wetlands are distinctive among the wetland areas chosen for this chapter, in that they are dominated by mangrove forests, which make up only about 2% of the world's wetlands (~150,000 km²; Spalding, Kainuma, and Collins 2010). Two of the core mangal families that are represented in both the Indo-West Pacific and Atlantic-East Pacific regions of mangrove distribution (Acanthaceae and Rhizophoraceae; Table 2.4) have representatives in both the Sundarbans and the Florida Everglades (*Rhizophora* and *Avicennia* species; Table 3.5). These two wetland areas also share very similar climates and latitudes (Figures 3.1 and 3.2); however, there are key differences between them that likely lead to the significant differences in overall plant species composition. First, the Everglades are primarily freshwater-fed and herbaceous dominated, with trees occurring most commonly on tree islands between inland sloughs (Wetzel et al. 2005) and as mangrove forests on the periphery, along the Gulf of Mexico and Atlantic coasts. The Sundarbans occur in a large tidal delta, where both the tides and seasonal tropical storms have a strong influence on the dominant vegetation.

We see nine families shared between the other two forested wetland areas, the Kolkheti Lowlands and the US Bottomland Forests: Betulaceae, Ebenaceae, Fagaceae, Juglandaceae, Moraceae, Oleaceae, Rosaceae, Sapindaceae, and Ulmaceae (Table 3.5). Among these, there were no shared species, but seven genera were represented at both locations: *Carpinus* (hornbeam, Betulaceae), *Diospyros* (ebony or persimmon, Ebenaceae), *Quercus* (oak, Fagaceae), *Morus* (mulberry, Moraceae), *Fraxinus* (ash, Oleaceae), *Acer* (Sapindaceae), and *Ulmus* (elm, Ulmaceae). As with the Sundarbans and Everglades, these two locations are at similar latitudes and have very similar temperature profiles (Figures 3.1 and 3.24), which could account for much of the similarity in tree taxa between the two.

3.4 GLOBAL THREATS TO WETLAND INTEGRITY

We will close this chapter by noting that wetlands face much the same ecological threats, regardless of where they are in the world. With humans having directly modified about three-fourths of the ice-free land on earth (Ellis and Ramankutty 2008), such factors as urbanization and land use conversion are common threats to the world's wetlands. These were specifically cited as threats to eight of the wetland areas we looked at in this chapter, and water diversion or withdrawals related to land use were given as threats in three of the remaining areas. Resource extraction (e.g., oil and gas mining, forest harvest) were given as threats to about half of the wetland areas, and pollution of water supplies to wetlands was mentioned for more than a third of them. All of these threats are symptoms of resource needs for our growing global human population, and all of these will need to be addressed if we are to continue to receive the benefits provided by ecologically intact wetland ecosystems.

Finally, and to aid our transition into the next chapter, we saw in the previous sections that another key problem facing wetlands is altered water supply, whether that results directly from human withdrawals (usually for agricultural irrigation), from drainage of wetlands prior to conversion to other land use, from climate change, or from upstream diversions of rivers or streams that feed the wetlands. Frequency and duration of the presence of water are critically important for wetland plants, and thus for wetland ecological integrity. For this reason, hydrology is the first

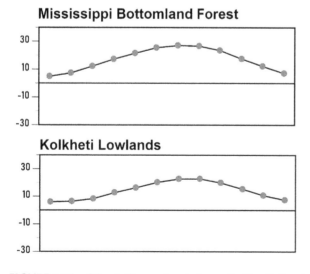

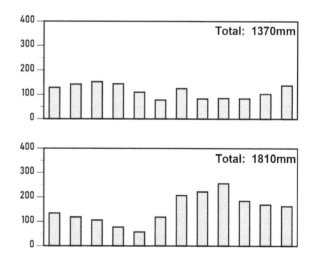

FIGURE 3.24 Climate Comparison between the Two Bottomland Forest Areas. Data were Obtained from the Worldclimate Database (www.worldclimate.com/). Mississippi Bottomland Forest Precipitation Data were from an On-Site Weather Station; Temperature Data were from a Station Approximately 30 Km away. Kolkheti Data were Provided for a Station Approximately 100 Km from the Wetland Area.

of the environmental factors that we will explore in detail in this text, and we do so in the next chapter.

3.5 IN SUMMARY

There have been several classifications or categorizations of wetland types, all of which are built around a common suite of wetland features. In no particular order, these are:

- Does the wetland accumulate peat (partially decomposed organic matter)?
- Is the wetland dominated by woody or by herbaceous vegetation?
- Is the wetland influenced primarily by freshwater or by saltwater?

In this chapter, I have used these major wetland features to define a few broad categories of wetlands within which several specific wetland types were described. Some of the more commonly encountered wetland types include marshes, wet meadows, swamps (forested wetlands), salt marshes, and peatlands (bogs and fens). Site-level factors that result in a distinction among specific wetland types include hydrology, underlying geology, and the morphology of the wetland basin itself; these factors collectively are referred to as the site's hydrogeomorphology (referring to the hydrologically modified **geomorphology** of the site).

When we look across wetlands at a broad geographic scale, we see patterns of plant diversity that reflect the taxonomic representation of aquatic and wetland plants that was discussed in the previous chapter. Wetland plant diversity is heavily characterized by the Nymphaeales, a few common monocot families (Poaceae, Cyperaceae, and those of the Alismatales), and some of the more taxonomically diverse families of eudicots (e.g., Asteraceae, Fabaceae). We also see some interesting geographic patterns in the relative frequency of monocot versus non-monocot taxa, analogous to general patterns of these groups across habitat types in general. Finally, with respect to patterns of taxonomic diversity across geographic regions, we find that areas with similar climates tend to harbor similar suites of wetland plant taxa.

A somewhat disheartening finding in taking a global look at wetland ecosystems is that there are very similar threats to their ecological integrity essentially everywhere we look. The most common threats encountered are impacts of anthropogenic disturbance (land use and land use change), introduction of non-native species, and climate change. One may note here that all three of these categories, although representing different types of ecological impact, are anthropogenic in origin. That is, humans have had a disproportionate influence on land cover change, species distributions, and recent rates of climate change. As we delve further into the inner workings of the ecological function of plants in aquatic and wetland ecosystems, the implications of these human-derived impacts should become more evident, as will the need for human intervention to ensure the continued provision of critical ecological services that wetlands provide for society.

3.6 FOR REVIEW

1. What are some characteristics that may be used to describe types of wetlands?
2. How are some of the wetland categories influenced by the dominant plant species they harbor?
3. What are the two major types of boreal peatland wetlands?
4. How do we distinguish between the wetland types from the previous question, and what drives those distinguishing features?
5. What are some common saltwater wetland types?
6. What are some ways that climate change is impacting wetlands around the world?
7. What are some global patterns encountered in wetland plant diversity?
8. Why might we see similar groups of plant species in widely separated regions of the globe?
9. What are some frequently encountered ways that human land use impacts wetlands?

3.7 REFERENCES

Abbott, Benjamin W., Jeremy B. Jones, Edward A. G. Schuur, F. Stuart Chapin, William B. Bowden, M. Syndonia Bret-Harte, Howard E. Epstein, et al. 2016. "Biomass Offsets Little or None of Permafrost Carbon Release from Soils, Streams, and Wildfire: An Expert Assessment." *Environmental Research Letters* 11 (3). https://doi.org/10.1088/1748-9326/11/3/034014.

Alahuhta, J., M. Lindholm, L. Baastrup-Spohr, J. García-Girón, M. Toivanen, J. Heino, and K. Murphy. 2021. "Macroecology of Macrophytes in the Freshwater Realm: Patterns, Mechanisms and Implications." *Aquatic Botany* 168: 103325.

Alho, C. J. R. 2009. "Biodiversity of the Pantanal: Response to Seasonal Flooding Regime and to Environmental Degradation." *Brazilian Journal of Biology* 68 (4 suppl): 957–66. https://doi.org/10.1590/s1519-69842008000500005.

Alonso, L. E., and L. Nordin. 2003. *A Rapid Biological Assessment of the Aquatic Ecosystems of the Okavango Delta, Botswana: High Water Survey*. Washington, DC: Conservation International Center for Applied Biodiversity Science.

Ashton, P. J., L. Nordin, and L. E. Alonso. 2003. "Introduction to the Okavango Delta and the AquaRap Expedition." In *A Rapid Biological Assessment of the Aquatic Ecosystems of the Okavango Delta, Botswana: High-Water Survey. RAP Bulletin of Biological Assessment*, edited by Leeanne E. Alonso and Lee-Ann Nordin, pp. 29–37. Washington, DC: Conservation International Center for Applied Biodiversity Science.

Berlanga-Robles, César Alejandro, Arturo Ruiz-Luna, and Guadalupe de la Lanza Espino. 2008. "Esquema de clasificación de los humedales de México." *Investigaciones Geográficas, Boletín del Instituto de Geografía, UNAM* (6): 25–46.

Bernal, Blanca, and William J. Mitsch. 2012. "Comparing Carbon Sequestration in Temperate Freshwater Wetland Communities." *Global Change Biology* 18 (5): 1636–47. https://doi.org/10.1111/j.1365-2486.2011.02619.x.

Borkhataria, Rena R., Paul R. Wetzel, Hiram Henriquez, and Stephen E. Davis. 2017. "The Synthesis of Everglades Restoration and Ecosystem Services (SERES): A Case Study for Interactive Knowledge Exchange to Guide Everglades Restoration." *Restoration Ecology* 25 (October): S18–26. https://doi.org/10.1111/rec.12593.

Bridgham, Scott D., John Pastor, Jan A. Janssens, Carmen Chapin, and J. Malterer Thomas. 1996. "Multiple Limiting Gradients in Peatlands: A Call for a New Paradigm." *Wetlands* 16 (1): 45–65.

Brinson, M. M. 1993. "A Hydrogeomorphic Classification for Wetlands." US Army Corps of Engineers Wetlands Research Program Technical Report WRP-DE-4.

Camill, Philip. 1999. "Patterns of Boreal Permafrost Peatland Vegetation across Environmental Gradients Sensitive to Climate Warming." *Canadian Journal of Botany* 77 (5): 721–33. https://doi.org/10.1139/cjb-77-5-721.

Clarkson, Beverley R., Louis A. Schipper, and Anthony Lehmann. 2004. "Vegetation and Peat Characteristics in the Development of Lowland Restiad Peat Bogs, North Island, New Zealand." *Wetlands* 24 (1): 133–51. https://doi.org/10.1672/0277-5212(2004)024[0133:vapcit]2.0.co;2.

Connor, Simon E., Ian Thomas, and Eliso V. Kvavadze. 2007. "A 5600-Yr History of Changing Vegetation, Sea Levels and Human Impacts from the Black Sea Coast of Georgia." *Holocene* 17 (1): 25–36. https://doi.org/10.1177/0959683607073270.

Cowardin, L. M., V. Carter, F. C. Golet, and E. T. LaRoe. 1979. "Classification of Wetlands and Deepwater Habitats of the United States." Washington, DC: Office of Biological Services, Fish and Wildlife Service.

Crow, Garrett E. 1993. "Species Diversity in Aquatic Angiosperms: Latitudinal Patterns." *Aquatic Botany* 44 (2–3): 229–58. https://doi.org/10.1016/0304-3770(93)90072-5.

Dahl, T. E. 2011. *Status and Trends of Wetlands in the Conterminous United States 2004 to 2009,* 108 pp. Washington, DC: U.S. Department of the Interior; Fish and Wildlife Service.

Dargie, Greta C., Simon L. Lewis, Ian T. Lawson, Edward T. A. Mitchard, Susan E. Page, Yannick E. Bocko, and Suspense A. Ifo. 2017. "Age, Extent and Carbon Storage of the Central Congo Basin Peatland Complex." *Nature* 542 (7639): 86–90. https://doi.org/10.1038/nature21048.

Dargie, Greta C., Simon L. Lewis, Ian T. Lawson, Tim J. Rayden, Lera Miles, Edward T. A. Mitchard, Susan E. Page, Yannick E. Bocko, and Suspense A. Ifo. 2019. "Congo Basin Peatlands: Threats and Conservation Priorities." *Mitigation and Adaptation Strategies for Global Change,* 1–18. https://doi.org/10.1007/s11027-017-9774-8.

David, Peter G. 1996. "Changes in Plant Communities Relative to Hydrologic Conditions in the Florida Everglades." *Wetlands* 16 (1): 15–23. https://doi.org/10.1007/BF03160642.

Davis, Steven M., Daniel L. Childers, Jerome J. Lorenz, Harold R. Wanless, and Todd E. Hopkins. 2006. "A Conceptual Model of Ecological Interactions in the Mangrove Estuaries of the Florida Everglades." *Wetlands* 25 (4): 832–42. https://doi.org/10.1672/0277-5212(2005)025[0832:acmoei]2.0.co;2.

Denk, Thomas, Norbert Frotzler, and Nino Davitashvili. 2001. "Vegetational Patterns and Distribution of Relict Taxa in Humid Temperate Forests and Wetlands of Georgia (Transcaucasia)." *Biological Journal of the Linnean Society* 72 (2): 287–332. https://doi.org/10.1006/bijl.2000.0502.

Díaz Barradas, M. C., M. Zunzunegui, R. Tirado, F. Ain-Lhout, and F. García Novo. 1999. "Plant Functional Types and Ecosystem Function in Mediterranean Shrubland." *Journal of Vegetation Science* 10 (5): 709–16. https://doi.org/10.2307/3237085.

Díaz-Paniagua, Carmen, Margarita Florencio, Carola Gómez-Rodríguez, Alexandre Portheault, Rocío Fernández-Zamudio, Pablo García-Murillo, Laura Serrano, and Patricia Siljeström. 2010. "Temporary Ponds from Doñana National Park: A System of Natural Habitats for the Preservation of Aquatic Flora and Fauna." *Limnetica* 29 (1): 41–58. https://doi.org/10.13039/501100000780.

Duggan, Kathryn, Lucy Roberts, Mary Beech, Hugh Robertson, Matthew Brady, Michael Lake, Kerry Jones, Kevin Hutchinson, and Shannon Patterson. 2013. *Arawai Kākāriki Wetland Restoration Programme: Whangamarino Outcomes Report 2007–2011.* Wellington, New Zealand: Research and Development Group, New Zealand Department of Conservation.

Ellery, K., and W. Ellery. 1997. *Plants of the Okavango Delta.* Durban, South Africa: Tsaro PublishersDurban, South Africa.

Ellery, W. N., and Budzanani Tacheba. 2003. "Floristic Diversity of the Okavango Delta, Botswana." In *A Rapid Biological Assessment of the Aquatic Ecosystems of the Okavango Delta, Botswana: High Water Survey. RAP Bulletin of Biological Assessment 27,* edited by L. E. Alonso and L. Nordin. Washington, DC: Conservation International.

Ellis, Erle C., and Navin Ramankutty. 2008. "Putting People in the Map: Anthropogenic Biomes of the World." *Frontiers in Ecology and the Environment* 6 (8): 439–47. https://doi.org/10.1890/070062.

Ervin, G. N., L. C. Majure, and J. T. Bried. 2006. "Influence of Long-Term Greentree Reservoir Impoundment on Stand Structure, Species Composition, and Hydrophytic Indicators." *Journal of the Torrey Botanical Society* 133 (3). https://doi.org/10.3159/1095-5674(2006)133[468:IOLGRI]2.0.CO;2.

Finlayson, C. Max. 2005. "Plant Ecology of Australia's Tropical Floodplain Wetlands: A Review." *Annals of Botany* 96 (4): 541–55. https://doi.org/10.1093/aob/mci209.

Finlayson, C. Max, John Lowry, Maria Grazia Bellio, Suthidha Nou, Robert Pidgeon, Dave Walden, Chris Humphrey, and Gary Fox. 2006. "Biodiversity of the Wetlands of the Kakadu Region, Northern Australia." *Aquatic Sciences* 68 (3): 374–99. https://doi.org/10.1007/s00027-006-0852-3.

Fontana, José Luis. 2008. "Vegetación y Diversidad de Ambientes En La Reserva Natural Isla Apipé Grande, Provincia de Corrientes, Argentina." *Miscelánea INSUGEO* 17 (2): 407–24.

Gao, Jay, Xi lai Li, Gary Brierley, Alan Cheung, and Yuan wu Yang. 2013. "Geomorphic-Centered Classification of Wetlands on the Qinghai-Tibet Plateau, Western China." *Journal of Mountain Science* 10 (4): 632–42. https://doi.org/10.1007/s11629-013-2561-4.

García, L. V., T. Marañón, A. Moreno, and L. Clemente. 1993. "Above-ground Biomass and Species Richness in a Mediterranean Salt Marsh." *Journal of Vegetation Science* 4 (3): 417–24. https://doi.org/10.2307/3235601.

Gopal, Brij, and Malavika Chauhan. 2006. "Biodiversity and Its Conservation in the Sundarban Mangrove Ecosystem." *Aquatic Sciences* 68 (3): 338–54. https://doi.org/10.1007/s00027-006-0868-8.

Gorham, E. 1991. "Northern Peatlands: Role in the Carbon Cycle and Probable Responses to Climatic Warming." *Ecological Applications* 1: 182–95.

Gorham, Eville, and J. A. N. A. Janssens. 1992. "Concepts of Fen and Bog Re-Examined." *Acta Societatis Botanicorum Poloniae* 61 (1): 7–20.

Green, A. J., J. Bustamante, and G. F. E. Janss. 2016. "The Wetland Book." *The Wetland Book,* 1–14. https://doi.org/10.1007/978-94-007-6172-8.

Harun-or-Rashid, Sheikh, Shekhar R. Biswas, Reinhard Böcker, and Michael Kruse. 2009. "Mangrove Community Recovery Potential after Catastrophic Disturbances in Bangladesh." *Forest Ecology and Management* 257 (3): 923–30. https://doi.org/10.1016/j.foreco.2008.10.028.

Harvey, Jud W., Paul R. Wetzel, Thomas E. Lodge, Victor C. Engel, and Michael S. Ross. 2017. "Role of a Naturally Varying Flow Regime in Everglades Restoration." *Restoration Ecology* 25 (October): S27–38. https://doi.org/10.1111/rec.12558.

Hodges, John D. 1997. "Development and Ecology of Bottomland Hardwood Sites." *Forest Ecology and Management* 90 (2–3): 117–25. https://doi.org/10.1016/S0378-1127(96)03906-0.

Iftekhar, M. S., and M. R. Islam. 2004. "Degeneration of Bangladesh's Sundarbans Mangroves: A Management Issue." *International Forestry Review* 6 (2): 123–35. https://doi.org/10.1505/ifor.6.2.123.38390.

Iftekhar, M. S., and P. Saenger. 2008. "Vegetation Dynamics in the Bangladesh Sundarbans Mangroves: A Review of Forest Inventories." *Wetlands Ecology and Management* 16 (4): 291–312. https://doi.org/10.1007/s11273-007-9063-5.

Junk, Wolfgang J., Mark Brown, Ian C. Campbell, Max Finlayson, Brij Gopal, Lars Ramberg, and Barry G. Warner. 2006. "The Comparative Biodiversity of Seven Globally Important Wetlands: A Synthesis." *Aquatic Sciences* 68 (3): 400–14. https://doi.org/10.1007/s00027-006-0856-z.

Keddy, Paul A. 2000. *Wetland Ecology: Principles and Conservation.* Cambridge, UK: Cambridge University Press.

Li, Xilai, Zaipo Xue, and Jay Gao. 2016. "Environmental Influence on Vegetation Properties of Frigid Wetlands on the Qinghai-Tibet Plateau, Western China." *Wetlands* 36 (5): 807–19. https://doi.org/10.1007/s13157-016-0788-x.

Loveless, Charles M. 1959. "A Study of the Vegetation in the Florida Everglades." *Ecology* 40: 1–9.

McInnes, Lynsey, F. Andrew Jones, C. David L. Orme, Benjamin Sobkowiak, Timothy G. Barraclough, Mark W. Chase, Rafaël Govaerts, Douglas E. Soltis, Pamela S. Soltis, and Vincent Savolainen. 2013. "Do Global Diversity Patterns of Vertebrates Reflect Those of Monocots?" Edited by Diego Fontaneto. *PLoS ONE* 8 (5): e56979. https://doi.org/10.1371/journal.pone.0056979.

Mitsch, William J., Blanca Bernal, Amanda M. Nahlik, Ülo Mander, Li Zhang, Christopher J. Anderson, Sven E. Jørgensen, and Hans Brix. 2013. "Wetlands, Carbon, and Climate Change." *Landscape Ecology* 28 (4): 583–97. https://doi.org/10.1007/s10980-012-9758-8.

Mitsch, William J., and James G. Gosselink. 2000. *Wetlands.* 3rd ed. New York: John Wiley & Sons, Inc.

———. 2007. *Wetlands.* 4th ed. Hoboken, NJ: John Wiley & Sons, Inc.

———. 2015. *Wetlands.* 5th ed. Hoboken, NJ: John Wiley & Sons, Inc.

Mitsch, William J., and Maria E. Hernandez. 2013. "Landscape and Climate Change Threats to Wetlands of North and Central America." *Aquatic Sciences* 75 (1): 133–49. https://doi.org/10.1007/s00027-012-0262-7.

Neiff, J. J., S. L. Casco, A. Cózar, A. S. G. Poi de Neiff, and B. Ubeda. 2011. "Vegetation Diversity in a Large Neotropical Wetland during Two Different Climatic Scenarios." *Biodiversity and Conservation* 20 (9): 2007–25. https://doi.org/10.1007/s10531-011-0071-7.

Newnham, R. M., P. J. de Lange, and D. J. Lowe. 1995. "Holocene Vegetation, Climate and History of a Raised Bog Complex, Northern New Zealand Based on Palynology, Plant Macrofossils and Tephrochronology." *Holocene* 5 (3): 267–82. https://doi.org/10.1177/095968369500500302.

Norton, David A., and Peter J. De Lange. 2003. "Fire and Vegetation in a Temperate Peat Bog: Implications for the Management of Threatened Species." *Conservation Biology* 17 (1): 138–48. https://doi.org/10.1046/j.1523-1739.2003.01131.x.

Odum, W. E., C. C. McIvor, and T. J. Smith. 1982. *The Ecology of the Mangroves of South Florida: A Community Profile.* Washington, DC: US Fish and Wildlife Service.

Pacella, L. F., S. Garralla, and L. Anzótegu. 2011. "Cambios en la vegetación durante el Holoceno en la región Norte del Iberá, Corrientes, Argentina." *Revista de Biología Tropical* 59: 103–12.

Pangou, Serge V., Gema Maury-Lechon, and Antoine Moutanda. 2003. "Monodominant Forests of Aucoumea Klaineana and Terminalia Superba in the Chaillu Forest (SW Congo, Africa)." *Polish Botanical Journal* 48: 145–62.

Pott, A., A. K. M. Oliveira, G. A. Damasceno-Junior, and J. S. V. Silva. 2011. "Plant Diversity of the Pantanal Wetland." *Brazilian Journal of Biology* 71 (1 suppl 1): 265–73. https://doi.org/10.1590/s1519-69842011000200005.

Prance, Ghillean T., and George B. Schaller. 2006. "Preliminary Study of Some Vegetation Types of the Pantanal, Mato Grosso, Brazil." *Brittonia* 34 (2): 228. https://doi.org/10.2307/2806383.

Rahman, M. R., and M. Asaduzzaman. 2013. "Ecology of Sundarban, Bangladesh." *Journal of Science Foundation* 8 (1–2): 35–47. https://doi.org/10.3329/jsf.v8i1-2.14618.

Ramberg, Lars, Peter Hancock, Markus Lindholm, Thoralf Meyer, Susan Ringrose, Jan Sliva, Jo Van As, and Cornelis VanderPost. 2006. "Species Diversity of the Okavango Delta, Botswana." *Aquatic Sciences* 68 (3): 310–37. https://doi.org/10.1007/s00027-006-0857-y.

Reis, Vanessa, Virgilio Hermoso, Stephen K. Hamilton, Douglas Ward, Etienne Fluet-Chouinard, Bernhard Lehner, and Simon Linke. 2017. "A Global Assessment of Inland Wetland Conservation Status." *BioScience* 67 (6): 523–33. https://doi.org/10.1093/biosci/bix045.

Rodgers, Leroy, Christen Mason, Ryan Brown, Ellen Allen, Philip Tipping, Mike Rochford, Frank Mazzotti, et al. 2018. *2018 South Florida Environmental Report, Volume I, Chapter 7: Status of Nonindigenous Species*, Vol. I. West Palm Beach, FL: South Florida Water Management District.

Ross, M. S., J. F. Meeder, J. P. Sah, P. L. Ruiz, and G. J. Telesnicki. 2000. "The Southeast Saline Everglades Revisited: 50 Years of Coastal Vegetation Change." *Journal of Vegetation Science* 11 (1): 101–12. https://doi.org/10.2307/3236781.

Shumway, C., D. Musibono, S. Ifuta, J. Sullivan, R. Schelly, J. Punga, J-C. Palata, and V. Puema. 2003. "Biodiversity Survey: Systematics, Ecology and Conservation along the Congo River September-October, 2002." Congo River Environment and Development Project (CREDP) Report. https://doi.org/10.13140/RG.2.1.1389.0961.

Spalding, Mark, Mami Kainuma, and Lorna Collins. 2010. *World Atlas of Mangroves.* New York: Taylor and Francis.

Sun, Hang, Yang Niu, Yong Sheng Chen, Bo Song, Chang Qiu Liu, De Li Peng, Jian Guo Chen, and Yang. 2014. "Survival and Reproduction of Plant Species in the Qinghai-Tibet Plateau." *Journal of Systematics and Evolution* 52 (3): 378–96. https://doi.org/10.1111/jse.12092.

Thormann, Markus N., Suzanne E. Bayley, and Randolph S. Currah. 2001. "Comparison of Decomposition of Belowground and Aboveground Plant Litters in Peatlands of Boreal Alberta, Canada." *Canadian Journal of Botany* 79 (1): 9–22. https://doi.org/10.1139/cjb-79-1-9.

Thormann, Markus N., Randolph S. Currah, and Suzanne E. Bayley. 1999. "The Mycorrhizal Status of the Dominant Vegetation Along a Peatland Gradient in Southern Boreal Alberta, Canada." *Wetlands* 19 (2): 438–50.

Turetsky, Merritt R., Evan S. Kane, Jennifer W. Harden, Roger D. Ottmar, Kristen L. Manies, Elizabeth Hoy, and Eric S. Kasischke. 2011. "Recent Acceleration of Biomass Burning and Carbon Losses in Alaskan Forests and Peatlands." *Nature Geoscience* 4 (1): 27–31. https://doi.org/10.1038/ngeo1027.

Turetsky, Merritt R., and Vincent L. St. Louis. 2006. "Disturbance in Boreal Peatlands." *Boreal Peatland Ecosystems* 188: 359–79. https://doi.org/10.1007/978-3-540-31913-9_16.

Úbeda, Bárbara, Adrian S. Di Giacomo, Juan José Neiff, Steven A. Loiselle, Alicia S. Guadalupe Poi, José Ángel Gálvez, Silvina Casco, and Andrés Cózar. 2013. "Potential Effects of Climate Change on the Water Level, Flora and Macro-Fauna of a Large Neotropical Wetland." *PLoS ONE* 8 (7): 1–9. https://doi.org/10.1371/journal.pone.0067787.

Wetzel, Paul R., Arnold G. van der Valk, Susan Newman, Dale E. Gawlik, Tiffany Troxler Gann, Carlos A. Coronado-Molina, Daniel L. Childers, and Fred H. Sklar. 2005. "Maintaining Tree Islands in the Florida Everglades: Nutrient Redistribution Is the Key." *Frontiers in Ecology and the Environment* 3 (7): 370–76. https://doi.org/10.1890/1540-9295(2005)003[0370:MTIITF]2.0.CO;2.

Wetzel, Robert G. 2001. *Limnology: Lake and River Ecosystems.* San Diego, CA: Academic Press.

Xu, Danghui, Honglin Li, Xiangwen Fang, Jinhua Li, Haiyan Bu, Wenpeng Zhang, Jingjing Wang, and Xiaolin Si. 2015. "Responses of Plant Community Composition and Eco-Physiological Characteristics of Dominant Species to Different Soil Hydrologic Regimes in Alpine Marsh Wetlands on Qinghai–Tibetan Plateau, China." *Wetlands* 35 (2): 381–90. https://doi.org/10.1007/s13157-015-0627-5.

Xue, Zhenshan, Xianguo Lyu, Zhike Chen, Zhongsheng Zhang, Ming Jiang, Kun Zhang, and Yonglei Lyu. 2018. "Spatial and Temporal Changes of Wetlands on the Qinghai-Tibetan Plateau from the 1970s to 2010s." *Chinese Geographical Science* 28 (6): 935–45. https://doi.org/10.1007/s11769-018-1003-1.

Zhao, Zhilong, Yili Zhang, Linshan Liu, Fenggui Liu, and Haifeng Zhang. 2015. "Recent Changes in Wetlands on the Tibetan Plateau: A Review." *Journal of Geographical Sciences* 25 (7): 879–96. https://doi.org/10.1007/s11442-015-1208-5.

4 Hydrology

4.1 OVERVIEW OF WETLAND HYDROLOGY

Hydrology is the most important ecological feature driving wetland processes. The depth, duration, frequency, and chemical composition of water on a site interact to determine whether that location can support a wetland ecosystem, as well as the structure and function of the wetland, if one is present. The study of these aspects of water, as well as the movement of water into and out of a wetland, whether via air or soil, all collectively fall under the broad umbrella of wetland hydrology (Davie 2008; Figure 4.1).

In this chapter, we will simultaneously look at how hydrology is measured in wetlands and how hydrology interacts with aquatic and wetland plants. With regard to the latter, we will look at ways that various aspects of hydrology affect plants, as well as some of the ways that plants impact hydrology. In considering hydrological measurements, we will look relatively broadly at methods used to investigate several major aspects of hydrology, but we also will look in much more depth at a few specific examples of hydrologic methods as part of examples that illustrate interactions between wetland hydrology and wetland plants.

An important concept with regard to wetland hydrology (or hydrology of any type of ecosystem) is the idea of a **water budget**, or the concept of water balance. The most basic water balance equation looks like (Davie 2008):

$$\Delta S = P \pm Q - ET \qquad (4.1)$$

where ΔS refers to the change in **water storage** in the wetland (can be positive or negative); P refers to precipitation (of all types); Q represents flow of water into or out of the wetland (flows into the wetland are assigned positive values); and ET (**evapotranspiration**) is the combination of evaporation of water from the wetland and **transpiration** of water through the plants. In cases where ET and outflows exceed precipitation and inflows, storage of water (S) within the wetland will decline, and the wetland can be thought of as running at a water deficit. Water surpluses occur when the opposite condition is true because the store of water within the wetland increases.

A somewhat more complicated visual depiction of the water balance equation is shown in Figure 4.2. The equation illustrated by Figure 4.2 would look like this (variables as defined in the figure):

$$\Delta S = P_n \pm Q_O \pm Q_{S/G} - ET \qquad (4.2)$$

In this equation, ΔS represents the change in the amount of water stored within the wetland (**storage**), P_n is the net precipitation arriving within the wetland, $\pm Q_O$ represents **overland** (i.e., surface) flows of water into and out of the wetland, $\pm Q_{S/G}$ are subsurface and groundwater flows into and out of the wetland, and ET represents the combination of evaporation and transpiration of water out of the wetland. Each of these will be discussed in greater detail later in the chapter.

It is important to note here that flows (Q from Equation 4.1) include overland flows such as runoff and stream discharge (labeled as "Q_O" in Figure 4.2), as well as flows belowground that we cannot see (Q_S and Q_G). We will see later that the "invisible" losses associated with evapotranspiration and subsurface flows sometimes account for most (or all) of the losses of water from a wetland. I have combined the two subsurface flows in Equation 4.2 and Figure 4.2 for simplicity, but that is also what is usually done in practice because of the difficulty in differentiating among different belowground flows. The difference between soil water and groundwater flow (as defined by Davie 2008) is whether the flows occur in the saturated (Q_G) or unsaturated (Q_S) portions of the underlying wetland basin. This is not something that we need to spend a lot of effort thinking about to appreciate the interactions that will be discussed in this chapter, but distinctions between the two will be pointed out when they are particularly relevant for a particular example.

To give a better feel for what these pieces of a wetland water budget look like in the real world, Figure 4.3 illustrates Equation 4.2 in a hypothetical wetland setting. One item in particular that may be more clearly depicted in Figure 4.3 than in Figure 4.2 is the difference between precipitation (P) and net precipitation (P_n). The latter of these represents the water that arrives in the wetland storage pool (even if briefly), after passing through, or being intercepted by, the wetland vegetation. Some of the intercepted precipitation will evaporate from the plant surfaces, while the remainder will enter the storage pool via **stemflow** (water that flows down the plants' stems) or **throughfall** (falling directly off the leaves, stems, etc.). Another important note here is that the Interception & Evaporation component of Figure 4.3 is, in reality, part of the water that is evaporated away from the wetland. Thus, it technically is part of the overall total ET for a wetland, even though it never truly arrives in the wetland storage; thus, Interception & Evaporation might be handled differently in an overall wetland budget, depending on which precipitation value is being used as an input to storage (P or P_n). For example, if we consider only P_n, then **interception** would already have been accounted for and would not be included in our overall ET estimate in the budget (although it would be important to recognize this contribution to ET separately from the budget).

DOI: 10.1201/9781315156835-4

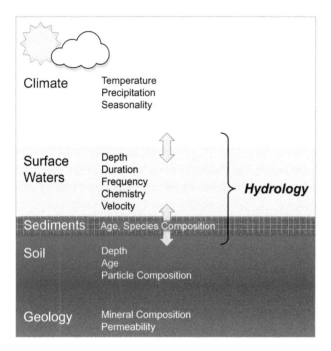

FIGURE 4.2   Components of a Wetland Water Budget. Overland Flows here can Include Discrete Stream Flows or more Dispersed Flows, such as Runoff from the Surrounding Landscape. Overland Flows also May Include Tidal Influences on the Wetland, Which Typically are Depicted in Figures such as this as both Inflows and Outflows (Or Net Tidal Influence) on the Right Side of the Diagram.

FIGURE 4.1   Hydrology Essentially Deals with all Aspects of Water in a Natural System: Depth, Duration, Frequency, Chemistry, and Movement into and Out of the System of Interest. Movement into and Out of the Wetland is Influenced by Characteristics of the Medium through which the Water Moves, Whether we are Thinking about Movement through the Air (Precipitation and Evaporation) or through Soil, Sediment, or other Material underlying the Wetland Basin.

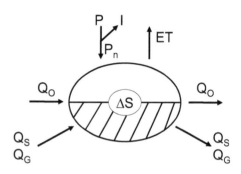

P   = precipitation
I   = interception; precipitation intercepted by vegetation
P_n = net precipitation, after passing through plant canopy
Q_O = overland flow; flow over the soil surface
Q_S = flow through the unsaturated soil; "soil water"
Q_G = flow through the saturated groundwater zone
ET = evapotranspiration
ΔS = change in wetland water storage

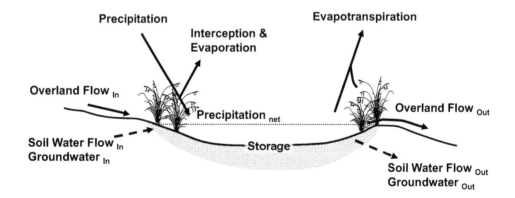

FIGURE 4.3   Illustration of Water Budget from Figure 4.2 from the Perspective of a Hypothetical Wetland. Two Items to Note here are that (1) Interception from Figure 4.2 Assumes that the Intercepted Precipitation Evaporates from Plant Surfaces, Rather than Flowing into the Wetland, and (2) Evapotranspiration Includes Losses of Water from Plant Surfaces Along with Losses from the Surface of any Standing Water in the Wetland.

Now that we have a basic understanding of the parts of a wetland water budget, we can look more closely at how some of the parts of the budget are measured.

4.2   QUANTIFYING HYDROLOGY WITH A HYDROLOGIC BUDGET

The precision needed in developing a hydrologic budget may vary widely, depending on one's objective. If the goal

is to obtain coarse estimates of maximum depth, duration, and timing of standing water in the wetland, less precision will be needed than if one is interested in mitigating a specific amount of incoming nutrients in a wastewater treatment wetland. As a result, the methods and instrumentation available and used in constructing a water budget vary widely across the spectrum of wetland management objectives. The following overview is not meant to cover all available methods, but it should give an appreciation for how one would measure major components of wetland hydrology and what those measurements tell us about the hydrology and its interaction with the plants.

4.2.1 PRECIPITATION AND INTERCEPTION

Precipitation is perhaps the most visible inflow component to wetland water budgets. Stream inflows can be directly observed as well, but not every wetland receives clearly defined stream inflows; however, every wetland at some point will receive contributions from precipitation. Even the Okavango Delta, the river delta in the Kalahari Desert that we saw in the last chapter, receives approximately 465 mm of seasonal rainfall per year (Figure 3.2).

As important as precipitation is, it also can be one of the simplest water budget components to estimate (Figure 4.4). Coarse estimates of total precipitation can be made with a simple graduated cylinder-type rain gauge, but such simple devices are subject to a few potentially important sources of error. Open cylinder rain gauges are susceptible to evaporation, which can result in losses of up to 100% if gauges are

left exposed to sunlight and not monitored with sufficient frequency. Evaporation can be reduced by using modified rain gauges, such as in Figure 4.4B–D, where the captured water is stored in a reservoir that minimizes loss to solar heating of the water surface (Davie 2008). Evaporation also can be reduced by adding a small amount of mineral oil to an open cylinder gauge, which will form a film atop the collected water, physically limiting evaporative losses (Mitsch, Dorge, and Wiemhoff 1977).

If rain gauges are placed at ground level, so that the captured precipitation reflects the amount of water reaching the level of the wetland surface, precipitation estimates are subject to overestimation because of splashing into the gauge from the surrounding areas or overflow of accumulated runoff (Davie 2008). To compensate for this potential source of error, the gauge may be surrounded by a mesh grid to reduce splashing, or it may be elevated above the ground surface. However, elevation above ground level subjects the gauge to increased levels of wind turbulence, as air flows around and over the gauge during precipitation events, blowing rain or snow away from the rain gauge

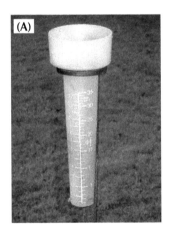

FIGURE 4.4 Three Common Types of Rain Gauge. (A) Simple Graduated Cylinder-Type Rain Gauge. (B) Standard Rain Gauge with Catch Funnel (Left) and Sun-Blocking Overflow Catch Basin (on the Right). (C) Exterior and (D) Interior Views of Tipping Bucket Rain Gauges; the Silver "Bucket" Behind the Funnel in (D) Tips Back and Forth Each Time One Side or the other Fills with Water, to Dump Water into a Collection Vessel, and Each Cycle can be Counted Automatically to Record Total Precipitation. Attributions: (A) © Kolling, Used under CC BY-SA 3.0 License, (B and C) Provided By Cambridge Bay Weather, and (D) © G43, Used under CC BY-SA 3.0 License (creativecommons.org/licenses/by-sa/3.0/deed.en).

opening (Davie 2008). Underestimation error resulting from turbulence can be another sizeable source of error, underestimating precipitation by at least 46% in some cases (Mekonnen et al. 2015).

Sophisticated tipping bucket rain gauges (Figure 4.4C–D) can be used to minimize evaporative losses and error due to splash (by enclosing the collection vessels, elevating the gauge above the ground surface, and integrating digital monitoring of captured precipitation with analog validation of amounts within the calibrated collection vessel). Estimates of precipitation from these gauges then can be combined with measurements of local wind speed to correct for potential turbulence-induced underestimates to yield the most precise estimates of total precipitation at a given point (Mekonnen et al. 2015), but even elaborate systems such as these require periodic calibration to ensure accuracy is maintained over time (Stransky, Bares, and Fatka 2007).

Interception, as depicted in Figure 4.3, is precipitation that is intercepted by the plant canopy and evaporated before it reaches the wetland storage pool. Note, as mentioned earlier, that there will be some precipitation that is temporarily intercepted but that eventually finds its way into the storage pool. This occurs via flow of the water down plant stems into the wetland or throughfall through the canopy once sufficient moisture has collected on the canopy to saturate the plant surfaces (Davie 2008). The simplest and most common method for estimating interception is to measure precipitation above and below the canopy, and then calculate interception as the difference between the two. However, we do sometimes see studies where the various individual components of precipitation-related inputs are measured. For example, collars may be placed around stems (trunks) of trees to collected stemflow, while rain gauges or other instruments may be used to capture throughfall beneath the canopy (Davie 2008). Above-canopy precipitation usually will be estimated by collecting precipitation data in an open (non-canopied) area either within or near the wetland.

4.2.2 Overland Flows

There are two primary forms of overland flow: streamflow and runoff. **Streamflow**, conceptually, is somewhat simple to measure or estimate. We typically see streamflow represented as a volume of water passing a particular point on the stream over some duration of time (e.g., m³ per second). Breaking down these units of measure, we can see that streamflow is a product of the velocity of the water (m per second) and the cross-sectional area of the stream at a given point (m²). However, because the velocity of water is rarely uniform across the width of a stream channel, estimation of flow is usually accomplished by dividing the stream width into some number of smaller sections, as in Figure 4.5, and measuring velocity at the center of each section, along with obtaining an estimate of the cross-sectional area of each section. The streamflow at that point along the stream is then calculated by summing the individual products of these areas and velocities. In some cases, if long-term

measurements of streamflow into or out of a wetland, or other water body, are desired, a weir or flume may be constructed for this purpose. These are structures constructed with a known cross-sectional area (flumes are usually rectangular or trapezoidal in cross-section; weirs often have a V-shaped or trapezoidal notch through which the water flows) where a single measurement of depth and velocity may be taken for estimation of flow.

We can obtain much more precise estimates of total streamflow in unaltered streams (i.e., those without a weir or flume), by attempting to include the volume of water that is passing this same point within the streambed (the hyporheic zone, or zone below the stream) or by using tracer dye methods to estimate volume by measuring the dilution of a measured amount of dye over time (and distance). On the other hand, if we lack access to instruments to measure velocity within the stream channel, we can estimate velocity for a particular stream reach by measuring the time it takes a floating object to travel a measured distance downstream. Each method will be subject to some degree of error or inaccuracy, and trade-offs among cost, effort, and accuracy should be weighed based on one's objective.

In contrast to streamflow, **runoff** is much more difficult to measure accurately. In principle, the volume of water entering a wetland as runoff could be calculated based on the depth of runoff flows and the velocity with which the water enters the wetland. In practice, however, measuring the depth of runoff is no simple task, and this depth would vary considerably around the periphery of a wetland, based on slope, microtopography, leaf and other litter from the vegetation, soil composition, and so on. When researchers have an interest in measuring runoff, they may install collection troughs at the base of a slope into which the water will accumulate and where the volume can be measured (Davie 2008). Ideally, multiple troughs would be installed at different locations, representing the diversity of topography and other features that may influence rates of runoff into the wetland, and the area draining into each would be

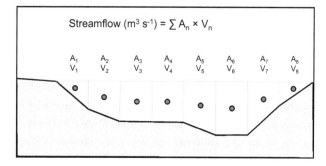

FIGURE 4.5 Diagrammatic Example of How One Would Estimate Total Discharge at a Given Point Along a Stream, by Integrating the Volume Passing that Point in Discrete Cross-Sections across the Point of Interest Along the Stream. Each Individual Area would be Estimated as the Area of a Triangle or Trapezoid, and Velocity Would be Measured at the Centroid of Each Area (Represented by the Dots) by Some Type of Streamflow Meter.

used to estimate total runoff from the surrounding basin into the wetland.

More often, however, runoff will be estimated based on the cumulative change in **stage**, or water depth, for the wetland over time (e.g., monthly or during a storm). The simplest means of doing this is to read depth from a staff gauge periodically (Figure 4.6). There are at least two ways this method would impact the accuracy of runoff estimates. First, as with any process, the frequency of measurement will impact accuracy, and it would be difficult to capture every instance of maximum depth associated with precipitation or other instances that may contribute runoff to the wetland. Second, the precision of the depth measurement is likely to be much lower with manual stage readings from a staff gauge than from more technologically sophisticated approaches.

One approach that was devised more than 100 years ago for continuous monitoring of stage height is the implementation of a mechanical water-level recorder (Figure 4.7).

This device consists of a counter-weighted float, attached to a gear system that "drives" the position of a pencil or other marker, and a mechanical clock-regulated spool on which a roll of recording paper is mounted. The clock controls the rate at which the recording paper moves past the marker, and the float controls the position of the marker on the paper. The result is a cumulative graph of water height, as shown at the bottom of Figure 4.7. A more modern approach uses electronic devices that measure pressure of the water (Figure 4.8), which increases with increasing depth and, in combination with a measure of local barometric pressure, provide a continuous estimate of changing water depth over time.

An important factor to bear in mind with the use of changes in water depth as a surrogate for runoff is that increases in water depth are a composite of all inflows to the wetland. Thus, one would need to account for all other inflows and then subtract those from the total change in depth in order to arrive at runoff estimates.

FIGURE 4.6 Staff Gauge and Water Control Structures from Ponds that had been Renovated to Increase the Area of Littoral Wetlands around their Perimeter. (A) and (D) Water Control Structures, with Wooden "Stoplogs" Used to Reduce Discharge from the Impoundments. (B) and (C) Staff Gauge Used to Monitor Water Level in the Pond from Photo (A).

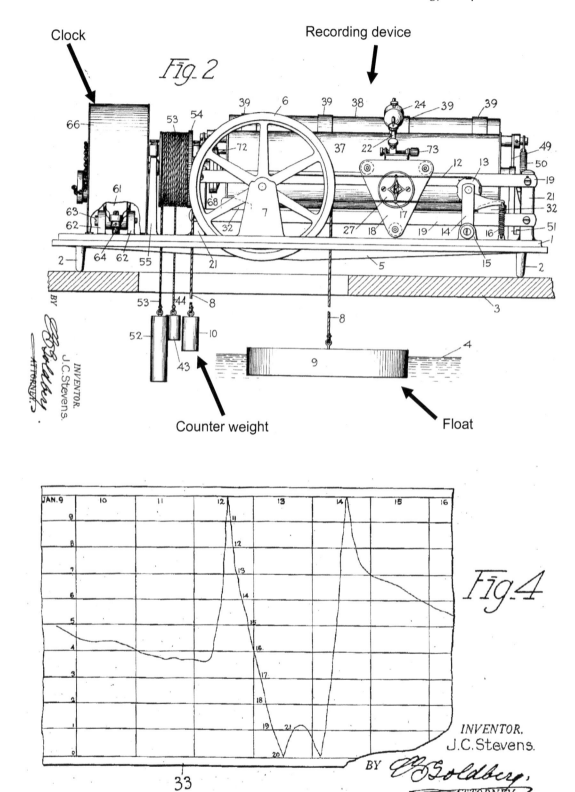

FIGURE 4.7   Diagrams of a "Stevens-Type" Water-Level Recorder, from Patent Documents Filed by J. C. Stevens on April 10, 1922. The Recording Device is a Roller, on Which a Spool of Paper is Mounted to Record Changes in Water Levels, as Shown in the Bottom Panel, Labeled here as "Fig. 4." the Mechanical Clock Controls the Rate at Which the Paper Rotates, and the Counter-Weighted Float Controls the Position of the Marker on the Roll of Paper. Water Levels from January 09 to 16 are Illustrated in the Bottom Panel. from Stevens (1924).

FIGURE 4.8   Electronic Water-Level Loggers can be Deployed for Months at a Time, Recording Water Levels at Very Short Intervals, to Produce Highly Detailed Wetland Hydroperiod Data. Shown here are the Location of a Deployed Water-Level Logger (Large Photo) and Close-up Photos of a Recorder Deployed to Track Barometric Pressure for Use in Correcting Data from the Submersed Recorders.

4.2.3   Soil Water Flow and Groundwater Flow

Movement of water through the soil (**soil water** flows) or into and out of aquifers or saturated zones within the soil (**groundwater** flows) is considerably more complicated to estimate than is precipitation, streamflow, or runoff. Because it is impractical to directly measure movement of water through the soil, through **aquifers** (i.e., rock or other geologic material that is capable of storing water), or through other porous media, these flows usually are estimated based on physical relationships between flow and measurable parameters of soil water or groundwater. In addition to the impracticability of placing instrumentation in the soil or underlying rock to measure the movement of water, the placement of such instruments could greatly alter the flows, such that the measurements would be quite unrepresentative of the flows in which we are interested.

The principles upon which soil water and groundwater hydrology are based were worked out by a 19th-century water-supply engineer from Dijon, France, by the name of Henri Darcy (Davie 2008). Darcy's Law, which is used to calculate flows of soil water and groundwater, is very similar to the calculation for streamflow we saw earlier, and is represented as:

$$Q = -K_{sat} \times A \times (\delta h/\delta x) \qquad (4.3)$$

where K_{sat} is the hydraulic conductivity of the soil or other water-conducting medium, A is the cross-sectional area of the layer in which we are interested, and $\delta h/\delta x$ is the hydraulic gradient across the layer of interest (Davie

2008). **Hydraulic conductivity** is the rate at which the soil or other medium conducts water, usually measured in distance per time (e.g., cm per day). The cross-sectional area is just that, the area of a representative cross-section of the soil or other layer in which we are interested. The **hydraulic gradient** is the change in **hydraulic head** (the total "pressure" on the water at that point, represented by h) over the distance (x) that the water would travel through the layer of interest. Change in hydraulic head here is a combination of the change in height of the water as it passes from one end of the layer to the other (elevational head) plus any change in pressure exerted on the water from overlying geological layers, if such are present (pressure head). The similarity between Darcy's Law and the calculation of streamflow is that Darcy's Law includes a cross-sectional area and two components of a velocity K_{sat} and $(\delta h/\delta x)$. These flow components of Darcy's Law amount to a potential velocity (the conductivity), weighted by the force pushing water through the layer (the hydraulic gradient). In terms of units, note that hydraulic conductivity is measured in distance per time, area is expressed as area (distance squared), and hydraulic gradient is unitless because its derivation divides one distance (hydraulic head) by another distance (distance the water moves). This leaves us with a flow value in units of volume per time, just as we saw earlier with streamflow.

The parameters that need to be measured, then, are the hydraulic conductivity of the material through which the water is passing, the cross-sectional area of the layer, and the change in head as water moves across the layer. The area may often be calculated based on the depths at which the other parameters are measured. That is, if we are measuring water moving through the upper meter of soil, we have one meter of depth, multiplied by the width of the location at which we are measuring flow. When resources are available to map the underlying geology of the wetland, area also could be measured based on the dimensions of the actual layers through which the water is moving. Hydraulic conductivity (rate of movement of water through the layer of interest) can be measured for small samples by passing water through cores of the materials in a lab setting (e.g., Mann and Wetzel 2000a) or by measurements within the wetland via well pumping tests (Davie 2008).

The final of the three parameters used to estimate subsurface flows, the hydraulic gradient, is usually measured with the assistance of one or more types of **wells** (Figure 4.9; Jackson 2006). "Wells," strictly speaking, allow access to subsurface water along all, or most, of their length, as with the monitoring well in Figure 4.9. Wells can be used to observe the height of the water table (the surface of the **saturated zone** or groundwater), which is the point at which the pressure head is equal to zero. Thus, at that point, the height of the water table above a reference elevation will equate to both the elevation head and the total hydraulic head at that point (Davie 2008; Jackson 2006). **Piezometers**, in contrast to wells, allow access to subsurface water only at or near the bottom end of the device (Figure 4.9). Thus, piezometers are somewhat more sensitive, in that they give a measure of the hydraulic head at a particular point within the soil water or groundwater. **Lysimeters**, which were designed to provide access to finely filtered soil water for analysis of solute concentrations, also can be used in the same way that

piezometers are used to measure hydraulic head at a particular depth belowground (Figure 4.9).

The ability to measure hydraulic head at a particular depth with piezometers or lysimeters makes these devices useful for determining fine-scale flow patterns in wetlands (or other ecosystems). Nested sets of piezometers can be installed, with the tip of each tube inserted at different depths belowground, to provide measurement of hydraulic head at different depths at a given location (Figure 4.10). Because water tends to move from higher to lower hydraulic head (i.e., from higher to lower pressure), the relative head at different depths will reveal the direction water should move at that location. For example, if the hydraulic head is greatest at the shallowest depth, water will tend to move away from that point, towards deeper areas with lower head, resulting in downward flow at that point. Thus, by installing a grid of nested sets of piezometers or lysimeters across a wetland, one can determine, in three dimensions, the subsurface flow paths (or flow networks) of water through a wetland (Jackson 2006). Note that the path along which water tends to flow will be perpendicular to lines along which hydraulic head is equal (Figure 4.11). Those lines of equal hydraulic head are referred to as equipotential lines, and they are analogous to topographic contour lines that are used to map the land's surface. For example, if we were to walk across a series of neighboring contour lines on a topographic map, we would be going either uphill or downhill. The closer the lines are to one another, the steeper will be the slope of the hill. Similarly, in a map of hydraulic potential, when we see the equipotential lines closer to one another, there will be a steeper hydraulic gradient, which will result in a faster flow of water. The exception to this would be in a soil where there is a correspondingly opposing gradient in hydraulic conductivity of the material through which the water flows, per Equation 4.3.

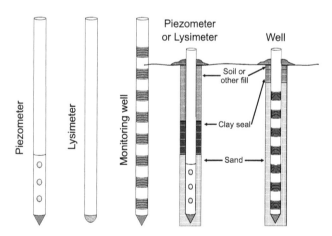

FIGURE 4.9 Three Different Types of Sampling Wells. The Piezometer and Lysimeter Types are Designed to Sample Water at Relatively Specific Depths, Through Openings at the Bottom End of the Sampler. Monitoring Wells, Usually Referred to Simply as Wells, Allow Access to Subsurface Water Along all or most of their Length. Illustrations on the Right Show Typical Installations for the Different Types of Water Sampling Devices. Note the Clay Seal in the Installation Drawings that is Used to Prevent or Reduce Contamination of the Layer of Interest by Water from above the Lateral Openings. Drawings are Based on those of Sprecher (1993).

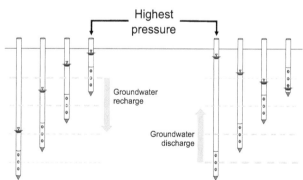

FIGURE 4.10 Examples of Nested Piezometers Used to Determine Localized Patterns of Hydraulic Head and Direction of Water Flow. Water Level in Each Piezometer Shown by an Inverted Triangle. The Highest Water Level Among the Four Piezometers Indicates the Piezometer with Highest Hydraulic Head, and Water will Tend to Flow away from the Depth being Sampled by that Piezometer, Toward Points of Lower Head. For Example, in the Set on the Left, Highest Head is Present at the Shallowest Depth, So Water Tends to Flow Downward. Drawings are Based on those of Sprecher (1993).

At a broader spatial scale, flow networks can be constructed to show movement of water among nearby wetlands, between streams and adjacent habitats, or between wetlands and the groundwater (Figure 4.12). In cases where we see that water moves from the wetland into the groundwater, we refer to this type of wetland as a **recharge wetland**—because the wetland essentially is recharging the groundwater supply (Figure 4.12). When the reverse is true

and the wetland is a net recipient of groundwater, we refer to this type of wetland as a **discharge wetland**—because the groundwater discharges into the wetland. The units of measure for hydraulic head are units of distance of the measured water level inside the well or piezometer above some reference elevation; mean sea level is often used as the reference for this purpose.

4.2.3.1 Interaction of Vegetation and Subsurface Hydrology in an Irish Bog

Clara Bog is a 650 ha bog wetland in central Ireland (Figure 4.13), 460 ha of which has been designated as a Ramsar site (ramsar.org). In addition to its international designation as a Ramsar wetland, Clara Bog was also designated nationally as a Nature Reserve in 1987, after purchase from a private company that had intended to mine the wetland for its peat reserves (Crushell et al. 2008). The basin within which Clara Bog formed was exposed roughly 11,500 years ago, following deglaciation of this area of Ireland (Crushell et al. 2008). Based on geological evidence, the bog appears to have begun as a shallow lake, surrounded by fen vegetation that gradually encroached

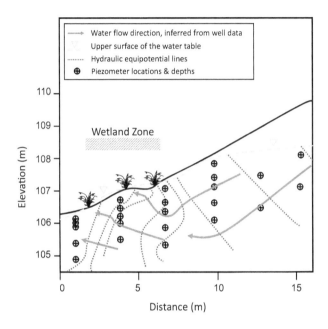

FIGURE 4.11 Results of a Study near Toronto, Canada, Aimed at Determining Patterns of Water Movement in a Small Headwater Stream Riparian Wetland. Twelve Sets of Nested Piezometers at Depths from 25 to 250 Cm Were Used to Map the Water Table and Subsurface Flows in and around the Wetland. Water Levels in the Piezometers are Used to Interpolate the Lines of Equivalent Hydraulic Pressure (Equipotential Lines), and Flows of Water Will Tend to be Perpendicular to those Lines. Based on Data in Waddington, Roulet, and Hill (1993).

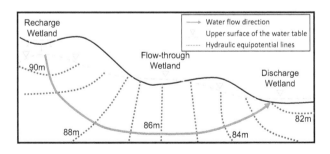

FIGURE 4.12 Hypothetical Regional Map of Subsurface Hydraulic Equipotential Lines. Networks of these Hydraulic Maps, Referred to as Flownets, Can be Used to Infer Broad-Scale Patterns of Water Flow and Potential Hydrologic Connections Among nearby Wetlands. In this Example, Subsurface Water Leaving the Wetland on the Far Left Would Eventually Find its Way into the Wetland on the Right, by Flowing down the Path of Declining Hydraulic Head. Based on Illustrations in Richardson, Wilding, and Daniels (1992).

FIGURE 4.13 (A) Clara Bog, in Central Ireland, is One of the Last Remaining Semi-Intact Raised Bog Peatlands in Ireland. This is Contrast to most other Nearby Peatlands, such as that in (B), Which Sits Only 5 Km to the West of Clara Bog and Shows Substantial Evidence of Peat Mining. Aerial Imagery from Esri (2022) "World Imagery" www.arcgis.com/home/item.html?id=1 0df2279f9684e4a9f6a7f08febac2a9.

upon the lake. This led to conversion of the lake to a fen, and eventually into a bog, as the water source shifted to precipitation dominated and vegetation shifted towards acidophilic raised bog species over a period of about 4,000 to 4,500 years (Crushell et al. 2008). Today, Clara Bog is the largest relatively intact bog ecosystem in Ireland, in contrast to the heavily mined bogs present on the surrounding landscape (Figure 4.13).

Flynn (1990) conducted a detailed study of the hydrology of Clara Bog, including detailed analyses of the geological layers underlying the wetland (Figures 4.14 and 4.15). The wetland has undergone some subsidence in recent decades but retains characteristics of a raised bog ecosystem and is populated by *Sphagnum* species, with a few relatively small permanently inundated areas, known as soaks. These ponded areas can originate through processes of **subsidence** that cause depressions within the raised peat platform or from inputs of mineral enriched groundwater through cracks or conductive "windows" in non-water-conducting layers (referred to as aquicludes) above confined aquifers; both of these likely have occurred in the Clara Bog system (Daly and Johnston 1994). Vegetation in the soaks is generally less acidophilic ("acid-loving") and includes species such as *Carex rostrata*, *Juncus effusus*, *Nuphar lutea*, and shrubs/trees from the genera *Vaccinium*, *Betula*, and *Salix* (Crushell et al. 2006). The periphery of these areas, however, is populated by multiple acidophilic *Sphagnum* species (e.g., *S. magellanicum*, *S. capillifolium*, *S. fallax*), with these mosses representing 80% to 100% of the plant cover in surveys at the margins of the soak habitats (Crushell et al. 2006).

Flynn (1990) used a network of wells and piezometers, along with his own geological studies of the wetland, to map three-dimensional flows of water through Clara Bog (Figures 4.14 and 4.15). He found that the former lakebed, which produced a confining layer of clay deposits beneath the bog, separated subsurface flows into two distinct compartments. Above the clays, the physical and conductive properties of the raised peat surface resulted in water moving generally outward, toward the bog edges, and away from the highest points of elevation in the center of the bog (Figures 4.14 and 4.15). The pH values in waters within the upper layers of peat ranged from about 4.0 to 5.0 in the central areas of the bog but were in the range of 6.4 to 7.0 along the bog's periphery and along roadways. Beneath the lake clays, water flowed primarily southward toward the Silver River, which passes a few kilometers to the southeast of the bog (Figure 4.15). Within these gravel and limestone aquifers beneath the lakebed clays, pH was in the range of 6.1 to 6.7.

Daly and Johnston (1994) presented an annual water budget for Clara Bog, developed from data in an unpublished thesis from the University of Wageningen, in the Netherlands (Figure 4.16). The water budget reflects the **ombrogenous** nature of this wetland, where inputs are dominated by or arise exclusively from precipitation ("ombro" = cloud), and this sets up conditions conducive to bog formation. Similarly, outputs from the bog reinforce

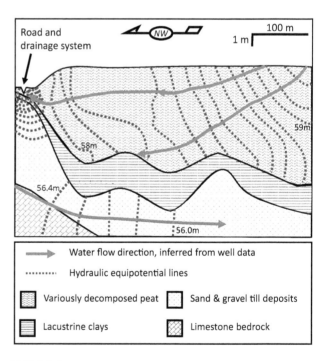

FIGURE 4.15   Mapped Equipotential Lines and Predicted Flow Paths of Subsurface Water in the Eastern Half of Clara Bog, Mapped on a Cross-Section of the Wetland, Along the Flowline Illustrated in Figure 4.14. Note that the Flows are Separated into Compartments above and below the Lacustrine Clays Layer. Data from Flynn (1990).

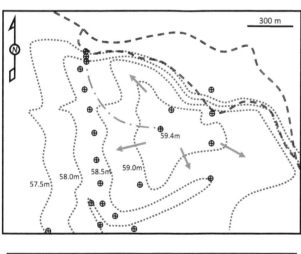

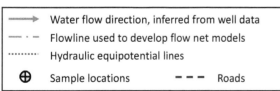

FIGURE 4.14   Mapped Equipotential Lines and Predicted Flow Paths of Subsurface Water in the Eastern Half of Clara Bog, Mapped from above. Data from Flynn (1990).

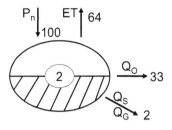

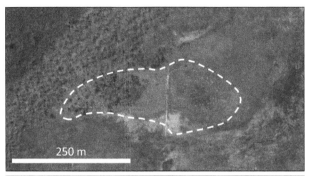

FIGURE 4.16 Water Budget for Clara Bog. Values are Percent of Inflows and Do not Sum Exactly to 100 Because of Rounding of Individual Values.

the relatively closed nature of the wetland basin, with only 2% of the outflows occurring as subsurface flows (through the cracks or windows in the clay layer mentioned earlier). Evapotranspiration, which will be discussed in detail shortly, was the dominant route by which water left the wetland, with 64% of incoming water leaving via evapotranspiration.

Thus, the hydrology of this wetland set up perfect conditions for the development of the bog-type vegetation that characterizes Clara Bog and that led to the raised nature of the peat surface, further reinforcing the boglike nature of this ecosystem (Bridgham et al. 1996). The lack of significant inputs of groundwater, coupled with accumulations of decaying organic matter (detritus), led to low-nutrient, acidic conditions to which a relatively narrow suite of species are adapted (e.g., Figure 3.16). These conditions also usually result in reduced rates of decomposition of the accumulated detritus, which will be further exacerbated in cold conditions, as at Clara Bog (which sits just above 53° north latitude), eventually leading to the development of a raised wetland platform. The increased elevation of the wetland surface then impacts hydrology by causing greater hydraulic head within the higher areas of the wetland, pushing water away from those high elevation areas, as we see in Figures 4.14 and 4.15.

4.2.3.2 Interaction of Vegetation and Subsurface Hydrology in a Great Lakes Coastal Fen

Cowles Bog Wetland is a complex of peatland and non-peat-accumulating wetlands covering approximately 80 ha in Indiana, USA (Figure 4.17; Shedlock et al. 1993; Wilcox, Shedlock, and Hendrickson 1986). The wetland is part of the Indiana Dunes National Lakeshore, and 22 ha of fen within the wetland was designated in 1966 as the Cowles Bog National Natural Landmark, to commemorate Dr. Henry Cowles use of the site for teaching. The wetland is situated very near a large port facility (Burns Harbor, Port of Indiana), as well as other industrial facilities to the west, bordered by a small highway to the south, beyond which lie residential areas and a large interstate highway, and is situated just south of a low-density coastal residential area on Lake Michigan (Figure 4.17). Numerous descriptions of

FIGURE 4.17 Aerial Imagery of the Central Mound Area of Cowles Bog Wetland, in Northwestern Indiana (USA), from June 2020. Top Photo Shows an Aerial View of the Area Depicted in Figure 4.18, with the Old Water Line (Thick White Dashed Line) and Topographic Contours (Faint White Dashed Lines) Illustrated. Bottom Photo Shows the Location of the Mound Relative to nearby Human Development (Port, Industrial Areas, Residential Areas, Highways). Aerial Imagery from Esri (2022) "World Imagery" www.arcgis.com/home/item.html?id=10df2279 f9684e4a9f6a7f08febac2a9.

the area were published in the 1910s and 1920s, and the US Geological Survey conducted research in the wetland in the 1970s to assess potential impacts of the industrial activities just to the west of the wetland. None of those studies, however, evaluated the impact that the site's hydrology had on the vegetation of Cowles Bog (Wilcox, Shedlock, and Hendrickson 1986).

During 1980–1984, Wilcox, Shedlock, and Hendrickson (1986) carried out detailed studies of the geology, hydrology, and vegetation of the Cowles Bog Wetland. In particular, they focused on a 4 ha raised mound (~1.4 m height) within the wetland, owing to its unique topography and plant species composition, including the only stand of northern white cedar (*Thuja occidentalis*) in the state of Indiana (west end of the central mound in Figure 4.18).

Wilcox, Shedlock, and Hendrickson (1986) installed two transects of wells, crossing the central mound, for monitoring groundwater levels and sample collection for water chemistry analyses. Each transect comprised five 5 cm diameter wells, ranging from 2.3 to 6.5 m deep. Additional wells were installed atop the mound, at four different depths, to sample water of differing depths in this central portion of the study area. The underlying geological strata

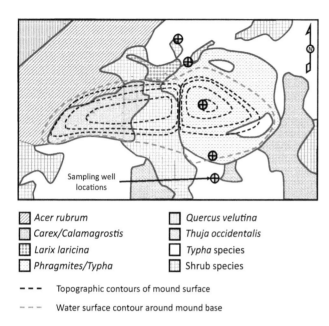

Sampling well
locations

▨ *Acer rubrum*   ▢ *Quercus velutina*

▨ *Carex/Calamagrostis* ▨ *Thuja occidentalis*

▨ *Larix laricina*   ▢ *Typha* species

▨ *Phragmites/Typha*   ▨ Shrub species

– – – Topographic contours of mound surface

– – – Water surface contour around mound base

FIGURE 4.18 Map of Vegetation on and around the Central Mound Area of Cowles Bog Wetland, Along with Water Level and Mound Topography, as in Wilcox, Shedlock, and Hendrickson (1986).

were mapped through the use of cores collected during the Wilcox, Shedlock, and Hendrickson (1986) study and a separate sedimentological study of the region.

Stratigraphic analyses of the cores showed that the central mound was composed primarily of variously decomposed

peat, underlain with a sand and marl layer embedded with carbonate shells (Figure 4.19). Below the sand layer was a glacial till clay layer, as we saw earlier with Clara Bog, but in this case, there is a sizeable breach in the clay, immediately below the central wetland mound. Analyses of the well data, in particular the nested set of wells atop the mound, indicated upwelling of water from the aquifer below the clay layer into the groundwater perched above the clay (Figure 4.19). Water chemistry indicated high concentrations of minerals in water near the mound, originating from material in the aquifer below the clay layer, and pH values at the mound were relatively high, at 7.03 to 7.16. Water sampled from a well installed ~450 m north of the mound had a pH measured at about 5.5 (Figure 4.19).

Many of the plant species identified at the Cowles Bog mound were considered to be fen-type species, including *Betula papyrifera*, *Thuja occidentalis*, *Cornus canadensis*, *Potentilla palustris*, and *Calamagrostis canadensis*. *Larix laricina* and *Sarracenia purpurea*, the latter of which was found within the *Typha* marsh, may be encountered in either minerotrophic (fen) or ombrotrophic (bog) wetlands. *Typha* and *Phragmites* are believed to have established in the Cowles Bog Wetland in association with human disturbances in and around the wetland, rather than as a natural part of wetland succession (Wilcox, Apfelbaum, and Hiebert 1985). The *Typha* marsh and shrub species associations of the mound area both were thought to be gradually invading the *Carex/Calamagrostis* marsh areas; thus, the more natural species assemblages of this wetland were thought to be the sedge-grass assemblage, along with the other four

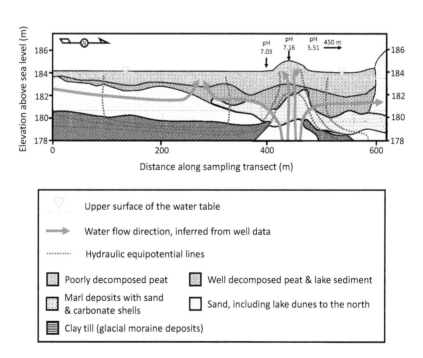

Upper surface of the water table

Water flow direction, inferred from well data

Hydraulic equipotential lines

▨ Poorly decomposed peat   ▨ Well decomposed peat & lake sediment

▨ Marl deposits with sand & carbonate shells   ▢ Sand, including lake dunes to the north

▤ Clay till (glacial moraine deposits)

FIGURE 4.19 Map of Equipotential Lines, Water Flow Paths, and Subsurface Stratigraphy of Cowles Bog Wetland, as in Wilcox, Shedlock, and Hendrickson (1986). The Large Opening in the Clay till Layer at Approximately 400–450 M Along the Transect was Hypothesized to have Strongly Influenced Vegetation in the Wetland, Potentially Driving Development of the Central Mound as a Result of its Impact on the Vegetation.

minerotrophic species mentioned earlier. The driving force for the development of this fen-type plant assemblage atop a raised peat mound wetland (rather than an ombrotrophic, bog-type assemblage) seems to have been the upwelling of higher pH waters from the underlying constrained aquifer (Wilcox, Shedlock, and Hendrickson 1986).

4.2.4 EVAPOTRANSPIRATION

The next piece of the water budget puzzle we will examine is evapotranspiration. Because of the extreme difficulty in teasing apart the contributions of evaporation from the wetland surface and transpiration through the plants, these two paths are lumped together in studies of hydrology. An important environmental parameter associated with both evaporation and transpiration is **water potential**. Water potential is a measure of the free energy of some sample of water, relative to pure water, and gives an indication of the tendency for that water to move along a gradient (Fitter and Hay 2002; Ray 1960). As materials, such as mineral ions and sugars, become dissolved within water or as water converts from liquid to gas through evaporation, the measured water potential will become more negative. Differences in water potential between two compartments of an ecosystem are a determining factor in the direction of water movement, with water moving from higher to lower potential, similar to what we just saw in the prior discussion of subsurface flows. Units used to measure water potential are Pascals (Pa), the standard metric unit for pressure; values for water potential in studies such as these usually will be given in either KPa or MPa, depending on the range of values being considered, and water potential is represented by the symbol ψ (Fitter and Hay 2002).

Total water potential in a particular compartment of the soil-plant system will integrate four distinct components (Fitter and Hay 2002; Taiz and Zeiger 2006):

1. **Pressure potential** (ψ_p): a positive potential exerted by hydrostatic pressure within plant tissues (an effect of the rigid cell walls);
2. **Solute potential** (ψ_s): negative pressure induced by dissolved substances in water;
3. **Matric potential** (ψ_m): negative pressure resulting from matric attractive forces between water and surfaces, such as soil particles and cell wall fibers; and
4. **Gravitational potential** (ψ_g): gravitational potential, which usually is very small and ignored for purposes such as ours.

Typical ranges of water potentials along a continuum from the soil or water, through the plant, and into the atmosphere (along the **soil-plant-air continuum**) are given in Figure 4.20. Water potentials approximate 0 MPa for freshwater wetlands, even within the saturated soil or sediment, and −2.7 MPa for saltwater systems (Cronk and Fennessy 2001; Taiz and Zeiger 2006; other references in

Figure 4.20). Most values for water potential will be represented by negative values, suggesting essentially a suction force that pulls water from areas of higher (less negative) potential to lower (more negative) potential. The one exception to this, as this topic pertains to plants, is the pressure force exerted within cells that helps the plant maintain **turgor**, owing to the rigid plant cell wall. In spite of those positive hydrostatic pressures inside the cell, however, we still see overall negative values for water potential within the plant, becoming more negative as we move from the roots, through the stems, and finally passing through the leaves into the atmosphere (Figure 4.20).

Note that there is a relatively large difference in potential between the water (or saturated soil) and the atmosphere—along the order of −100 MPa in terrestrial systems. In aquatic and wetland systems, however, this difference may be substantially less, owing to the relatively high humidity maintained above the water column, within the plant canopy. Nevertheless, as long as the water potential of the atmosphere is less than that within the air spaces of the leaf, water will tend to move out of the leaf and into the atmosphere. This movement of water vapor through the plant into the atmosphere is called transpiration, and it benefits the plant in numerous ways, including the facilitation of transport of nutrients and other dissolved substances through the plant and as a component of thermoregulation of leaf surfaces. In the context of water budgets, transpiration serves to move water out of the wetland storage pool and into the atmosphere.

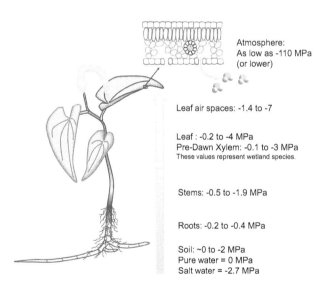

Atmosphere:
As low as -110 MPa
(or lower)

Leaf air spaces: -1.4 to -7

Leaf : -0.2 to -4 MPa
Pre-Dawn Xylem: -0.1 to -3 MPa
These values represent wetland species.

Stems: -0.5 to -1.9 MPa

Roots: -0.2 to -0.4 MPa

Soil: ~0 to -2 MPa
Pure water = 0 MPa
Salt water = -2.7 MPa

FIGURE 4.20 Approximate Water Potentials Encountered Along the Soil-Plant-Air Continuum. Water Moves from Higher (Less Negative) to Lower (More Negative) Potentials Along this Path, Regulated by the Stomatal Openings on Leaves but Influenced by Solute Concentrations within Plant Tissues Along the Way. Based on Data from Cronk and Fennessy (2001), Donovan, Linton, and Richards (2001), Johnson et al. (2013), Jones and Muthuri (1984), Taiz and Zeiger (2006), and Touchette et al. (2009, 2010).

There is a long list of physical factors that influence rate of evapotranspiration (ET), all centered in one way or another on water vapor density in the atmosphere and the movement of air. Among these factors are (Davie 2008):

1. Incoming solar radiation, which provides the heat/energy input to drive evaporation;
2. Water vapor pressure deficit of the air (difference between actual water vapor pressure and vapor pressure at saturation), which is a measure of the amount of water that can be added to the air via evapotranspiration;
3. Density of the air, which also will influence the potential water vapor capacity of the air;
4. Temperature of the air, yet another determinant of the capacity for water vapor to enter the atmosphere (cooler air can hold less water than warm air);
5. Aerodynamic resistance of air overlying the vegetation to mixing of moist and dry air, which can affect the velocity and turbulence of air currents, both of which are positively correlated with movement of water out of plants and into the overlying atmosphere;
6. Wind speed above the plant canopy, where faster velocities can accelerate conduction of moisture through the plants; and
7. Height at which ET is being estimated (and thus, at which measurements are being made), which will affect wind speed and turbulence, among other factors.

Measuring all of these variables, especially with precision and throughout daily cycles of warming and cooling, is a complicated task and requires fairly sophisticated instrumentation (Figure 4.21). Even with instrumentation installed to collect these data, the precision of these ET estimates will decline with distance from the point of measurement, along with corresponding variation in plant species, canopy cover, and so on. That is not to say that data collected through these methods are not much more precise, in general, than ET estimates made without this level of detail, but they do still have somewhat limited applicability, and the instrumentation is quite expensive at US\$40,000 or more for one station.

In the absence of the technology to measure the aforementioned variables, estimates of ET are still possible. These estimates rely on more readily available data related to the outlined factors, such as potential evapotranspiration (measured with evaporation pans, as in Figure 4.21), wind speed, temperature, precipitation, elevation, and latitude (Cronk and Fennessy 2001; Davie 2008; Fitter and Hay 2002; Mitsch and Gosselink 2000). Two of these are the Thornthwaite and Linacre formulas, given next, as in Dolan et al. (1984).

$$\text{Thornthwaite ET} = 1.6 \times \left(10 \times T / I\right)^{\alpha} \qquad (4.4)$$

where T = mean monthly temperature; I = annual heat index, calculated by summing, across the 12 months:

$$\text{monthly i} = \left(T / 5\right)^{1.514} \qquad (4.5)$$

FIGURE 4.21 Meteorological Instrumentation Used in Estimating Various Components of a Wetland Water Budget. (A) Meteorological Station in a Formerly Ponded Area of the Talladega Wetland Ecosystem (Twe), Discussed in a Later Section of this Chapter. (B) and (C) Instrumentation Atop an Eddy Flux Tower in the Northern Portion of the Twe, Used to Develop Fine-Scale Estimates of Evapotranspiration for the Wetland. Numbered Components are Used to Measure (1) Light, (2) Temperature and Humidity, (3) Wind Speed and Direction, (5) **Potential Evaporation**, (6) Light, (7) Wind Speed and Direction, (8) Humidity, and (9) Three-Dimensional Wind Speed and Direction. Solar Power Panels (4) and (10) Help to Power the Unit, and Battery and Data Storage Sit to the Left of (4).

And

$$\alpha = 6.75 \times 10^{-7} \times I^3 - 7.71 \times 10^{-5} I^2 + 1.792 \times 10^{-2} I$$
$$+ 0.49239 \qquad (4.6)$$

The other formula is:

$$\text{Linacre ET} = \left[\frac{500 \times (T + 0.006\,h)\, /}{(100 - A) + 15 \times (T - T_d)} \right] / (80 - T) \quad (4.7)$$

where T = monthly mean temperature; h = elevation above sea level; A = latitude; and T_d = mean dew point (an index of the water-holding capacity of the air). The Linacre formula is a simplification of yet another ET formula that is sometimes used (Penman formula, see, e.g., Rosenberry et al. [2004] for the much more complicated Penman formula).

Rosenberry et al. (2004) compared the Penman and Thornthwaite formulas, along with another 11 ET calculations, to ET values obtained via an energy budget formula based on wetland-specific measurements of temperature, humidity, wind speed, and transfers of radiant energy into, out of, and within a prairie wetland in North Dakota. They found that the Penman formula performed well in their study, but not quite as well as two somewhat simpler formulas: the Priestley-Taylor and the deBruin-Keijman methods (Rosenberry et al. 2004). However, all three of the Penman, Priestley-Taylor, and deBruin-Keijman equations provided estimates within 0.1 mm per day of ET measured via the energy budget method. It was noted that the Thornthwaite formula (called the Mather formula by Rosenberry et al.) was the simplest of the formulas tested and provided ET estimates within 0.5 mm per day of those obtained from the energy budget calculation. A major advantage of the Thornthwaite equation is that one only needs monthly temperatures for the wetland of interest to estimate ET for the wetland. Again, as with any of the methods discussed so far, there are trade-offs between the increased simplicity of the calculations and the local accuracy of the resulting estimates.

Another approach to estimating ET that relies solely on local data, and thus is highly applicable to local conditions, is the "diurnal method," which uses measured changes in wetland water level to estimate ET for a particular wetland. In this method, water levels within the wetland are recorded over time and used to estimate ET (Figure 4.22). This method assumes ET to be at or very near zero at night, so in the absence of precipitation, nighttime changes in water level are assumed to be solely from surface and subsurface flows into and out of the wetland (Cronk and Fennessy 2001; Dolan et al. 1984; Mitsch and Gosselink 2000). Nighttime changes in observed water level (usually midnight to 4AM or 6AM) are converted to an hourly rate and used to determine expected change over a 24-hour period in the absence of evapotranspiration. The observed change in water level across a 24-hour period is then subtracted from the expected 24-hour change, and this difference represents losses attributable to ET (Figure 4.22).

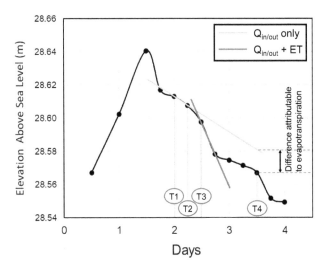

FIGURE 4.22 Water Levels Recorded Over a Four-Day Period from a Marsh Located 40 Km West of Orlando, Florida (USA). These Data were Used to Derive Evapotranspiration Estimates Using the Diurnal Method. Water Levels at Time Points T1–T4 can be Used in Equation 4.8 to Estimate Et. Data from Dolan et al. (1984).

Despite this method relying on much less detailed measurements than those discussed earlier, ET calculated in this way can be quite sensitive to local changes in environmental conditions. For example, Wilcox (2019) examined data collected in a mountain fen in North Carolina (USA) during a total solar eclipse on August 21, 2017, and found evidence of the eclipse in water-level changes during the eclipse totality (when incoming solar radiation was almost completely blocked). Evapotranspiration was found to be significantly lower during the eclipse and for almost an hour afterward, with ET values estimated from three groundwater wells dropping by about 36% during that period.

The difference between observed and expected water-level change based on this method is approximately equal to total ET only when standing water is present in the wetland. This is because these parameters are measured as changes in height of the water surface, but when the water table level is below the surface of the soil, each unit of change in height of the water surface within the soil includes not only the volume of water present in a given area of wetland, but also the soil (or organic matter, etc.) through which the water moves. This property of the soil-water system is referred to as specific yield (abbreviated as S_y), described as the "fillable porosity" of the soil (Maréchal et al. 2006) or the relative yield of drainable water from a volume of soil (or sediment, peat, etc.). If, for example, we have a volume of soil that is 42% "soil material" and 58% pore space, the specific yield of this volume of soil would be 0.58 (Figure 4.23), and a 1 cm drop in water level in this soil would really equate to a loss of 0.58 cm of water from the wetland. The formula for calculating ET based on the data in Figure 4.22 would be:

$$\text{ET} = S_y \times \left[24 \times ((T2 - T1)\,/\,6) \right] - (T4 - T3) \quad (4.8)$$

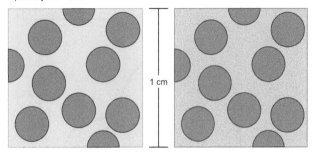

FIGURE 4.23 Specific Yield is the Proportion of the Change in Height of Water Level that is Attributable to the Volume of Water in a Unit of Soil. In this Example, the Water itself (Blue in the Left Panel) Accounts for Only 58% of the Volume of the Soil-Water System. A 1 Cm Drop in Water Level here Equates to a Vertical Loss of Only 0.58 Cm of Water, as the Remaining Soil Particles Occupied the other 42% of Volume.

where (T2–T1)/6 gives us the hourly nighttime change in water level, which is multiplied by 24 to give the expected 24-hour change in water level mentioned earlier, and T4–T3is the 24-hour change in observed water level from noon on day 2 until noon on day 3.

In the study from which the data in Figure 4.22 were obtained, specific yield was measured throughout the upper 50 cm of the organic peat layer (Dolan et al. 1984). Then, for each day that ET was estimated using this approach, the appropriate S_y for that day's water table level was used for the calculations in Equation 4.8. Dolan et al. (1984) found that ET estimated from the diurnal method described earlier was very closely correlated with the product of the monthly plant biomass and the atmospheric water-saturation deficit for the wetland (regression coefficient, $r^2 = 0.85$; Figure 4.24), as one might expect, given that ET is driven to a large degree by transpirational evaporation from the plant surfaces (Figure 4.20). Because of this relationship and the fact that it was very stable across the year, regardless of season or biomass accumulation at any point during the year, Dolan et al. were able to use data on plant biomass and daily water-saturation deficit to calculate ET for days on which the diurnal method could not be used (e.g., days with rain or periods of widespread flooding).

Dolan and colleagues also compared ET values based on their diurnal measurements with estimates from three other established methods of ET estimation: pan evaporation (using an evaporation pan like that shown in Figure 4.21) and calculations based on the Thornthwaite and Linacre formulas (Equations 4.4 and 4.7). They found that the Thornthwaite formula generally gave monthly ET estimates that were close to calculations based on biomass and daily water-saturation deficits, and the annual summed ET for these two differed by less than one percent. The Linacre method, on the other hand, overestimated ET (compared with the Thornthwaite and the diurnal calculations) by

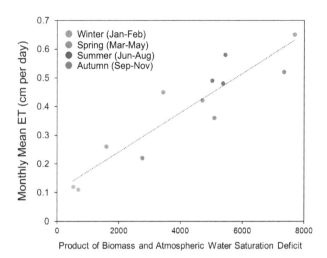

FIGURE 4.24 Relationship Between Et, Estimated by the Diurnal Method, and the Product of Biomass and Water-Saturation Deficit, a Measure of How Much Additional Moisture the Atmosphere can Hold. Biomass and Atmospheric Water-Saturation Deficit Combined to Predict 85% of the Variation in Monthly Mean Estimated Et. Data from Dolan et al. (1984).

about 27%, and ET estimates from the pan evaporation method overestimated by 50%.

4.2.4.1 Interaction of Vegetation and Hydrology in a Riparian Wetland in Alabama (USA)

As a graduate student, I worked in a small riparian wetland ecosystem (~15 ha total area) in the Talladega National Forest in Alabama (USA; Figure 4.25). Since about 1992, this wetland, referred to as the Talladega Wetland Ecosystem (TWE), has been the site of many collaborative research projects aimed at understanding the ecosystem-level functioning of wetlands. Research has focused on carbon and energy transfers (i.e., food web or trophic dynamics), ecology of plants, microbes, insects, and fish, and, of course, hydrology (Chaubey and Ward 2006). Some of the more detailed, fine-scale investigations of hydrology were conducted by Mann and Wetzel (1999, 2000a, 2000b).

The focal area of research in the TWE was an approximately 1.5 ha beaver pond that had been created in the late 1940s (Chaubey and Ward 2006). The pond provided an open-water habitat with populations of *Nymphaea odorata*, *Proserpinaca palustris*, and other aquatic species until around 1996–1997, when the beaver dam collapsed, draining the ponded area. The periphery of the wetland had been occupied by a dense population of *Juncus effusus*, along with other assorted wetland species, such as *Dulichium arundinaceum*, *Rhexia virginica*, and *Alnus serrulata*. After the pond drained, *J. effusus* began to colonize the exposed sediments and encroached rapidly into the new wet meadow habitat. One focus of the work by Mann and Wetzel was the interaction between *J. effusus* and subsurface hydrology in the TWE.

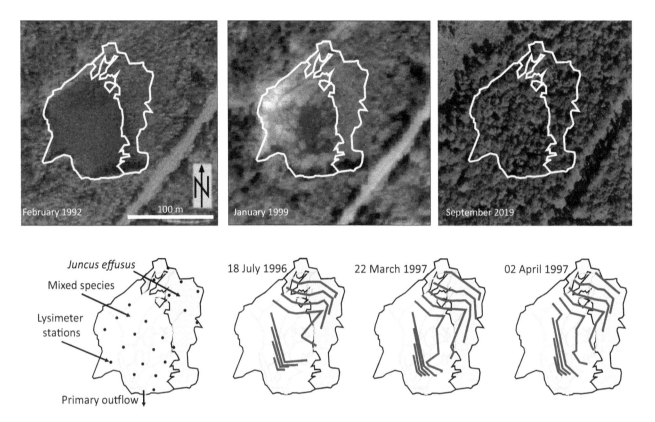

FIGURE 4.25 Top Panels: Time Series of Aerial Photos Looking Down on the Ponded Portion of the Twe (1992) and after Drainage of the Beaver Pond (1999 and 2019). Bottom Panels: Map of Lysimeter Grid and Subsurface Equipotential Lines at 20 Cm Depth, Mapped in 1996 and 1997. A Depth of 20 Cm was the Typical Rooting Depth of *Juncus Effusus* Reported by Wetzel and Howe (1999). Imagery for 1992 and 1999 Obtained from US Geological Survey Earth Explorer (https://earthexplorer.usgs.gov/); 2019 Imagery Obtained from Esri (2022) "World Imagery" www.arcgis.com/home/item.html?id=10df2279f9684e4a9f6a7f08febac2a9. Lower Maps Based on Data in Mann and Wetzel (2000a, 2000b).

Mann and Wetzel installed a network of 22 sets of nested lysimeters (e.g., Figures 4.9 and 4.10 and Figure 4.25) to measure three-dimensional subsurface flows of water through the TWE. At each location, lysimeters were installed at depths of 20, 50, 75, and 100 cm, and water levels in the lysimeters were measured weekly at each location for about 16 months (Mann and Wetzel 2000a). In addition, they measured the hydraulic conductivity of sediments every 10 cm, from 10 cm to 80 cm depth, at 56 locations throughout the wetland, to determine not only the directions of water flows, but also the velocities (Figure 4.26; Mann and Wetzel 2000b).

Mann and Wetzel were also interested in the role of evapotranspiration in subsurface hydrology, especially given the predominance of *J. effusus* in the wetland. Other work in the TWE had shown that this plant can produce as much as 10 kg of biomass per year (above- and below-ground biomass combined), about 70% of which is aboveground tissues (Wetzel and Howe 1999). Because of the very close correlation between plant biomass and ET (e.g., Figure 4.24), it was expected that *J. effusus* might have a substantial influence on subsurface movement of water via the plant's role in ET in the wetland. To test this specifically,

Mann and Wetzel conducted a greenhouse experiment where the movement of water through different types of soil material (clay, sand) was measured via a chemical tracer, bromide, relative to the position of *J. effusus* in greenhouse tanks (Mann and Wetzel 2000a).

Recall that lines of equivalent hydraulic head (**equipotential lines**) are very similar to topographic contour lines. That is, the closer the equipotential lines are to one another, the steeper the gradient of hydraulic potential per distance water travels. Thus, if hydraulic conductivity is similar between two areas, we expect that water will move more quickly through areas where equipotential lines are closer to one another. In Figure 4.25, water flows generally from the north to the south, leaving the formerly ponded area at the stream outflow in the bottom center of the wetland. We see that equipotential lines were farthest apart in the northern and eastern portions of the wetland, in the area dominated by *J. effusus*, and they become closer together as water neared the wetland outflow.

In contrast to patterns of hydraulic potential, Mann and Wetzel (2000b) found that hydraulic conductivity of the sediments was greatest in the area dominated by *J. effusus* (Figure 4.26). They also found that the soil and sediments

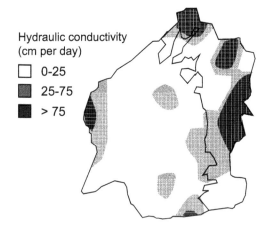

Hydraulic conductivity
(cm per day)

☐ 0-25
▨ 25-75
■ > 75

FIGURE 4.26 Hydraulic Conductivity of Sediments in the Formerly Ponded Area of the Talladega Wetland Ecosystem, at 20 Cm Depth, Based on Analyses of 56 Sediment Cores Distributed across the Wetland. Higher Hydraulic Conductivity of the Soil Suggests the Potential for Faster Subsurface Water Velocities. Based on Data in Mann and Wetzel (2000a).

there had greater porosity, especially in the upper 20 cm (which is what is represented in Figures 4.25 and 4.26). This was the result of a greater accumulation of organic matter in the sediments of the *J. effusus* area, as well as the **porosity** caused by penetration of roots through the soil and sediments. In the greenhouse experiment, they further found that, in the more porous material (sand), the dissolved bromide tracer tended to move toward *J. effusus* plants, presumably pulled by transpiration through the plants. Thus, the dominant plant species in this wetland was found to exhibit multiple mechanisms for influencing subsurface water movement within the wetland. The increased porosity and conductivity of sediments could have accelerated movement of subsurface water through the wetland, while transpiration of the plants themselves may have slowed the lateral movement of water while generating transpirational losses of water from the wetland to the atmosphere.

4.2.5 CHANGE IN STORAGE

In a perfect world, once we have measured all the aforementioned components of a wetland's water budget, we can calculate the difference between total inputs and total outputs to determine the change in wetland water storage over a specified interval of time. Several examples of this approach are shown in Figure 4.27. One aspect of this figure to note is that only one of the budgets includes all of the components that we looked at earlier (Figure 4.27). Additionally, some of the other budgets in Figure 4.27 include elements that we did not cover, such as tidal flows and riverine flooding.

The cypress swamp that was used for the water budget in Figure 4.27 was a cypress-tupelo swamp (a forested wetland dominated by *Taxodium* and *Nyssa* species) situated near the Cache River in southern Illinois. William Mitsch and

colleagues (1977) developed this water budget as part of a comprehensive study of the wetland aimed, in part, at quantifying some of the ecosystem services it provided (e.g., nutrient retention). In this study, rather than calculating change in wetland water storage by difference, they estimated change in water storage by changes in water level in the wetland and solved the following equation for unmeasured subsurface flows:

$$\Delta S = P_n + Q_O + Q_{S+G} - ET - \text{Weir} \qquad (4.9)$$

where Q_{S+G} = net subsurface inflow/outflow balance and "Weir" = streamflow measured via a weir installed on the major outflow of the wetland.

Net precipitation (P_n; throughfall) was measured as precipitation beneath the tree canopy using nine rain gauges distributed across the 35-hectare (90-acre) wetland. Evaporation from the gauges was reduced by adding mineral oil, and a measured quantity of ethylene glycol (1.25 cm) was added during winter months to prevent freezing of water in the gauges (which would damage the gauge). Total precipitation was measured in a non-canopied area within 1 km of the wetland.

Overland flows (runoff) were estimated based on changes in the water level in the wetland, which was measured with a water-level recorder like that in Figure 4.7. Throughfall was subtracted from these values to yield an estimate of runoff associated with rainstorms. Evapotranspiration was estimated in this study by a combination of the diurnal method and the pan evaporation method.

Some of the subsurface flows were measured for the wetland through the use of groundwater wells (Figure 4.9) installed on two sides of the swamp. These wells were used to estimate flows out of the wetland into the Cache River and flows into the wetland from adjacent upland areas. However, not all subsurface flows were measured; the remainder were estimated by difference from Equation 4.9.

The result of these measurements was the finding that inflows approximated outflows from the wetland; that is, storage changed very little over the course of an annual cycle (Figure 4.27B). The majority of inputs came in the form of overland flow (runoff) from the surrounding landscape, and the majority of outflow was via streams the drained the wetland. Evapotranspiration accounted for about 29% of the water leaving the wetland, between intercepted precipitation (9% of total budget) and ET occurring between precipitation events (20% of the annual budget).

Keeping in mind that our main focus in this book is the plants that occur in these ecosystems, ET is of considerable importance, as it indicates a direct involvement of plants in the hydrology of the wetland. Across the examples in Figure 4.27, we see a variety of patterns, from tropical wetlands where ET represents less than 10% of the budget (owing to very large surface inflows and outflows) to more hydrologically isolated wetlands where ET represents more than 80% of the total budget. It is useful to also consider

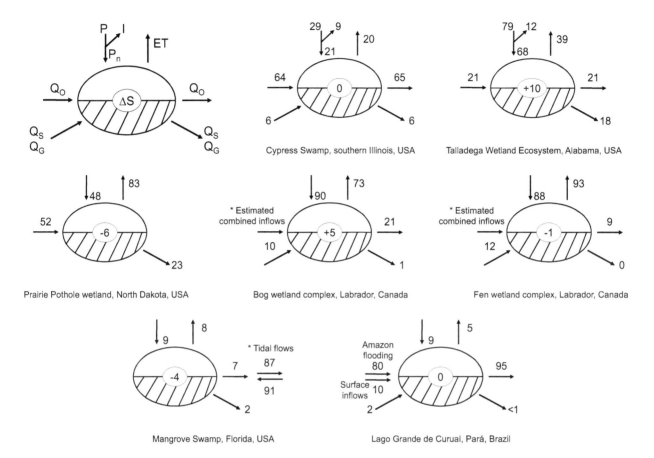

FIGURE 4.27 Water Budgets from a Selection of Wetlands Representing a Variety of Potential Hydrologic Conditions. Values are Percent of Inflows and Do not Sum Exactly to 100 Because of Rounding of Individual Values. Inflows for E and F were Estimated from the Difference between Precipitation Inputs and Δs Values for the Canadian Wetlands. Data Sources: (B) Mitsch, Dorge, and Wiemhoff (1977), (C) Chaubey and Ward (2006), (D and G) Mitsch and Gosselink (2000), (E and F) Price and Maloney (1994), and (H) Bonnet et al. (2008).

that the budgets in Figure 4.27 represent annual sums in each of the water budget components, but those individual components vary over time, based on local climate, yearly variations in weather, plant growth, and all of the other factors we have discussed earlier. To illustrate some of this temporal variation, we can examine data from the TWE wetland in closer detail.

Chaubey and Ward (2006) provided monthly data from the years 1994 and 1995 in their hydrologic budget for the TWE wetland in Figure 4.27. Those data are broken out by budget components in Figure 4.28. Here, we can see the annual cycle of ET, which increases with plant growth during spring (March through May), peaks during summer (June through August), and declines through autumn and into winter. As it happens, this wetland, in Alabama (USA), typically receives most of its precipitation during winter (December through February), although some years have a fairly wet tropical storm season (September through November). We also see that losses resulting from ET exceeded throughfall inputs during much of the active growing season in both years. In fact, evapotranspirative losses contributed 60% of total outflows in 1994 and 53% of outflows in 1995 (Chaubey and Ward 2006).

4.2.6 HYDROPERIOD INTEGRATES THE HYDROLOGY ACROSS TIME

Measuring all the parameters needed to construct a water budget is a laborious task (Figure 4.27). Even those who are interested in developing these budgets find it to be overwhelming, as there are few examples where all the components were measured, or even estimated. A shortcut to the process, for those who are less interested in the intricacies of how and where water moves into and out of the wetland, is simply to measure the temporal patterns of inundation in the wetland. The integration of these patterns over time is referred to as the **hydroperiod** (Mitsch and Gosselink 2000), and it can tell us much about the conditions plants will face within the wetland and, as a consequence, the types of aquatic or wetland plants we would expect to see in the wetland. Because the hydroperiod represents water depth over time, the methods used to develop a hydroperiod are those associated with water level in the wetland, sometimes including subsurface water. The instruments, then, that one would use for describing a wetland hydroperiod would be some type of water-level recorder and wells of some sort, if subsurface water is of interest.

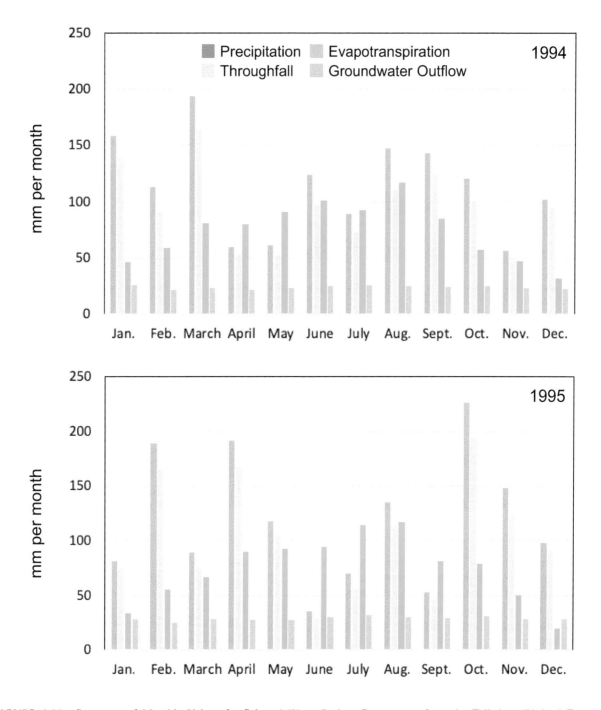

FIGURE 4.28 Summary of Monthly Values for Selected Water Budget Components from the Talladega Wetland Ecosystem (Alabama, USA). A few Items of Note here are the Increasing Evapotranspiration Rates during the Peak Growing Season (May–August), the Somewhat Inverse Relation Between Precipitation and Evapotranspiration, and the More or Less Uniform Groundwater Outflows throughout Both Years. Data Taken from Chaubey and Ward (2006).

In some of my previous work, one of my students (Cory Shoemaker) was interested in determining hydroperiods for a set of wetlands he was studying during his doctoral research. In this project, we were interested in the vegetation of wetlands established and managed under the US Department of Agriculture's Wetland Reserve Program (WRP) and how they compared with nearby unmanaged wetlands. In some instances (e.g., Figure 4.29), WRP wetlands were situated very near non-managed wetlands, facilitating comparison of the two wetland types by reducing much of the extraneous variation in soils, topography, or precipitation that one might see between wetlands separated by long distances. Hydroperiods for the two wetlands in Figure 4.29 are given in Figure 4.30, along with precipitation data from two meteorological stations within 25 km of the wetlands.

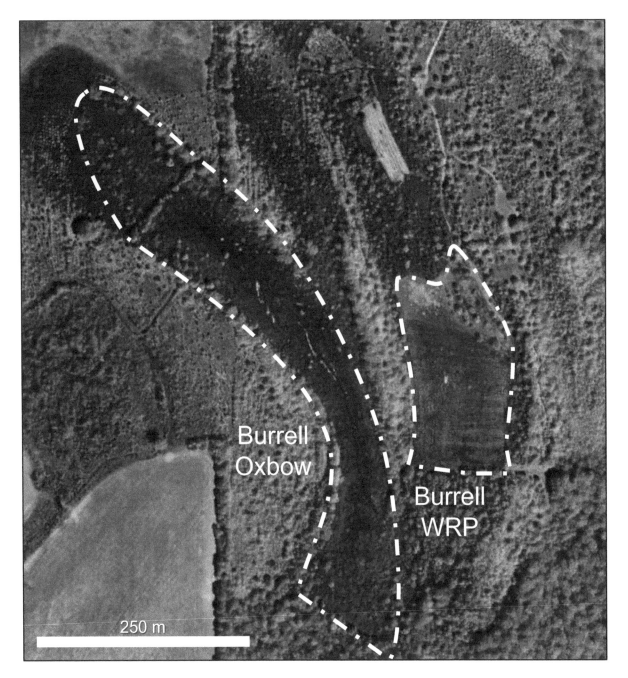

FIGURE 4.29   Two Wetlands Included in the Dissertation Studies of Cory Shoemaker at Mississippi State University, Located in Northwest Mississippi (USA). Burrell Oxbow is a Remnant Oxbow Wetland from an Abandoned River Channel in the Upper Yazoo-Mississippi River System. Burrell Wrp is a Wetland Restoration Site Managed as Winter Waterfowl Habitat as Part of the Usda Natural Resources Conservation Service Wetland Reserve Program. Aerial Imagery from Esri (2022) "World Imagery" www.arcgis.com/home/item. html?id=10df2279f9684e4a9f6a7f08febac2a9.

One fairly noticeable difference between these two wetland hydroperiods is that the oxbow wetland was flooded to a depth greater than 5 cm on every day throughout the year, whereas the managed WRP wetland had no standing water on 177 of the 372 days measured (47.5%). The reason for the flood-free days in the WRP wetland is that it was being managed to encourage growth of annual plant species, with the expectation that they will produce an abundance of seeds for overwintering waterfowl. Thus, the water levels

in the WRP wetland were regulated with control structures such as those shown in Figure 4.6 so that soil was exposed during the growing season to allow seed production. Later, water levels were raised in the autumn to provide stopover habitat for migrating birds. The results of these manipulations, along with the associated differences in hydrology, were quite evident in the plant data for these sites. Surveys of the WRP site recorded 45 plant species in summer of 2015, 18 of which were grasses, sedges, or rushes, versus

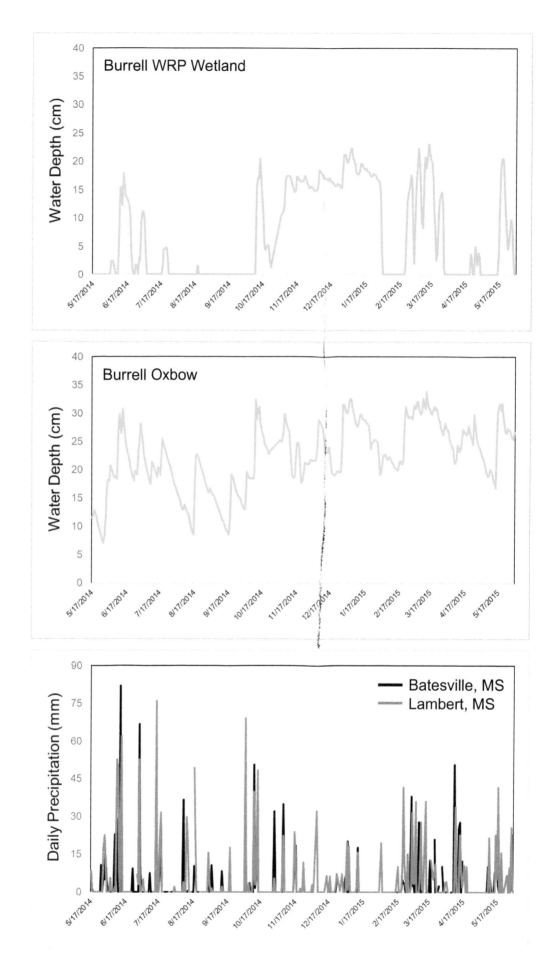

FIGURE 4.30 **Hydrographs** for the Burrell Wrp Site and Burrell Oxbow, Measured with Water-Level Loggers as Shown in Figure 4.8 (Cory Shoemaker, Unpublished Data). Precipitation Data in the Bottom Panel Obtained from the US National Oceanic and Atmospheric Administration's National Centers for Environmental Information (www.ncdc.noaa.gov/cdo-web).

only 22 plant species in the non-managed oxbow wetland, half of which were wetland tree species.

4.3   IN SUMMARY

The depth, duration, frequency, and chemical composition of water on a site, the last of which will be discussed in Chapters 5 and 7, have enormous impacts not only on whether a given site will become a wetland, but also on what type of wetland will occur or can be maintained on a given site. In fact, hydrology was one of the three requisite features defining wetlands in Chapter 1, and as a result, for determining whether a given species of plant would be considered a wetland plant or an upland plant.

The water balance approach is used to integrate inflows and outflows of water for a given wetland (or other ecosystem) into a water budget for that system. The "balance" aspect of this is the reconciliation of inflows versus outflows, under the principle that the outflows plus any storage within the wetland cannot exceed the total inflows to a system. A complete water budget will consider precipitation inputs and surface and subsurface inflows, along with surface and subsurface outflows and outflows back to the atmosphere via evapotranspiration (ET). However, some of these attributes of a water budget are difficult to measure (and sometimes have a small impact on one's objective), and they may be estimated via alternative approaches or even ignored under some circumstances.

We saw in this chapter that, although ET may be directly and significantly influenced by plants (via the transpiration component of ET), all components of a water budget can have significant relationships with plants, depending on the type of wetland and circumstances under which the hydrology is being studied. In some cases, plant-driven evapotranspiration can account for 90% or more of water losses from a wetland. However, the presence and/or chemical composition of subsurface inflows can determine the specific suites of plant species that may successfully establish within a wetland, which in turn can determine potential rates of transpiration of that water back out of the system. Thus, the hydrology of a wetland is not only a complex wetland attribute in and of itself, but it also can have even more complex interactions with the plants occupying the site, with some very interesting feedbacks existing between the living and nonliving components of the system.

4.4   FOR REVIEW

1. Summarize a hydrologic budget for a wetland.
2. What are the basic principles behind methods for measuring the different components of a wetland water budget (net precipitation, streamflow, runoff, subsurface flows, evapotranspiration, storage)?
3. What is the force that allows plants to carry out evapotranspiration, and what is the soil-plant-air continuum?

4. Measured evapotranspiration rates often are *less* than potential rates suggested from measuring water loss from evaporation pans. What are some possible reasons for this difference?
5. What are some ways that hydrology influences/affects plants in wetlands?
6. What are some ways that wetland plants can affect hydrology in wetlands?

4.5   REFERENCES

Bonnet, M. P., G. Barroux, J. M. Martinez, et al. 2008. "Floodplain Hydrology of an Amazon Floodplain Lake (Lago Grande de Curuaí)." *Journal of Hydrology* 349: 18–30.

Bridgham, Scott D., John Pastor, Jan A. Janssens, Carmen Chapin, and J. Malterer Thomas. 1996. "Multiple Limiting Gradients in Peatlands: A Call for a New Paradigm." *Wetlands* 16 (1): 45–65.

Chaubey, I., and G. M. Ward. 2006. "Hydrologic Budget Analysis of a Small Natural Wetland in Southeast USA." *Journal of Environmental Informatics* 8 (1): 10–21. https://doi.org/10.3808/jei.200600073.

Cronk, Julie K., and M. S. Fennessy. 2001. *Wetland Plants: Biology and Ecology*. Boca Raton, FL: CRC Press.

Crushell, P. H., M. G. C. Schouten, A. J. P. Smolders, J. G. M. Roelofs, and P. S. Giller. 2006. "Restoration of Minerotrophic Vegetation within an Irish Raised Bog Soak System." *Biology and Environment: Proceedings of the Royal Irish Academy* 106B (3): 371–85. https://doi.org/10.3318/BIOE.2006.106.3.371.

Crushell, Patrick, Andrew Connolly, Matthijs Schouten, and Fraser J. G. Mitchell. 2008. "The Changing Landscape of Clara Bog: The History of an Irish Raised Bog." *Irish Geography* 41 (1): 89–111. https://doi.org/10.1080/00750770801915596.

Daly, Donal, and Paul Johnston. 1994. "The Hydrodynamics of Raised Bogs: An Issue for Conservation." In *The Balance of Water—Present and Future*, edited by T. Keane and E. Daly, 105–21. Dublin, Ireland: AGMET Group.

Davie, Tim. 2008. *Fundamentals of Hydrology*. 2nd ed. New York: Routledge.

Dolan, Thomas J., Albert J. Hermann, Suzanne E. Bayley, and John Zoltek. 1984. "Evapotranspiration of a Florida, U.S.A., Freshwater Wetland." *Journal of Hydrology* 74: 355–71.

Donovan, L. A., M. J. Linton, and J. H. Richards. 2001. "Predawn Plant Water Potential Does Not Necessarily Equilibrate with Soil Water Potential under Well-watered Conditions." *Oecologia* 129: 328–35.

Fitter, Alastair H., and Robert K. M. Hay. 2002. *Environmental Physiology of Plants*. 3rd ed. San Diego: Academic Press, Inc.

Flynn, Raymond M. 1990. "Clara Bog: A Hydrogeological Study." Master's thesis, University of Birmingham, UK.

Jackson, C. Rhett. 2006. "Wetland Hydrology." In *Ecology of Freshwater and Estuarine Wetlands*, edited by Darold P. Batzer and Rebecca R. Sharitz, 43–81. Berkeley, CA: University of California Press.

Johnson, Mark S., Eduardo G. Couto, Osvaldo B. Pinto, Jr., et al. 2013. "Soil CO_2 Dynamics in a Tree Island Soil of the Pantanal: The Role of Soil Water Potential." *PLOS One* 8: e64874.

Jones, M. B., and F. M. Muthuri. 1984. "The Diurnal Course of Plant Water Potential, Stomatal Conductance and Transpiration in a Papyrus (*Cyperus papyrus* L.) Canopy." *Oecologia* 63: 252–55.

Mann, Carroll J., and Robert G. Wetzel. 1999. "Photosynthesis and Stomatal Conductance of *Juncus effusus* in a Temperate Wetland Ecosystem." *Aquatic Botany* 63: 127–44.

———. 2000a. "Hydrology of an Impounded Lotic Wetland—Wetland Sediment Characteristics" *Wetlands* 20: 23–32.

———. 2000b. "Hydrology of an Impounded Lotic Wetland—Subsurface Hydrology." *Wetlands* 20: 33–47.

Maréchal, J. C., B. Dewandel, S. Ahmed, L. Galeazzi, and F. K. Zaidi. 2006. "Combined Estimation of Specific Yield and Natural Recharge in a Semi-Arid Groundwater Basin with Irrigated Agriculture." *Journal of Hydrology* 329 (1–2): 281–93. https://doi.org/10.1016/j.jhydrol.2006.02.022.

Mekonnen, G. B., S. Matula, F. Doležal, and J. Fišák. 2015. "Adjustment to Rainfall Measurement Undercatch with a Tipping-Bucket Rain Gauge Using Ground-Level Manual Gauges." *Meteorology and Atmospheric Physics* 127 (3): 241–56. https://doi.org/10.1007/s00703-014-0355-z.

Mitsch, William J., Carol L. Dorge, and John R. Wiemhoff. 1977. "Forested Wetlands for Water Resource Management in Southern Illinois." Report to the University of Illinois Water Resources Center, Urbana, Illinois. https://doi.org/10.1017/CBO9781107415324.004.

Mitsch, William J., and James G. Gosselink. 2000. *Wetlands*. 3rd ed. New York: John Wiley & Sons, Inc.

Price, J. S., and D. A. Maloney. 1994. "Hydrology of a Patterned Bog-fen Complex in Southeastern Labrador, Canada." *Nordic Hydrology* 25: 313–30.

Ray, Peter M. 1960. "On the Theory of Osmotic Water Movement." *Plant Physiology* 35 (6): 783–95.

Richardson, J. L., L. P. Wilding, and R. B. Daniels. 1992. "Recharge and Discharge of Groundwater in Aquic Conditions Illustrated with Flownet Analysis." *Geoderma* 53: 65–78.

Rosenberry, Donald O., I. Stannard David, Thomas C. Winter, and Margo L. Martinez. 2004. "Comparison of 13 Equations for Determining Evapotranspiration from a Prairie Wetland, Cottonwood Lake Area, North Dakota, USA." *Wetlands* 24: 483–97.

Shedlock, Robert J., Douglas A. Wilcox, Todd A. Thompson, and David A. Cohen. 1993. "Interactions between Ground Water and Wetlands, Southern Shore of Lake Michigan, USA." *Journal of Hydrology* 141: 127–55.

Sprecher, Steven W. 1993. "Installing Monitoring Wells/Piezometers in Wetlands." US Army Corps of Engineers Technical Report HY-IA-3.1. US Army Corps of Engineers Waterways Experiment Station, Vicksburg, MS.

Stevens, John C. 1924. "Water-level Recorder." United States Patent number 1,494,034.

Stransky, D., V. Bares, and P. Fatka. 2007. "The Effect of Rainfall Measurement Uncertainties on Rainfall-Runoff Processes Modelling." *Water Science and Technology* 55 (4): 103–11. https://doi.org/10.2166/wst.2007.100.

Taiz, Lincoln, and Eduardo Zeiger. 2006. *Plant Physiology*. 4th ed. Sunderland, MA: Sinauer Associates, Inc.

Touchette, B. W., L. R. Iannacone, G. Turner, and A. Frank. 2010. "Ecophysiological Responses of Five Emergent-wetland Plants to Diminished Water Supply: An Experimental Microcosm Study." *Aquatic Ecology* 44: 101–12.

Touchette, B. W., G. A. Smith, K. L. Rhodes, and M. Poole. 2009. "Tolerance and Avoidance: Two Contrasting Physiological Responses to Salt Stress in Mature Marsh Halophytes *Juncus roemerianus* Scheele and *Spartina alterniflora* Loisel." *Journal of Experimental Marine Biology and Ecology* 380: 106–12.

Waddington, J. M., N. T. Roulet, and A. R. Hill. 1993. "Runoff Mechanisms in a Forested Groundwater Discharge Wetland." *Journal of Hydrology* 147: 37–90.

Wetzel, Robert G., and Michael J. Howe. 1999. "High Production in a Herbaceous Perennial Plant Achieved by Continuous Growth and Synchronized Population Dynamics." *Aquatic Botany* 64: 111–29.

Wilcox, Douglas A., Steven I. Apfelbaum, and Ronald D. Hiebert. 1985. "Cattail Invasion of Sedge Meadows Following Hydrologic Disturbance in the Cowles Bog Wetland Complex, Indiana Dunes National Seashore." *Wetlands* 4: 115–28.

Wilcox, Douglas, Robert Shedlock, and William Hendrickson. 1986. "Hydrology, Water Chemistry, and Ecological Relations in the Raised Mound of Cowles Bog." *Journal of Ecology* 74: 1103–17.

Wilcox, Jeffrey D. 2019. "Total Solar Eclipse Effects on Evapotranspiration Captured by Groundwater Fluctuations in a Southern Appalachian Fen." *Hydrological Processes* 33: 1538–41.

5 Critical Features of the Aquatic Environment

5.1 KEY FEATURES OF WATER AS A MEDIUM FOR PLANT LIFE

In Chapter 2, we looked at the progression of traits that were accumulated by plants as they evolved from an aquatic lifestyle into the early land plants, then tracheophytes, spermatophytes, and eventually angiosperms. Along the way, plants acquired and retained characters such as a waxy cuticle, stomata, and production of physically supportive, water-conducting vascular tissues. All these characteristics facilitated survival in the water-limited terrestrial habitats that plants colonized over a period of some 400 to 500 million years. As we also saw, however, while most groups of plants continued along a path of terrestrial existence, certain groups of these plants, from diverse nodes within the evolutionary tree, occasionally branched out to recolonize the aquatic habitat, facing new challenges and acquiring new traits as a result. In this chapter, we will take an in-depth look at some of the challenges that plants faced in their transition back to the aquatic environment. In the following chapter, we will see some of the diverse ways that flowering plants have adapted to the challenges of the aquatic environment in which they now thrive.

To build a basis for this chapter's discussion, we first look at the core metabolic processes of *photosynthesis* and *respiration* (Figure 5.1; Table 5.1); these energy metabolizing pathways set the stage for the adaptations we will see in the next chapter. The following reaction summarizes the process of photosynthesis, in which plants use light energy from the sun to convert inorganic carbon (in the form of carbon dioxide) into carbohydrates, releasing oxygen in the process. The four aspects of the aquatic habitat that are the focus of this chapter are all represented in this reaction: carbon dioxide, water, and light, as inputs, and oxygen, as one of the two products.

$$CO_2 + H_2O + Light \rightarrow CH_2O + O_2 \qquad (5.1)$$

This equation is a summary of two complex steps commonly referred to as the "light-dependent" and "light-independent" reactions of photosynthesis (top leftmost part of Figure 5.1). In the light-dependent reactions, the water on the left side of Equation 5.1 is split into protons, electrons, and oxygen, using solar energy captured by the chlorophyll molecule. The capture of light by chlorophyll takes place on the **thylakoid** membrane, inside the chloroplasts, while the splitting of water itself occurs inside the **lumen** (internal space) of the thylakoid, resulting in an accumulation of protons that drives the formation of **ATP**. The light-independent reactions, in contrast, take place in the **stroma** (the fluid inside the chloroplast but outside the thylakoid; Figure 5.1). Electrons and ATP from the light-dependent reaction are used to synthesize carbohydrates by adding the electrons to carbon dioxide. We refer to this process of CO_2 reduction as **carbon fixation**. As depicted in Figure 5.1, there are numerous similarities between photosynthesis and respiration. Aside from the overall summary reactions being essentially the opposite of one another, both processes use similar molecules to shuttle electrons from place to place, use proton gradients to produce ATP, and rely on that ATP as a temporary energy currency to drive some of the reactions. We will revisit these in later sections of this chapter.

The harvesting of light and the incorporation of CO_2 into carbohydrates take place inside the chloroplasts, but there are many obstacles that reduce the amount of light that can be captured within the chloroplast. The first several of these include the plant's cell walls, the cuticle that surrounds the leaf (if present), and the water within which the leaf may be submersed (Figure 5.2). Within the water, there will be dissolved and/or suspended (i.e., non-dissolved) mineral and organic matter, all of which can absorb, reflect, or scatter light as it passes through the water. There also will be suspended organisms, including photosynthetic algae and cyanobacteria, which absorb light of the same wavelengths used by aquatic plants. Finally, there may also be deposits of organic or mineral matter, along with entire communities of microorganisms, including algae and/or cyanobacteria, directly on the surfaces of submersed leaves (Figure 5.2). The result of this for a **submersed plant** (or submersed parts of other plants) is an array of obstacles reducing the amount of light that ultimately arrives at the thylakoid membrane to drive the production of carbohydrates, in addition to the stresses imposed on aerobic organisms living beneath the water. In many ways, this presents submersed plants with challenges like those faced by plants growing in the shade of other, taller plants, such as in the forest understory. We will see in Chapter 6 that some aquatic and wetland plants handle this challenge in much the same way as understory plants do, namely by altering the morphology of their leaves for better resource acquisition.

We will begin to dissect the nature of these obstacles presented by the aquatic environment by first exploring some of the key characteristics of water, then look more specifically at the availability of light, carbon dioxide, and oxygen for aquatic and wetland plants.

DOI: 10.1201/9781315156835-5

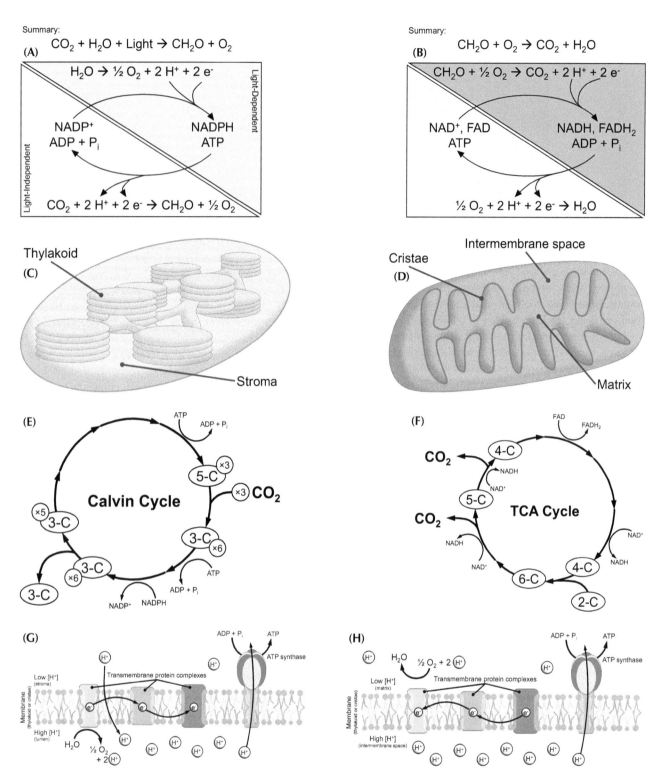

FIGURE 5.1 Photosynthesis and Respiration are Two Core Metabolic Processes in Plants, and Key Aspects of Each Process Set the Conditions under Which Plant Adaptations are Required for Existence in Aquatic and Wetland Habitats. Here, I've Highlighted some of the more Salient Features of these Processes to Aid in the Mechanistic Understanding of Concepts Covered in this Chapter. (A) Both Processes can be Considered as Combinations of Biochemical Reduction-Oxidation Reactions in Which Intermediary Molecules Serve to Shuttle Protons, Electrons, and Energy Among Compartments within the Cell (See Table 5.1 for More Detail). The Two Processes also Very Nearly Mirror One Another, in Terms of the Reactants and Products and the Extraction Vs. Storage of Energy. (B) the Cellular Organelles and their Most Relevant Components for these Processes. (C) Summaries of the Pathways by Which Carbohydrates are Formed (Calvin Cycle) and Broken Down (Tricarboxylic Acid Cycle; TCA). Note here the Intake of CO_2 in the Calvin Cycle Vs. The Release of CO_2 from the TCA Cycle. Terms such As 3-C Refer to the Number of Carbon Atoms in the Compounds Resulting from the Various Steps in These Cycles. The Calvin Cycle must be Run Three Times to Yield One Three-Carbon Carbohydrate for Use in other Cellular Processes (Bottom Left of the Calvin Cycle). See Table 5.1 for Descriptions of Other Abbreviations in these Cycles. (D) Both of these Processes are Accompanied by or Partnered with Electron Transport Chains on Their Intra-Organellar Membranes (Thylakoid in Photosynthesis; Cristae in Mitochondrial Respiration). These Electron Transport Chains Drive the Production of an Electrochemical Proton Gradient across the Membrane that is Responsible for the Generation of Atp Via a Membrane-Associated Protein Complex Referred to as an Atp Synthase. For More Detail on these Processes, Falkowski and Raven (1997) is an Excellent Resource, as are Taiz and Zeiger (2006) and Zubay (1998).

TABLE 5.1

Carriers and currency of biochemical energy transfers in photosynthesis and respiration. Other forms of cellular respiration found among bacteria also share these features, although some of the specific electron carriers may differ. For more detail, Gottschalk (1986), Taiz and Zeiger (2006), and Zubay (1998) are excellent references.

Category	Description
Electron carriers	NADP⁺, NAD⁺, and FAD are molecules referred to as coenzymes that serve to move electrons during a variety of biochemical processes (NADP⁺: nicotinamide-adenine dinucleotide phosphate; NAD⁺: nicotinamide-adenine dinucleotide; FAD: flavin-adenine dinucleotide).
Energy currency	ATP (adenosine triphosphate) is an energy-carrying molecule in biological systems formed from the combination of an ADP molecule (which also carries energy) and an inorganic phosphate molecule (P_i).
Proton gradient	Some of the protons (H⁺) liberated during photosynthesis and respiration become compartmentalized either within the lumen (innermost compartment) of the thylakoid membrane in chloroplasts or in the intermembrane space of mitochondria. Protons accumulate in those spaces, creating an electrochemical gradient that is used to generate ATP, in a manner somewhat like the accumulation of water behind a hydroelectric dam for use in generating electricity.

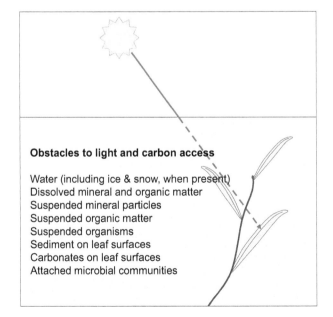

Obstacles to light and carbon access

Water (including ice & snow, when present)
Dissolved mineral and organic matter
Suspended mineral particles
Suspended organic matter
Suspended organisms
Sediment on leaf surfaces
Carbonates on leaf surfaces
Attached microbial communities

FIGURE 5.2 Aquatic Plants Face Multiple Levels of Obstacles in Acquiring Adequate Quantity and Quality of Sunlight to Carry Out Photosynthesis, Including Dissolved and Suspended Materials in the Water Column and Even the Water itself.

5.1.1 TEMPERATURE AND DENSITY RELATIONS OF WATER

Water is a **polar molecule** (Figure 5.3) because of the difference in **electronegativity** of the hydrogen (slightly positive) and oxygen atoms (slightly negative), which causes an uneven sharing of electrons within the molecule. This polar nature of water results in the formation of weak hydrogen bonds between adjacent water molecules. We can see in Figure 5.3 that each oxygen atom has the potential to form **hydrogen bonds** with four other nearby oxygen molecules, through the sharing of hydrogen atoms among them. Although these bonds individually are weak in nature, the sheer numbers of these within a volume of water result in collectively strong cohesive forces. It is these forces that provide the matric potential used to move water upward through xylem tissues, as we saw in Chapter 4.

The nature of these hydrogen bonds also give water its characteristic behavior regarding temperature. As temperatures cool, the rate of vibration of atoms within the molecules and the movement of the molecules themselves decrease (see, e.g., Wetzel 2001). This decreased movement allows a bit of a stabilization of the hydrogen bonding, bringing the molecules closer to one another and increasing the density of the liquid water (Figures 5.3 and 5.4). At a temperature of approximately 3.98 °C (~39–40 °F), water reaches its maximum density (Figure 5.4). As temperatures continue to cool, movement continues to slow until finally, at 0 °C (32 °F), the molecular vibrations slow by five to seven orders of magnitude, allowing formation of a stabilized **lattice** structure similar to that in Figure 5.3C (Millero 2013; Wetzel 2001). It is this property that gives solid water a lower density than liquid water and allows ice to float. Were this not the case, ice would sink as it formed, and bodies of water would gradually freeze solid from the bottom up. This would be bad news for aquatic plants and other forms of aquatic life.

Another important temperature-related property of water is its ability to resist changes in temperature, or its **specific heat** (amount of heat required to increase one gram of water by 1 °C). Water has one of the highest specific heat values of any commonly encountered substance (Millero 2013; Wetzel 2001), giving it the ability to stabilize aquatic environments, and adjacent terrestrial environments, against changes in temperature. Thus, we often see milder, less fluctuating climates in coastal areas. The high specific heat of water also means that large amounts of energy must be lost to the surrounding air in order for liquid water to freeze into ice, and likewise, large amounts of energy are required to melt ice into liquid water (Wetzel 2001). As with other properties of water, its high specific heat results from the multitude of hydrogen bonds that draw the molecules to one another.

5.1.2 WATER AS A SOLVENT

The polar nature of the water molecule also results in water readily associating with dissolved ionic materials (e.g.,

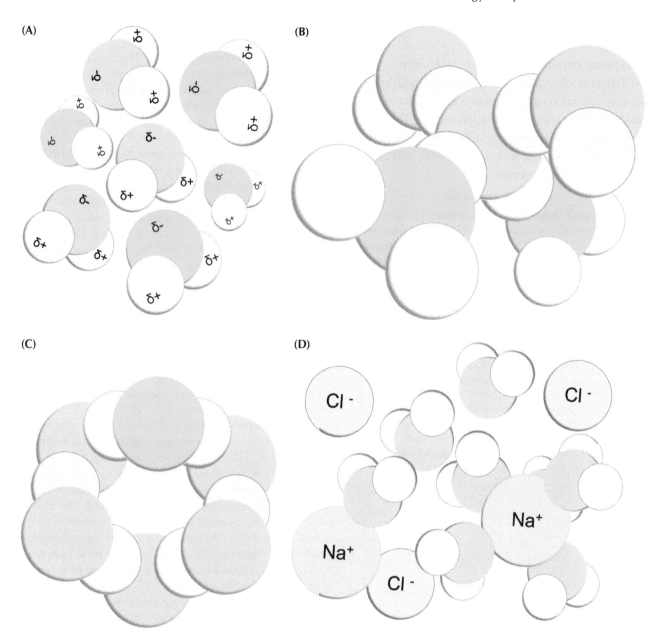

FIGURE 5.3 Water is a Polar Molecule (A) Capable of Forming Hydrogen Bonds with up to Four Neighbors (B). The Nature of These Bonds Allows Water to Form a Lattice Structure of Repeating Units, Similar in Configuration to (C). The Space Formed by the Lattice Rings in this Configuration Gives Water a Lower Density in Solid Form Vs. Liquid at its Freezing/Melting Point of 0 °C. Dissolved Ions, such as Sodium (Na⁺) and Chloride (Cl⁻) from Table Salt, Can Disrupt the Bonding Among Water Molecules and Affect the Behavior of the Water. Based on Illustrations and Data in Belch, Berkowitz, and Mccammon (1986), Millero (2013), and Wetzel (2001).

sodium chloride, NaCl; Figure 5.3), as the water molecules readily dissociate into protons (H^+) and hydroxide ions (OH^-; Millero 2013). In fact, the solvent properties of water are as much as four to 40 times greater than some common organic solvents (Millero 2013). Furthermore, the attractional forces between ionized water and dissolved solutes are so strong that some of the key properties of pure water are substantially altered in the process of forming an ionic solution with water (Millero 2013). Examples are:

1. The freezing point of the solution is lower than that of pure water;

2. The boiling point of the solution is higher than that of pure water;

3. The electrical conductivity of the solution is higher than that of pure water; and

4. The temperature at which the solution reaches maximum density is lower than pure water.

These effects that the solute has on the freezing and boiling points of water are the result of disruption of the typical molecular structure of the water by the dissolved ionic solute (Millero 2013). The dissolved ions interact strongly with the water molecules, preventing them from escaping as

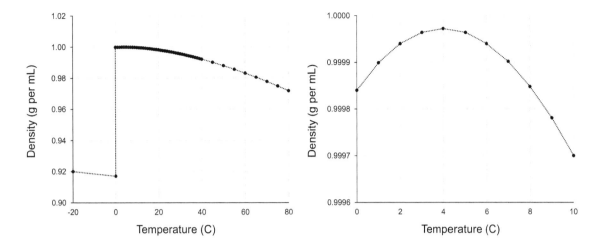

FIGURE 5.4   Density of Air-Free Pure Water at Standard Pressure (1 Atmosphere). Right Panel is Focused on the Differences in Density from 0 to 10 °C, to Emphasize the Highest Density at ~4 °C. Data from Kohlrausch (1996).

vapor during heating or from forming a lattice during freezing, and this is in part responsible for the altered critical temperatures of the solution, relative to pure water.

5.2   PATTERNS OF LIGHT AVAILABILITY

As pointed out in Figure 5.2, there are many potential obstacles as light travels from the sun to the chlorophyll molecules in any given plant. First off, gaseous and particulate materials in the atmosphere absorb, reflect, and scatter a substantial amount of incoming sunlight, particularly in the wavelengths of visible light (400–700 nm), where absorption of energy for photosynthesis takes place (Figure 5.5). In those wavelengths, we see about 30% of the incoming

solar radiation absorbed before it reaches the earth (at sea level). At longer wavelengths (> 700 nm, in the **infrared**), we see up to 100% of the incoming radiation absorbed by atmospheric water vapor, with lesser fractions (in fewer bands of wavelengths) absorbed by atmospheric oxygen and carbon dioxide. For a much more detailed overview of the behavior of solar radiation in the atmosphere, Gates (1963) provides an excellent review, relative to energy availability in natural ecosystems.

If we look globally at the amount of energy successfully passing through the atmosphere, we see considerable variation in the seasonal or annual patterns of energy arriving at a given point (Figure 5.6). Seasonal maxima and minima become much more pronounced as we move toward the

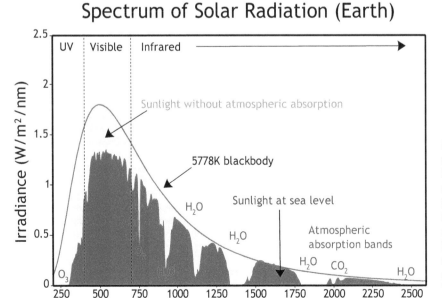

FIGURE 5.5   Solar Radiation Spectrum at the Top of the Atmosphere (Upper Boundary of Yellow/Lighter Bars) and at Sea Level (Red/Darker Bars). The Solid Curve Represents the Idealized Spectrum of Light Emitted from the Sun (Labeled "5778k Blackbody"). Various Regions of Absorption by Water, O_2, O_3 (Ozone), and CO_2 in the Atmosphere are Indicated on either Side of the Visible Region. © Nick84, Used under CC BY-SA 3.0 License (creativecommons. org/licenses/by-sa/3.0/deed.en).

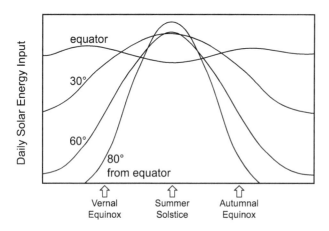

FIGURE 5.6 Relative Comparison of Approximate Total Daily Solar Radiation Impinging upon a Horizontal Surface as a Function of Season and Latitude. Redrawn from Wetzel (2001).

poles from the equator, reflective of differences in the ways that seasons are expressed in the temperate versus tropical latitudes. Near the poles, we see very small energy inputs during the long, dark winters. For example, at 70° N latitude, there is no official sunrise between about November 24 until January 18 (years 2019–2020; US National Oceanic and Atmospheric Administration [NOAA] Earth System Research Lab, www.esrl.noaa.gov).

To get an idea of the total energy inputs experienced by a leaf, Figure 5.7 summarizes many sources of energy impinging upon a horizontal surface in nature. This considers not only the direct sunlight hitting the surface, but also sunlight reflected from below and sunlight scattered by gases and particles as it passes through the atmosphere. The latter can be quite important in some ecosystems, where low-lying cloud cover is a common occurrence. For example, in spruce-fir (*Picea rubens* and *Abies fraseri*) forests of the southern Appalachian mountains (North Carolina, USA), low-lying clouds frequently immerse the tree canopy in water vapor, with cloud immersion observed on ~70% of days, accounting for as much as 45% of the total annual water input (Johnson and Smith 2006), and plant species in these forests are clearly adapted to these light conditions. For example, seedlings of *A. fraseri* exhibited three times higher rates of photosynthesis during cloud immersion of the forest understory than on days with clear skies.

In addition to light energy, Figure 5.7 considers heat inputs to the leaf surface, which also are important drivers of plant physiology, influencing rates of transpiration, metabolism, and so on. We can see here that the heat energy received by the leaf can be much greater than the reflected sunlight, potentially equaling the total direct sunlight and skylight received in some ecosystems. Heat energy also tends to fluctuate less throughout the day, as heat continues to be radiated from the soil, water, and other surfaces during nighttime (recall the prior discussion of the heat retention capacity of water, which is usually present in soil, especially in and around wetlands). Comparing the total

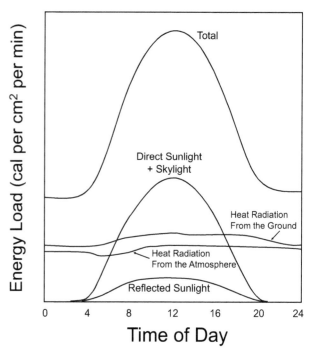

FIGURE 5.7 Daily Cycle of Estimated Light and Heat Load (Total Radiation Load) on a Horizontal Surface, Based on Measurements Taken on June 5, 1954, at Hamburg, Germany (~55.5° N Latitude). Redrawn from Gates (1963).

energy inputs among a tropical rainforest, an alpine tundra ecosystem, and in the Alaskan interior (65° N) on the summer solstice reveals that the total amount of energy received by the surface throughout the day can be much higher in the tundra and arctic habitats on days of peak solar input (Figure 5.8). This highlights the need to consider climate extremes, rather than the mean or median, when thinking about environmental physiology.

5.2.1 INFLUENCE OF WATER ON LIGHT

Most of us are familiar with the refractive properties of water; that is, the ability of water to differentially **transmit** (i.e., allow the penetration of) light of differing wavelengths. This is true for visible light, which we see evidenced in the form of rainbows, but also for the ultraviolet (UV) and infrared portions of the light spectrum (Figures 5.5 and 5.9). In Figure 5.9, we see that water selectively absorbs in the UV range (wavelengths shorter than 400 nm) and in the red regions (also infrared, from Figure 5.5). Wavelengths in the range of about 450 to 500 nm (blue region of the visible spectrum) tend to be absorbed the least by pure water, and, thus, these wavelengths tend to penetrate to the greatest depths (Figure 5.10). The blue region of the spectrum also has a high capacity to be backscattered by pure water (Figure 5.11). The combined low absorption and high backscatter of the blue portion of the visible light spectrum combine to give us beautiful blue waters of the clear ocean and deep lakes, such as Crater Lake (Oregon, USA), Lake

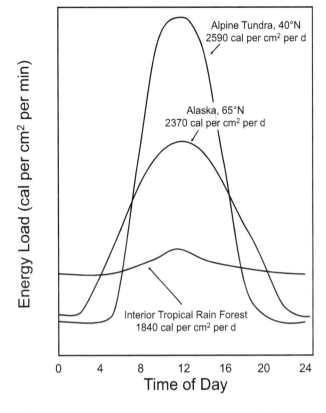

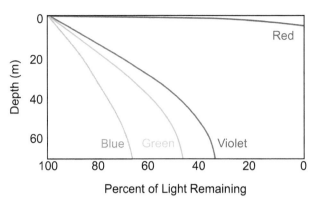

FIGURE 5.10   Relative Depth Penetration in Distilled Water by Selected Bands in the Visible Light Spectrum. Redrawn from Wetzel (2001).

FIGURE 5.8   Daily Cycles of Estimated Total Radiation Load on a Horizontal Surface at Summer Solstice, Estimated from Observed Values for Sunlight and Skylight at the Given Locations, Along with Estimates of Heat Radiation Based on those Light Values. Redrawn from Gates (1963).

Baikal (eastern Russia), lakes Malawi (Figure 5.12) and Tanganyika (eastern Africa rift valley), and Lake McKenzie (Australia). Even in much shallower waters, we can see evidence of this high transmission and backscatter of blue light, in the absence of phytoplankton and other particulate materials that tend to interfere with passage of light through water (e.g., the geyser pool shown in Figure 5.12).

These differences in the degree to which water absorbs or transmits different wavelengths of light is critically important for photosynthesis of all plants, but especially aquatic species. The physiological impact of these differences in absorbance results from the absorption

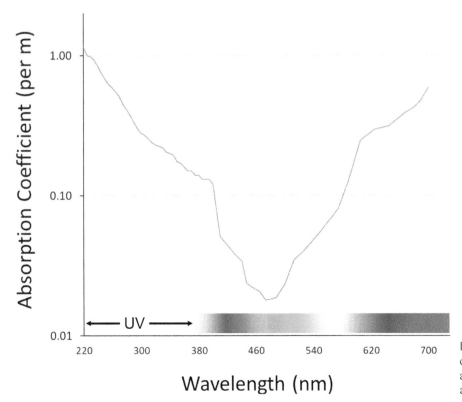

FIGURE 5.9   Light Absorption Spectrum of Pure Water. Based on Data in Lenoble and Saint-Gully (1955), Morel (1974), and Wetzel (2001).

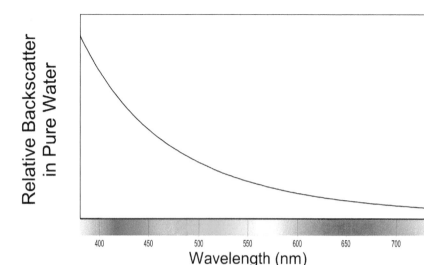

characteristics of the chlorophyll pigments used in light capture (Figure 5.13). There are several different forms of chlorophyll, as well as other photosynthetic and accessory pigments found among photosynthetic organisms. Green plants produce chlorophylls *a* and *b*, whose absorption spectra are shown in Figure 5.13, and chlorophylls *c* and *d* are found in cyanobacteria and some microalgal taxa, such as diatoms and dinoflagellates (Falkowski and Raven 1997; Taiz and Zeiger 2006). The strongest absorption of the plant chlorophylls is in the blue and red portions of the visible spectrum (Figure 5.13). As we have just seen, blue wavelengths are much more available in pure water with increasing depth, so this high absorption in the blue region is a plus for aquatic plants!

On the other hand, chlorophyll *c*, which is produced by some of the microalgae that grow attached to submersed surfaces (e.g., diatoms), has a very similar absorption spectrum to that of the chlorophylls *a* and *b*. The peak blue absorbance for chlorophyll *c* lies roughly between that of chlorophyll *a* and *b*, roughly approximating the magnitude of the chlorophyll *a* blue peak (Falkowski and Raven 1997). This poses a problem for submersed macrophytes, in that any attached algae living on leaf surfaces have the first opportunity at the incoming sunlight, intercepting some of the specific wavelengths of energy that would have been available for use by the plant. In Chapter 10, we will see that some plant species have evolved mechanisms for dealing with some of these microbial competitors.

5.2.2 INFLUENCE OF DISSOLVED AND SUSPENDED MATERIALS ON LIGHT

In addition to the effects that the water itself has on light, there are impacts of the materials within the water as well. We already saw in the previous section that other photosynthetic organisms can reduce the penetration of certain wavelengths of light through the water, as a consequence of their using similar photosynthetic pigments. As alluded

to in Figure 5.2, there are many other types of materials within the water column, in both dissolved and suspended form.

Vähätalo, Wetzel, and Paerl (2005) investigated the differences in light absorption among three components of freshwater and estuarine ecosystems in the Neuse River of North Carolina (USA; Figure 5.14). They examined numerous spectral absorbance characteristics of phytoplankton, non-algal particulate matter, and "chromophoric" (i.e., light-absorbing) dissolved organic matter (CDOM), in comparison with the water itself. They did this for water sampled from four cypress-tupelo swamps, six reservoirs, 11 estuarine wetlands, and 22 streams, from across the 16,000 km² Neuse River drainage basin. Curiously, they found that the phytoplankton absorbed a relatively small but consistent fraction of visible light (never more than 30%), whereas the dissolved organic matter frequently was found to absorb the greatest fraction of light (usually > 30%), from among the four components examined (Figure 5.14). Another interesting finding of their study was that the patterns among phytoplankton, non-algal particles, and CDOM were relatively consistent among the four different types of ecosystems examined.

The consistently high light absorption by CDOM is not restricted to solar radiation in the visible portion of the spectrum. Dissolved organic matter also absorbs a great deal of radiation in the UV wavelengths (Figure 5.15). Absorption of UV by dissolved carbon is so high that almost all the radiation in the range of 305–320 nm disappears within 2 m of depth in waters with as little as 1–2 mg of dissolved organic carbon per liter. Thus wetlands, with their often high concentrations of dissolved organic carbon, which is released during decomposition of the high amounts of plant material present, may experience very high rates of **UV-induced photodegradation** of dissolved organic material (physical degradation of organic matter resulting from exposure to high-energy UV radiation; e.g., Vähätalo and Wetzel 2004).

FIGURE 5.12 Top: Lake Malawi, Mozambique (Africa), Seen from Likoma Island. Bottom: One of the Many Dozens of Geyser Pools in Yellowstone National Park (Wyoming, USA). The High Temperatures and High Dissolved Salt Content of these Geyser Pools Restricts Growth of Algae that Might otherwise Obscure Light Penetration through their Waters. Photo of Lake Milawai © Worldtraveler/CC BY-SA 3.0, https://commons.wikimedia.org/w/index.php?curid=486719.

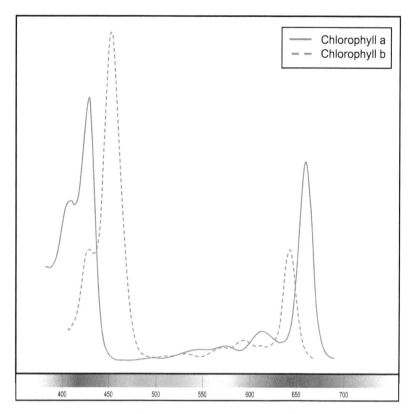

FIGURE 5.13 Visible Region Absorption Spectra for Chlorophylls a and *B*, Extracted in a Solution of Acetone and Calcium Carbonate, Followed by an Extraction in Ether, and Subsequently Washed in Ether to Separate the Pigments from other Materials in the Filtered Extract. Absorption Was Measured for the Final Ether Extract. Redrawn from Comar and Zscheile (1942).

5.3 CARBON DIOXIDE AVAILABILITY IN THE WATER

Having ended the previous section by looking at **dissolved** *organic* carbon, now is a great time to examine the **dissolved** *inorganic* **carbon** that is used by plants in photosynthesis. As indicated in Figure 5.1, plants capture carbon, in the form of CO_2, during photosynthesis and "fix" that carbon into carbohydrates. In the process of doing this, they are harnessing solar energy captured by chlorophyll in the form of chemical bonds within the carbohydrates being produced (via the Calvin cycle; Figure 5.1). Specifically, the energy is stored in the carbon-hydrogen bonds within the carbohydrates. Ecologically, this is an important concept to understand, as those carbon-hydrogen bonds represent the currency by which energy is transferred within ecosystems, after its initial capture by the plants. Incidentally, this fact that the plants are first to produce biomass from the sun's energy is the reason they are referred to as **primary producers**. The breaking of those carbon-hydrogen bonds, which takes place within the mitochondria during aerobic respiration, releases the energy so that it may be exchanged to another currency (ATP) for use within cells (the right-hand side of Figure 5.1). We revisit this later in this chapter and again in the two chapters that follow, along with information on anaerobic respiration.

Where do the plants obtain the CO_2 that is captured in the Calvin cycle? For those plants whose photosynthetic parts are exposed to the atmosphere, the CO_2 comes directly from the air, entering the leaves through the stomata. Plants with submersed leaves, on the other hand, rely largely on dissolved forms of CO_2. Although water presents aquatic plants with an obstacle to obtaining the atmospheric CO_2, this gas is relatively soluble in water (about 200 times more soluble than oxygen, for example; Wetzel 2001), and water usually contains a considerable amount of dissolved inorganic carbon. Furthermore, the amount of CO_2 within the water column, generally speaking, is on the rise. According to NOAA (www.climate.gov), in 2021, the global atmospheric CO_2 concentration was approximately 415 parts per million (ppm) and was increasing at a rate of about 2 ppm per year. However, the specific form in which dissolved carbon is most abundant varies strongly with the water's pH.

The relationship between CO_2 dynamics and pH is the result of the carbonate buffering system illustrated in Figure 5.16. As CO_2 enters the water, the rate at which it becomes **hydrated** (i.e., fully dissolved into an aqueous form) depends on the pH of the water. At pH values below about 8, the hydration of CO_2 is relatively slow (Wetzel 2001), but at higher pH, where OH^- ions are more readily available, hydration is more rapid, owing to the reaction of CO_2 and OH^- to form bicarbonate ions (HCO_3^-; Figure 5.16). Once in solution, CO_2 will enter the buffering reactions shown in Figure 5.16, and the ratios among forms of available inorganic carbon will lie close to the ratios depicted for a particular pH in that figure. For example, at pH below

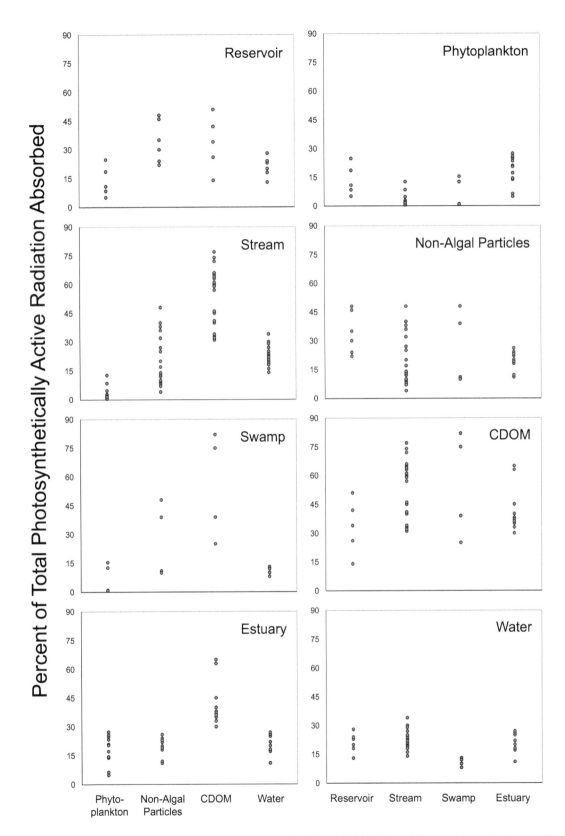

FIGURE 5.14 Comparison of Light Absorption at Wavelengths from 400–700 Nm (Photosynthetically Active Radiation), Among Different Aquatic and Wetland Ecosystems and by Different Dissolved and Suspended Materials in the Water. Cdom = Chromophoric (Or Light-Absorbing) Dissolved Organic Matter. Based on Data in Vähätalo, Wetzel, and Paerl (2005).

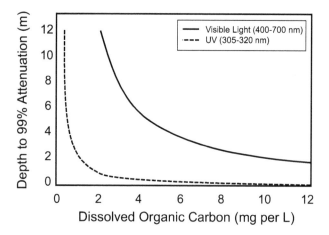

FIGURE 5.15 Increased Attenuation of Ultraviolet (Uv) Wavelengths by Dissolved Organic Carbon, in Comparison with Visible Light. The Attenuation Curve for Uv Wavelengths is an Averaged Curve for Wavelengths of 305 and 320 Nm. Redrawn from Wetzel (2001).

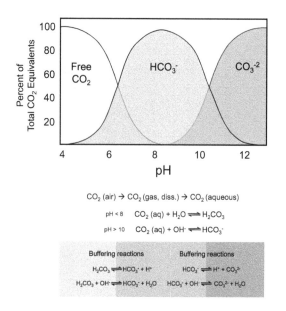

FIGURE 5.16 Relationship between Ph and Relative Concentration of Different Forms of Inorganic Carbon in Freshwaters, at 15 °C. Redrawn from Golterman (1969) and Wetzel (2001). Equations in the Lower Half of the Figure Demonstrate the Various Ph-Dependent, Bidirectional Transitions Among Inorganic Carbon Species, with Equilibria at Ph = 6.4 and 10.4 at a Temperature of 15 °C (Represented by the Differential Shading from Left to Right; Millero 2013; Wetzel 2001). The Specific Equilibrium Points are Strongly Influenced by Temperature and the Presence of Dissolved Ions Other than those Indicated.

about 6.4 (which is the equilibrium pH value for the conversion between carbonic acid [H_2CO_3] and HCO_3^- + H^+), free CO_2 will begin to dominate, becoming even more prevalent with declining pH, until, around pH of about 4, dissolved inorganic carbon will exist almost exclusively as free CO_2

and H_2CO_3 (Allan 1995; Wetzel 2001). Similar patterns are seen at pH of roughly 10.4 and above (equilibrium pH of HCO_3^- with CO_3^{2-} and H^+), where carbonate (CO_3^{2-}) would begin to dominate the inorganic carbon pool.

At pH values of approximately 6.4 and 10.4 (again, subject to effects of temperature and other dissolved materials), the equilibrium buffering reactions will release and consume protons (H^+) and hydroxyl ions (OH^-). This, in conjunction with the tendency for water itself to dissociate into protons and hydroxyl ions, tends to offset any significant changes in pH values that might result from diffusion of CO_2 into (or out of) the water or from uptake or release of CO_2 via photosynthesis and respiration of the aquatic biota. Given these relationships between inorganic carbon and pH, we might imagine that submersed plants growing at pH values higher than 6.4 would face a challenge in acquiring free CO_2 to use in photosynthesis, and we would be correct in that assumption.

The pH of surface waters, absent of human intervention, falls somewhere in the range of 2 to 12 (Wetzel 2001). You will likely recognize pH values of 2 and 12 as being rather extreme, and these are quite rare in natural systems, occurring in areas with a strong influence of mineral inputs from volcanic material on the low end and **calcareous** (calcium-rich) geological formations on the high end. Recall from Chapter 3 that bogs often have pH values below 4, with minima around 3.3 to 3.5 (Figure 3.16, but see also Wetzel 2001). At the other extreme, many of the more basic lake ecosystems (and their marginal wetland areas) have pH values at or above 8, and typical *maximum* values for natural surface waters are around pH of 10 (Cronk and Fennessy 2001; Wetzel 2001). Furthermore, many truly aquatic plant species across North America occur within a pH range of 6.3 to 9.0 (Hutchinson 1975), placing them well within the range of pH where HCO_3^- is the dominant form of inorganic carbon (Figure 5.16). Note also that the carbonate buffering system (the reactions at the bottom of Figure 5.16) help to maintain pH values within this range. We examine in Chapter 6 how submersed plants deal with situations where inorganic carbon comes predominantly in the form of HCO_3^-. I should also point out that pH above this range is relatively uncommon, with none of Hutchinson's 68 examples falling above pH = 9.0, and Golterman (1969) indicating that situations where carbonate (CO_3^{2-}) dominates (i.e., waters with pH > 10.4) are essentially negligible when considering freshwaters as a whole.

Another piece of this puzzle is that during photosynthesis, when plants (or other primary producers) remove CO_2 from the water, they can begin to shift the equilibrium among the various forms of dissolved inorganic carbon. We saw earlier that at low pH, when CO_2 becomes hydrated, it forms carbonic acid (H_2CO_3), which dissociates to bicarbonate (HCO_3^-) and a proton (H^+), thereby reinforcing the low pH (but buffered by an equilibrium conversion of HCO_3^- and water to carbonic acid and hydroxyl ions; Figure 5.16):

$$CO_2 + H_2O \rightarrow H_2CO_3 \rightarrow HCO_3^- + H^+ \qquad (5.2)$$

However, this is a reversible process, the opposite of which results from the removal of CO_2 from the water by photosynthesizing plants as the buffering system replaces CO_2 that is removed by photosynthesis. This shift in the equilibrium effectively "pulls" the reactions to the left (Equation 5.3). The cascading series of reactions resulting from this disruption of carbon equilibria will tend to increase the water's pH because it results in protons being removed from solution as they combine with HCO_3^- to form H_2CO_3:

$$CO_2 + H_2O \leftarrow H_2CO_3 \leftarrow HCO_3^- + H^+ \qquad (5.3)$$

Fluctuations in pH resulting from daily patterns of photosynthesis can be as large as 3 to 4 pH units, but more commonly are on the order of 1 to 2 units, owing to the buffering capacity of carbonate buffering reactions in Figure 5.16 (Cronk and Fennessy 2001; Maberly 1996). Furthermore, the carbonate buffering system can act to restore pH values during the nighttime, when plant uptake of CO_2 has ceased (for most species). Similarly, cellular respiration of the plants or the other groups of organisms in their surroundings releases CO_2 into the water and can lead to a decline in pH; this can be quite high in wetlands, where large amounts of dead biomass accumulate and subsequently decompose within the water and sediments. The CO_2 resulting from cellular respiration and other forms of energy metabolism also can be collected by the plants for use in photosynthesis, and sometimes may even be recycled within the plants themselves. We will revisit this idea in Chapter 6.

5.4 OXYGEN AVAILABILITY IN THE WATER AND SEDIMENTS

As we have seen, oxygen is one of the products of photosynthesis (Equation 5.1, Figure 5.1), but oxygen also is required to carry out **cellular respiration** in aerobic organisms, including plants, and this can be represented as:

$$CH_2O + O_2 \rightarrow CO_2 + H_2O \qquad (5.4)$$

where carbohydrates are broken down, with the aid of oxygen, yielding carbon dioxide and water. We also saw in Figure 5.1 that the protons and electrons removed from carbohydrates on the left side of this equation are used to generate reducing potential that aids in the production of ATP in the mitochondria, prior to the eventual formation of the water on the right side of the equation.

Aerobic respiration generates a considerable amount of ATP that the organism (a plant, for example) uses in carrying out other metabolic functions to support survival, reproduction, growth, and defense. Thus, the availability of oxygen within the water and the flooded soil and sediments is important biologically and ecologically for plants, animals, and aerobic microorganisms that form the communities living in aquatic and wetland habitats. Keep in mind also that although oxygen is generated through photosynthesis, this takes place in the light-dependent reactions and,

thus, is strictly limited by the availability of light of the relevant wavelengths.

5.4.1 PATTERNS OF OXYGEN IN THE WATER AND SEDIMENTS

The potential oxygen-holding capacity of water (i.e., the concentration when water is saturated with dissolved oxygen) is influenced by both atmospheric pressure and water pressure, as well as water temperature (Figure 5.17). Oxygen concentrations along the curve in Figure 5.17 represent water that is 100% saturated with oxygen; however, it is possible for water to be supersaturated with greater concentrations of oxygen. For example, moving waters may become supersaturated because of the capture of turbulence-induced air bubbles, and still waters may become supersaturated because of high rates of oxygen generation through photosynthesis in dense canopies of submersed vegetation. Experimentally, we also can supersaturate water with oxygen by bubbling oxygen gas into the water.

Patterns of oxygen availability in wetlands are incredibly complex and heterogeneous in space and time, because of the following factors:

- Daily fluctuations in dissolved oxygen concentrations driven by photosynthesis in the water column (increased $[O_2]$ in the day, decreased $[O_2]$ at night);
- Consumption of oxygen in flooded soils and sediments by aerobic microbial metabolism;
- Release of oxygen into the soil and sediment from plants (discussed in Chapter 6);
- Inputs of moving water by rainfall or overland surface flow into the wetland; and
- Exposure of soil and sediment during times of water deficit in the wetland (e.g., during drought).

In an experimental examination of small-scale patterns of oxygen availability within the upper surface of wetland

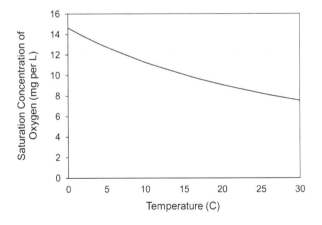

FIGURE 5.17 Relationship between Temperature and Saturating Levels of Oxygen in Freshwaters. Based on Data Given in Wetzel (2001).

sediments, Carlton and Wetzel (1988) constructed small experimental chambers containing 8 to 12 mm thick layers of sediment and the naturally attached microalgae growing on the upper surface of those sediments. In their experiments, they were interested specifically in the effects of algal photosynthesis on oxygen availability in the water and in the thin layer of sediments. They also looked at the effect of photosynthesis on nutrients dissolved in the water, and we will talk about that in Chapter 7.

One interesting finding from their work was that in chambers not exposed to light, where no photosynthesis had taken place for at least six hours, oxygen concentration declined to zero within the upper one millimeter of sediments because of microbial metabolism within those sediments (Figure 5.18). This was despite the water above the sediments having oxygen levels at saturation for the ambient temperatures during the experiments. In that same experiment, chambers were evaluated after exposure to either two or six hours of light to facilitate algal photosynthesis. Although oxygen levels were increased in the upper millimeter of sediment in response to algal photosynthesis, even in the chambers that had been exposed to light for six hours, oxygen levels reached zero at 4 mm into the sediments. These results speak to the rapid rates with which microbial metabolism consumes oxygen in wetland sediments, as well as the difficulty that respiring plant roots would face attempting to acquire the oxygen they need in that environment, in the absence of the adaptations that we will look at in Chapter 6.

Relative to some of the discussion earlier in this chapter about the interactions between photosynthesis and pH in the water, Carlton and Wetzel (1988) also measured pH

in their experimental chambers (Figure 5.19). As expected, they found that pH in the sediments increased in response to six hours of light exposure (and photosynthesis), but it still dropped rapidly within the sediments, declining by almost two pH units within the upper two millimeters of sediments. This decline was likely driven by the release of carbon dioxide via the metabolic activities of the sediment microbes.

Carlton and Wetzel (1988) also looked at longer-term patterns of oxygen concentrations in the sediments, imposing cycles of light and dark on the microbial communities (Figure 5.20). Again, as expected, they found that oxygen concentrations in the upper millimeter of sediment increased after exposure to light, approximating levels in the oxygen-saturated water above the sediments, and when lights were turned off, sediment oxygen levels returned to zero. As a further examination of rapid water column oxygen

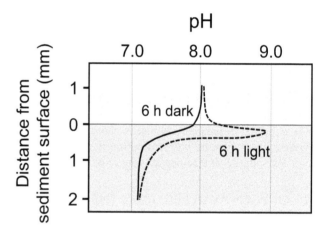

FIGURE 5.19 Profiles of Ph in Experimental Microcosms Used in the Studies Depicted in Figure 5.17. Microcosms here were either Held in the Dark for the Duration of the Measurements (6 H Dark) or Exposed to Light for Six Hours. Redrawn from Carlton and Wetzel (1988).

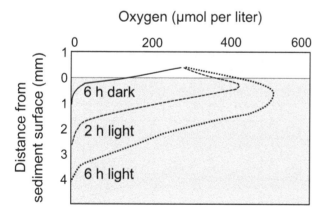

FIGURE 5.18 Profiles of Oxygen Concentration in an Experimental Microcosm Study. Microcosms were either Held in the Dark for the Duration of the Measurements (6 H Dark) or Exposed to Light for either Two or Six Hours. Sediments were Collected from the Lake Floor at a Water Depth of Eight Meters in Lawrence Lake, Michigan (USA); the Upper 8–12 Mm of Sediments, with Attached Microalgae and Associated Microbes, were Used in the Experiments. Oxygen Concentration was Measured using a Microelectrode (Outer Diameter 15–40 Mm) at 0.2 Mm Intervals within the Sediments. Redrawn from Carlton and Wetzel (1988).

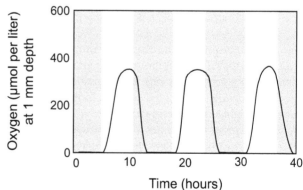

FIGURE 5.20 Profiles of Oxygen Concentration in Response to Dark-Light Cycles, at 1 Mm Depth within the Sediments, in Experimental Microcosms Used in the Studies Depicted in Figure 5.17. Shaded Areas of the Figure Correspond to Periods of Darkness. Redrawn from Carlton and Wetzel (1988).

depletion by sediment microbes, Carlton and Wetzel experimentally supersaturated the water in their experimental chambers to levels at 375% of capacity (Figure 5.21). In this experiment, they again found that microbial metabolism in the sediments drew down the oxygen concentration to zero within the upper 4 mm of sediments, emphasizing the degree to which sediment microbial metabolism can influence resource availability in wetlands (in this case O_2).

The effects of this rapid consumption of oxygen in the sediments are not restricted to these very small spatial scales, however. In a larger-scale study carried out in a Colorado (USA) lake, Buscemi (1958) investigated patterns of dissolved oxygen in stands of the submersed plant *Elodea canadensis* (from the Hydrocharitaceae). He found that in early and mid-summer, when light availability and water circulation were limited by dense growth of submersed *Elodea*, water column oxygen concentrations within the plant canopy began to decline rapidly as far as 40 to 60 cm above the sediments (Figure 5.22). In that

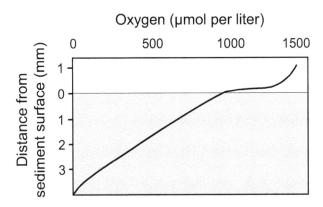

FIGURE 5.21 Profile of Oxygen Concentration along a Depth Gradient in Microcosm Sediments, Where Overlying Water was Oxygenated to 375% of Ambient Saturation. Redrawn from Carlton and Wetzel (1988).

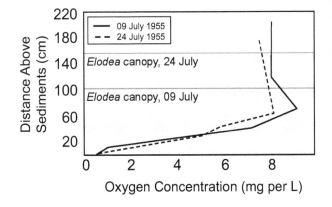

FIGURE 5.22 Depth Profiles for Oxygen in Stands of *Elodea Canadensis* as Measured in a Colorado (USA) Lake, at 2,499 M Elevation, in July 1955. The Plant Canopy was Approximately 50 Cm Taller on July 24, as Indicated by the Two Different Horizontal Lines. Redrawn from Buscemi (1958).

study, Buscemi noted that oxygen concentrations beneath the submersed *Elodea* canopy eventually increased during late summer, when the lowermost leaves of the plant had died and decomposed, allowing deeper light penetration and potentially reducing respiratory demands of attached microbes on the leaves (and possibly the lowermost, shaded leaves themselves). Thus, the combination of low rates of photosynthesis in the deeper, more heavily shaded areas of the *Elodea* stand, the reduced circulation of oxygenated water into the stand of vegetation, and the respiratory demands of the sediment microbes led to very low oxygen concentrations throughout the early parts of the summer.

Similarly, Turner, Cholak, and Groner (2010) found that the canopy of leaves from a large stand of *Nelumbo lutea* in Pennsylvania (USA) substantially reduced the concentrations of dissolved oxygen, compared to adjacent areas of open water (Figure 5.23). The differences were not substantial in early summer (June), but, as shown in Figure 5.23, the daily maximum and minimum oxygen concentrations were lower under the shade of the *Nelumbo* canopy in July and August, with frequent **hypoxia** (oxygen concentrations below 2.5 mg L^{-1}) observed in the warmest month (August). These two studies highlight the importance of balance between light availability to drive photosynthesis and **heterotrophic** oxygen consumption in aquatic and wetland habitats in maintaining oxygen concentrations in the water column within these ecosystems.

5.4.2 REDOX CHEMISTRY

Another important influence that oxygen has on the environmental conditions within wetlands touches on the topic of reduction-oxidation (redox) chemistry. This often is a difficult concept to grasp, but we can use photosynthesis and respiration as examples. In reduction-oxidation reactions, a **reduction** is the gain of one or more electrons, which results in a decrease (or reduction) in the electrical charge of the atom receiving those electrons. Thinking about this as a reduction in electrical charge helps to make it an easier concept to remember. In biological systems, a donated electron is almost always accompanied by a proton, equating to the transfer of a hydrogen atom, H (H = H^+ + e^-). Furthermore, a reduction of one atom (or molecule) is always accompanied by the **oxidation** of another, which is the loss of electrons, resulting in an increase in the charge of the atom that has given up the electron(s).

We can illustrate this with a simple redox equation:

$$DH + R \rightarrow RH + D \tag{5.5}$$

In Equation 5.5, we have a donor atom, D, that is giving up an electron to a recipient atom, R. So, in Equation 5.5, we have atom D being oxidized while atom R is being reduced, and the source of the reducing power is the hydrogen that is being transferred from D to R.

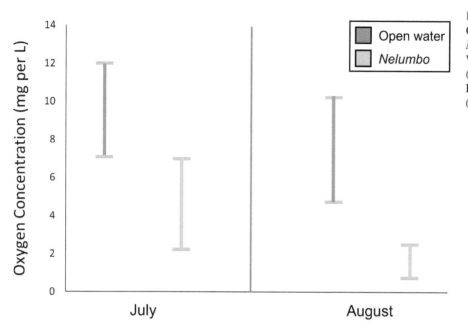

FIGURE 5.23 Ranges of Measured Oxygen Concentrations in a Stand of *Nelumbo Lutea*, and in Nearby Open Water, in a Reservoir in Pennsylvania (USA) in Summer 2007. Based on Data in Turner, Cholak, and Groner (2010).

Now, let us take another look at photosynthesis and respiration in the context of redox chemistry. Recall from Figure 5.1 (and Equation 5.1) that we can represent photosynthesis as:

$$CO_2 + H_2O + Light \rightarrow CH_2O + O_2 \qquad (5.1)$$

which can be broken into separate oxidation and reduction steps, as in Figure 5.1. In the first step:

$$H_2O + Light \rightarrow 2H^+ + 2e^- + \tfrac{1}{2}O_2 \qquad (5.6)$$

water is oxidized by the removal of electrons (and protons, as discussed earlier). Then, in the second step:

$$CO_2 + 2H^+ + 2e^- \rightarrow CH_2O + \tfrac{1}{2}O_2 \qquad (5.7)$$

carbon dioxide is reduced through the addition of the electrons that were removed from water in Equation 5.6.

Similarly, we have seen that respiration can be summarized as:

$$CH_2O + O_2 \rightarrow CO_2 + H_2O \qquad (5.4)$$

Here also, we can consider separate oxidation and reduction steps that essentially mirror the prior Equations 5.6 and 5.7. The first step:

$$CH_2O \rightarrow CO_2 + 2H^+ + 2e^- \qquad (5.8)$$

illustrates the oxidation of a carbohydrate molecule, such as that formed during photosynthesis, releasing electrons (and protons). As with photosynthesis, the electrons (and protons) are picked up by **electron carriers**, and in the pro-

cess are used to generate the **proton gradient** in the mitochondrion that drives the production of ATP (Figure 5.1). In aerobic respiration, the electron carriers ultimately transfer those protons and electrons to oxygen, in a reduction reaction that yields water (Equation 5.9).

$$O_2 + 2H^+ + 2e^- \rightarrow H_2O \qquad (5.9)$$

Equation 5.9 also highlights the role of oxygen as a **terminal electron acceptor** in aerobic respiration. What this means is that oxygen sits at the end of a chain of electron transfers along the membranes of the cristae in the mitochondrion, using the carrier molecules mentioned in Figure 5.1 and Table 5.1 to generate the proton gradient that drives production of ATP. Electrons are carried from the carbohydrates that are being oxidized to one of several membrane proteins, and the transfers of the electrons drive the movement of protons into the intermembrane space of the mitochondrion, where they accumulate (Figure 5.1D). Those protons then are used to drive the production of ATP from ADP, using another complex of membrane proteins (the ATP synthase) that moves the protons back across the membrane into the matrix, as ATP is formed.

The amount of energy generated by this movement of electrons is dependent upon the number of protons that can be moved across the membrane, which depends on the electrochemical distance those electrons are carried. When cells use oxygen as the final electron acceptor, this provides the greatest difference in reduction-oxidation potential between the terminal acceptor (oxygen) and the source of the electrons (carbohydrates, in this case). This is illustrated in Figure 5.24, which gives the redox potential of the pairs of reactions involved in some of the more common bacterial carbohydrate metabolism pathways. As illustrated,

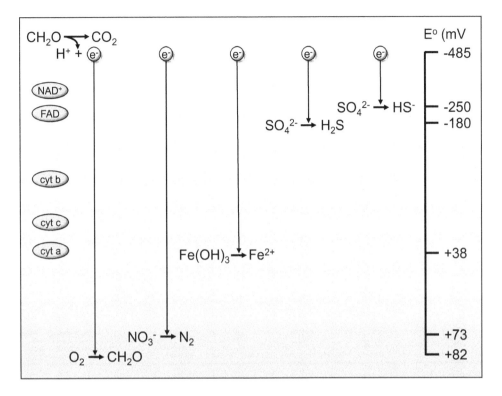

FIGURE 5.24 Reduction Potential (E°) of Redox Half-Reactions Involving Selected Electron Acceptors Relevant in Wetland Sediments, and of Selected Electron Carriers. Reduction Half-Reactions are at the Bottom of Each Downward-Pointing Arrow. Energy Yield for the Oxidation of Carbohydrates Decreases with Successive Electron Acceptors from Left to Right, as the Difference in E° between the Oxidation (Top) and Reduction (Bottom) Half-Reactions Decreases. Part of the Added Energy Comes from the Transfer of Additional Electrons as More Carriers are Added to the Transfer Chain. Cytochromes (Cyt A, B, and C) are Associated with the Mitochondrial Electron Transport Chain Shown in Figure 5.1d. Data from Gottschalk (1986), Reddy and Delaune (2008), Seager, Schrenk, and Bains (2012), Weber, Achenbach, and Coates (2006), and Zubay (1998).

"dropping" the electrons onto oxygen from a carbohydrate like glucose yields a greater "fall" for the electrons (and thus, greater energy yield) than any of the other electron acceptors listed, with relative differences in energy yield for common bacterial electron acceptors given in Figure 5.25.

The strong electronegativity of the oxygen molecule gives it a high capacity for oxidation (or removal of electrons from other compounds; Reddy and DeLaune 2008), which gives it a correspondingly high redox potential (approximately 820 millivolts), relative to other potential electron acceptors. Recall that the strong electronegativity of oxygen is also one of the things that gives the water molecule many of its biologically important characteristics.

5.4.3 Anaerobic Microbial Metabolism

As has been mentioned frequently in this and previous chapters, the decomposition of plant-derived organic matter by microorganisms in wetland sediments leads to the rapid consumption of dissolved oxygen in the water column and in the soil and sediments within wetlands and aquatic habitats. In the absence of adequate supplies of oxygen, the biota shift from aerobic to **anaerobic metabolism**, to the extent possible, either individually or as a community. What I mean by shifting as a community is that, within the

sediments, we find communities of microorganisms (largely bacteria and Archaea) comprising suites of species adapted to use a wide variety of terminal electron acceptors, such as those in Figures 5.24 and 5.25. As time passes after the soil becomes flooded or as depth increases within flooded soil, we tend to find shifts in the terminal electron acceptor that is most abundant and that yields the greatest amount of metabolic energy to soil microorganisms (Figure 5.25 and 5.26). This shift in the abundance of electron acceptors is accompanied by a shift of the dominant microbial species, because of the corresponding abilities of those species to obtain energy under the prevailing sediment conditions.

Once oxygen has been depleted, the commonly encountered electron-accepting molecules progress generally in the order of nitrogen (in the form of nitrate, NO_3^-), manganese (as MnO_2), iron (as $Fe(OH)_3$), and sulfate (SO_4^{2-}; Figure 5.26). As supplies of sulfate are exhausted, the dominant microbial metabolism may switch to various simple carbon-containing molecules as electron acceptors, generating methane in the process (this is called **methanogenesis**). There are many different types of microbes (largely archaeans) that carry out methanogenesis, and this will be covered in more detail in Chapter 7.

Unfortunately, the use of some of these alternate electron acceptors comes with negative consequences for the

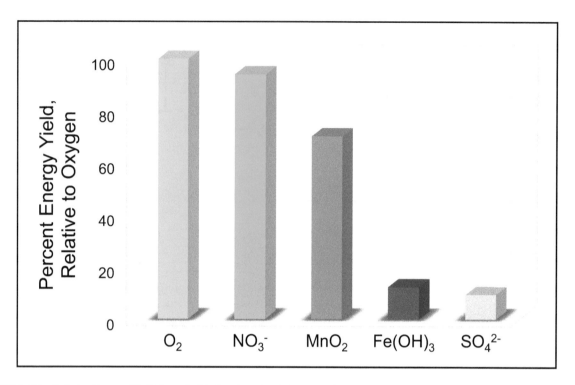

FIGURE 5.25   Relative Energy Yield from Oxidation (I.e., Breakdown) of Carbon-Containing Compounds, Using Selected Electron Acceptors Relevant in Wetland Sediments. Values are the Average Relative Energy Yield from Oxidation of Glucose, Acetate, and Benzoate, Using Selected Electron Acceptors Relevant in Wetland Sediments. Glucose and Acetate are Common Carbon Compounds Used as an Energy Source in Wetland Sediments, and Benzoate is Representative of Numerous Organic Contaminants often Polluting Industrial or other Wastewaters. Data from Reddy and Delaune (2008).

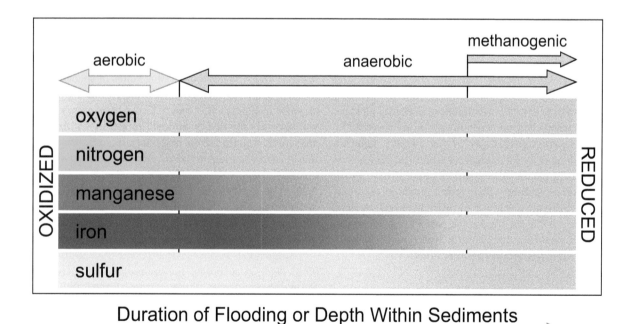

FIGURE 5.26   Time (Or Depth) Sequence in the Chemical Reduction of Common Electron Acceptors in Wetland Sediments. Oxygen, at the Top of the Figure, is First to become Fully Reduced in Flooded Sediments, Followed by the others in Sequence from Top to Bottom. As Sulfur becomes Fully Reduced, Microbial Metabolism may Switch to Processes that Result in the Release of Methane, Termed Methanogenesis. Redrawn from Reddy and Delaune (2008).

TABLE 5.2

Reduced forms of biogeochemically important elements in flooded wetland soil and sediments and their potential effects on wetland plants. Based on information in Cronk and Fennessy (2001), Mitsch and Gosselink (2007), Pezeshki (2001), and Pezeshki and DeLaune (2012).

Element	Reduced Forms	Potential Impact on Plants
Nitrogen	N_2, N_2O, NH_4^+	No direct negative impact at natural concentrations. Conversion to N_2O or N_2 can result in loss of nitrogen from the system.
Manganese (Mn^{4+})	Mn^{2+}	Reduced manganese can interfere with protein structure, including enzymes and membrane transport proteins. Natural concentrations of reduced manganese in wetlands can result in toxicity.
Iron (Fe^{3+})	Fe^{2+}	Reduced iron, like other metals, is more water-soluble than the oxidized form. Concentrations often found in wetlands can impact photosynthesis and respiration. Oxidized iron near the roots also can immobilize phosphorus and can reduce the efficiency of nutrient uptake.
Sulfur	S^{2-}, HS^-, H_2S	Reduced sulfur can react with metals in the wetland forming insoluble metal sulfides, reducing the availability of those minerals for plants. When taken up by plants, reduced sulfur can interfere with proteins involved in photosynthesis and in aerobic and anaerobic energy metabolism. Sulfide toxicity is more common in coastal saltwater and brackish marshes because of the high sulfur content of seawater.
Carbon	CH_4	No direct negative effects of methane on plants. The gas is transported through the plant to the atmosphere.

plants (Table 5.2). Recall from the earlier overview of redox chemistry that the addition of electrons to an atom (such as oxygen when it is serving as a terminal electron acceptor) is a reduction reaction. Thus, the form of any of these electron acceptors after they have received those electrons is their reduced form. With nitrogen, for example, we see multiple potential reduced forms arising from this anaerobic metabolism (Table 5.2). At typical concentrations of these molecules, we see no direct negative impacts on the plants, but reduction to N_2 gas can result in the nitrogen leaving the wetland into the atmosphere (a process called **denitrification**, owing to the effect of removing nitrogen from the ecosystem). However, in the case of manganese, iron, and sulfur, the reduced forms can have direct harmful effects on plants by binding to and interfering with the function of plant proteins (Table 5.2). These reduced elements can (1) impact proteins involved in photosynthesis and respiration (thereby impacting energy metabolism), (2) bind to essential nutrients in the water and make them unavailable to the plants, and (3) bind to the surfaces of roots, decreasing the efficiency with which plants take up nutrients and water (even in aquatic and wetland habitats). The toxicity arising from accumulation of these reduced elements in flooded soil makes oxygen depletion in the sediments a twofold problem for plants. Initially, they may suffer from oxygen deprivation, leading eventually to death, but even if they do not succumb to hypoxia (the low oxygen concentrations), they may be impacted by toxicity from reduced forms of iron, manganese, or sulfur around their roots.

Now that we have seen some of the more critical challenges that hydrophytes face in acquiring basic necessities for survival (light, carbon, and oxygen), we will look at how these plants manage to survive in aquatic and wetland environments. In the next chapter, we will examine some of the more important characteristics that hydrophytes possess that have enabled them to survive in the aquatic environment. We will examine these specifically as they relate to the features of the aquatic environment that we have covered here in Chapter 5: oxygen, carbon, and light.

5.5 IN SUMMARY

Survival of plants in aquatic environments is built around satisfying the needs of basic energy metabolism. Four critical components of the aquatic habitat (light, carbon, oxygen, and water) are all represented in the chemical reaction for photosynthesis:

$$CO_2 + H_2O + Light \rightarrow CH_2O + O_2 \quad (5.1)$$

All of these except for light are also represented in the complementary reaction for aerobic respiration:

$$CH_2O + O_2 \rightarrow CO_2 + H_2O \quad (5.4)$$

In this chapter, we have taken an in-depth look at these four components of the aquatic environment from chemical, physical, and biological perspectives to build a framework within which to investigate key plant adaptations in the following chapter.

The fundamental chemical nature of the water molecule sets the stage for this framework by influencing the behavior (and thus the availability) of the light, carbon, and oxygen in wetlands and aquatic habitats. The strong electronegativity of the oxygen atom gives water its polar nature, which drives the properties of water as a solvent, as a medium for gas exchange, and as a physical environment within which plants and other aquatic organisms live.

The absorptive capacity of water for different wavelengths of solar radiation also strongly influences the suitability of aquatic environments as habitat for hydrophytes, through both heat-retention capacity and ability to transmit light of wavelengths (and thus, energy) suitable for positive net rates of photosynthesis. At the same time, however, patterns of water density afford buoyancy to organic and inorganic sediments, along with phytoplankton and other suspended organisms (including the aquatic plants themselves), all of which have the potential to reduce light availability for submersed photosynthetic plant surfaces.

The chemical properties of water also influence the availability of inorganic carbon for photosynthesis, through what is known as the carbonate buffering system. The interplay among properties of water, pH of surface waters, and chemistry of the underlying geology strongly influence the dominant form of available carbon in surface waters. This, in turn, influences the rates at which plants can capture carbon, thereby locking away solar energy, even when sunlight is abundant.

Finally, all of the aforementioned factors interact to influence the availability of oxygen in aquatic and wetland habitats. Oxygen is a requisite for plants to gain the greatest benefit from the energy locked away through photosynthesis. However, the accumulation of organic matter in flooded sediments results in rapid consumption of oxygen by wetland microorganisms, leading to the use of alternative metabolic pathways within the microbial communities. These alternative pathways use nitrogen, manganese, iron, and sulfur as alternative electron acceptors. The latter three of these can result in the accumulation of toxic reduced metals in the soil or sediments surrounding plant roots, creating a significant selective force for the adaptive features that will be covered in the following chapter.

5.6   FOR REVIEW

1. How are the processes of photosynthesis and respiration interrelated?
2. What are ways in which the aquatic environment provides obstacles to plants' obtaining adequate amounts of light?
3. How do chemical properties of the water molecule influence aquatic life?
4. How is photosynthesis "tuned" to the behavior of light in the atmosphere and in water?
5. How does the carbon buffering system of freshwaters influence availability of carbon for photosynthesis?
6. Why is oxygen important for wetland plants?
7. What are patterns of oxygen distribution in wetlands? What factors influence these?
8. What is the typical order in which alternative electron acceptors are used in wetland microbial metabolism? Why do we see this pattern?

5.7   REFERENCES

Allan, J. David. 1995. *Stream Ecology: Structure and Function of Running Waters*. London: Chapman & Hall.

Belch, Alan C., Max Berkowitz, and J. A. McCammon. 1986. "Solvation Structure of a Sodium Chloride Ion Pair in Water." *Journal of the American Chemical Society* 108: 1755–61.

Buscemi, Philip A. 1958. "Littoral Oxygen Depletion Produced by a Cover of *Elodea canadensis*." *OIKOS* 9: 239–45.

Carlton, Richard G., and Robert G. Wetzel. 1988. "Phosphorus Flux from Lake Sediments: Effect of Epipelic Algal Oxygen Production." *Limnology and Oceanography* 33 (4): 562–70. https://doi.org/10.4319/lo.1988.33.4.0562.

Comar, C. L., and F. P. Zscheile. 1942. "Analysis of Plant Extracts for Chlorophylls a and b by a Photoelectric Spectrophotometric Method." *Plant Physiology* 17: 198–209.

Cronk, Julie K., and M. S. Fennessy. 2001. *Wetland Plants: Biology and Ecology*. Boca Raton, FL: CRC Press.

Falkowski, Paul G., and John A. Raven. 1997. *Aquatic Photosynthesis*. Malden, MA: Blackwell Science.

Gates, David M. 1963. "The Energy Environment in Which We Live." *American Scientist* 51 (3): 327–48.

Golterman, H. L. 1969. *Methods for Chemical Analysis of Fresh Waters*. Edinburgh: Blackwell Scientific Publications.

Gottschalk, Gerhard. 1986. *Bacterial Metabolism*, 2nd ed. New York: Springer-Verlag.

Hutchinson, G. Evelyn. 1975. *A Treatise on Limnology: Volume III, Limnological Botany*. New York: John Wiley & Sons, Inc.

Johnson, Daniel M., and William K. Smith. 2006. "Low Clouds and Cloud Immersion Enhance Photosynthesis in Understory Species of a Southern Appalachian Spruce-Fir Forest (USA)." *American Journal of Botany* 93 (11): 1625–32. https://doi.org/10.3732/ajb.93.11.1625.

Kohlrausch, Friedrich. 1996. *Praktische Physik*. Stuttgart: B. G. Teubner.

Lenoble, Jacqueline and M. Bernard Saint-Gully. 1955. "Sur l'absorption du rayonnement ultraviolet par l'eau distillée." *Comptes Rendus Hebdomadaires des Séances de l'Académie des Sciences* 240: 954–55.

Maberly, S. C. 1996. "Diel, Episodic and Seasonal Changes in pH and Concentrations of Inorganic Carbon in a Productive Lake." *Freshwater Biology* 35 (3): 579–98.

Millero, Frank J. 2013. *Chemical Oceanography*. 4th ed. Boca Raton, FL: CRC Press.

Mitsch, William J., and James G. Gosselink. 2007. *Wetlands*. 4th ed. New York: John Wiley & Sons, Inc.

Morel, A. 1974. "Optical Properties of Pure Water and Pure Sea Water." In *Optical Aspects of Oceanography Symposium*. London: Academic Press.

Pezeshki, S. R. 2001. "Wetland Plant Responses to Soil Flooding." *Environmental and Experimental Botany* 46: 299–312.

Pezeshki, S. R., and R. D. Delaune. 2012. "Soil Oxidation-reduction in Wetlands and Its Impact on Plant Functioning." *Biology* 1: 196–221.

Reddy, K. Ramesh and Ronald D. DeLaune. 2008. *Biogeochemistry of Wetlands: Science and Applications*. Boca Raton, FL: CRC Press.

Seager, Sara, Matthew Schrenk, and William Bains. 2012. "An Astrophysical View of Earth-based Metabolic Biosignature Gases." *Astrobiology* 12: 61–82.

Taiz, Lincoln and Eduardo Zeiger. 2006. *Plant Physiology*. 4th ed. Sunderland, MA: Sinauer Associates, Inc.

Turner, Andrew M., Emily J. Cholak, and Maya Groner. 2010. "Expanding American Lotus and Dissolved Oxygen Concentrations of a Shallow Lake." *The American Midland Naturalist* 164 (1): 1–8. https://doi.org/10.1674/0003-0031-164.1.1.

Vähätalo, Anssi V., and Robert G. Wetzel. 2004. "Photochemical and Microbial Decomposition of Chromophoric Dissolved Organic Matter during Long (Months-Years) Exposures." *Marine Chemistry* 89 (1–4): 313–26. https://doi.org/10.1016/j.marchem.2004.03.010.

Vähätalo, Anssi V., Robert G. Wetzel, and Hans W. Paerl. 2005. "Light Absorption by Phytoplankton and Chromophoric Dissolved Organic Matter in the Drainage Basin and Estuary of the Neuse River, North Carolina (U.S.A.)." *Freshwater Biology* 50 (3): 477–93. https://doi.org/10.1111/j.1365-2427.2004.01335.x.

Weber, Karrie A., Laurie A. Achenbach, and John D. Coates. 2006. "Microorganisms Pumping Iron: Anaerobic Microbial Iron Oxidation and Reduction." *Nature Reviews Microbiology* 4: 752–64.

Wetzel, Robert G. 2001. *Limnology: Lake and River Ecosystems*. San Diego: Academic Press.

Zubay, Geoffrey. 1998. *Biochemistry*. 4th ed. Dubuque, IA: Wm. C. Brown Publishers.

6 Adaptations for Life in the Aquatic Environment

6.1 CRITICAL LIMITATIONS FOR LIFE IN THE WATER

We saw in the previous chapter that plants in aquatic and wetland habitats face many challenges to acquiring the basic necessities for survival. Light, carbon dioxide, and oxygen can be in very limited supply, especially for submersed species that find all of their resource-acquiring tissues surrounded by water and potentially coated with inorganic deposits and communities of microbes that are competing for the same resources. Additionally, oxygen deficiency can make uptake of water itself difficult, even when the plant finds itself bathed in water, and hypoxia can also result in the accumulation of toxins in the root zone (Pezeshki 2001). Nevertheless, we find that certain species of plants do exceptionally well in aquatic and wetland habitats, owing to adaptations that we will discuss in the present chapter.

6.2 ANAEROBIC/HYPOXIC PHYSIOLOGY

The saturation concentration for oxygen in water at temperatures of 8 °C to 10 °C averages approximately 9 mg L^{-1}, or 0.0009%, in contrast to a concentration of about 21% in the atmosphere (Figure 5.16; Wetzel 2001). Furthermore, the rate at which oxygen diffuses into the water is "several thousand times" slower than for diffusion of oxygen in the air (Sculthorpe 1967, but many sources cite a value of 10,000 times slower). Thus, even for plants that find themselves temporarily underwater because of storm-induced flooding, water presents a significant barrier to obtaining oxygen needed for **aerobic metabolism**. Much like humans, even short periods under water without being properly equipped to access oxygen can be detrimental, or even fatal, for plants.

6.2.1 "ENERGY CRISIS" METABOLISM

When oxygen levels become inadequate for aerobic respiration to continue in plants, one or more anaerobic metabolic processes may be initiated (Figure 6.1). These processes allow the plant to do three physiologically important things: (1) continue to extract energy from stored carbohydrates (and/or lipids), albeit at a lower level and for a limited time, (2) continue to dissipate electrons that become energized via light absorption by chlorophylls, and (3) scavenge/ deactivate potentially harmful oxygen molecules that can no longer be effectively used as electron acceptors. I will briefly discuss details of the second and third processes, before spending a bit more time on the first.

When light energy remains available under conditions of suboptimal oxygen supply (termed hypoxia for low oxygen or **anoxia** for the absence of oxygen), light-absorbing pigments continue to absorb light energy, which energizes electrons and splits water, releasing an excess of oxygen. Under these conditions, there are numerous proteins and protein complexes that will bind to oxygen to prevent it from oxidizing membrane components or other proteins that are necessary for cellular metabolism (Czarnocka and Karpiński 2018; Greenway and Armstrong 2018; Armstrong et al. 2019). When these processes are insufficient to lock away all of the excess oxygen, electrons, and other biochemically oxidizing atoms that begin to accumulate in the cell (collectively termed **reactive oxygen species**), the plant will begin to experience cell and tissue damage. Paradoxically, then, inadequate access to external oxygen supplies can result in the accumulation of harmful and unusable oxygen within plant cells.

Regarding the earlier first process, extraction of energy under hypoxic conditions, there are two primary mechanisms by which this is known to occur (Figure 6.1B and C), and these may begin within minutes of the plant experiencing hypoxic conditions (Armstrong et al. 2019). In general, the inability of the plant to use oxygen as the terminal electron acceptor in the electron transport chain within the mitochondrion (aerobic respiration, Figure 5.1) results in shutdown of the TCA cycle (Taiz et al. 2015; Zubay 1998), which leads rapidly to an energy deficit within the plant and a suite of anaerobic physiological processes that Greenway and Armstrong (2018) refer to as "**energy crisis metabolism**." One of the early physiological changes that occurs when cells switch to energy crisis metabolism is the production of lactic acid via **fermentation** (Figure 6.1B), which allows the plant to continue to extract limited energy from carbohydrates in the absence of oxygen, at approximately 6% the rate provided via aerobic respiration. Production of lactic acid leads to a decrease in cellular pH, and this shift in pH away from the optimal pH for the lactic acid–producing enzyme (lactate dehydrogenase; LDH) leads to production of ethanol via alcohol dehydrogenase (ADH; Figure 6.1C). Ethanol fermentation yields the same amount of energy as lactic acid fermentation, but both yield substantially less energy than aerobic respiration.

The reduced amount of energy available through these fermentative processes leads to a reduction in all energy-requiring processes within the plant cells. One of these processes is the pumping of excess hydrogen ions (that is, protons) into the cells' **vacuoles** (these are the large, central organelles that make up most of the interior volume

DOI: 10.1201/9781315156835-6

A. Glycolysis

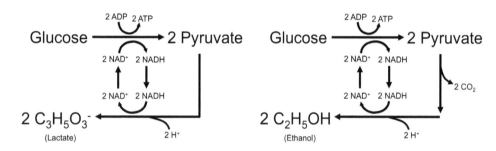

B. Lactic acid fermentation

C. Ethanol fermentation

Zubay and
Brooker et al.

D. Phytoglobin-nitric oxide cycle

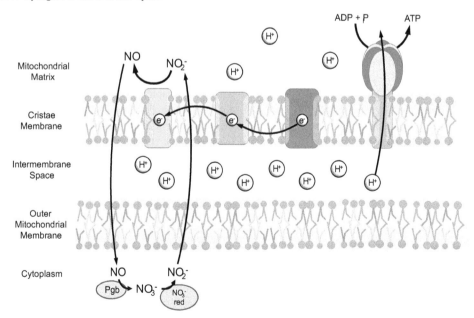

FIGURE 6.1   Glycolysis and Alternative Anaerobic Metabolic Pathways in Plants Experiencing Hypoxia. Portions Redrawn from Brooker et al. (2017), Igamberdiev and Hill (2009), and Zubay (1998). The Products from Glycolysis (A) will Find their Way into the TCA Cycle as Needed under Aerobic Conditions. Note that the Reaction that Transitions into the Gray Shaded Area in (A) Results in Two of the Three-Carbon Molecules; thus, everything in the Shaded Area should be Doubled to Obtain the Net Yields Indicated. Within a few Minutes of Hypoxia, However, the Pyruvate Generated Through Glycolysis will begin to be Consumed Via Fermentative Processes, Generating First Lactic Acid (B), then Ethanol (C), Once Cellular Ph Drops Sufficiently to Alter Relative Enzyme Activities Involved in those Processes. Another Route by Which Plants may Generate Atp under Hypoxic Conditions, Albeit in Limited Quantities, is by Using Nitrite as a Terminal Electron Acceptor and then Recycling Nitrite Via the Phytoglobin-Nitric Oxide Cycle (D). Phytoglobin (Pgb) and Nitrate Reductase (NO_3^- Red) are Cytoplasmic Proteins Mediating the Recycling of Nitrite, Which Returns to the Mitochondrial Matrix to again Serve as a Terminal Electron Acceptor.

of plant cells). With a reduced ability to move protons into the vacuoles, where they are isolated from the rest of the cell's metabolic processes, the cationic content of the cytoplasm increases and the pH of the cytoplasm drops (i.e., the acidity increases), leading to disruption of **ionic gradients** within the cells, disruption of cellular metabolism, and damage to cells and plant tissues (Czarnocka and Karpiński 2018; Taiz and Zeiger 2002). In plant roots, which are the organs most likely to experience hypoxia in wetland plants, these processes lead to reduced **permeability** of the roots to water and to overall water deficit in the plant, which usually results in the plant closing its stomata, in an effort to conserve water (even though the plant is sitting with its roots in saturated soil). This closing of the stomata then results in further reduced rates of water uptake through the roots, reduced diffusion of oxygen into the plant through the leaves, and accumulation of CO_2 and other gases within the plant. This situation is similar to the oxidative stress described earlier, where physiological responses to hypoxia resulted in the accumulation of unusable oxygen inside the plant, except here, we see abundant water *outside* the plant that the plant is incapable of accessing for normal metabolic uses.

Another, relatively recently recognized physiological process that takes place in plant tissues experiencing hypoxia-induced energy deficits is mediated by plant proteins called **phytoglobins** (short for "phyto-hemoglobins;" Armstrong et al. 2019; Igamberdiev and Hill 2009). One function of phytoglobins is the regeneration of nitrite (NO_2^-) in the cytoplasm, as part of a *phytoglobin-nitric oxide cycle*. The regenerated nitrite is transported into the mitochondrial matrix where it can serve as a terminal electron acceptor when oxygen levels are low (Figure 6.1D; Armstrong et al. 2019; Igamberdiev and Hill 2009). Recall from Figures 5.23 and 5.24 that the reduction potentials and energy yields for oxygen and nitrogen are quite similar, with use of nitrogen as a terminal electron acceptor yielding approximately 94% as much energy as with oxygen. Armstrong et al. (2019) indicated that although this process can result in the generation of some ATP, as in aerobic metabolism (compare Figures 5.1D and 6.1D), this is likely

only significant in a narrow band of root tissue between the innermost vascular tissues and the outer root cortex. Thus, it has only limited capacity to extend the metabolism of plants experiencing flood-induced hypoxia. The phytoglobin-nitric oxide cycle also serves important roles in scavenging nitric oxide from within the mitochondrion (where it would tend to displace oxygen from the proteins in the electron transport chain) or scavenging excess oxygen, as previously discussed (Armstrong et al. 2019).

6.2.2 Ethylene Production and Signaling

The metabolic processes in the previous section can be observed within a few minutes of exposure to hypoxic conditions, and they lead to corresponding responses in altered cellular metabolism and hormone signaling among plant parts. Many of the short-term responses to hypoxia (i.e., those occurring within a few hours) are also observed in plants experiencing drought conditions, or water deficits. These include closure of the stomata, reduced leaf expansion (in new leaves), **senescence** (i.e., systematic, controlled death) of older leaves, and downward curvature of leaf petioles, referred to as **epinasty** (Vartapetian and Jackson 1997). All of these responses can impact transpiration rates, leading to a reduced loss of water through leaves, but they also have the result of reducing movement of oxygen and carbon dioxide between internal leaf air spaces and the atmosphere. This latter effect is critical in some of the key plant adaptations that we will see later in this chapter.

Another metabolic response to flooding is the increased production of the enzyme ACC synthase and its product, 1-aminocyclopropane-1-carboxylic acid (ACC), which is the precursor compound for the plant hormone *ethylene* (Figure 6.2). Increased production of ACC synthase can continue for up to 48 hours after onset of hypoxia in flooded root systems (Vartapetian and Jackson 1997), accompanied by continued production of ACC in the flooded roots. This ACC then may be transported through the xylem tissues into the shoots of the plant where, in the presence of sufficient oxygen concentrations (e.g., from photosynthesis or transport into the leaves via stomata), it will be converted

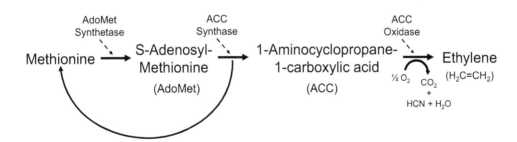

FIGURE 6.2 Simplified Biosynthesis Pathway for Ethylene, from the Sulfur-Containing Amino Acid, Methionine. Based on Taiz and Zeiger (2002) and Wang, Li, and Ecker (2002). The Three Steps Shown here are Catalyzed by the Enzymes Adomet Synthetase, Acc Synthase (Which is Produced in Response to Flooding), and Acc Oxidase. Two of These, Acc Synthase and Acc Oxidase, are Discussed in the Text, Regarding their Role in Ethylene Signaling and Plant Responses to Flooding and Hypoxia. Note That Acc Oxidase Requires Oxygen as a Substrate for the Production of Ethylene.

TABLE 6.1

Important classes of plant hormones and other signal molecules involved in plant responses to the environment. Information taken from Graham, Graham, and Wilcox (2006) and Taiz and Zeiger (2002).

Signal Type	Precursors	Effects on Plant
Abscisic acid	Carotenoid pigments	Enhances salinity tolerance; Promotes seed development; Promotes dormancy; Regulates stomata opening
Auxins	Tryptophan (an amino acid)	Maintains apical dominance and growth; Regulates phototropism; Promotes root development; Stimulates ethylene synthesis
Cytokinins	Adenine (a precursor of adenosine, in ATP)	Promotes cell division; Regulates vascular tissue development; Delays aging
Ethylene	1-aminocyclopropane-1-carboxylic acid (ACC)	Induces aerenchyma formation; Promotes fruit ripening; Induces leaf and flower aging; Mediates responses to pathogen attack
Gibberellins	Terpenoids	Promotes cell division and cell enlargement; Stimulates seed germination; Induces flowering
Jasmonic acid	Fatty acids	Mediates resistance to multiple environmental stresses, including herbivory
Salicylic acid	Phenolic acids	Mediates responses to pathogen attack
Systemin	Amino acids	Serves as a signal of physical wounding

to ethylene (Taiz and Zeiger 2002; Vartapetian and Jackson 1997). This conversion to ethylene cannot take place in hypoxic roots because it is an oxygen-consuming reaction (Figure 6.2); thus, severely hypoxic or anoxic tissues are incapable of producing ethylene.

Ethylene, which may be produced constitutively in some plant species or upon exposure to various stresses in other species, is involved in numerous physiological responses of plants to stress, often in conjunction with other plant hormones, such as auxins, gibberellins, or abscisic acid (Table 6.1). Especially relevant for aquatic and wetland plants is the involvement of ethylene in stem elongation, formation of adventitious roots, and the formation of air spaces or channels, referred to as aerenchyma, within plants. The specific mechanisms by which ethylene signaling takes place remain under investigation. Some of these involve the phytoglobin-nitric oxide cycle discussed in the previous section (Figure 6.1D) and may be mediated via components of that pathway interacting with hormones that influence the production or accumulation of ethylene within plant tissues (Manac'h-Little, Igamberdiev, and Hill 2005; Kapoor et al. 2018). Some of these interactions lead to the accumulation of reactive oxygen species (mentioned in an earlier section) that can initiate a process known as programmed cell death, which is involved in aerenchyma formation in some plants (Seago et al. 2005). Some of these processes are discussed further in the relevant sections later.

6.3 AERENCHYMA

It was emphasized earlier that energy crisis metabolism is only a temporary solution to conditions of hypoxia or anoxia. The **catabolic** processes that allow plants to break down and extract energy from carbohydrates under oxygen-deficient conditions yield substantially less energy than do

aerobic processes and, thus, they cannot sustain the plant indefinitely. As a result, aquatic and wetland plants must have a means of maintaining sufficient internal oxygen concentrations to allow the persistence (or recovery) of aerobic metabolism under conditions of saturated soils or standing water. One of the key anatomical features that allows this is the presence of internal tissues that form air passageways throughout the plant, allowing diffusion of oxygen from areas of higher concentration into areas of oxygen demand or deficiency (Figure 6.3). These tissues are called **aerenchyma**.

Formation of aerenchyma has been best studied in root tissues, and published descriptions of aerenchyma tissue date back at least to the mid-1800s, when Schleiden (1849) described "air canals" and "air cavities" in various species, including *Nymphaea* and species from the Poaceae, Asteraceae, and Apiaceae (the "carrot" or "parsley" family). Shortly thereafter, DeBary (1877) devoted roughly half of a chapter about "intercellular spaces" to a discussion of aerenchyma, described as "air-containing chambers and canals." In his treatment of these tissues, DeBary classified aerenchyma into two broad categories, based on the mechanism of its formation. The first of these was **schizogeny**, a process wherein cells split apart from one another to form openings that later fill with gases, and the second is **lysigeny**, where cells die, and their collapse leads to the formation of voids within the tissue (Figure 6.4). Recently, Seago et al. (2005) described a third mechanism of aerenchyma formation, termed **expansigeny**, wherein division and growth of cells surrounding a pre-existing intercellular space result in expansion of that space, leading to formation of aerenchyma tissue (Figure 6.4).

Recent investigations of aerenchyma by Jung, Lee, and Choi (2008) and Takahashi et al. (2014) mentioned the work by Seago and colleagues (2005) but chose to restrict

FIGURE 6.3 Petioles of the Free-Floating Aquatic Plant *Eichhornia Crassipes* Contain Extensive Aerenchyma Tissue that Provides Air Spaces that Aid in Floatation but also Allow Movement and Storage of Gases within the Plant. In the Upper Photo, the Cross-Section on the Left was Taken near the Base of the Petiole, while the One on the Right was Taken Mid-Petiole. Note also the Horizontal Diaphragms that Dissect the Hollow Central Lacuna. These Diaphragms are Typical of Longer Aerenchyma Channels in Multiple Species and can be Seen in Some of the Photos of Shoot Tissues in Figure 6.7.

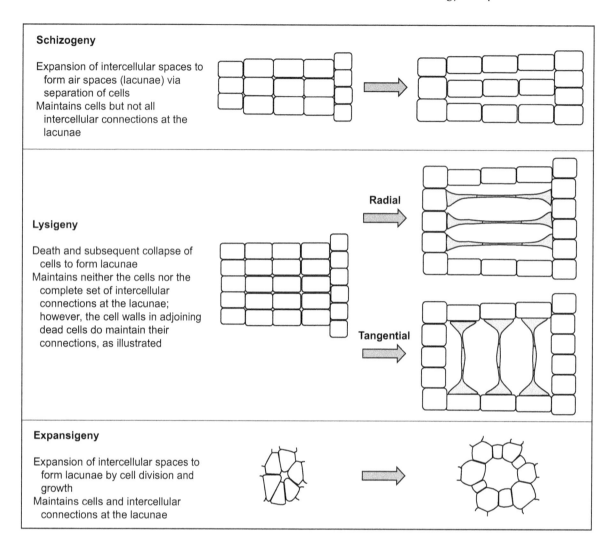

Schizogeny

Expansion of intercellular spaces to form air spaces (lacunae) via separation of cells
Maintains cells but not all intercellular connections at the lacunae

Lysigeny

Death and subsequent collapse of cells to form lacunae
Maintains neither the cells nor the complete set of intercellular connections at the lacunae; however, the cell walls in adjoining dead cells do maintain their connections, as illustrated

Radial

Tangential

Expansigeny

Expansion of intercellular spaces to form lacunae by cell division and growth
Maintains cells and intercellular connections at the lacunae

FIGURE 6.4   Generalized Aerenchyma-Forming Processes. Adapted from Illustrations in Jung, Lee, and Choi (2008) and Seago et al. (2005), with Additional Input from Dr. James Seago. Both Published Sources Provide Considerable Detail on these General Categories of Aerenchyma Formation, as well as Combinations and Specific Subsets of Each.

their treatment of aerenchyma formation to the processes of schizogeny and lysigeny. An important consideration in reconciling these three aerenchyma-producing processes (schizogeny, lysigeny, and expansigeny) is where and when the process takes place. In my prior description, I mentioned that expansigeny involves the expansion of pre-existing intercellular spaces. As noted by Seago et al. (2005; and personal communication with Dr. James Seago), those spaces themselves arise through schizogeny in or near the root apical meristem (RAM). The RAM is the region of the root tip just inside the root cap, where mitotic cell divisions give rise to new cells of the growing root tip (Figure 6.5). The expansion of air spaces that characterizes expansigenous aerenchyma then takes place once tissues grow beyond the RAM, into the region of cell elongation, differentiation, and maturation (Figure 6.5).

Because all three forms of aerenchyma initiate through schizogeny in (or near) the RAM, we need to think about

distinguishing among the processes occurring *outside* the RAM that lead to formation of air passageways that characterize mature aerenchyma tissues. At that point, we can consider two key attributes of the resulting aerenchyma to distinguish among the processes. The first is whether the cells associated with the *lacunae* (the air spaces) retain their original connections once the air spaces are fully formed, and the second is whether the lacunae are formed via the death of cells (Figure 6.4).

Although we have been aware of the presence and general function of aerenchyma since the 1840s, we still have a relatively poor understanding of the biological and genetic mechanisms involved in the formation of these tissues. Lysigenous aerenchyma is the best understood of these processes, with some of the molecular signals having been discovered in recent years (Takahashi et al. 2014). It is clear that ethylene accumulation in plant tissues (often resulting from submergence, as discussed previously) plays a prominent

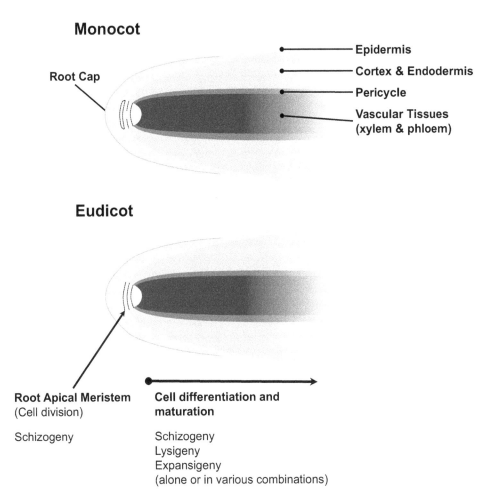

Monocot

Root Cap

Epidermis

Cortex & Endodermis

Pericycle

Vascular Tissues
(xylem & phloem)

Eudicot

Root Apical Meristem
(Cell division)

Schizogeny

**Cell differentiation and
maturation**

Schizogeny
Lysigeny
Expansigeny
(alone or in various combinations)

FIGURE 6.5   Generalized Anatomy of a Growing Root Tip in Monocots and Eudicots. The Root Apical Meristem (Ram) Region (The White Zone in Each Diagram) Can Be Divided into Three or Four Series of Meristematic Cells. In Monocots (Top): (1) the Apical-Most Tier are the Root Cap Initials and Give Rise to the Apical and Lateral Tissues of the Root Cap, With the Lateral Root Cap Tissues Arising from Meristematic Cells Around the Edges of This Region; (2 and 3) the Epidermal and Cortical Initials Give Rise to the Epidermis and Cortex (+ Endodermal) Tissues (The Endodermal Tissues are Located Just Outside the Pericycle), and (4) the Procambium, or Stellar Initials, from Which the Pericycle and Vascular Tissues are Derived. In Eudicots, the Epidermal Initials are Part of the Apical Layer of Cells, Collectively Giving Rise to the Apical Root Cap, Later Root Cap, and Epidermal Tissues. Surrounding the Basalmost Portions of the Ram is a Small Zone of Slowly Dividing Cells (~10× More Slowly than Meristematic Cells), Termed the Quiescent Center, that Regulates Meristematic Activity in the Ram Region. Based on Illustrations and Information in Della Rovere et al. (2016), Seago and Marsh (1989), and Scheres, Mckhann, and Van Den Berg (1996), with Considerable Additional Direct Input from Dr. James Seago.

role in inducing lysigeny in plants exhibiting this form of aerenchyma. In addition to the review by Takahashi et al. (2014), Evans (2003) and Drew, He, and Morgan (2000) also reviewed some of the molecular pathways involved in lysigenous aerenchyma formation, focusing specifically on the process of **programmed cell death** in producing the lacunae in lysigenous aerenchyma. Programmed cell death is a process whereby some environmentally induced cell signaling pathway becomes activated, resulting in the death of a region of cells within plant tissues. Evans (2003) indicated that ACC synthase (involved in ethylene production; Figure 6.2) is produced within a few hours of exposure to hypoxia and that, in some plant species, production of cellular receptors for ethylene increases within a half

hour of submergence. Those receptors are later involved in transmitting signals associated with ethylene-induced cell death (Takahashi et al. 2014). Evans (2003) and Takahashi et al. (2014) also indicated a role for nitric oxide (NO) and reactive oxygen species (ROS) in programmed cell death in some plant species (the latter is well known in the field of plant response to pathogens). Recall the previous discussion on the potential for both ROS and NO concentrations to increase in plant tissues during periods of oxygen deficiency.

While lysigeny has received considerable interest, we know much less about the molecular and genetic mechanisms bringing about schizogeny, and even less about expansigeny, as it was only introduced to the literature in

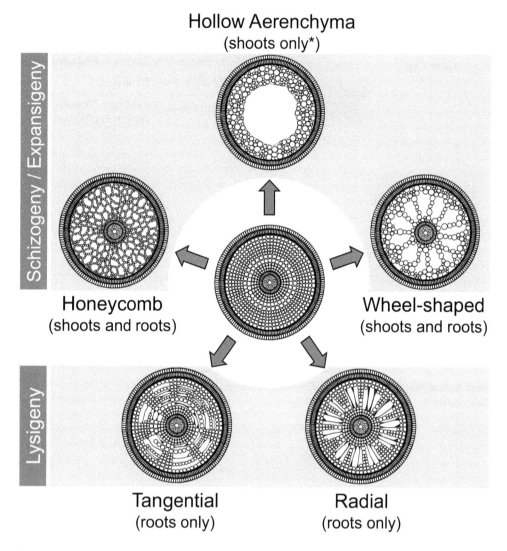

Hollow Aerenchyma
(shoots only*)

Schizogeny / Expansigeny

Lysigeny

Honeycomb
(shoots and roots)

Wheel-shaped
(shoots and roots)

Tangential
(roots only)

Radial
(roots only)

FIGURE 6.6   Some of the most Common Patterns of Aerenchyma Formation Among Aquatic and Wetland Plants, Based on Data, Drawings, and Micrographs in Jung, Lee, and Choi (2008), Justin and Armstrong (1987), and Seago et al. (2005). * Among the Angiosperms Covered by Jung, Lee, and Choi (2008), Hollow Aerenchyma only Appeared in Shoots. However, they Indicate that Hollow Aerenchyma also Can Be Found in Roots of *Isoetes* Species (Quillworts), Among the Lycophyta; Shoots of those Species Exhibit the Wheel-Shaped Aerenchyma.

2005. Whereas lysigeny appears to be an **inducible** process (i.e., it may be initiated in response to some environmental signal), schizogeny appears to be a **constitutive** process, occurring as a pre-programmed aspect of development, although it may be more prevalent under conditions of hypoxia (Evans 2003). This constitutive nature seems to be especially true in species exhibiting expansigeny, where the differential growth and expansion of air passages occur following an initial schizogeny within the RAM (Figure 6.5; Seago et al. 2005).

With the knowledge that there is an underlying genetic component to all forms of aerenchyma production, we might expect to find a signal of this in the systematic relationships among plant taxa. Jung, Lee, and Choi

(2008) investigated this, using 110 species of aquatic and wetland plants (representing 65 genera, across 42 families and 21 orders). These species included lycophytes (*Isoetes* species) and monilophytes (in the genera *Equisetum* and *Marsilea*), but the majority were angiosperms (105 species). There were indeed some interesting patterns, summarized in Figure 6.6 and Table 6.2. Among these were:

- Aerenchyma in shoot tissues (stems and leaves) seemed to arise predominantly through schizogeny and expansigeny;
- Except in the Poales, shoot aerenchyma typically exhibited the honeycomb structural pattern;

- In the Poales, shoot aerenchyma exhibited either the hollow or leafy pattern (but note that hollow aerenchyma appears in cylindrical shoots, which might resemble leafy aerenchyma if compressed; see Figure 6.7 for a comparison); and
- Development of root aerenchyma demonstrated no consistent relationship with taxonomy, exhibited evidence of both schizogeny and lysigeny, and appeared to be quite diverse, even within orders (e.g., species in the Alismatales display at least four different patterns of root aerenchyma including combinations of schizogeny, expansigeny, and lysigeny).

I should note here that even though the species examined by Jung, Lee, and Choi (2008) spanned a breadth of angiosperms, the 105 species examined represent only about 2% of the aquatic and wetland angiosperm genera mentioned in Table 2.3. As a result, although the prior list summarizes trends among the findings of Jung, Lee, and Choi (2008), with the incorporation of expansigeny as a form of

aerenchyma development, there is plenty of room for new findings or amendments to this summary. A very recent publication by Schweingruber (2020) includes detailed micrographs of 400 species of macrophytes and provides an opportunity for further examination of these patterns.

Before moving on to other adaptations, it is important to look briefly at the function of aerenchyma. As discussed earlier, these tissues are thought to provide aeration for oxygen-deprived tissues, especially roots, during periods of saturated soil or sediments (which is a constant stress for the floating-leafed, submersed, and free-floating species). There also is evidence that production of aerenchyma, even constitutively produced types, is enhanced under saturated soil conditions. Justin and Armstrong (1987) conducted an experiment on root responses to flooding, using 91 species of plants representing non-wetland species (29 species), wetland species (42), and species with intermediate habitat affinities (20). Porosity of roots in the wetland species increased substantially in response to flooding (water 10–20 cm above soil surface), whereas the strongest response measured for non-wetland species was mortality

TABLE 6.2

Aerenchyma patterns among selected major plant taxa examined by Jung, Lee, and Choi (2008). Numbers given represent the numbers of species in which the given aerenchyma trait appeared (numerator), out of the number of species for which data or observations were available (denominator). "Leafy aerenchyma" refers to leaves or photosynthetic culms wherein the spongy interior mesophyll tissues of the leaves are permeated with large lacunar air canals, as in *Typha*, or various species of grasses (Poaceae) and sedges (Cyperaceae).

Group	Order	Families Included	Roots	Shoots
Basal angiosperms	Nymphaeales	Cabombaceae, Nymphaeaceae	Honeycomb 5/5 species	Honeycomb 4/5 species
Monocots	Alismatales	Alismataceae, Hydrocharitaceae, Juncaginaceae, Potamogetonaceae, Ruppiaceae	Radial Lysigeny 10/19	Honeycomb 27/31
	Poales	Sparganiaceae, Typhaceae	Radial Lysigeny 6/6	Leafy Aerenchyma 5/6
		Juncaceae	Radial Lysigeny 2/2	Hollow Aerenchyma 2/2
		Cyperaceae	Tangential Lysigeny 14/14	Leafy Aerenchyma 9/14
		Poaceae	Radial Lysigeny 6/6	Hollow + Leafy Aerenchyma 2/2
Eudicots	Ranunculales	Ranunculaceae	Radial Lysigeny 2/4	Honeycomb or Hollow Aerenchyma 3/4
	Myrtales	Lythraceae, Onagraceae	Honeycomb 2/2	Honeycomb 5/5
	Caryophyllales	Polygonaceae	Honeycomb 3/3	Honeycomb 2/3
	Asterales	Asteraceae, Campanulaceae, Menyanthaceae	Radial Lysigeny 3/4	Honeycomb 5/6
	Lamiales	Lamiaceae, Lentibulariaceae, Pedaliaceae, Phrymaceae, Scrophulariaceae	Radial Lysigeny 8/8	Honeycomb 7/9

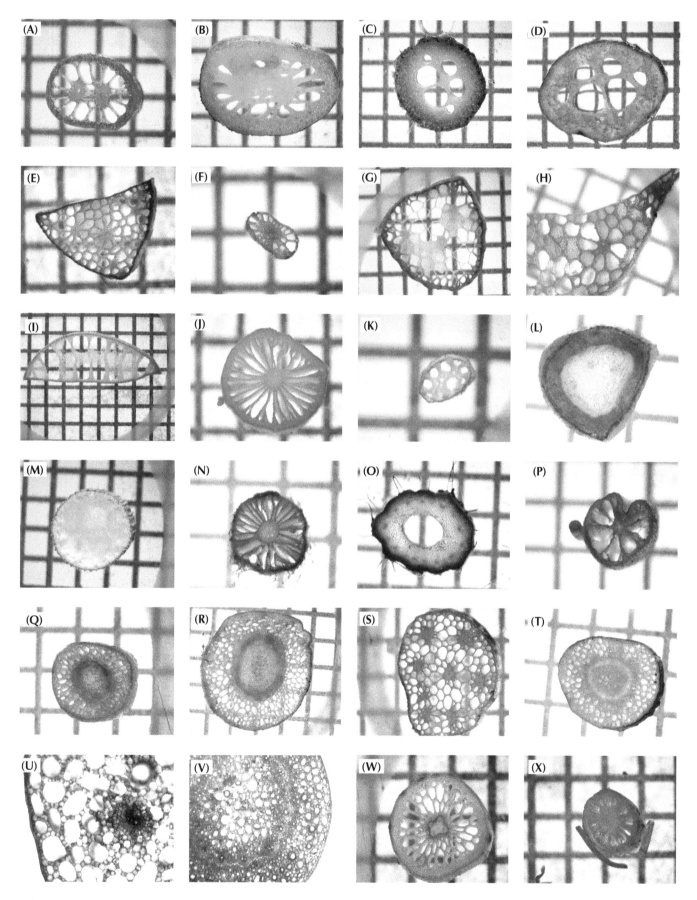

FIGURE 6.7 Examples of Aerenchyma from Species across the Angiosperms. (A–D) Nymphaeales; (E–H) Alismatales; (I–N) Poales; (O–P) Ranunculales; (Q–R) Myrtales; (S–V) Apiales; and (Y–Z) Asterales. Aerenchyma Patterns Follow Closely those Expected Based on Information Compiled for Table 6.2. Grid Squares all Equal 1 Mm². *Hydrocotyle Ranunculoides* Images Provided Courtesy of Dr. James Seago, Stained with Toluidine Blue O; Petiole (U) Photographed at 10×, Stolon (V) at 5×.

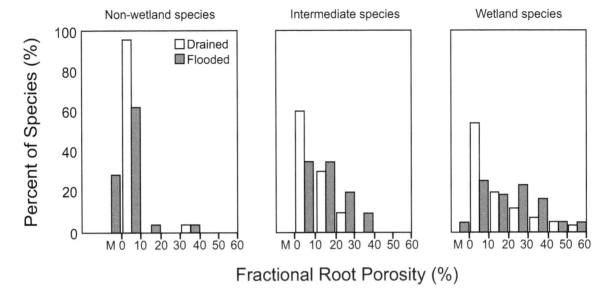

FIGURE 6.8 Root Porosity is More Prone to Shift in Wetland-Adapted Plant Species (Far Right) and those Species Capable of Tolerating Wet or Dry Soil Conditions, than Among Species Typically Restricted to Drier Soils (Left). The Non-Wetland-Adapted Species were much more Likely to Experience Mortality (M Category on the Horizontal Axis) than the Intermediate or Wetland-Adapted Species. Data from Justin and Armstrong (1987).

of almost 30% of those species (Figure 6.8). These increases in root porosity for the wetland species were accompanied by a higher mean soil redox potential for the wetland than non-wetland species. Approximately 35% of wetland species maintained soil redox above −100 mV, and 10%–15% maintained redox above +100 mV, in contrast with non-wetland species, fewer than 5% of which maintained soil redox above −100 mV (Justin and Armstrong 1987).

The higher soil redox values maintained by wetland plants, with their greater root porosity, likely result from oxygen leakage from the roots into the surrounding soil (Vartapetian and Jackson 1997), which we will talk about further in section 6.5. That oxygen, however, is transported to the roots either from photosynthetic tissues or from the atmosphere itself, via the networks of aerenchyma throughout the plant (Cronk and Fennessy 2001; Vartapetian and Jackson 1997). Visser et al. (1997) conducted a series of experiments in which they measured diffusion of oxygen and of ethylene through adventitious roots of two *Rumex* species (Polygonaceae). They found that aerenchyma networks in these plants allowed diffusion of oxygen into and ethylene out of roots grown in oxygen-deprived conditions, avoiding growth inhibition that ordinarily would be caused by oxygen deficiency and ethylene accumulation.

In *Typha latifolia*, Constable, Grace, and Longstreth (1992) found that leaf aerenchyma occupied more than 50% of the total leaf volume (vs. only 6% in the rhizomes, or belowground stems). During the warmest month of summer, those air spaces held more than 6,300 μL CO_2 per L of air space at 06.00h (6:00AM), whereas the atmospheric CO_2 concentration at that time was just over 600 μL L^{-1}.

In other words, the overnight accumulation of CO_2 in the leaf aerenchyma amounted to more than ten times the concentration in the air just outside the leaves! This CO_2 was assumed to have resulted from a combination of cellular respiration in the plant tissues themselves (both above- and belowground) and microbial respiration in the sediments surrounding the plants' roots and rhizomes (Constable, Grace, and Longstreth 1992). As the day progressed, internal leaf [CO_2] decreased, until it roughly matched that of the surrounding air at 14.00h (2:00PM), when photosynthetic rates were highest. Later work by Constable and Longstreth (1994) showed that this internally stored CO_2 can be a significant source of carbon for photosynthesis in *T. latifolia*., suggesting that aerenchyma may serve not only to facilitate aerobic respiration and root metabolism, but may enhance photosynthesis as well.

6.4 AERATION SYSTEMS

The basic means by which gases move through aerenchyma within a plant is by **passive diffusion** (Figure 6.9; Cronk and Fennessy 2001). For example, in the case of CO_2 storage described earlier for *Typha*, CO_2 will diffuse passively within the aerenchyma at night, until concentrations are more or less equivalent throughout the plant (but still much higher than in the surrounding air). Once sunlight is available, however, photosynthesis will begin to consume that CO_2, and within the plant it will diffuse towards the areas of highest demand (e.g., the leaves) until concentrations in the aerenchyma are roughly equal to those outside the plant. Similarly, oxygen will diffuse through the aerenchyma

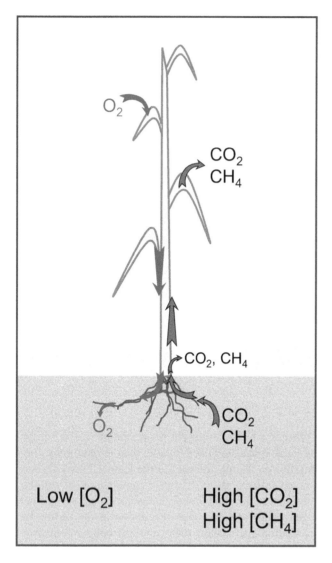

FIGURE 6.9 Passive Diffusion of Gases in Aquatic and Wetland Plants. Passive Diffusion Simply Moves Gases from Areas of High Concentration to Areas of Low Concentration (of the Respective Gas). In this Case, the High Demand of Oxygen in the Sediments Results in Very Low Concentrations there, as has been Discussed Earlier in the Chapter, and this Drives Diffusion of Oxygen through the Aerenchyma Network into the Sediments (Left Side of the Figure). Similarly, Heterotrophic Microbial Metabolism in the Sediments Releases Considerable amounts of Carbon Dioxide and Methane. The High Concentrations of these Gases in the Sediments Result in their Diffusion into the Plant, Through the Aerenchyma Channels, and into the Water and/or Atmosphere (Right Side). Redrawn from Cronk and Fennessy (2001).

towards the roots along a diffusion gradient, having been either liberated via photosynthesis in the aboveground tissues or diffused into the plant from the atmosphere.

An important note here is that these and other gases are not restricted to travel within the plant. In the case of oxygen diffusion into the belowground structures, if there are sufficient oxygen concentrations and potential routes of escape through the roots or rhizomes, oxygen may continue to diffuse out of the plant and into the soil or sediments. Carbon dioxide similarly may escape from the plant before it arrives at the stomatal openings in leaves if sufficiently porous tissues are encountered as it diffuses through the plant. Both of these alternative routes can function towards the benefit of the plant, as we will see later in the chapter.

6.4.1 AERATION VIA POSITIVE PRESSURE

In addition to passive diffusion, plants benefit from certain structural characteristics that facilitate more active movement of gases through their aerenchyma networks. In these processes, physical characteristics of the plant's leaves, stems, roots, and aerenchyma combine with physical properties of gases and water to enhance ventilation throughout the plant (Figure 6.10). Some of the key structural and developmental characteristics at play here are related to differences in leaves and stems as they mature, age, and then senesce (i.e., die, but in a manner that allows redistribution of resources within the plant), as described by Grosse, Büchel, and Tiebel (1991). In the youngest newly emerged leaves, intercellular spaces have not yet developed, and the leaves tend to be highly resistant to gaseous diffusion. However, as the leaves expand, intercellular gas spaces and small-pored aerenchyma develop, making the leaves permeable to gases but still capable of holding pressure that accumulates in response to temperature or humidity gradients. As the leaves age, air spaces expand further and the tissues become more permeable to gases, resulting in a decreased ability to maintain internal pressures. Once the leaves begin to senesce, they will be even more permeable and often will serve as areas of efflux for gases from other parts of the plant, including belowground structures.

This general phenomenon was first recorded by Professor Raffeneau-Delile, of the Montpellier Faculty of Medicine (mid-1800s), re-examined by the Japanese botanist Ohno (early 1900s), and followed up with experiments by numerous other European scientists during the late 1800s and early 1900s (Grosse, Armstrong, and Armstrong 1996). These scientists collectively showed that the general process of gas transport in species of *Nelumbo* is driven by physical processes influenced by temperature and humidity gradients and that the air spaces through the plant are more or less continuous with one another across multiple individual leaves, via the rhizome (e.g., Figure 6.10C). Apparently, interest in this phenomenon was lost during most of the 1900s, as Grosse, Armstrong, and Armstrong (1996) reported that there was "no substantial work" published on this topic between about 1912 and 1979.

A series of papers published by Dacey and colleagues during 1979–1982 renewed interest in the mechanisms of active gas transport in aquatic plants, and a summary of those results is shown in Figure 6.10A (Dacey 1981). Dacey's work focused on *Nuphar luteum* (probably more correctly *N. advena*, since the plants were from a lake in Kalamazoo County, Michigan, USA). We first saw *Nuphar*

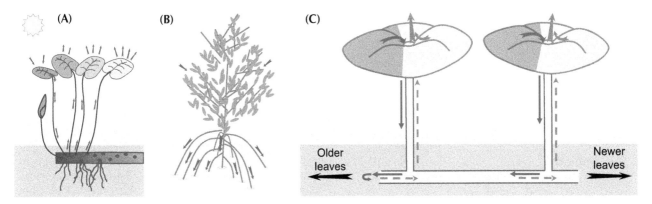

FIGURE 6.10 Movement of Gases in Aquatic and Wetland Plants as a Result of Thermo-Osmotically Driven Positive Pressure Within Leaves. (A) Young Leaves of *Nuphar* (and other Water Lilies) Tend to Build Higher Pressures than Do Older Leaves, Causing Directional Movement of Gases from Younger Leaves Towards (and then out of) Older Leaves, Via Aerenchyma in the Petioles and Rhizomes; (B) Similarly, Leaves of some Mangrove Species (E.g., *Rhizophora* Species) have been Shown to Use Thermo-Osmotic Pressure within the Leaves to Move Gases into the Flooded Root Systems, with Efflux Venting through Lenticels on Pneumatophores; (C) *Nelumbo* Species Use a Complex System of Air Channels to Move Oxygen-Rich Air through the Blue-Shaded "Half" of Leaf Blades Down through the Petioles and into the Rhizomes, Towards Older Rhizome Segments. Upon Reaching Obstructed Channels, the Air is Redirected Towards the Rhizome Apex, but some of the Air Leaves the Plant Along the Way Via a Separate Set of Channels that Exit at the Center of the Leaf Blade. More Detail on Each is Provided in the Text. (A) Redrawn from Dacey (1981); (C) Redrawn from Vogel (2004).

advena in Chapter 2, in the simplified example of how plant phylogenies are assembled (Figure 2.3). Air spaces in the leaf petioles and in the roots and rhizomes of this species can make up 40%–60% of the volume of those tissues, providing substantial aeration capacity to provide oxygen to belowground organs (Dacey 1981). Using a somewhat elaborate system of "intravenous" gas and temperature measurements, Dacey and colleagues mapped the movement and distribution of gases through the interconnected leaves and rhizomes of individual *Nuphar* plants, such as that depicted in Figure 6.10.

In these studies, they used inert ethane gas as a tracer through the plants, as well as measuring concentrations of methane escaping leaves from the apex of the rhizome, working back toward the base of the plant. In both cases, they found that gases moved from the younger (but expanded) leaves, through the rhizomes, and out the older leaves (Dacey 1981). As indicated in Figure 6.10, rates of efflux of methane from the leaves increased along a gradient from younger to older, with no gas escaping directly from the young leaves to the atmosphere during daytime and rates of methane release increasing from 1 mL/h at the second leaf sampled (next-to-youngest leaf) to 3 mL/h at the sixth leaf (7 cm down the rhizome from the youngest sampled leaf). Measurements of internal pressurization of the leaves showed that pressure increased through the day until around 14.00h–16.00h and then declined, corresponding with daily patterns of sunlight availability. In fact, there was a very close relationship between light intensity and leaf pressure (virtually a 1:1 correlation). Curiously, they also observed a slight increase in internal leaf pressure in response to air currents moving across the sampled leaves, in spite of the presumed cooling effect this might have on leaf temperature.

This latter effect of wind movement was further investigated by Grosse and colleagues (Grosse, Büchel, and Tiebel 1991; Grosse 1996), using 15 species of aquatic plants from the Nymphaeaceae (11 species), Alismataceae (1), Nelumbonaceae (1), and Menyanthaceae (2). The observed phenomenon is referred to as humidity-induced convection and, in contrast to thermally induced process described earlier, requires no difference in temperature between the plant and air. The movement of air over the leaf results in a decreased humidity directly above the leaf, in contrast to the unchanged internal humidity within the aerenchyma tissues of the leaf (Figure 6.11). This decrease in the concentration of water vapor above the leaf simultaneously increases the relative concentration of other gases in the air, such as oxygen and nitrogen. The result is that effective concentrations of oxygen outside the leaf increase while the concentrations inside the leaf remain unchanged. This establishes what is, in a manner of thinking, an artificial oxygen concentration gradient between the air and leaf that causes oxygen to enter the leaf. This and the thermally induced convection described in the previous paragraph are depicted in Figure 6.11.

Both of these processes require a porous, but somewhat diffusion-resistant, partition to create the gradients necessary to drive diffusion of oxygen into the leaves. Schröder, Grosse, and Woermann (1986) discovered such a layer in *Nuphar lutea*, situated between the upper **palisade parenchyma** layer (the column-like layer of mesophyll cells just inside the epidermis) and the inner aerenchyma. In *N. lutea*, stomata have diameters of about 2.4 μm, and spaces among the palisade cells average around 15 μm, both of which are too large to set up the necessary diffusion gradients (Schröder, Grosse, and Woermann 1986). Spaces between the cells of the barrier layer, however, are on the order of 1

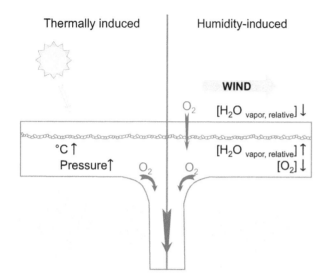

FIGURE 6.11 Positive Pressure Development Via Thermo-Osmotic Processes. On the Left, Sunlight Heats the Leaf, Resulting in Pressurization of the Gases inside the Lacunar Spaces of the Aerenchyma. The Result is Movement of Gases Away from the Younger Leaves, Which Tend to Maintain Highest Pressurization. On the Right, Air Movement across the Leaf Surface Results in a Humidity Gradient across the Upper Epidermis. The Differential in Effective Water Vapor Concentration (with Drier Air on the Outside of the Leaf) Results in an Opposing Gradient of Gases (Including Oxygen). Thus, When the Relative Water Vapor Pressure Increases Inside the Leaf, Oxygen and Other Gases Tend to Move into the Leaf, Where the Relative Concentrations of these Gases (Among the Sum of Water Vapor Plus other Gases) are Lower. Diffusion within the Plant then Results in Movement of Oxygen Towards Areas of Highest Consumption or Loss. A Layer of Densely Packed Cells (Diagrammed) Between the Upper Palisade Layer of Cells and the Internal Aerenchyma Makes these Pressure and Diffusion Gradients Possible in *Nuphar* Species (Grosse, Büchel, and Tiebel 1991; Schröder, Grosse, and Woermann 1986).

FIGURE 6.12 Lenticels on Stems of *Betula Nigra*, *Morella* (*Myrica*) *Cerifera*, and *Cephalanthus Occidentalis*. The Lenticels are the Tiny, Lighter-Colored Spots on the Stems of these and Many other Woody Species. Lenticels are Areas where the Outer Waterproof Bark is Interrupted, Allowing Exchange of Gases with Underlying Tissues. Typically, Tissues Just inside the Lenticels are Characterized by Numerous Intercellular Air Spaces, Further Facilitating Gas Exchange (Graham, Graham, and Wilcox 2006).

μm or smaller, which is small enough to induce the diffusion gradients that drive **thermo-osmotic convective airflows** in aquatic plants (Figure 6.11).

Pressurized ventilation is not restricted to herbaceous plants; it has been documented in a number of woody species including members of *Alnus*, *Salix*, *Betula*, *Populus*, *Taxodium* (Grosse, Armstrong, and Armstrong 1996), and species of mangroves in the genus *Rhizophora* (Evans, Okawa, and Searcy 2005; Evans, de Leon, and Sai 2008; Evans, Testo, and Cerutti 2009). Evans and colleagues (Evans, Okawa, and Searcy 2005; Evans, de Leon, and Sai 2008) studied internal airflow in *Rhizophora mangle* and *R. stylosa* and found continuous routes of air transport, using pressurized air tests, beginning at structures referred to as

"cork warts" on the leaves, traveling down through leaf petioles, into stems, through the prop roots, and finally into the root system (Figure 6.10B). Within the leaves, the air moves through a system of air spaces on the upper (**adaxial**) surface of the leaves, and within the roots, the oxygenated **influx** air moves through an inner system of aerenchyma. **Efflux** (outgoing) gases, on the other hand, move through aerenchyma in the outer portions of the root cortex, eventually exiting the plant through **lenticels** on roots or pneumatophores (see Figure 6.12 for a general description of lenticels). Further work by Evans, Testo, and Cerutti (2009) showed consistent patterns of this pressurized air transport among 11 species of mangroves sharing similar characteristics of cork wart density on leaves, volume of aerenchyma in leaves, and ratio of stem aerenchyma to number of leaves per stem. Species that had higher values for these (or, more specifically, the product of multiplying these three values) were more efficient at internal aeration.

As diagrammed in Figure 6.10C, pressurized aeration in *Nelumbo* species is somewhat more complicated than in the *Nuphar* species studied by Dacey and Grosse and colleagues. Following up on the work by European and Japanese scientists of the 1800s and early 1900s (mentioned earlier), Vogel (2004) carried out a detailed series of studies to map the air circulation system in *Nelumbo nucifera*. He found that the aeration system in leaves in this species are divided into four sectors. As indicated in Figure 6.10C, there are a basal and **distal** half, each of which is itself somewhat segregated into a left and right half. The basal half (shaded blue in Figure 6.10) is that half towards the "base" of the plant, or towards the older portions of the plant, and it generally is responsible for airflow into the whole-plant aeration system. Within the plant's rhizomes, there are three pairs of aerenchyma channels that carry airflow towards older portions of the plant and another two channels that move air towards the plant apex (newer parts and growing point). This fourth pair of channels is also connected to upward channels that lead to a central area on the upper surfaces of leaves (termed a central "plate" by Vogel), where air can exit the plant through pores that are opened and closed to control airflow. The distal half of each leaf blade (unshaded portion in the figure) appears to function primarily in moving air through that half of the leaf and out of the central plate, even though the aerenchyma channels in that half of the leaf are connected with the channels that flow down through the petiole (Vogel 2004).

6.4.2 Aeration via Negative Pressure

The previous examples, generally speaking, rely on the accumulation of a positive pressure in leaves to "push" air into other parts of the plant. The humidity-induced flows in the right half of Figure 6.11 might be considered to be driven, at least in part, by a diffusion gradient that "pulls" oxygen into the leaves. However, even there, the pressure that can accumulate on the inside of the barrier will also have a role in moving oxygen and other gases throughout the plant, along with the diffusion gradient created by the increased oxygen concentration. There are other means of gas transport in hydrophytes that rely on a negative pressure (essentially a vacuum) to pull air or oxygen through or into the plant.

The wetland grass *Phragmites australis* (common reed) has been used quite extensively to understand aeration systems (Armstrong et al. 1996; Brix, Sorrell, and Schierup 1996), and the general processes for active aeration in this plant are shown in Figure 6.13. As was shown in the water lilies, humidity gradients develop between air outside the plant and the internal spaces of the leaf or culm, just inside stomatal openings (Armstrong et al. 1996). This humidity gradient sets up an oxygen diffusion gradient in the opposite

direction, leading to diffusion of oxygen into the plant, as in the right-hand side of Figure 6.11. Once inside the leaf or culm, stomatal resistance to the exit of gases results in a pressurization, pushing oxygen down the culm, through the rhizomes, and into the root system, where demands are highest. Other gases are eventually vented out of the plant through older tissues, following the same directional movement of gases that is established in the pressurized culms. This form of humidity-induced airflow also has been demonstrated in the horsetail species *Equisetum telmateia* (Armstrong and Armstrong 2009), suggesting the anatomical features supporting this process have quite ancient roots among the vascular plants.

In contrast to this humidity-induced pushing of gases through the plant, differential movement of wind across a stand of *Phragmites* can create a negative pressure in taller dead and/or broken culms, which will experience high wind velocities because of their height, relative to other culms (Figure 6.13, right-hand side). This negative pressure, which is proportional to the speed of wind passing across the culms (Armstrong, Armstrong, and Beckett 1992; Armstrong et al. 996), results in air being pulled through the plant, entering via older, broken, and shorter culms. The result is that oxygenated air can be pulled into and circulated throughout portions of the plant's rhizome system, oxygenating the root system and the surrounding sediments via oxygen leakage from the roots (Armstrong, Armstrong, and Beckett 1992; Armstrong et al. 1996).

In the previous section, we saw a study by Evans, Testo, and Cerutti (2009) that identified a suite of structural characteristics responsible for pressurized air transport in mangroves. Those investigators found that fewer than half of the species in their study used positively pressurized transport to aerate their root systems; *Avicennia marina* was among those species not using positive pressurization. *Avicennia marina* and other mangrove species have been shown to benefit from negative pressurization that pulls oxygen into **pneumatophores** and subsequently the belowground structures following inundation of the pneumatophores by high tides (Figure 6.14). Scholander, van Dam, and Scholander (1955) examined the influence of tides on gas concentrations and internal pressures in roots of *A. germinans* in the Bahamas to determine the role of pneumatophores on root system aeration in this species. They found that submergence of the pneumatophores by the incoming tides resulted in a decline of both internal oxygen concentrations and internal gas pressures, with the lowest pressures (greatest suction) and lowest oxygen concentrations observed in the most deeply inundated roots. Based on measurements of changes in gas concentrations and internal pressures, Scholander, van Dam, and Scholander (1955) concluded that respiration within the roots, along with dissolution of respired CO_2 out of the root system into the surrounding water, accounted for the declines in internal pressures. That is, the loss of CO_2 into the water column, without an accompanying influx of gases to replace it, left a gaseous void that was responsible for the negative pressures in the

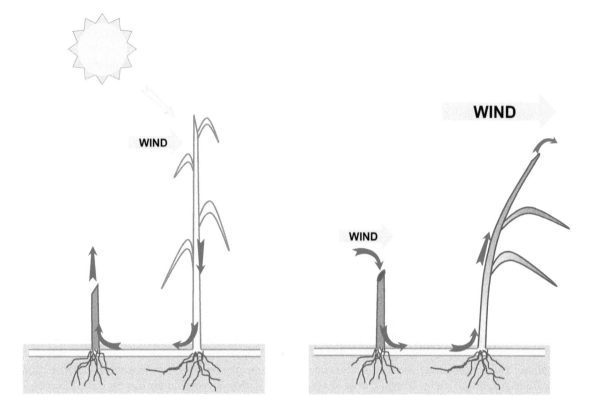

FIGURE 6.13   Contrasting Airflow Driven by Positive Pressures (Left) with that Driven by Negative Pressures (Right) in *Phragmites*. On the Left, as in Figure 5.36, Humidity-Induced Processes Create a Gradient of Oxygen Diffusion into the Culm, after Which Positive Pressure Forces air Down, through Aerenchyma Channels, into the Porous Rhizomes and Out Through Older, Senescent Shoots. On the Right, Differential Wind Speeds, with Faster Winds Passing across Taller, Broken Shoots Creates a Negative Pressure, Resulting in Air being Pulled into Shorter, Fragmented Senescent Shoots, and through the Plant, Eventually Leaving Via the Taller Shoots. This Process is Referred to as Venturi-Induced Convective Airflow. Redrawn from Figures in Armstrong et al. (1996).

root system. Similar experimental studies were conducted in Sydney, Australia, by Skelton and Allaway (1996), using *Avicennia marina*, with the same findings and conclusions.

The accumulated negative pressure (and low oxygen concentration) in the submerged root and pneumatophore system of *Avicennia* mangroves causes oxygen to quickly diffuse back into the root system upon aeration of the pneumatophores during falling tides (Figure 6.14). Pre-flooding oxygen concentrations typically are restored before arrival of the next high tide (Scholander, van Dam, and Scholander 1955), and Curran, Cole, and Allaway (1986) showed that there is sufficient capacity within the root aerenchyma system to store enough oxygen to tolerate up to six hours of tidal inundation.

Dissolution of carbon dioxide into the water column has also been shown to drive aeration in herbaceous wetland species. A thin film of air/gases surrounds submersed tissues in hydrophytes (on the order of 0.05 mm; Beckett et al. 1988), enhancing aeration by increasing the surface area of the plant through which gases can enter (Figure 6.15). Carbon dioxide is roughly 140 times more soluble in water than is oxygen (Raskin and Kende 1985), and this leads to a rapid loss of CO_2 from the plant when diffusion gradients exist between the internal air spaces and surrounding water or sediments. Raskin and Kende (1985) showed

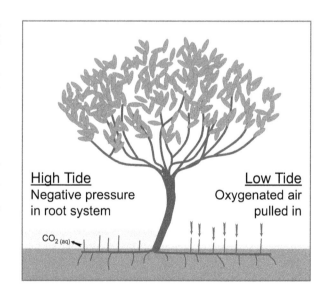

FIGURE 6.14   Root Aeration in Mangroves Driven by Negative Pressures. In some Mangrove Species (E.g., *Avicennia Marina*), during High Tide, When Pneumatophores are Submersed, the Dissolution of Carbon Dioxide from Roots and Pneumatophores into the Water Column Results in a Vapor Deficit in the Root System, Creating a Negative Pressure there. as the Tide Falls and the Pneumatophores are Exposed, Oxygen-Rich Air is Pulled into the Root System, Restoring Pre-Flood Oxygen Levels (Skelton and Allaway 1996).

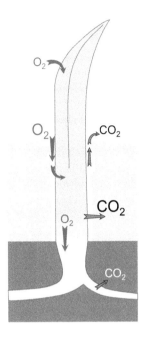

FIGURE 6.15 A Thin Gas Film, on the Order of 0.05 Mm, Surrounds Submersed Portions of Hydrophyte Leaves (If the Leaf Apex Protrudes above the Water Line), Increasing the Effective Surface Area Across Which Oxygen Can Enter the Plant During Mass-Flow Aeration Driven by CO_2 Loss. Carbon Dioxide that Exits the Leaf Can Enter the Water Surrounding this Film by Dissolution into the Water, as in Figure 5.15, or it can Continue to Diffuse Upward, Out of the Film and into the Atmosphere. Oxygen will be Pulled into Emergent Tissues and, Via the Film, into Submersed Tissues, Filling the Void Created by CO_2 Leaving the Plant. Carbon Dioxide Similarly Can leave the Root System by Dissolution into the Saturated Soil, Further Increasing the Gas Deficit within the Plant. Based on Information in Beckett et al. (1988) and Raskin and Kende (1985).

evidence for this process in deep-water rice but also indicated that their results had been observed in other rice varieties. Rates of gas intake in their work were higher during night than day because some of the internal CO_2 was used in photosynthesis during daylight and thus did not dissolve into the water column. Influx of gases also was correlated with an acidification of the surrounding water, as would be expected with increased CO_2 dissolution (see Figure 5.15). Koncalova, Pokorny, and Kvet (1988) also reported this CO_2-driven mass flow aeration in the sedge *Carex gracilis*.

6.5 RADIAL OXYGEN LOSS

As mentioned earlier, if supplies are adequate, some of the oxygen that is circulated throughout a plant's aerenchyma network will leak out of the root system into the surrounding soil and/or sediments. The oxygen leakage is driven by the significant oxygen deficits that develop in sediments largely because of the combination of microbial metabolism and soil flooding (Pezeshki 2001). This leakage of oxygen from root systems, termed **radial oxygen loss (ROL)**, can be sufficient to elevate soil redox status (e.g., Justin and Armstrong 1987) and oxidize some of the otherwise toxic reduced minerals that we saw in section 5.4.3 (Table 5.2). Similar to the patterns we saw in Figure 5.17, the elevated oxygen levels typically are restricted to the first millimeter or so of the surrounding soil, sediments, or other medium (Figure 6.16), but this is adequate to provide some physiological protection to the roots at the point of leakage (Colmer 2003).

Armstrong et al. (2000) set up an experimental system with *Phragmites australis* to measure oxygen gradients at 10 μm to 50 μm intervals within roots and the deoxygenated

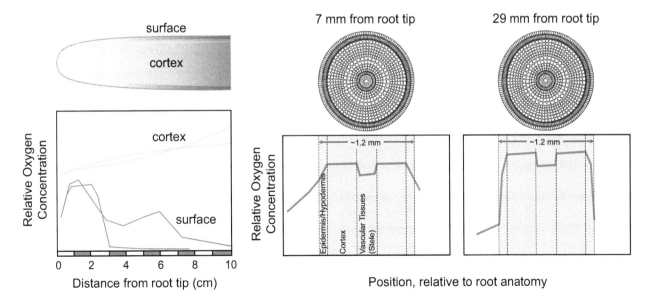

FIGURE 6.16 Radial Oxygen Loss from Roots of *Phragmites Australis*, Demonstrating Longitudinal and Lateral Patterns of Oxygen Leakage. Oxygen Release from the Roots is Greatest at, or Just Past, the Tip and Decreases as the Root Matures, Even though Oxygen Concentrations become Progressively Higher within the Root Cortex as One Moves away from the Root Tip. Longitudinal Transect Data on the Left were Obtained from Two Different *P. Australis* Roots. Based on Data in Armstrong et al. (2000).

rooting medium in which they were grown (Figure 6.16). They measured oxygen profiles longitudinally along the length of the roots and laterally along cross-sections within and outside roots at different distances from the root tips. They found elevated oxygen levels within the **rhizosphere** (the area of medium immediately surrounding and metabolically affected by the roots), but oxygen levels outside the roots were highest near the root tip and declined relatively rapidly, within the first 10 cm from the root tip (Figure 6.16). Inside the root, oxygen levels were highest within the cortex, that is, the tissues between the epidermis-hypodermis and the **stele** (vascular tissues and pericycle), and concentrations within the roots displayed a pattern opposite that of those outside the root, increasing with distance from the root tip (Figure 6.16). Recall from section 6.3 that maturation of aerenchyma tissues, which occurs within the cortex tissues, also increases with distance from the root tip. As a result, oxygen concentrations within the cortex would be highest in the region of the root where aerenchyma tissues are most developed, and they would decline as oxygen moves towards the root tip, where epidermal permeability and ROL tend to be highest (among wetland-adapted plants).

The reduced permeability of roots to oxygen that we see as we move basally away from the root tip typically is attributed to the development of some form of epidermal or hypodermal barrier to the movement of gases (and potentially other substances) across the root epidermis. Armstrong et al. (2000) mentioned that this decline in oxygen permeability had potential connections with development of anaerobic conditions or the presence of toxins, but indicated that the mechanism was, at the time, unknown. Colmer (2003) discussed ROL barriers at some length, but still indicated that the signaling mechanisms responsible for development of these barriers was unknown. More recent follow-up work by Colmer and colleagues (Watanabe et al. 2017) has revealed a chromosomal segment in *Zea nicaraguensis* (a wild, flood-tolerant congener of cultivated maize, *Zea mays*) that is linked with development of an ROL barrier and with reduced solute transport across the root epidermis and hypodermis. However, the specific molecular mechanism was still under investigation at the time of publication of that research.

Colmer (2003) also discussed potential adaptive benefits of this ROL barrier in hydrophytes. First among these was the enhancement of longitudinal oxygen transport brought about by the external barrier. That is, the lack of an escape route along most of the root length forces oxygen all the way to the root tip before it has an opportunity to escape. A second potential benefit of the barrier is to prevent entry of toxins (see again Table 5.2) along much of the root length. Although the root tip, where there is no barrier, might be susceptible to such toxins, heightened ROL at the root tip reduces this risk by oxidizing many of the potential toxins outside that region of the root. Jensen et al. (2005) echoed this hypothesis following their work with the seagrass *Zostera marina* (Alismatales, Zosteraceae). They

found ROL within the first 5 mm beyond root tips and suggested that barriers to ROL beyond that point in this species may prevent the entry of hydrogen sulfide (H_2S) into roots in the marine environment, where sulfur concentrations can be quite high. It was mentioned in that paper that other researchers had discovered **suberized** hypodermal cells in *Z. marina*, which likely help in creating the ROL barrier. **Suberin** is a complex of fatty acids that is deposited in cell walls of roots (and other tissues) that aids in protection against water loss and pathogen entry (Taiz and Zeiger 2002). Thus, the interplay between ROL and the presence of an external barrier to ROL (and passage of water and solutes) seems to act in some ways as a "tuning device," as it were, whereby the combination of these in an individual species helps to determine where they might fit along a flooding continuum. In fact, Pi et al. (2009) suggested that these two factors, along with the degree of aerenchyma development, helped to determine where each of a suite of eight mangrove tree species occurred along an inundation gradient. The species that had the least developed ROL barrier and the highest aerenchyma porosity appeared to be most suited to flooded areas and were considered "pioneering mangrove species." The species that had the tightest barrier to ROL along with the least developed root aerenchyma was the least tolerant of flooded conditions among the eight species examined.

Radial oxygen loss has been demonstrated in species of all four wetland plant growth forms, as well as in various wetland tree species (Carpenter, Elser, and Olson 1983; Cronk and Fennessy 2001; Pi et al. 2009; Sand-Jensen, Prahl, and Stokholm 1982; Shu et al. 2015; Smits et al. 1990; Visser et al. 2000). The units of measurement vary in published reports, and many papers do not directly measure oxygen but instead use indicator dyes or other approaches. However, when amounts of released oxygen are available, ranges for values typically are fairly consistent across growth forms or other groupings. For example, Shu et al. (2015) gave a range of roughly 0.10 to 0.32 μg O_2 per mg of root per hour (mg^{-1} h^{-1}) for two cultivars of *Eichhornia crassipes* (a free-floating species) grown under a variety of conditions. Carpenter, Elser, and Olson (1983) gave a value of 1.91 μg O_2 mg^{-1} h^{-1} for the submersed *Myriophyllum verticillatum* and a range of 0.8 to 9.8 μg O_2 mg^{-1} h^{-1} for ROL rates of emergent species published elsewhere. Sand-Jensen, Prahl, and Stokholm (1982) reported ROL rates for a variety of submersed species ranging from 0.08 to 5.4 μg O_2 mg^{-1} h^{-1}. Visser et al. (2000) reported ROL rates in units of μg O_2 released per area of root (instead of mass), per time, and found that values within 2 cm of the root tip ranged from roughly 1.2 to 4.8 μg O_2 cm^{-1} h^{-1}, for diverse group of five emergent species. This is in comparison with the aforementioned submersed *M. verticillatum*, which had an ROL rate of 1.07 μg O_2 cm^{-1} h^{-1} (Carpenter, Elser, and Olson 1983).

It seems, then, that rates of ROL tend to fall within a relatively narrow range, with either submersed or free-floating species at the lower end of the spectrum. This latter detail should not be surprising, given that ROL is driven largely by a diffusion gradient from the root into the surrounding medium. Many studies have shown that ROL rates are higher when the demand for oxygen just outside the roots is greatest, or said another way, when concentrations of free oxygen outside the roots are lowest (Cronk and Fennessy 2001; Visser et al. 2000). In free-floating species, roots are suspended within the water column, which (generally speaking) will have a higher concentration of oxygen than the organic sediments in which other growth forms are rooted. Submersed species, by virtue of not being in contact with the atmosphere, have less direct access to oxygen than species with shoots in contact with the air. It also has been shown that rates of ROL tend to be higher during the daytime (or when plants are exposed to light), given that oxygen is a by-product of the light-dependent reactions of photosynthesis (Carpenter, Elser, and Olson 1983; Sand-Jensen, Prahl, and Stokholm 1982). Cronk and Fennessy (2001) also discussed a number of other patterns of ROL among species and growth forms, as well as plant traits, such as aerenchyma porosity, that have been shown to influence ROL among hydrophytes.

6.6 ROOT ADAPTATIONS

Adventitious roots (Figure 6.17) are those that form at (or near) the interface of the air and either the water column or inundated soil, typically at some point along the stem (usually at nodes) where roots would not typically be encountered (Parent et al. 2008). Sculthorpe (1967) touched very briefly on adventitious rooting in aquatic plants, comparing the adventitious roots of herbaceous plants (he termed them "air-roots") to the pneumatophores of cypress (*Taxodium* species) and mangroves. He mentioned that they were thought to serve as a sort of direct "short circuit" to provide access to atmospheric oxygen, potentially increasing transport of oxygen to submersed organs, in comparison to transport via the lacunar aerenchyma networks.

However, as was discussed previously, flooding results in a substantial decrease in the energy available to submersed organs, especially roots, which are surrounded by oxygen-demanding soil and sediments. One result of this, if oxygen supplies are not quickly restored, is death of some of the affected tissues. Thus, another function of adventitious roots is to replace deeper roots that have been lost to flooding-induced hypoxia, mitigating some of the lost water and nutrient uptake capacity (Parent et al. 2008; Sauter 2013; Vartapetian and Jackson 1997; e.g., Figure 6.17D).

The timing of adventitious root production varies widely among plant species, with some species constitutively forming adventitious root primordia before experiencing flooding or hypoxia (rice, for example) and others delaying production of adventitious roots as much as a few days after initiation of flooding (Steffens and Rasmussen 2016). Other factors that influence when adventitious roots may be produced include developmental stage at which flooding is experienced, water temperatures, and depth and duration of flooding (Steffens and Rasmussen 2016). Depth and duration of flooding and water temperature likely will influence the rate at which critical biochemical and physiological signals are produced, transported, and detected within the plant, as well as how rapidly the plant can respond to those signals. Ethylene and auxins are two of the more important signal molecules (Table 6.1), each of which seems to initiate different aspects of adventitious root production and emergence (Parent et al. 2008; Sauter 2013; Steffens and Rasmussen 2016). Auxins appear to be involved in the production and/or growth of the adventitious roots themselves, whereas ethylene seems to be involved in the death of epidermal cells that permits the new root to emerge from the plant into the surrounding soil or water (Steffens and Rasmussen 2016). This particular role of ethylene is thought to be mediated through reactive oxygen species that accumulate during hypoxic conditions (Steffens and Rasmussen 2016).

Because of the stresses imposed on roots that are produced in deeper layers of the soil or sediments, most wetland and aquatic plant species will have relatively shallow root systems. Cronk and Fennessy (2001) mentioned that most roots of herbaceous species will be found in the upper 10 cm or so of soil, and even wetland trees, such as *Taxodium distichum* (Figure 6.17D), produce a very small percentage of roots below about 30 cm depth. Recall also from one of the examples used in Chapter 4 that the typical rooting depth for *Juncus effusus* growing along the margins of an Alabama beaver wetland was approximately 20 cm. Shallow rooting of wetland trees often results in "tip-ups" (Figure 6.18), which sometimes create canopy gaps that can increase environmental heterogeneity and plant species diversity in forested wetlands.

6.7 RAPID SHOOT ELONGATION

Much of the anatomical adaptations that have been presented thus far have dealt with providing oxygen to roots once they are exposed to hypoxic conditions (or with ventilating other gases out of the root system). Aerenchyma increases the permeability of the plant tissues to internal gases, pressurized ventilation (both positively and negatively pressurized) helps to actively move those gases within the plant, and adventitious roots help provide a more direct conduit for gases to reach the root system. What happens, however, when the shoot systems of plants in the emergent or floating-leaved growth forms becomes submersed because of flooding?

In these cases, which may occur seasonally or because of unexpected flooding from anthropogenic activities, the plant must have a mechanism for providing relatively rapid access of the shoot to atmospheric oxygen. This access is provided through a process called **rapid shoot elongation**. Sculthorpe (1967) referred to this process as "accommodation of the petiole," in reference to the plant's need to extend the petiole as a means of accommodating the increased water depth. Early experiments illustrated by Sculthorpe

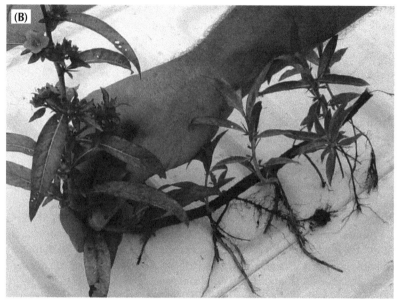

FIGURE 6.17   A few Examples of Adventitious Rooting in Wetland Plants. (A) Massive Adventitious Rooting in *Ludwigia Decurrens* Growing Along a Lakeshore, (B) Adventitious Rooting at Nodes of *Hydrolea Uniflora*, (C) Two Different Forms of Adventitious Rooting at Nodes of *Ludwigia Peploides*, and (D) Adventitious Roots Produced at Multiple Elevations Along the Base of a *Taxodium Distichum* Tree Growing in the Shallow Portion of a Managed Reservoir. Note also the Pneumatophores ("Knees") at the Upper Level of the Root System in D.

showed that petiole elongation takes place by a combination of cell division and cell elongation and could occur at rates of almost 2 cm per hour (Figure 6.19). Sculthorpe (1967) credited increased petiole length with at least two important benefits in wetland plants. The first of these was exposure of photosynthetic tissues to sunlight at the water's surface, as this is critical to plant survival. The second was an ability of elongated petioles in floating-leafed species to move in response to wind or water currents, thus reducing damage to the leaf surface resulting from water movement. Another important benefit is the gas exchange afforded by

direct contact with the atmosphere. In fact, gas exchange (or the lack thereof) likely is one mechanism regulating the process of rapid shoot elongation in these plants.

Arber (1920) addressed rapid shoot elongation briefly through descriptions of experiments with the free-floating *Hydrocharis morsus-ranae* (European frogbit) and with species of *Ranunculus* and *Marsilea*. In reference to experiments conducted in the 1870s (by A. B. Frank) and 1880s (by G. L. Karsten), Arber pointed out that the cessation of petiole elongation had been shown to be independent of light availability at the water's surface and also was

FIGURE 6.18 Tip-Up of an American Beech (*Fagus Grandifolia*), Showing the Shallow Root System Developed as a Consequence of Growing Near a Beaver Wetland.

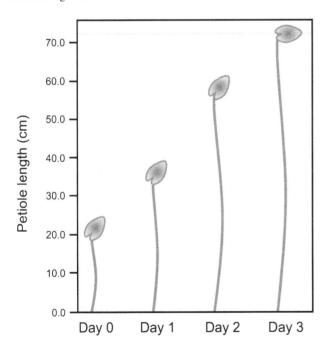

FIGURE 6.19 Elongation of Petioles in *Nymphoides Peltata* Following Transplantation into Deeper Water (I.e., After Submergence of the Plants). Petioles Elongated by as much as 1.7 Cm Per Hour during the Experiment, and Elongation Tended to be Fastest in the Middle to Upper Third of the Petiole Length. Based on Data Given in Sculthorpe (1967).

unaffected by reduced pressure on the leaf as it emerges from the water. Independence from light was demonstrated in experiments conducted in darkness, wherein petioles of *Hydrocharis* would only grow until the leaf reached the water surface, regardless of light availability. With respect to water pressure, Arber (1920) describes experiments conducted with species of *Hydrocharis*, *Ranunculus*, and *Marsilea* demonstrating that when tubes of oxygen-free air were placed over individual leaves, the associated petioles continued to grow even after leaves had reached the surface of the water. Arber concluded that contact with oxygen thus might have been a key factor in signaling when petioles should stop elongation.

More recent work by Voesenek et al. (2004, 2006) has helped to identify more specific anatomical and biochemical/physiological aspects of rapid shoot elongation in wetland plants (Figures 6.20 and 6.21). In their work with *Rumex palustris* (and other *Rumex* species), they found ethylene to be a key signal in initiating two separate, but interrelated, components of shoot elongation. The first step in the process involves combined signaling by ethylene and the physical angle of the leaf petiole, to initiate the activity of auxin on hyponastic adjustment of the leaf angle (Figure 6.20). **Hyponasty** refers to upward movement of the leaf blade resulting from elongation of cells on the petiole's

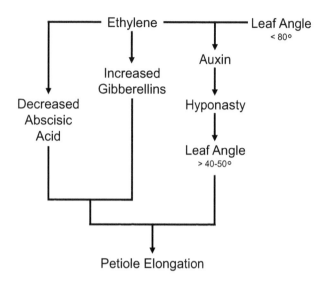

FIGURE 6.20 Network of Signals Involved in Rapid Petiole Elongation in *Rumex Palustris*. Leaf Angle Refers to the Angle between the Leaf Petiole and Horizontal. Redrawn from Voesenek et al. (2004).

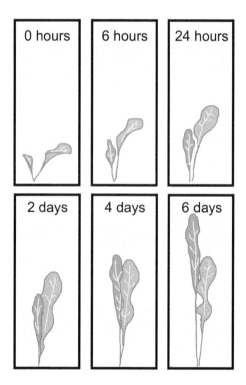

FIGURE 6.21 Time Series of Hyponastic Growth (The Rotation to Vertical) and Petiole Elongation in *Rumex Palustris* Plants Following Submergence. Here, One Younger (Left) and One Older Leaf (Right) were Followed through the Time Course, from Among the Six or so Leaves on the Individual Plant. Drawn from Photos in Voesenek et al. (2004).

lower surface, resulting in a "pressure" (-nasty) from the lower (hypo-) side of the petiole. In other words, the lower side of the petiole grows faster than the upper, resulting in upward movement of the leaf blade. Once the leaf angle reaches approximately 45° (40°–50° out of horizontal),

another signal cascade initiates elongation of cells on both the upper and lower surfaces of the petiole, resulting in the blade moving more vertically towards the water's surface. This progressive change in leaf angle, followed by petiole elongation, is illustrated in Figure 6.21. Petiole elongation itself involves an inhibition of abscisic acid and an accompanying stimulation in the production of gibberellins (Table 6.1), which decreases the ratio of abscisic acid to gibberellin within the plant tissues. Once this ratio and the leaf angle reach the necessary levels, the petiole begins to elongate, and the hyponastic movement continues until the petiole is growing approximately vertically (> approx. 80° out of horizontal; Voesenek et al. 2004). At the cellular level, petiole elongation is initiated by the production of proteins called **expansins** that are involved in loosening the chemical bonds among cell wall fibers. This loosening of the cell wall fibers allows expansion of the cells, and the production of expansins is influenced by the combination of ethylene concentration and the ratio of abscisic acid:gibberellins (Jackson 2008; Voesenek et al. 2004, 2006). An interesting note here is that all of the shoot elongation in *Rumex palustris* was found to have resulted from cell elongation, with no contribution at all from cell division (Voesenek et al. 2003).

As long as this combination of factors is satisfied, petiole elongation will continue until the leaf surface emerges from the water. Given this understanding of the role of ethylene in shoot elongation (and other processes we have studied involving ethylene), it is quite possible that Arber's hypothesis about the involvement of oxygen in cessation of petiole elongation might be at least partially replaced with one suggesting that ventilation of ethylene from the emerging leaf is a key factor in the termination of shoot elongation. That is, the available evidence suggests that elongation is driven, in part, by ethylene, and that elongation ceases once the leaf can vent accumulated ethylene to the atmosphere. Accumulation of ethylene in submersed plant tissues is clearly a critical factor governing responses of these plants to changes in their physical environment.

6.8 HETEROPHYLLY

Heterophylly refers to the presence of two or more leaf types or morphologies on an individual plant ("hetero" = different; "phyll" = leaf). Generally speaking, for aquatic and wetland plants, heterophylly is the production of submersed leaves that tend to be either long and ribbonlike or highly dissected and fanlike in structure (Figures 6.22 and 6.23), and often being only two to four cell layers thick (Figure 6.24). Aerial or floating leaves (those that are in contact with the atmosphere), on the other hand, tend to have entire margins, thicker cuticles, more stomata, and a greater thickness than the submersed leaves (Figures 6.24 and 6.25).

In the earlier literature on this topic with respect to hydrophytes, heterophylly and a related term, heteroblasty, were used almost interchangeably. **Heteroblasty** ("blasty" = embryonic or early developmental stage) refers to the condition wherein young, or juvenile, structures

FIGURE 6.22 Early Depictions of Heterophylly in Species of *Potamogeton* (Left) and *Ranunculus* (Right), Taken from Dodoens (1557). Note the Two Morphologically Distinct Leaf Types on Each Species, where the Submersed Leaves are either Long and Thin (*Potamogeton*) or Finely Dissected (*Ranunculus*). The Phenomenon is Common in Aquatic and Wetland Plants, Especially those Well Adapted to the Fluctuating Water Levels that are Characteristics of Wetland Ecosystems.

differ morphologically from those of older, mature, or adult structures or individuals. In the case of its relation to heterophylly, submersed leaves were considered a reversion to a "juvenile" condition, while floating or aerial leaves were considered to represent the "adult" form.

In some of the early writing on this topic, Goebel (1900) discussed the tendency for leaves of aquatic and wetland plants to revert to a "primary" or "juvenile" morphology when the plant is "unfavourably influenced." He further indicated that it was possible to bring about this reversion to juvenility (to use his terminology) by changing the medium in which plants were growing or "any other limiting cause" that might influence leaf development, even in plants that had already produced leaves of what was considered to be an adult morphology. Arber (1920), Sculthorpe (1967), and Hutchinson (1975) all referenced the juvenile versus adult condition for leaves in their discussions of heterophylly, even though Sculthorpe (1967) specifically wrote, "there is no *a priori* reason to assume that the ribbon-shaped or dissected submerged leaves . . . are juvenile, or . . . floating or aerial leaves necessarily adult." This juvenile versus adult concept of heterophylly persisted, for the most part, into the 1990s, when experimental work began to demonstrate that environmental conditions could influence leaf development well after leaves had emerged from the leaf bud. Jones (1995), for example, showed that leaves of cucumber did not begin to respond to environmental conditions until they had emerged from the bud and grown to a length between 1.2 and 2.0 mm. In the aquatic genera *Ranunculus* and *Hippuris*, Bruni, Young, and Dengler (1996) showed that many components of heterophyllous leaf development could be induced by a change in the leaf environment even after

individual leaves had reached half their final size. These and similar studies suggest that environmentally induced heterophylly is not driven by a developmental reversion to juvenile status, as was long considered to be the case.

6.8.1 FACTORS INFLUENCING HETEROPHYLLY

In her discussion of heterophylly, Arber (1920) identified numerous environmental conditions that could influence leaf morphology in heterophyllous species. Among these were damage to the plant, poor nutrition, low light availability, and other "adverse conditions." Similarly, Sculthorpe (1967) and Hutchinson (1975) described **photoperiod** (the length of daylight in the diurnal cycle), temperature, depth, light intensity, and underlying geology (which influences the relative availability of calcium and potassium) as factors that can influence the morphology of leaves in some aquatic species. Wells and Pigliucci (2000) conducted an extensive review on heterophylly in aquatic plants and found eight factors that consistently influenced leaf morphology in heterophyllous species (Table 6.3). These included photoperiod, temperature, and light intensity (three factors that had been previously noted by the authors mentioned earlier), along with osmotic stress (i.e., water stress), relative humidity, red to far-red light ratios, and concentrations of abscisic acid or gibberellic acid (see again Table 6.1 for information on these plant hormones).

The first five environmental factors given in Table 6.3 could be used to clearly distinguish an aerial or terrestrial environment from an aquatic environment. Within the water, we expect temperature, light intensity, and osmotic stress to be lower, as has been discussed in Chapters 4 and 5. Higher relative humidity is also an expected condition underwater. The influence of water on photoperiod relates to the absorption of light by water, where we might expect the total daily duration of available light for photosynthesis to be shorter within the water column, owing to extinction of light as it passes through water (e.g., Figure 5.11). The degree to which photoperiod might be reduced, then, would be influenced strongly by the depth at which the plants were growing.

The remaining three factors may not be so clearly linked to the plant's (or leaf's) ability to detect, or ability to respond to, its position above or below the water. We will look first at the ratio of red to far-red light. Plant responses to this factor are regulated by a special group of photoreceptive pigments called phytochromes. **Phytochromes** are protein-based pigments that are regulated primarily by absorption of red (650–680 nm wavelength) or far-red light (710–740 nm). Their influence on plant development began to be discovered in the 1930s, and one intriguing discovery at that time was that the effect of phytochrome on plant development could be reversed by switching between exposure to red light and far-red light (Taiz and Zeiger 2002). Because sunlight consists of both red and far-red light (Figure 5.9), the pool of phytochrome within plant tissues will consist of both the red-activated and far-red-activated

FIGURE 6.23 Examples of Heterophylly from across the Hydrophytic Angiosperms. (A–B) *Nuphar Advena* (Nymphaeaceae), Showing Submersed (S) and Floating (F) Leaves. Floating Leaves are Noticeably Thicker and Tougher than Submersed Leaves. (C) Floating and Submersed Leaves of *Cabomba Caroliniana* (Cabombaceae). (D) Aerial (A) and Submersed Leaves of *Proserpinaca Palustris* (Haloragaceae). (E–F) Aerial, Submersed, and Intermediate (Int) Leaf Forms of *Myriophyllum Aquaticum* (Haloragaceae). (G–H) Aerial and Submersed Leaves of *Rotala Rotundifolia* (Lythraceae). (I) Submersed and Floating Leaves of *Callitriche Heterophylla* (Plantaginaceae).

forms of phytochrome, in different ratios, owing to the relative proportions of photons in the two wavelength ranges. As a result, the ratio of red to far-red light (or R:FR ratio) is an important determinant of the outcome of plant developmental responses to phytochrome activity. Recall from Figure 5.9 that light in the far-red region is absorbed more strongly than light in the red region. Wetzel (2001) gave specific extinction coefficients of approximately 0.5 per m for red light and 1.5 per m for far-red light; that is, far-red light is absorbed roughly three times as quickly as red light when passing through pure water. The result of this difference in absorption by water is that the R:FR ratio will always tend to increase with depth through the water column (because the FR light disappears more quickly). Looking back at Table 6.3, we see that higher R:FR ratios tend to lead to the production of submersed leaf forms, as would be expected based on the relative absorption of red

and far-red light with depth, in combination with the activity of phytochrome.

Finally, we will examine the effects of the plant hormones abscisic acid (ABA) and gibberellic acid (GA; a gibberellin). Abscisic acid is well known as a drought-associated plant hormone (Wells and Pigliucci 2000), and we see in Table 6.1 that ABA acts to enhance salinity tolerance. It does this via a number of specific mechanisms, including regulation of stomatal opening. As it relates to heterophylly, ABA usually promotes the development of aerial or terrestrial type leaves. This should make sense, given that plants generally will be more susceptible to drought stress in terrestrial environments. This relationship among ABA, drought, and the production of the aerial leaf type also makes sense in light of the effects of temperature, light intensity, relative humidity, and osmotic stress on leaf type in heterophyllous species (Table 6.3).

A. *Elodea canadensis*

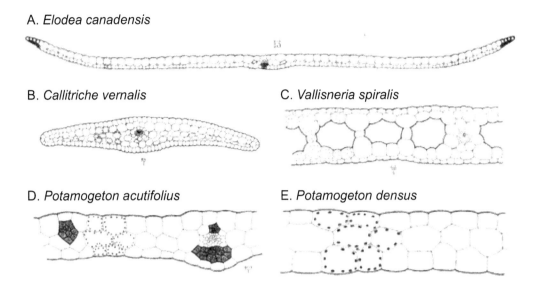

B. *Callitriche vernalis*

C. *Vallisneria spiralis*

D. *Potamogeton acutifolius*

E. *Potamogeton densus*

FIGURE 6.24 Cross Sections of Selected Submersed Leaves. Note the Absence of Stomata in these Cross-Sections. (A) *Elodea Canadensis*; (B) *Callitriche Vernalis*, Where the Epidermis Lacked Both Stomata and Chlorophyll); (C) *Vallisneria Spiralis*, with Numerous Parallel Aerenchyma Channels; (D) *Potamogeton Acutifolius*, with Regularly Distributed Vascular Bundles; (E) *Potamogeton Densus*, Cross-Section, Between Vascular Bundles. Taken from Schenk (1886).

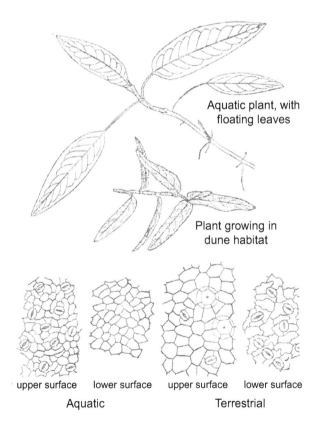

Aquatic plant, with floating leaves

Plant growing in dune habitat

upper surface   lower surface   upper surface   lower surface

Aquatic   Terrestrial

FIGURE 6.25 Upper: General Form of *Polygonum Amphibium* Growing in the Water or on Lakeshore Dunes. Note the Adventitious Rooting on the Aquatic Plant and Substantially Hairier Leaves of the Terrestrial Plant. Lower: Presence/Absence of Stomata on Leaves of Aquatic and Terrestrial Forms of *P. Amphibium*. The Aquatic Form Produced Stomata Only on the Upper Surface, Which was in Contact with the Atmosphere, whereas the Terrestrial Form Produced Stomata on Both Surfaces, with Fewer Produced on the Upper Surface (Which would have been more Directly Exposed to Sunlight and more Susceptible to Water Loss). Images Taken from Arber (1920), Where they were Credited to J. Massart (1910, Esquisse De La Géographie Botanique De La Belgique, Original Art not Seen).

TABLE 6.3

Factors affecting the morphology of leaves in heterophyllous aquatic and wetland plant species. Conditions given are those that were shown to promote the expression of either aerial or submersed type leaf morphology, regardless of where the leaf was actually formed. Information from Iida et al. (2016), Momokawa, Kadono, and Kudoh (2011), and Wells and Pigliucci (2000).

Factor	Aerial Leaf Morphology	Submersed Leaf Morphology
Temperature	High	Low
Light Intensity	High	Low
Osmotic Stress	High	Low
Photoperiod	Long	Short
Relative Humidity	Low	High
Red:Far-Red Ratio (R:FR)	Low	High
Abscisic Acid (ABA)	Induced aerial type	
Gibberellic Acid (GA)		Induced submersed type

Unfortunately, the role of GA in the expression of heterophylly is not quite so clear-cut (Wells and Pigliucci 2000). In many species, GA tends to promote production of submersed type leaves, but this is not always the case. *Eichhornia crassipes* and *Proserpinaca palustris* both show the opposite effect of GA application under experimental conditions, producing aerial-type leaves (Wells and Pigliucci 2000). Wells and Pigliucci (2000) also suggested that GA activity might be influenced by ABA concentration or by photoperiod. This latter factor makes sense, given the role of GA in fruit ripening and seed germination (Table 6.1). Taiz and Zeiger (2002) also indicated that GA activity might be influenced by photoperiod, but that its influence on developmental events such as fruit ripening and flowering also tend to vary from species to species. The specific role of GA in the production of submersed type leaves thus is a bit of a mystery, but likely involves other hormones (ABA) and some degree of response to photoperiod, which has been connected experimentally with the expression of heterophylly in hydrophytes.

6.8.2 Benefits of Heterophylly

Because heterophylly is characterized (usually) as the expression of different morphologies of leaves submersed within the water column, in contrast with those exposed to the air, we might think about potential benefits of heterophylly in the context of the contrasting aerial versus submersed leaf types. Generally speaking, submersed leaves are characterized by a suite of characteristics that we also find in plants grown under shade, in contrast to those

exposed to full sunlight. These include (Wells and Pigliucci 2000):

- Thinner leaves (i.e., fewer layers of cells; Figure 6.24);
- Greater degree of lobing or dissection, or having a greater leaf perimeter per unit leaf area in the case of longer linear submersed leaves (i.e., narrower leaf cross-sections);
- Greater surface area per unit mass;
- Lower density of stomata or potential lack of stomata;
- Fewer veins/vascular bundles; and
- Thinner cuticle.

Typically, this suite of characteristics in aquatic and wetland plants is thought to enhance photosynthetic activity of submersed leaves by enhancing capture of light and increasing the supply of inorganic carbon for each photosynthetic cell. This has, in fact, been demonstrated for some species, such as *Potamogeton* species (five species; Frost-Christensen and Sand-Jensen 1995) and *Batrachium peltatum* (Ranunculaceae; Nielsen and Sand-Jensen 1993), for which submersed-type leaves had equivalent or higher photosynthetic performance underwater than did aerial- or floating-type leaves of the same species.

The production of leaves with fewer layers of cells places each individual cell in closer proximity to the surrounding water, thereby reducing the physical distance required for diffusion of inorganic carbon from the surrounding water into cells for photosynthesis. The thinner cuticle, or total lack of cuticle, also enhances the diffusion of carbon into the leaves. The combination of thinner leaves and reduced or absent cuticle precludes the need for stomata to enhance uptake of inorganic carbon, and the fact that the leaves are submersed precludes the need for stomatal regulation of transpiration. The greater overall surface area per unit leaf mass increases the area of the water column within which light can be captured, as well as the volume of water from which carbon can be extracted (Arber 1920).

The narrower leaf cross-sectional area that results from increased dissection of submersed dicot leaves or increased linear form of submersed monocot leaves enhances diffusion into the leaves by increasing the leaf surface area per unit volume. However, there are other benefits to this aspect of heterophylly. The reduced width of individual leaf segments also impacts dynamics of water movement around the leaves; there are two potential benefits of this for the plant. First, the turbulent nature of water movement is influenced in part by the length of surface across which the water moves, with longer distances resulting in greater turbulence, all else being equal (Allan 1995). Thus, narrower leaf segments should experience reduced turbulence, reducing the likelihood of physical damage to the narrower submersed leaves (Wetzel 2001). Second, the width of leaves and leaf segments influences the thickness of the still **boundary layer** that surrounds the leaf (and all other objects within the water). Somewhat non-intuitively, given the previous relation of turbulence and leaf width, when water (or any other fluid) travels greater distances along a surface, the still boundary layer lying immediately adjacent to that surface increases in thickness (Abernathy 1970). As discussed in Chapter 5, diffusion of gases in water, including CO_2, is as much as 10,000× slower than in air. Gas exchange within the water is enhanced greatly by movement of the water, but it is reduced by lack of movement. Thus, within the still boundary layer that surrounds the leaves, plants must rely solely on diffusion for the uptake of inorganic carbon; the thicker this layer is, the more difficult it becomes to obtain carbon for photosynthesis. The boundary layer around submersed leaves often may be the greatest obstacle to obtaining carbon for photosynthesis, accounting for as much as 90% of the resistance to CO_2 diffusion (Madsen 1984). So, owing to the physical properties of fluid dynamics, narrower leaves both enhance carbon uptake and reduce the likelihood of physical damage caused by moving waters.

When all of these factors are considered collectively, we can see many potential benefits of heterophylly for plant species that tend to occupy habitats with fluctuating water levels, as is the case in natural wetlands. When the plants find themselves suddenly submersed (or suddenly exposed to the air), they have mechanisms to detect this via environmental cues, to respond via phytochrome and certain hormones, and to react by producing new leaves that are better suited to the new habitat conditions. The result is that these species can modify their **phenotypes** (that is, their physical form) to optimize resource capture as conditions change throughout the year or from year to year within a given aquatic or wetland ecosystem.

6.9 MODIFICATIONS TO PHOTOSYNTHETIC METABOLISM

We saw early in Chapter 5 that photosynthesis comprises two sets of biochemical reactions commonly referred to as the light-dependent and light-independent reactions (Figure 6.26). In the light-dependent reactions (Figure 6.26C), water is split into protons, electrons, and oxygen, using solar energy captured by the chlorophyll molecule. The capture of light by chlorophyll takes place on the thylakoid membrane, inside the chloroplasts, while the splitting of water takes place inside the thylakoid, in association with protein-pigment complexes bound to and within the thylakoid membrane itself (refer to Figure 5.1 for detail on involvement of the membrane-bound components). The light-independent reactions take place in the stroma (the fluid inside the chloroplast but outside the thylakoid; Figure 6.26B). The light-independent reactions collectively are referred to as the Calvin cycle, and it is within this pathway that CO_2 is physically captured, or "fixed," into the sugars that plants use for other aspects of their metabolism. However, there is an alternative preceding pathway used by some plants, under particular circumstances, which serves in the initial capture of CO_2 prior to it being shunted to

the Calvin cycle for incorporation into the plant's carbohydrate metabolism. This alternative pathway and the different ways that it is represented among aquatic and wetland plants will be discussed in the following section.

The Calvin cycle makes use of the enzyme ribulose bisphosphate carboxylase-oxygenase (**RuBisCO**) to attach CO_2 molecules to pre-existing ribulose bisphosphate molecules within the stroma of the chloroplast. Ribulose bisphosphate (RuBP) is the 5-carbon molecule in Figure 6.26D that is combined with the incoming CO_2 molecules to form the six 3-carbon molecules shown in the Calvin cycle diagram that we are using here. Because this first step in the Calvin cycle results in the formation of molecules containing three carbons, we frequently refer to this predominant mode of photosynthesis as C_3 photosynthesis. It is important to note this because we will see two other modes of photosynthesis that will be distinguished from C_3 photosynthesis in part by the number of carbon atoms in the molecule resulting from initial fixation of inorganic carbon.

The enzyme responsible for initial fixation of atmospheric CO_2 in C_3 plants, RuBisCO, has more than one biochemical function, as indicated by the inclusion of two "-ases" in its name: ribulose bisphosphate carboxyl*ase*-oxygen*ase*. We can often discern the function of an enzyme by having a look at the term to which the "-ase" in its name is attached. In this case, we note that RuBisCO can function to carboxylate ribulose bisphosphate, but also to oxygenate this same molecule. One of the factors that influences an enzyme's activity is the concentration of the molecules upon which it acts (i.e., its **substrates**). For an enzyme with two potential substrates, as is the case with RuBisCO, the relative abundance of the two substrates will be an important factor influencing which of the reactions the enzyme catalyzes. In the case of RuBisCO, then, the relative concentrations of CO_2 and O_2 are critical factors influencing the rate of photosynthesis in C_3 plants.

If RuBisCO is provided equal concentrations of CO_2 and O_2, the rate of carboxylation in C_3 angiosperms, on average, will be approximately 82× to 90× the rate of oxygenation (Taiz et al. 2015). This tendency to carry out the carboxylase function when CO_2 is abundant certainly works to the advantage of C_3 plants, as they rely on RuBisCO for all of their carbon capture. However, water in equilibrium with the air has a CO_2:O_2 ratio of around 0.04 (or approximately 25 O_2 molecules for each CO_2 molecule), and under these conditions, the rate of carboxylation averages approximately 3× the rate of oxygenation (Taiz and Zeiger 2002). Now, consider a chloroplast inside a plant cell, in a submersed leaf, which may be covered in algae or cyanobacteria, which themselves are surrounded by a boundary layer of still water within which the concentration of free CO_2 is strongly influenced by the pH of the water and the biogeochemistry of the underlying wetland basin. Not only does the plant now face substantial obstacles to accessing CO_2 for photosynthesis, but also during the daytime, the light-dependent reactions of photosynthesis will *release* O_2 into the stroma of the chloroplast as CO_2 is consumed! All of

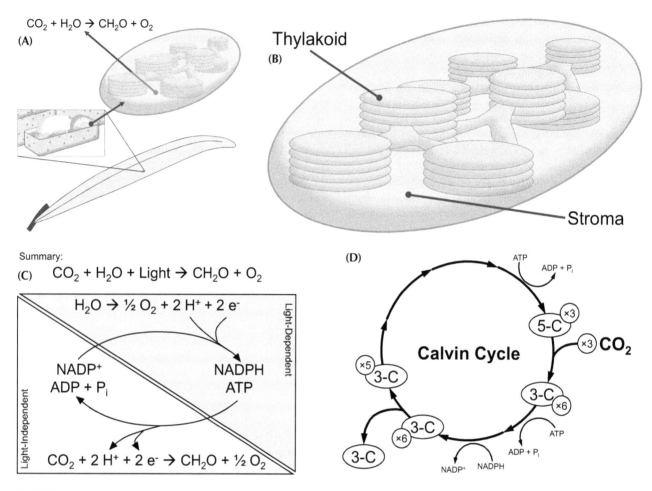

FIGURE 6.26 A General Depiction of Photosynthesis, as it is Carried Out by Most Aquatic and Wetland Plant Species. (A) Photosynthesis Takes Place within the Chloroplasts, in Cells of Photosynthetic Tissues; this is not Restricted to Leaves Alone. (B) Light Capture Takes Place on the Thylakoid Membrane, Which is Organized into Stacks Called "Grana," and Splitting of Water Occurs inside the Internal Thylakoid Membrane Space, or Lumen (Figure 5.1), in Very Close Association with Thylakoid Membrane-Bound Pigments and Proteins. The Remainder of the Photosynthetic Process Illustrated here Occurs within the Stroma of the Chloroplasts. (C) Conceptual Organization of the Light-Dependent and Light-Independent Reactions of Photosynthesis, Organized into Oxidation and Reduction Stages. (D) Generalization of the Calvin Cycle, wherein Co_2 is Reduced to Organic Sugars, the First of which Contains Three Carbons; As a Result, the Process Illustrated in this Figure is Usually Referred to As "C_3 Photosynthesis."

these challenges set up a scenario where the oxygenating ability of RuBisCO, termed **photorespiration**, begins to predominate, leading to a loss of some of the CO_2 that was previously captured by the plant (Figure 6.27).

There is a cost to photorespiration, in that two of the carbons that had been previously fixed via the Calvin cycle (with an expenditure of ATP energy) are removed from the RuBP substrate. By way of a complex series of reactions in the **peroxisome** (a cellular organelle responsible for highly oxidative reactions via the enzyme catalase) and the mitochondrion, some of that carbon may be recycled in the form of 3-carbon compounds that can be reused in the Calvin cycle. Note in Figure 6.27 (lower left) that there are 2-carbon molecules lost from RuBP via photorespiration. Those molecules can be shunted to the peroxisome, where they are partially oxidized before being sent to the mitochondrion. In the mitochondrion, two of these 2-C molecules are processed into one 3-C molecule, with the loss of one CO_2, and the resulting 3-C

molecule is cycled back through the peroxisome on its way to be reused in the chloroplast. This recycling can result in a net loss of only 25% of the carbons that are removed from RuBP by photorespiration. However, this process requires energy and reducing potential (NADH), and none of the original ATP used to fix those carbon atoms is recovered (Gurevitch, Scheiner, and Fox 2006; Taiz et al. 2015).

Given the potential cost of photorespiration, how is it that RuBisCO is so important in plants and has been suggested by some to be the most abundant protein on earth (Bar-On and Milo 2019; Raven 2013)? Because this enzyme is so abundant, there must be some benefit to the oxygenase function of RuBisCO that offsets this cost. As a thought experiment, we might think a bit about when the oxygenase function would be more prevalent for C_3 plants, or more specifically stated, when we might see carbon dioxide levels decline and oxygen levels increase. As mentioned earlier, these conditions might be expected during the daytime, when plants are actively

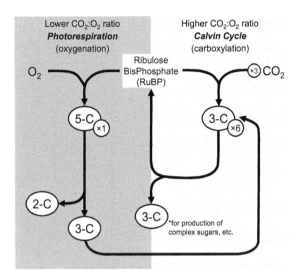

FIGURE 6.27 The Enzyme Responsible for the Initial Capture of Co_2 in the Calvin Cycle, Rubisco, is Capable of Carboxylation and Oxygenation of the Sugar Ribulose Bisphosphate (Rubp). The Relative Prevalence of these Two Processes in Plants Depends on the Ratio of $Co_2{:}O_2$ within the Photosynthetic Tissues. When the Concentration of O_2 is High, Relative to that of Co_2 (I.e., a Low $Co_2{:}O_2$ Ratio), Photorespiration May Dominate, Leading to the Gradual Loss, Via Mitochondrial Respiration, of some of the Accumulated Carbon (see the 2-Carbon Molecule Leaving the Cycle in the Lower Left). As Discussed in the Text, this Process has been Shown to Provide Some Photoprotective Benefits to Plants, Offsetting the Energetic Costs.

photosynthesizing, that is, when they are consuming CO_2 and releasing O_2. For the terrestrial ancestors that gave rise to aquatic and wetland species (or for hydrophyte species living in moist-soil wetland margins), we might also expect this to be further exacerbated on days when the plants close their stomata to conserve water. Closure of the stomata would tend to restrict entry of CO_2 into the leaves while simultaneously inhibiting the escape of O_2 generated during photosynthesis. Under these conditions, plants have difficulty continuing photosynthesis but have no means to avoid the continued capture of light energy. In section 6.2, we discussed harmful reactive oxygen species and the damage they can cause; accumulation of excess light energy without the ability to dissipate that energy can similarly degrade proteins and lipids within plant cells. Some have hypothesized that the ability to continue to dissipate this excess energy by recirculating some of the previously captured CO_2 (via photorespiration) serves a protective function during times that C_3 plants must "choose" between water conservation or continued photosynthesis, and some experimental evidence has supported this hypothesis (Gurevitch, Scheiner, and Fox 2006; Taiz and Zeiger 2002).

6.9.1 C_4 Photosynthesis

As we just saw, the key enzyme responsible for carbon fixation in C_3 photosynthesis (RuBisCO) presents an energetic cost under conditions that make CO_2 less available,

such as when terrestrial plants or plants inhabiting shallow wetland margins must close their stomata to conserve water. Plants growing along wetland margins may have fewer concerns about water conservation (outside of drought periods), but recall that these species sometimes close their stomata in response to flooded soil conditions. This stomatal closure can affect CO_2 and O_2 concentrations in wetland plants in the same way that closure would affect terrestrial plants. For truly aquatic plants, especially the submersed species, the water itself presents an obstacle to CO_2 access, potentially placing them under conditions conducive to photorespiration. Given that the critical factor regulating the relative carboxylation versus oxygenation function of RuBisCO is the ratio of CO_2 to O_2 within the plant's photosynthetic tissues, a solution to this problem would be an ability to deliver adequate supplies of CO_2 to chloroplasts during daylight hours. This solution is usually called a **carbon-concentrating mechanism**, and there are two primary categories of these mechanisms found among plants, including those of aquatic and wetland habitats: the **C_4 photosynthetic pathway** and **crassulacean acid metabolism** (CAM, sometimes also referred to as aquatic acid metabolism [AAM] in aquatic plants; Figures 6.28–6.31).

I mentioned earlier that C_3 photosynthesis gets its name from the production of 3-carbon molecules after initial fixation of CO_2 in the Calvin cycle. The first variation on photosynthesis that we will look at is C_4 photosynthesis, named such because the molecule formed from initial CO_2 fixation contains four carbon atoms (Figures 6.28–6.29).

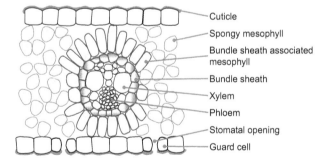

FIGURE 6.28 One Configuration of the Characteristic Sheathing of the Vascular Bundles in Most Plant Species Known to Exhibit C_4 Photosynthesis. The "Wreath-Like" Arrangement of Bundle Sheath and Mesophyll Cells is often Referred to as Kranz Anatomy, from the German Word for "Wreath." the Arrangement of these Cells Allows Capture of Co_2 in the Mesophyll Cells, Using the Enzyme Pep Carboxylase, Followed by Shuttling of the Resulting 4-Carbon Compounds into the Bundle Sheath Cells, Where a Carbon is Removed (in the Form of Co_2) for Use by Rubisco in the Calvin Cycle. The Remaining 3-Carbon Molecule is then Sent Back to the Mesophyll Cell for Reuse in C_4 Carbon Fixation. The Immediate Adjacency of the Bundle Sheath Cells and Vascular Tissues Allows Direct Transfer of Metabolites into the Phloem for Distribution within the Plant. More Details are Provided in Figure 6.29. Based on Micrographs and Diagrams Provided in Sage and Monson (1999).

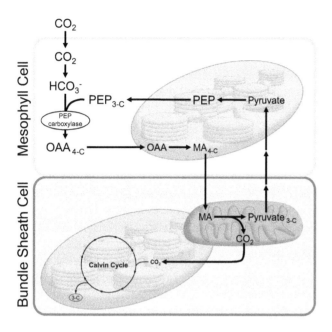

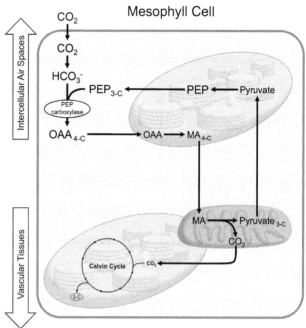

FIGURE 6.29 One Pair of Cells from the Sheathing Structure in a Typical C₄ Plant. The Arrangement of these Cells Allows Capture of Co₂ in the Mesophyll Cells, Which are Situated Adjacent to the Leaf's Internal Air Spaces, Using the Enzyme Pep Carboxylase. The Resulting 4-Carbon Compounds then are Shuttled into the Bundle Sheath Cells, Where a Carbon is Removed (In the Form of Co₂) for Use by Rubisco in the Calvin Cycle. The Remaining 3-Carbon Molecule is then Sent Back to the Mesophyll Cell for Reuse in C₄ Carbon Fixation. The Specific 4-Carbon and 3-Carbon Molecules Shuttled Between the Mesophyll and Bundle Sheath Cells Vary by Plant Species, as Does the Specific Location of the Decarboxylation of the Shuttled 4-Carbon Compound. Pep = Phosphoenolpyruvate; Oaa = Oxaloacetate; Ma = Malic Acid. Based on Information in Sage and Monson (1999) and Taiz et al. (2015).

FIGURE 6.30 Single-Cell C₄ Photosynthesis is Very Similar to Dual-Cell C₄ Photosynthesis. In the Form Illustrated Here, Carbon Fixation Occurs in the Upper End of the Mesophyll Cell, Which is Adjacent to the Intercellular Air Spaces and the External Co₂ Supply. Mitochondrial Decarboxylation Occurs at the Opposite End of the Cell, Where Mitochondria Cluster Around Chloroplasts Carrying Out the Calvin Cycle. In this Type of Single-Cell C₄ Photosynthesis, the Calvin Cycle is Carried Out at the End of the Cell that Lies Adjacent to the Vascular Bundles (The Bottom Here), Analogous to the Spatial Organization of Bundle Sheath Cells in Plants with Kranz-Type Leaf Anatomy. Based on Information in Lung, Yanagisawa, and Chuong (2012) and Taiz et al. (2015).

Hatch (2002), one of the two scientists after which the C₄, or Hatch-Slack, pathway was named, discussed a number of factors observed in some tropical grasses during the mid-1900s that suggested there might have been something other than ordinary C₃ photosynthesis operating in those species. These species included the grass crops maize, sugarcane, and sorghum (Hatch 2002). Among the noteworthy characteristics were high water use efficiency, high growth rates at higher temperatures and light levels, two distinct types of chloroplasts within individual cells, and a distinct internal leaf anatomy characterized by a clustered sheath of photosynthetic cells surrounding vascular bundles (Figure 6.28).

The key enzyme that makes the C₄ pathway possible is **phosphoenolpyruvate (PEP) carboxylase**, which catalyzes the reaction of bicarbonate with phosphoenolpyruvate, a 3-carbon molecule (Figure 6.29). Note here that PEP carboxylase does not have an oxygenase function; therefore, it serves only to capture carbon into organic compounds, specifically converting PEP and HCO_3^- to the 4-carbon compound oxaloacetate (OAA). The bicarbonate for this reaction is supplied to PEP carboxylase in C₄ plants

through the conversion of CO_2 by the enzyme carbonic anhydrase after CO_2 enters the cytoplasm of the mesophyll cells (Sage and Monson 1999). This CO_2 could have been delivered to the mesophyll cells through partially open stomata, through the aerenchyma network after diffusing into the plant from the soil or sediments, or even from cellular respiration within the plant itself.

In some species, the OAA produced by PEP carboxylase is converted to malic acid (MA) or aspartic acid prior to being shuttled to the bundle sheath cell for **decarboxylation** (removal of one carbon, in the form of CO_2). Because PEP carboxylase activity is independent of relative concentrations of CO_2 and O_2, C₄ plants have greater flexibility in managing water balance in hot, dry conditions, which improves their water use efficiency and allows them to outperform C₃ plants under those conditions, in terms of both net photosynthetic rate and water management. For example, at any given degree of stomatal openness, the rate of photosynthesis for terrestrial C₄ plants is in the range of 2.3× to 2.6× higher than for C₃ species (Hetherington and Ian Woodward 2003). As discussed earlier, water use efficiency generally is not a problem for aquatic plants, and

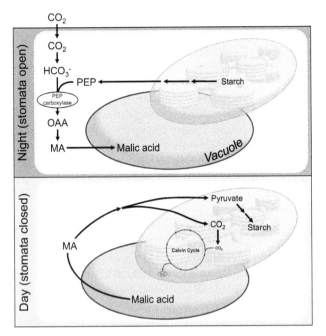

FIGURE 6.31 Typical Crassulacean Acid Metabolism (Cam) Involves a Day/Night Separation of Carbon Capture from Carbon Usage. The Enzyme Pep Carboxylase is Used to Capture CO_2 into 4-Carbon Oaa Molecules at Night, Which are Converted to Malic Acid and Stored in Vacuoles (Large Intracellular Storage Organelles) Until Daytime. The Closure of Stomata during the Daytime Allows Significant Water Conservation, but Usually Comes at a Substantial Cost in Terms of Photosynthetic Productivity. Liberation of CO_2 from the Stored Ma Allows Periods of Carbon Concentration in the Chloroplasts Early in the Day that Helps Minimize Photorespiration (Until Ma Supplies are Exhausted). Most Aquatic and Wetland Species Exhibiting Cam Photosynthesis are Submersed and Lack Stomata, But the General Timing of When Carbon is Captured Vs. Used in the Calvin Cycle Remains the Same. Based on Information in Taiz et al. (2015).

most aquatic plants lack stomata, so the benefit of C_4 photosynthesis in aquatic and wetland plants is focused primarily on mitigating the difficulty some species face in maintaining consistently high supplies of CO_2 to offset photorespiration, or simply to compete effectively for inorganic carbon against other photosynthetic species with which they share the water column.

We saw earlier in this chapter that most submersed aquatic plants produce leaves that are only two to four cells layers thick (Figure 6.24). Given this spatial constraint, it would be difficult to construct leaves that make use of the wreath-like Kranz anatomy illustrated in Figures 6.28 and 6.29. Nevertheless, many of these species face the potential of restricted access to free CO_2 within the water column that could result in unfavorable CO_2:O_2 balance for photosynthesis, and many species have employed a modified approach to the C_4 pathway to accomplish this, carrying out C_4 photosynthesis within individual cells (Figure 6.30). There are at least seven aquatic species that have been reported to carry out single-cell C_4 metabolism: *Egeria densa*, *Hydrilla verticillata*, and *Ottelia alismoides* of the Hydrocharitaceae,

Sagittaria subulata (Alismataceae), *Eleocharis acicularis* (Cyperaceae), and the grasses *Orcuttia californica* and *Orcuttia viscida* (Poaceae; Bowes 2011; Han et al. 2020). Note that all of these are monocot species.

Single-cell C_4 photosynthesis is very similar to the dual-cell (Kranz-type) C_4 photosynthesis discussed earlier. The mechanism for concentrating CO_2 in species exhibiting this form of photosynthesis results from an intracellular spatial segregation of different chloroplast types and mitochondria within mesophyll cells. In one form of single-cell C_4 photosynthesis, the initial fixation of carbon via PEP carboxylase occurs in the outer periphery of the cytoplasm, all around the outer region of the photosynthesizing cells. Decarboxylation of the 4-carbon malic acid (to release free CO_2 for photosynthesis) occurs in the central region of the cell, where mitochondria cluster around chloroplasts carrying out the Calvin cycle. In the other form, illustrated in Figure 6.30, carbon fixation occurs in the outermost end of the mesophyll cell (adjacent to the intercellular air spaces), and mitochondrial decarboxylation occurs at the opposite end of the cell, where mitochondria cluster around chloroplasts carrying out the Calvin cycle. In the type of single-cell C_4 photosynthesis illustrated in Figure 6.30, the Calvin cycle is carried out at the end of the cell that lies adjacent to the vascular bundles; thus, the spatial organization of these cells resembles that of bundle sheath cells in plants with Kranz-type leaf anatomy (Figure 6.29).

Evolutionarily speaking, C_4 photosynthesis is somewhat rare, with only about 8,150 angiosperms known to exhibit any form of C_4 photosynthesis (Sage 2017), accounting for fewer than 3% of all angiosperm species. No land plant groups other than the angiosperms are known to possess C_4 photosynthesis. However, there are a few algal species known to exhibit single-cell C_4 metabolism, including a marine diatom and a marine macroalga (Bowes 2011). Among the angiosperms, C_4 species are very unevenly distributed across 20 families, with 80% of them coming from the monocots (Bowes 2011; Sage 2017). Roughly 5,000 of those are grasses (Poaceae), and 1,300 or so species are sedges (Cyperaceae), with about 500 of those species belonging to the genus *Cyperus*, which happens to contain the most C_4 species among all angiosperm genera. These two families of monocots, then, contain about 77% of all known C_4 species, with the remaining 23% scattered across two families of monocots and 16 families of eudicots. We have seen previously that both the grasses and sedges are well represented among wetland flora, and some genera that contain both C_4 species and wetland species are *Andropogon*, *Digitaria*, *Echinochloa*, *Panicum*, *Paspalum*, *Saccharum*, *Setaria*, and *Spartina*, among the grasses, and the sedge genera *Bulbostylis*, *Cyperus*, *Eleocharis*, *Fimbristylis*, and *Rhynchospora* (Sage 2017).

6.9.2 CAM or AAM Photosynthesis

The other primary carbon-concentrating mechanism found among plants is generally referred to as crassulacean acid metabolism (CAM), although we sometimes see it referred to as aquatic acid metabolism (AAM) in aquatic plants (Cockburn 1985; Cronk and Fennessy 2001). The differences between CAM and C_4 photosynthesis typically are summarized in terms of "where" versus "when" carbon is captured by incorporation into 4-carbon compounds. As we saw with C_4 metabolism, there is a spatial segregation between the initial capture of CO_2 and its liberation for use in the Calvin cycle, whether we are talking about species with Kranz anatomy or those that make use of single-cell C_4 metabolism (Figure 6.29 and 6.30). In CAM photosynthesis, on the other hand, we see a temporal segregation of CO_2 capture versus its use in the Calvin cycle (Figure 6.31), with capture of CO_2 occurring at night and liberation of CO_2 for use in the Calvin cycle happening during the day. The capture, storage, and subsequent use of the CO_2 all take place in the same cell, in contrast to dual-cell C_4 metabolism.

In CAM metabolism, the same enzyme that was used to capture CO_2 in C_4 plants (PEP carboxylase) is used to capture CO_2 and form oxaloacetic acid (OAA). Again, as with C_4 plants, the OAA is converted into malic acid (MA), but in CAM plants, the MA is stored in the vacuole until it can be used during the day. In the daytime, MA is released from the vacuole, and it diffuses to the chloroplasts, where it is decarboxylated, releasing CO_2 (for the Calvin cycle) and pyruvate, that is converted to sugars for storage in starch (until it is needed for carbon capture the following night). So, we see in CAM plants that the capture and storage intermediates (MA and pyruvate) are essentially shuttled back and forth between the vacuole and chloroplasts, where in C_4 plants they were shuttled between mesophyll cells and bundle sheath cells for the same purposes. The end result of both alternative forms of photosynthesis is that CO_2 is concentrated near the site of the Calvin cycle, mitigating potential energetic costs of photorespiration by RuBisCO.

The distinction between "CAM" and "AAM" really boils down to a technicality of how terms are defined. If CAM photosynthesis is defined strictly as employing carbon capture while stomata are closed, then species that lack stomata (such as submersed aquatic species) cannot, technically speaking, exhibit CAM metabolism. Thus, when Cockburn (1985) described all the different forms taken by species using the general CAM metabolism blueprint, he used a general term of photosynthetic acid metabolism, or PAM, for all of the different forms. He then split all PAM species into those that have stomata (stomatal CAM, and yes, he called these "SCAM" species) and those that lack stomata (astomatal PAM). Those without stomata (the APAM species) then were divided into terrestrial species, with terrestrial astomatal acid metabolism (TAAM), and aquatic species, which exhibit what he called aquatic acid metabolism. For simplicity's sake, I am simply going to refer to this means of carbon concentration as CAM metabolism.

The advantage of CAM photosynthesis for terrestrial plants is that plants can open their stomata at night to capture CO_2 that diffuses into the plant, store that carbon in the form of MA in the vacuole, and then close the stomata

during the day to reduce water loss via transpiration. This is a more extreme means of conserving water than we saw in C_4 plants, and we typically see CAM metabolism in plant species characteristic of very arid ecosystems (or very dry microhabitats within other ecosystems). Cacti are typical examples of terrestrial CAM plants.

As with C_4 plants, we might wonder why we would see CAM metabolism in aquatic plants, given the ready availability of water. In fact, all known CAM species among the aquatic and wetland plants are not only aquatic, but also are submersed aquatic species! These include at least two species of *Crassula* (from the Crassulaceae, after which CAM photosynthesis was named), two or more species of *Vallisneria* (Hydrocharitaceae), at least one species each from *Sagittaria* (Alismataceae) and *Littorella* (Plantaginaceae), and 30 or so species of *Isoetes*, the lycophyte genus mentioned in this context in Chapter 2 (Keeley 1998b; Silvera et al. 2010). Some authors have included additional genera among those with CAM metabolism, but Keeley (1998b) suggested that most of those lack one or more of the key requirements for inclusion as true CAM species. Some of them will be discussed later in section 6.9.4.

Cockburn (1985) discussed several potential advantages of CAM in aquatic species, and most are related to challenges that have been discussed here in Chapters 5 and 6. The first of these was mitigating the physical challenge of diffusion of CO_2 into and through the water by concentrating carbon (as MA in the vacuole) during the night, when respiratory release of CO_2 and lower temperatures will increase CO_2 concentration in the water. Another benefit suggested for CAM photosynthesis in aquatic plants was the reduction of photorespiration brought about by having concentrated CO_2 within the cells during the night. Finally, Cockburn (1985) suggested that, because carbon capture by PEP carboxylase can take place at relatively low CO_2 concentrations, CAM photosynthesis (or C4 metabolism, for that matter) can create fairly steep CO_2 gradients to encourage diffusion of CO_2 toward sites of carbon fixation within the plants. This is somewhat reminiscent of some of the pressurized ventilation approaches we saw back in section 6.4.

6.9.3 Use of Bicarbonate

In Chapter 5, we saw that the predominant form of inorganic carbon in aquatic systems between pH 6.4 to pH 10.4 is bicarbonate (HCO_3^-), with free CO_2 declining rapidly once pH exceeds 6.4, approaching zero around pH 8 (Figure 5.15). Because the pH of most lakes is above 6.0 (Wetzel 2001), bicarbonate is frequently the most available form of inorganic carbon for submersed aquatic plants. As a result, many submersed species have adaptations for taking up and using HCO_3^-.

Three mechanisms are known by which bicarbonate is taken up by aquatic plants, including the charophyte macroalgae (Prins and Elzenga 1989; Wetzel 2001): co-transport of bicarbonate and hydrogen ions (HCO_3^- and

H$^+$); extracellular release of carbonic anhydrase enzyme, which converts bicarbonate to CO_2 immediately outside the cells (used largely by charophytes); and a complex system of extracellular acidification of bicarbonate accompanied by co-transport of cations, such as K$^+$ or Ca^{2+} through the **apoplast** (i.e., within the cell walls but outside the cell membranes; Figure 6.32).

The latter mode of bicarbonate usage is found in several submersed aquatic genera (*Elodea, Egeria, Hydrilla, Potamogeton*; all of these are monocots within the Alismatales), wherein bicarbonate uptake occurs in the exterior cell walls of the lower leaf epidermis only (Prins and Elzenga 1989). In these species, we sometimes find infoldings of the cell membranes of the lower epidermal cells, accompanied by ingrowths of the cell wall, where the acidification of bicarbonate is thought to take place, facilitated by light-activated proton pumps (Babourina and Rengel 2010; Krabel et al. 1995; Prins et al. 1982; Prins and Elzenga 1989; Figure 6.32). The pumping of protons out of the lower epidermis of these species is balanced by a loss of OH$^-$ ions through the upper epidermis, to maintain pH and ionic balance within the plant, but this also results in a pH gradient between the upper and lower surfaces of the leaves. In *Hydrilla verticillata*, for example, lower surface pH may be as low as 4, while pH at the upper surface reaches 10 (Bowes et al. 2002). At the upper surface, the elevated pH can result in deposition of calcium carbonate on the leaf surfaces (Wetzel 2001), while the significant decrease in pH on the lower surface can shift the carbon buffering system in favor of a dominance of CO_2, versus bicarbonate, during the daytime (Prins and Elzenga 1989).

Although this polar mechanism of bicarbonate usage has been studied most thoroughly in aquatic angiosperms, there are species of *Vallisneria* and *Myriophyllum* that use carbonate without any evidence of developing these polar gradients as a result (Prins and Elzenga 1989). Additionally, *Myriophyllum* is somewhat unusual in being a non-monocot angiosperm making direct use of bicarbonate as a carbon source. *Myriophyllum spicatum* (a noteworthy invasive aquatic plant) also has been found to simultaneously make use of different carbon sources on the same plant, with stems taking up CO_2 and while leaves take up HCO$_3^-$ (Prins and Elzenga 1989).

6.9.4 SWITCHING AMONG PHOTOSYNTHETIC PATHWAYS AND INTERMEDIATE PHENOTYPES

You may have noticed in the preceding subsections that some plant species appear to exhibit multiple forms of photosynthesis. One of the more prominent of these was *Hydrilla verticillata*, which has been shown to employ

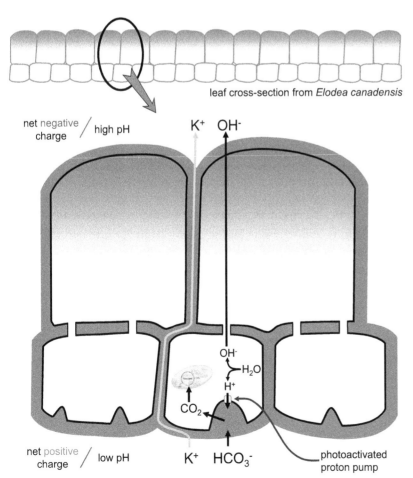

leaf cross-section from *Elodea canadensis*

net negative charge / high pH

K$^+$ OH$^-$

OH$^-$
H$_2$O
H$^+$
CO$_2$

net positive charge / low pH

K$^+$ HCO$_3^-$

photoactivated proton pump

FIGURE 6.32 Uptake of Bicarbonate (Hco$_3^-$) by Submersed Macrophytes, such as *Elodea Canadensis*, is often Coupled with Co-Transport of Cations, such as K$^+$, and Powered in Part by Photoactivated Proton Pumps on the Plasma Membrane of Cells of the Lower Epidermis. Acidification of Bicarbonate within the Cell Walls (as a Result of the Proton Pumping) Liberates Co$_2$, Which is Transferred into the Cells for Fixation Via the Calvin Cycle. To Maintain Ionic and Charge Balance within the Leaf, the Co-Transported Cations and Oh$^-$ Anions are Released Through Cells of the Upper Epidermis. This Shuttling of Ions and Acidification of the Lower Epidermal Cell Walls Results in Both a Charge Differential and a Ph Differential between the Upper and Lower Leaf Surfaces. Based on a Diagram from Prins et al. (1982), with Additional Information from Babourina and Rengel (2010), Krabel et al. (1995), and Prins and Elzenga (1989).

both C_3 and C_4 photosynthesis, while also having the capacity to access inorganic carbon from bicarbonate (Bowes et al. 2002). Species of *Hydrilla* and of *Egeria* are known to shift modes of photosynthetic metabolism when exposed to low CO_2 concentrations in the surrounding water (Bowes et al. 2002). Very recently, another species from the Hydrocharitaceae, *Ottelia alismoides*, has been suggested to employ C_4 and CAM metabolism, while also being able to utilize bicarbonate from the water column (Han et al. 2020). A number of the other species that are capable of using bicarbonate also show variation in this capacity, depending on local conditions, with variation having been shown for species of *Potamogeton* and *Elodea* among habitats, across the growing season, and even among individual leaves on a single plant (Prins and Elzenga 1989).

At least one strictly C_4 species (a grass, *Orcuttia californica*) demonstrates an interesting form of heterophylly. In this species, the terrestrial leaves exhibit the typical Kranz anatomy shown in Figure 6.28; however, aquatic leaves of this grass, although still exhibiting C_4 biochemistry, do not produce the Kranz-type internal leaf anatomy (Keeley 1998a). There are a few other wetland species that exhibit variation not only in the physical morphology of leaves, but also in photosynthetic metabolism that correlates with position of leaves, relative to the water. These species, thus, display a form of anatomical *and* physiological heterophylly. One of these, *Eleocharis baldwinii*, produces aerial culms (i.e., stems) with Kranz anatomy and clear C_4 metabolism (Uchino et al. 1995). However, submersed culms of this species produce substantially reduced bundle sheath cells and exhibit photosynthetic metabolism that is intermediate between C_3 and C_4, with some nighttime carbon capture, as in CAM photosynthesis. A congener of *E. baldwinii*, *Eleocharis vivipara*, displays a similar form of heterophylly in which the aerial culms produce the Kranz-type internal anatomy with C_4 metabolism, but submersed culms lack the Kranz anatomy and display C_3 photosynthetic metabolism (Ueno 2001). We will see another example of species capable of switching between C_3 and CAM photosynthesis in Chapter 9, as part of evidence used to determine patterns of species diversification among a group of *Sagittaria* species.

6.10 IN SUMMARY

In Chapter 5, we looked at the many ways in which water influences the availability of light, carbon, and oxygen for aquatic and wetland plants. In this chapter, we have taken an in-depth look at some of the more noteworthy adaptations present among these plants that allow them to not only survive, but also flourish in the face of the physical challenges presented by the aquatic environment. Submersed species, in particular, are presented with substantial obstacles to resource acquisition, having their resource-acquiring tissues surrounded by water and coated by microbes,

sediment, and inorganic carbon precipitates. In spite of these challenges, a few submersed species are well known globally as important invaders of aquatic habitats; this success is accomplished, in part, through suites of adaptations discussed in this chapter.

We began by looking at short-term responses of plants following submergence caused by sudden flooding. Plants are equipped with a number of physiological mechanisms to mitigate short periods of oxygen deficiency caused by flooding, including fermentative anaerobic metabolism and the use of nitric oxide as an alternative electron acceptor in cellular respiration. However, these metabolic processes cannot sustain the plant over longer durations of flooding, so species that are well adapted to aquatic and wetland environments must possess longer-term strategies for persistence in these ecosystems. It turns out that one of the indirect physiological by-products of anaerobic metabolism, ethylene, serves as a signal for production of several anatomical modifications used by hydrophytes for dealing with life under water or in continuously flooded soils.

Accumulation of ethylene, which occurs when flooding prevents the escape of this gas from plant tissues, serves as a signal to initiate elongation of shoots, production of adventitious roots, and formation of aerenchyma tissues, which themselves are critical components of ventilation mechanisms employed by aquatic and wetland species. Each of these three processes involves not only ethylene, but also other plant hormones, each of which has a set of specific, environmentally driven functions for plant development. Formation of aerenchyma is possibly the most important of these ethylene-induced responses to flooding, in that the aerenchyma networks within the plant are ultimately responsible not only for ventilation of the plant, which provides needed oxygen to all its tissues, but also for the escape of the ethylene gas itself.

Ventilation of plant tissues through the aerenchyma system takes many specific forms, but all of them are driven by some form of pressurization within the tissues, either a positive pressure that "pushes" air throughout the plant or a negative pressure, or vacuum, that pulls air through the plant. The direction of airflow is determined in large part by the architecture of the plant tissues themselves. Positive pressures accumulate in some species because of warming of younger tissues that are better capable of accumulating these pressures and moving air towards the older, less airtight tissues. In other species, there are specific ventilating structures (mangrove pneumatophores, for example) through which air escapes after having been pressurized in other parts of the plant, such as the leaves. Negative pressures can be established in plant tissues by wind movements across stems of differing heights or by dissolution of gases out of submersed tissues. The latter of these creates a gas deficit within the plant tissues that results in air being pulled into the plant when aerating structures are exposed to air.

While the negative pressures used to ventilate some species are generated in part by the dissolution of CO_2 out of submersed plant tissues, many hydrophyte species also experience a loss of O_2 from the root tissues. Although this seems to work in opposition to the purpose of the adaptations covered in this chapter, the loss of oxygen from roots, termed radial oxygen loss, serves an important function for wetland-adapted species. In Chapter 5, we discussed several disadvantages to the accumulation of reduced minerals within the plant root zone. Radial oxygen loss, which typically occurs only at the growing root apices, serves to counter this threat by providing a thin region of oxidized soil immediately outside the roots, mitigating potential risks of accumulation of reduced compounds in the rhizosphere.

In addition to production of aerenchyma, the accumulation of ethylene leads to production of adventitious roots, placing new roots in closer proximity to supplies of oxygen, and to rapid elongation of shoots, allowing leaves to rapidly reach the water surface for ventilation and photosynthesis. Another adaptation of plants to the changing water levels that are so characteristic of wetland habitats is the production of different leaf types specialized for aerial or for submersed conditions. This adaptation, termed heterophylly, is regulated by a suite of environmental factors that collectively inform new leaves about the environment within which they will be produced, allowing new leaves to be tailored to current conditions. The result is that the leaves are better fitted for capture of light, for uptake of carbon, and to resist the particular physical challenges that the new environment may pose (e.g., desiccation or stresses from moving waters).

The final group of adaptations covered in this chapter dealt with modifications to photosynthesis that enhance carbon capture for plants in habitats with low CO_2 concentrations, as is often the case underwater. Two of these, C_4 and CAM photosynthesis, are considered to be carbon-concentrating mechanisms, meaning they serve to ensure the concentration of CO_2 is as high as possible at the chloroplasts involved in the Calvin cycle itself (the light-independent reactions of photosynthesis). These two alternatives to the more common C_3 photosynthesis are distinguished primarily by what I have termed the "where" versus "when" aspect of carbon capture. Plants using C_4 photosynthesis concentrate carbon by capturing it into 4-carbon molecules in one area and then shuttling it to another area where the carbon is released for use in the Calvin cycle. This may involve segregating the capture and use into different cells (mesophyll vs. bundle sheath cells) or into different compartments of an individual cell, in single-cell C_4 metabolism. In contrast, CAM plants separate capture and use of carbon temporally, capturing the carbon at night and storing it in 4-carbon acids until the daytime, when it can be used in the Calvin cycle. Both of these carbon-concentrating mechanisms help the plant avoid potential losses of carbon and energy resulting from photo-respiration that may be carried out by the enzyme RuBisCO

under high O2:low CO_2 conditions. They also help the plants capture carbon when concentrations of CO_2 are relatively low, for example, because of competition with other aquatic primary producers or because of the very slow diffusion of gases within the water.

Another carbon capture adaptation employed by aquatic plants is the direct capture or use of bicarbonate, which often is the predominant form of inorganic carbon in the water column. There are three modes by which this is accomplished by aquatic primary producers. The mode that tends to be most common among aquatic angiosperms uses light-activated proton pumps to acidify the bicarbonate immediately outside cells of the leaf's lower epidermis, allowing access to the resulting CO_2 for photosynthesis. However, this approach requires co-transport of other ions, including hydroxyl anions, to balance electrical charge and ion content of the leaves. This co-transport results in substantial differences in pH between the upper and lower leaf surfaces, often leading to the accumulation of calcium carbonate on the upper leaf surface, potentially reducing net photosynthesis for these plants.

Finally, examples of species that can switch among C_3, C_4, CAM, and bicarbonate capture in response to environmental conditions serve as a reminder that wetlands are highly dynamic ecosystems. Plants have responded to this dynamism by evolving diverse combinations of ecological and physiological traits to ensure their growth, survival, and reproduction not only in spite of, but sometimes as a product of, that environmental variation. We will revisit this idea in Chapters 8 and 9.

6.11   FOR REVIEW

1. Why do plants need to have special adaptations for living in aquatic and wetland habitats?
2. In what ways does accumulation of ethylene under flooded conditions impact plant anatomy and physiology?
3. How do the three processes through which aerenchyma tissues develop differ from one another?
4. What are some key differences/similarities among passive diffusion, positive pressure-induced ventilation, and negative pressure-induced ventilation? Why do plants not simply rely on passive diffusion for aeration?
5. What is the benefit for wetland plants from radial oxygen loss?
6. What are advantages and disadvantages associated with the production of adventitious or shallow roots under flooded conditions?
7. How is the process of rapid shoot elongation regulated in plants experiencing flooding?
8. How does heterophylly benefit aquatic plants? That is, for which resource(s) is it an adaptation? How does heterophylly mitigate low availability of that resource/those resources?

9. What are some of the environmental factors that result in expression of heterophylly in aquatic plants? How are those factors related to the resource(s) for which heterophylly appears to be an adaptation?

10. What are some adaptations in aquatic and wetland plants that permit them to persist in areas with low CO_2 availability in the water?

6.12 REFERENCES

Abernathy, F. H. 1970. "Film Note for Fundamentals of Boundary Layers." Chicago, IL: Encyclopedia Britannica Educational Corporation.

Allan, J. David. 1995. *Stream Ecology: Structure and Function of Running Waters*. London: Chapman and Hall.

Arber, Agnes R. 1920. *Water Plants: A Study of Aquatic Angiosperms*. Cambridge, UK: Cambridge University Press.

Armstrong, J., and W. Armstrong. 2009. "Record Rates of Pressurized Gas-Flow in the Great Horsetail, *Equisetum telmateia*. Were Carboniferous Calamites Similarly Aerated?" *New Phytologist* (184): 202–15.

Armstrong, J., W. Armstrong, and P. M. Beckett. 1992. "*Phragmites australis*: Venturi- and Humidity-Induced Pressure Flows Enhance Rhizome Aeration and Rhizosphere Oxidation." *New Phytologist* 197–207.

Armstrong, J., W. Armstrong, P. M. Beckett, J. E. Halder, S. Lythe, R. Holt, and A. Sinclair. 1996. "Aquatic Botany Pathways of Aeration and the Mechanisms and Beneficial Effects of Humidity- and Venturi-Induced Convections in *Phragmites australis* (Cav.) Trin. Ex." *Aquatic Botany* 54: 177–97.

Armstrong, W., D. Cousins, J. Armstrong, D. W. Turner, and P. M. Beckett. 2000. "Oxygen Distribution in Wetland Plant Roots and Permeability Barriers to Gas-Exchange with the Rhizosphere: A Microelectrode and Modelling Study with *Phragmites australis*." *Annals of Botany* 86 (3): 687–703. https://doi.org/10.1006/anbo.2000.1236.

Armstrong, William, Peter M. Beckett, Timothy D. Colmer, Timothy L. Setter, and Hank Greenway. 2019. "Tolerance of Roots to Low Oxygen: 'Anoxic' Cores, the Phytoglobin-Nitric Oxide Cycle, and Energy or Oxygen Sensing." *Journal of Plant Physiology* 239 (March): 92–108. https://doi.org/10.1016/j.jplph.2019.04.010.

Babourina, Olga, and Zed Rengel. 2010. "Ion Transport in Aquatic Plants." In *Waterlogging Signalling and Tolerance in Plants*, edited by S. Mancuso and S. Shabala, 221–38. Berlin: Springer-Verlag. https://doi.org/10.1007/978-3-642-10305-6.

Bar-On, Yinon M., and Ron Milo. 2019. "The Global Mass and Average Rate of Rubisco." *Proceedings of the National Academy of Sciences of the United States of America* 116 (10): 4738–43. https://doi.org/10.1073/pnas.1816654116.

Beckett, P. M., W. Armstrong, S. H. F. W. Justin, and J. Armstrong. 1988. "On the Relative Importance of Convective and Diffusive Gas Flows in Plant Aeration." *New Phytologist* 110: 463–68.

Bowes, G., S. K. Rao, G. M. Estavillo, and J. B. Reiskind. 2002. "Comparisons with Terrestrial C4 Systems." *Functional Plant Biology* 29: 379–92.

Bowes, George. 2011. "Single-Cell C4 Photosynthesis in Aquatic Plants." In *C4 Photosynthesis and Related CO2 Concentrating Mechanisms*, edited by Agepati S. Raghavendra and Rowan F. Sage, 63–80. Dordrecht, The Netherlands: Springer Science + Business Media.

Brix, Hans, Brian K. Sorrell, and Hans-henrik Schierup. 1996. "Gas Fluxes Achieved by in Situ Convective Flow in *Phragmites australis*." *Aquatic Botany* 54: 151–63.

Brooker, Robert J., Eric P. Widmaier, Linda E. Graham, and Peter D. Stiling. 2017. *Biology*. 4th ed. New York: McGraw-Hill.

Bruni, Nadia C., Jane P. Young, and Nancy G. Dengler. 1996. "Leaf Developmental Plasticity of *Ranunculus flabellaris* in Response to Terrestrial and Submerged Environments." *Canadian Journal of Botany* 74: 823–37.

Carpenter, Stephen R., James J. Elser, and Karen M. Olson. 1983. "Effects of Roots of *Myriophyllum verticillatum* L. on Sediment Redox Conditions." *Aquatic Botany* 17 (3–4): 243–49. https://doi.org/10.1016/0304-3770(83)90060-8.

Cockburn, W. 1985. "Variation in Photosynthetic Acid Metabolism in Vascular Plants: CAM and Related Phenomena." *New Phytologist* 101 (1): 3–24.

Colmer, T. D. 2003. "Long-Distance Transport of Gases in Plants: A Perspective on Internal Aeration and Radial Oxygen Loss from Roots." *Plant, Cell and Environment* 26: 17–36.

Constable, J. V. H., James B. Grace, and David J. Longstreth. 1992. "High Carbon Dioxide Concentrations in Aerenchyma of *Typha latifolia*." *American Journal of Botany* 79 (4): 415–18.

Constable, J. V. H., and David J. Longstreth. 1994. "Aerenchyma Carbon Dioxide Can Be Assimilated in *Typha latifolia* L. Leaves." *Plant Physiology* 106: 1065–72.

Cronk, Julie K., and M. S. Fennessy. 2001. *Wetland Plants: Biology and Ecology*. Boca Raton, FL: CRC Press.

Curran, M., M. Cole, and W. G. Allaway. 1986. "Root Aeration and Respiration in Young Mangrove Plants (*Avicennia marina* (Forsk.) Vierh.)." *Journal of Experimental Botany* 37 (181): 1225–33.

Czarnocka, Weronika, and Stanisław Karpiński. 2018. "Friend or Foe? Reactive Oxygen Species Production, Scavenging and Signaling in Plant Response to Environmental Stresses." *Free Radical Biology and Medicine* 122 (October 2017): 4–20. https://doi.org/10.1016/j.freeradbiomed.2018.01.011.

Dacey, J. W. H. 1981. "Pressurized Ventilation in the Yellow Waterlily." *Ecology* 62 (5): 1137–47.

DeBary, A. 1877. *Comparative Anatomy of the Vegetative Organs of the Phanerogams and Ferns (English Translation)*. Lon: Oxford University Press.

Della Rovere, Federica, Laura Fattorini, Marilena Ronzan, Giuseppina Falasca, and Maria Maddalena Altamura. 2016. "The Quiescent Center and the Stem Cell Niche in the Adventitious Roots of *Arabidopsis thaliana*." *Plant Signaling and Behavior* 11: e1176660.

Dodoens, Rembert. 1557. *Histoire des plantes en laquelle est contenue la description entiere des herbes . . . non seulement de celles qui croissent en ce païs, mais aussi des autres estrangères qui viennent en usage de médecine*. N.p.: l'Imprimerie de Jean Loë.

Drew, M. C., C. J. He, and P. W. Morgan. 2000. "Ethylene-Triggered Cell Death during Aerenchyma Formation in Roots." *Symposia of the Society for Experimental Biology* 52 (3): 183–92.

Evans, David E. 2003. "Aerenchyma Formation." *New Phytologist* 161 (1): 35–49. https://doi.org/10.1046/j.1469-8137.2003.00907.x.

Evans, Lance S., Maryvic F. de Leon, and Erika Sai. 2008. "Anatomy and Morphology of *Rhizophora stylosa* in Relation to Internal Airflow and Attim's Plant Architecture." *The Journal of the Torrey Botanical Society* 135 (1): 114–25. https://doi.org/10.3159/07-ra-027r.1.

Evans, Lance S., Yuuya Okawa, and Dennis G. Searcy. 2005. "Anatomy and Morphology of Red Mangrove (*Rhizophora mangle*) Plants in Relation to Internal Airflow." *The Journal of the Torrey Botanical Society* 132 (4): 537–50. https://doi.org/10.3159/1095-5674(2005)132[537:aamorm]2.0.co;2.

Evans, Lance S., Zachary M. Testo, and Jonathan A. Cerutti. 2009. "Characterization of Internal Airflow within Tissues of Mangrove Species from Australia: Leaf Pressurization Processes." *The Journal of the Torrey Botanical Society* 136: 70–83.

Frost-Christensen, Henning, and Kaj Sand-Jensen. 1995. "Comparative Kinetics of Photosynthesis in Floating and Submerged Potamogeton Leaves." *Aquatic Botany* 51 (1–2): 121–34. https://doi.org/10.1016/0304-3770(95)00455-9.

Goebel, K. E. 1900. *Organography of Plants, Especially of the Archegoniata and Spermaphyta: General Organography—Part 2. Special Organography*. Oxford: Clarendon Press.

Graham, Linda E., James M. Graham, and Lee W. Wilcox. 2006. *Plant Biology*. 2nd ed. Upper Saddle River, NJ: Pearson Education, Inc.

Greenway, Hank, and William Armstrong. 2018. "Energy-Crises in Well-Aerated and Anoxic Tissue: Does Tolerance Require the Same Specific Proteins and Energy-Efficient Transport?" *Functional Plant Biology* 45 (9): 877–94. https://doi.org/10.1071/FP17250.

Grosse, W., J. Armstrong, and W. Armstrong. 1996. "A History of Pressurised Gas-Flow Studies in Plants." *Aquatic Botany* 54: 87–100.

Grosse, Wolfgang. 1996. "Pressurised Ventilation in Floating-Leafed Aquatic Macrophytes." *Aquatic Botany* 54: 137–50.

Grosse, Wolfgang, Hans Bernhard Büchel, and Helga Tiebel. 1991. "Pressurized Ventilation in Wetland Plants." *Aquatic Botany* 39: 89–98.

Gurevitch, Jessica, Samuel M. Scheiner, and Gordon A. Fox. 2006. *The Ecology of Plants*. 2nd ed. Sunderland, MA: Sinauer Associates, Inc.

Han, Shijuan, Stephen C. Maberly, Brigitte Gontero, Zhenfei Xing, Wei Li, Hongsheng Jiang, and Wenmin Huang. 2020. "Structural Basis for C4 Photosynthesis without Kranz Anatomy in Leaves of the Submerged Freshwater Plant *Ottelia alismoides*." *Annals of Botany* 125 (6): 869–79. https://doi.org/10.1093/aob/mcaa005.

Hatch, Marshall D. 2002. "C4 Photosynthesis: Discovery and Resolution." *Photosynthesis Research* 73: 251–56.

Hetherington, Alistair M., and F. Ian Woodward. 2003. "The Role of Stomata in Sensing and Driving Environmental Change." *Nature* 424: 901–8.

Hutchinson, G. Evelyn. 1975. *A Treatise on Limnology: Volume III, Limnological Botany*. New York: John Wiley and Sons, Inc.

Igamberdiev, Abir U., and Robert D. Hill. 2009. "Plant Mitochondrial Function during Anaerobiosis." *Annals of Botany* 103 (2): 259–68. https://doi.org/10.1093/aob/mcn100.

Iida, Satoko, Miyuki Ikeda, Momoe Amano, Hidetoshi Sakayama, Yasuro Kadono, and Keiko Kosuge. 2016. "Loss of Heterophylly in Aquatic Plants: Not ABA-mediated Stress but Exogenous ABA Treatment Induces Stomatal Leaves in *Potamogeton perfoliatus*." *Journal of Plant Research* 129: 853–62.

Jackson, Michael B. 2008. "Ethylene-Promoted Elongation: An Adaptation to Submergence Stress." *Annals of Botany* 101 (2): 229–48. https://doi.org/10.1093/aob/mcm237.

Jensen, Sheila Ingemann, Michael Kühl, Ronnie Nøhr Glud, Lise Bolt Jørgensen, and Anders Priemé. 2005. "Oxic Microzones and Radial Oxygen Loss from Roots of *Zostera marina*." *Marine Ecology Progress Series* 293: 49–58. https://doi.org/10.3354/meps293049.

Jones, C. S. 1995. "Does Shade Prolong Juvenile Development? A Morphological Analysis of Leaf Shape Changes in *Cucurbita argyrosperma* Subsp. *sororia* (Cucurbitaceae)." *American Journal of Botany* 82 (3): 346–59. https://doi.org/10.2307/2445580.

Jung, Jongduk, Seung Cho Lee, and Hong Keun Choi. 2008. "Anatomical Patterns of Aerenchyma in Aquatic and Wetland Plants." *Journal of Plant Biology* 51 (6): 428–39. https://doi.org/10.1007/BF03036065.

Justin, S. H. F. W., and W. Armstrong. 1987. "The Anatomical Characteristics of Roots and Plant Response To Soil Flooding." *New Phytologist* 106 (3): 465–95. https://doi.org/10.1111/j.1469-8137.1987.tb00153.x.

Kapoor, Karuna, Mohamed M. Mira, Belay T. Ayele, Tran Nguyen, Robert D. Hill, and Claudio Stasolla. 2018. "Phytoglobins Regulate Nitric Oxide-Dependent Abscisic Acid Synthesis and Ethylene-Induced Program Cell Death in Developing Maize Somatic Embryos." *Planta* 247 (6): 1277–91. https://doi.org/10.1007/s00425-018-2862-5.

Keeley, Jon E. 1998a. "C4 Photosynthetic Modifications in the Evolutionary Transition from Land to Water in Aquatic Grasses." *Oecologia* 116 (1–2): 85–97. https://doi.org/10.1007/s004420050566.

———. 1998b. "CAM Photosynthesis in Submerged Aquatic Plants." *The Botanical Review* 64: 121–75.

Koncalova, Hana, Jan Pokorny, and Jan Kvet. 1988. "Root Ventilation in *Carex gracilis* Curt.: Diffusion or Mass Flow?" *Aquatic Botany* 30: 149–55.

Krabel, Doris, Walter Eschrich, Yuri V. Gamalei, Jörg Fromm, and Hubert Ziegler. 1995. "Acquisition of Carbon in *Elodea canadensis* Michx." *Journal of Plant Physiology* 145 (1–2): 50–56. https://doi.org/10.1016/S0176-1617(11)81845-6.

Lung, Shiu-Cheung, Makoto Yanagisawa, and Simon D. X. Chuong. 2012. "Recent Progress in the Single-cell C4 Photosynthesis in Terrestrial Plants." *Frontiers in Biology* 7: 539–47.

Madsen, T. V. 1984. "Resistance to CO2 Fixation in the Submerged Aquatic Macrophyte *Callitriche stagnalis* Scop." *Journal of Experimental Botany* 35 (152): 338–47.

Manac'h-Little, Nathalie, Abir U. Igamberdiev, and Robert D. Hill. 2005. "Hemoglobin Expression Affects Ethylene Production in Maize Cell Cultures." *Plant Physiology and Biochemistry* 43 (5): 485–89. https://doi.org/10.1016/j.plaphy.2005.03.012.

Momokawa, Naoko, Yasuro Kadono, and Hiroshi Kudoh. 2011. "Effects of Light Quality on Leaf Morphogenesis of a Heterophyllous Amphibious Plant, *Rotala hippuris*." *Annals of Botany* 108: 1299–306.

Nielsen, Søren Laurentius, and Kaj Sand-Jensen. 1993. "Photosynthetic Implications of Heterophylly in *Batrachium peltatum* (Schrank) Presl." *Aquatic Botany* 44 (4): 361–71. https://doi.org/10.1016/0304-3770(93)90077-A.

Parent, Claire, Nicolas Capelli, Audrey Berger, Michèle Crèvecoeur, and James Dat. 2008. "An Overview of Plant Responses to Soil Waterlogging." *Plant Stress* 2 (1): 20–27.

Pezeshki, S R. 2001. "Wetland Plant Response to Soil Flooding." *Environmental and Experimental Botany* 46: 299–312.

Pi, N., N. F. Y. Tam, Y. Wu, and M. H. Wong. 2009. "Root Anatomy and Spatial Pattern of Radial Oxygen Loss of Eight True Mangrove Species." *Aquatic Botany* 90 (3): 222–30. https://doi.org/10.1016/j.aquabot.2008.10.002.

Prins, H. B. A., and J. T. M. Elzenga. 1989. "Bicarbonate Utilization: Function and Mechanism." *Aquatic Botany* 34: 59–83.

Prins, H. B. A., J. F. H. Snel, P. E. Zanstra, and R. J. Helder. 1982. "The Mechanism of Bicarbonate Assimilation by the Polar Leaves of Potamogeton and Elodea. CO2 Concentrations at the Leaf Surface." *Plant, Cell and Environment* 5: 207–14.

Raskin, Ilya and Hans Kende. 1985. "Mechanism of Aeration in Rice." *Science* 228 (4697): 327–29.

Raven, John A. 2013. "Rubisco: Still the Most Abundant Protein of Earth?" *New Phytologist* 198 (1): 1–3. https://doi.org/10.1111/nph.12197.

Sage, Rowan F. 2017. "A Portrait of the C4 Photosynthetic Family on the 50th Anniversary of Its Discovery: Species Number, Evolutionary Lineages, and Hall of Fame." *Journal of Experimental Botany* 68 (2): e11–28. https://doi.org/10.1093/jxb/erx005.

Sage, Rowan F., and Russell K. Monson. 1999. *C4 Plant Biology*. Edited by Rowan F. Sage and Russell K. Monson. San Diego: Academic Press.

Sand-Jensen, Kaj, Claus Prahl, and Hans Stokholm. 1982. "Oxygen Release from Roots of Submerged Aquatic Macrophytes." *Oik* 38 (3): 349–54.

Sauter, Margret. 2013. "Root Responses to Flooding." *Current Opinion in Plant Biology* 16 (3): 282–86. https://doi.org/10.1016/j.pbi.2013.03.013.

Schenk, H. 1886. *Die Biologie der Wassergewaechse*. Bonn: Verlag von Max Cohen and Sohn.

Scheres, Ben, Heather I. McKhann, and Claudia van den Berg. 1996. "Roots Redefined: Anatomical and Genetic Analysis of Root Development." *Plant Physiology* 111: 959–64.

Schleiden, J. M. 1849. *Principles of Scientific Botany (English Translation)*. London: Longman, Brown, Green, and Longmans.

Scholander, P. F., L. van Dam, and Susan I. Scholander. 1955. "Gas Exchange in the Roots of Mangroves." *American Journal of Botany* 42 (1): 92–98.

Schröder, Peter, Wolfgang Grosse, and Dietrich Woermann. 1986. "Localization of Thermo-Osmotically Active Partitions in Young Leaves of *Nuphar lutea*." *Journal of Experimental Botany* 37 (10): 1450–61. https://doi.org/10.1093/jxb/37.10.1450.

Schweingruber, Fritz H. 2020. *Anatomic Atlas of Aquatic and Wetland Plant Stems*. Cham, Switzerland: Springer.

Sculthorpe, C. D. 1967. *Biology of Aquatic Vascular Plants*. London: Edward Arnold Publishers Ltd.

Seago, James L., and Leland C. Marsh. 1989. "Adventitious Root Development in *Typha glauca*, with Emphasis on the Cortex." *American Journal of Botany* 76: 909–23.

Seago, James L., Leland C. Marsh, Kevin J. Stevens, Aleš Soukup, Olga Votrubová, and Daryl E. Enstone. 2005. "A Re-Examination of the Root Cortex in Wetland Flowering Plants with Respect to Aerenchyma." *Annals of Botany* 96 (4): 565–79. https://doi.org/10.1093/aob/mci211.

Shu, Xiao, Qi Deng, Quan Fa Zhang, and Wei Bo Wang. 2015. "Comparative Responses of Two Water Hyacinth (*Eichhornia crassipes*) Cultivars to Different Planting Densities." *Aquatic Botany* 121: 1–8. https://doi.org/10.1016/j.aquabot.2014.10.007.

Silvera, Katia, Kurt M. Neubig, W. Mark Whitten, Norris H. Williams, Klaus Winter, and John C. Cushman. 2010. "Evolution along the Crassulacean Acid Metabolism Continuum." *Functional Plant Biology* 37 (11): 995–1010. https://doi.org/10.1071/FP10084.

Skelton, Nicholas J., and William G. Allaway. 1996. "Oxygen and Pressure Changes Measured in Situ during Flooding in Roots of the Grey Mangrove *Avicennia marina* (Forssk.) Vierh." *Aquatic Botany* 54 (1996): 165–75.

Smits, A. J. M., P. Laan, R. H. Thier, and G. van der Velde. 1990. "Root Aerenchyma, Oxygen Leakage Patterns and Alcoholic Fermentation Ability of the Roots of Some Nymphaeid and Isoetid Macrophytes in Relation to the Sediment Type of Their Habitat." *Aquatic Botany* 38 (1): 3–17. https://doi.org/10.1016/0304-3770(90)90095-3.

Steffens, Bianka, and Amanda Rasmussen. 2016. "The Physiology of Adventitious Roots." *Plant Physiology* 170 (2): 603–17. https://doi.org/10.1104/pp.15.01360.

Taiz, Lincoln, and Eduardo Zeiger. 2002. *Plant Physiology*. 4th ed. Sunderland, MA: Sinauer Associates, Inc.

Taiz, Lincoln, Eduardo Zeiger, Ian Max Moller, and Angus Murphy. 2015. *Plant Physiology and Development*. 6th ed. Sunderland, MA: Sinauer Associates, Inc.

Takahashi, Hirokazu, Takaki Yamauchi, Timothy David Colmer, and Mikio Nakazono. 2014. "Aerenchyma Formation in Plants." In *Low-Oxygen Stress in Plants*, edited by J. T. VanDongen and F. Licausi, 247–65. Wien: Springer-Verlag.

Uchino, Akira, Muneaki Samejima, Ryuichi Ishii, and Osamu Ueno. 1995. "Photosynthetic Carbon Metabolism in an Amphibious Sedge, *Eleocharis baldwinii* (Torr.) Chapman: Modified Expression of C4 Characteristics under Submerged Aquatic Conditions." *Plant and Cell Physiology* 36 (2): 229–38. https://doi.org/10.1093/oxfordjournals.pcp.a078754.

Ueno, Osamu. 2001. "Environmental Regulation of C3 and C4 Differentiation in the Amphibious Sedge *Eleocharis vivipara*." *Plant Physiology* 127 (December): 1524–32. https://doi.org/10.1104/pp.010704.1524.

Vartapetian, Boris B., and Michael B. Jackson. 1997. "Plant Adaptations to Anaerobic Stress" 79: 3–20.

Visser, E. J. W., T. D. Colmer, C. W. P. M. Blom, and L. A. C. J. Voesenek. 2000. "Changes in Growth, Porosity, and Radial Oxygen Loss from Adventitious Roots of Selected Mono- and Dicotyledonous Wetland Species with Contrasting Types of Aerenchyma." *Plant, Cell and Environment* 23 (11): 1237–45. https://doi.org/10.1046/j.1365-3040.2000.00628.x.

Visser, E. J. W., R. H. M. Nabben, C. W. P. M. Blom, and L. A. C. J. Voesenek. 1997. "Elongation by Primary Lateral Roots and Adventitious Roots during Conditions of Hypoxia and High Ethylene Concentrations." *Plant, Cell, and Environment* 20: 647–53.

Voesenek, L. A. C. J., J. J. Benschop, J. Bou, M. C. H. Cox, H. W. Groeneveld, F. F. Millenaar, R. A. M. Vreeburg, and A. J. M. Peeters. 2003. "Interactions between Plant Hormones Regulate Submergence-Induced Shoot Elongation in the Flooding-Tolerant Dicot *Rumex palustris*." *Annals of Botany* 91: 205–11. https://doi.org/10.1093/aob/mcf116.

Voesenek, L. A. C. J., T. D. Colmer, R. Pierik, F. F. Millenaar, and A. J. M. Peeters. 2006. "How Plants Cope with Complete Submergence." *New Phytologist* 170 (2): 213–26. https://doi.org/10.1111/j.1469-8137.2006.01692.x.

Voesenek, L. A. C. J., J. H. G. M. Rijnders, A. J. M. Peeters, H. M. Van De Steeg, and H. De Kroon. 2004. "Plant Hormones Regulate Fast Shoot Elongation under Water: From Genes to Communities." *Ecology* 85 (1): 16–27. https://doi.org/10.1890/02-740.

Vogel, S. 2004. "Contributions to the Functional Anatomy and Biology of *Nelumbo nucifera* (Nelumbonaceae) I. Pathways of Air Circulation." *Plant Systematics and Evolution* 249: 9–25. https://doi.org/10.1007/s00606-004-0201-8.

Wang, Kevin L.-C., Hai Li, and Joseph R. Ecker. 2002. "Ethylene Biosynthesis and Signaling Networks." *The Plant Cell* (Suppl): S131–S151.

Watanabe, Kohtaro, Hirokazu Takahashi, Saori Sato, Shunsaku Nishiuchi, Fumie Omori, Al Imran Malik, Timothy David Colmer, Yoshiro Mano, and Mikio Nakazono. 2017. "A Major Locus Involved in the Formation of the Radial Oxygen Loss Barrier in Adventitious Roots of Teosinte *Zea nicaraguensis* Is Located on the Short-Arm of Chromosome 3." *Plant Cell and Environment* 40 (2): 304–16. https://doi.org/10.1111/pce.12849.

Wells, Carolyn L., and Massimo Pigliucci. 2000. "Adaptive Phenotypic Plasticity: The Case of Heterophylly in Aquatic Plants." *Perspectives in Plant Ecology, Evolution, and Systematics* 3: 1–18.

Wetzel, Robert G. 2001. *Limnology: Lake and River Ecosystems.* San Diego: Academic Press.

Zubay, Geoffrey L. 1998. *Biochemistry.* 4th ed. Dubuque, IA: Wm. C. Brown Publishers.

7 Plant Nutrition and Sediment Biogeochemistry

7.1 PLANT NUTRITION

In Chapters 5 and 6, we talked extensively about three elemental nutrients needed by plants: carbon, oxygen, and hydrogen. However, plants require much more than these to live, develop normally, and successfully reproduce. The essential nutrients that plants require can be broken into those that are required in relatively large amounts (termed **macronutrients**) and those that are required in usually much smaller amounts (**micronutrients**). These are given in Tables 7.1 and 7.2, along with typical concentrations in dried plant tissues, their biochemical or physiological roles within the plants, and some of the usual symptoms of nutrient deficiencies. Other nutrients may be required by some species, in much smaller, trace, concentrations, or by symbionts of the plants themselves. Symbiotic nitrogen-fixing bacteria that will be introduced later in the chapter, for example, require trace quantities of nickel and cobalt (Barbour et al. 1999; Taiz et al. 2015); although not a direct requirement of the plants, deficiency of these nutrients in the soil can impact health of plants that rely on the bacteria.

In Chapter 5 we also learned a bit about how conditions within wetland soils and sediments influence the availability of nitrogen, manganese, iron, and sulfur (e.g., Figure 5.25). We saw there that radial oxygen loss (ROL) is critical for reducing toxicity that would result from the reduced forms of some of these elements (Table 5.2). In the case of nitrogen and sulfur, however, some plant species preferentially take up oxidized forms (Barbour et al. 1999; Cronk and Fennessy 2001), making ROL important not only for reducing the toxic impacts of sediment hypoxia but also for enhancing access to certain macronutrients. In this chapter, we take a closer look at some of the more important essential nutrients, the means by which they are made available for plants, and a few adaptations plants have gained for dealing with nutrient deficiencies.

7.2 CARBON, HYDROGEN, AND OXYGEN AS ESSENTIAL NUTRIENTS

We have seen the specifics of plant acquisition of these nutrients in the previous two chapters, so only a brief review will be provided here, largely to contextualize these three elements as plant nutrients. Hydrogen, as we saw in Figure 5.1, is derived from the splitting of water during the light-dependent reactions of photosynthesis. Since water is formed from only two elements, hydrogen and oxygen, the splitting of water also releases oxygen, which goes on to become one of the end products of photosynthesis (O_2).

This process of oxygenic photosynthesis, in fact, was the ultimate source for virtually all of the oxygen in the atmosphere (Dismukes et al. 2001). We also saw in Chapter 5 that the hydrogen atoms released from splitting of water are used to generate ATP that helps to drive the Calvin cycle, after which some of those same hydrogen atoms are used to reduce the CO_2 that is captured during the Calvin cycle. Thus, within the carbohydrates produced during photosynthesis, the origins of the individual elemental nutrients are the following:

- Carbon and oxygen in plant-produced carbohydrates derive from CO_2 that is brought into the plants from various sources, including the atmosphere, heterotrophic respiration within the sediments, respiration within the plant itself, or dissolved inorganic carbon within the water column.
- Hydrogen in those carbohydrates comes from the splitting of water, which is taken up from the surrounding environment.

The carbohydrates produced during photosynthesis then are used by the plant for energy in cellular respiration, but also to form the skeleton of the biological molecules used in constructing the plant's cells, tissues, and organs (Table 7.1). In addition, those same tissues eventually find their way into other compartments of the wetland food web, through direct consumption by herbivores or following death of the tissues and consumption by members of the decomposer guild within the food web. Thus, photosynthesis of the individual plant becomes part of a larger cycle of carbon flow within the wetland (Figure 7.1), as well as between the wetland and the other ecosystems with which the wetland has connections.

We also see in the **carbon cycle** depicted in Figure 7.1 that the interconnection between photosynthesis and respiration illustrated in Figure 5.1 is really a microcosm of the larger interconnections between primary producers and consumers within the wetland, as carbon is taken up from the atmosphere via photosynthesis and later returned to the atmosphere after degradation of the resulting organic compounds by various consumers, decomposers, and so on. Another component of the carbon cycle that deserves special mention is the process of methanogenesis, production of methane via microbial metabolism. Methanogenesis was mentioned very briefly in Chapter 5 during the discussion of anaerobic metabolism, where we saw that it was the predominant means of microbial metabolism once the

DOI: 10.1201/9781315156835-7

TABLE 7.1

Essential plant macronutrients, with their biological functions and symptoms of deficiency. Data and information from Barbour et al. (1999), Hillel (2008), and Taiz et al. (2015).

Macronutrient	Approx. % in Dry Plant Tissues	Biochemical Roles	Symptoms of Deficiency
Carbon—C	45	Carbon, oxygen, and hydrogen form the skeleton of all organic biomolecules	Generally, we think of deficiencies only with respect to the mineral nutrients below because of the generally high availability of carbon and hydrogen relative to other macronutrients. Oxygen deficiency was covered extensively in Chapter 6.
Oxygen—O	45		
Hydrogen—H	6		
Nitrogen—N	2	Amino acids, proteins, nucleotides, nucleic acids, chlorophylls	Yellowing of older leaves (chlorosis), leaf abscission, purple coloration (anthocyanin accumulation)
Potassium—K	1	Maintains ionic balance and turgor pressure, regulates stomata, enzyme cofactor	Spotty chlorosis on tips and margins of older leaves, progressing toward leaf bases with death of tissue (necrosis)
Calcium—Ca	0.5	Component of cell walls, intercellular signaling, enzyme cofactor, involved in membrane permeability	Necrosis of younger, actively growing tissues, often with deformity of leaves or roots
Magnesium—Mg	0.2	Component of chlorophyll, enzyme cofactor	Chlorosis between leaf veins, especially in older leaves, eventual loss of all color in leaves followed by death of tissues
Phosphorus—P	0.2	Nucleic acids, ATP/ADP, cell membrane phospholipids	Stunted growth, dark green leaves with necrotic spots, delayed maturation, anthocyanin accumulation
Sulfur—S	0.1	Amino acids, proteins, coenzyme component	Chlorosis of younger leaves, stunted growth, anthocyanin accumulation
Silicon—Si	0.1	Component of cell walls	Susceptibility to falling over (lodging) and fungal infection

TABLE 7.2

Essential plant micronutrients, with their biological functions and symptoms of deficiency. Actual concentrations of these elements are much more variable, on a percentage basis, than with macronutrients, with some varying by as much as two orders of magnitude (100×). Data and information from Barbour et al. (1999), Hillel (2008), and Taiz et al. (2015).

Micronutrient	Approx. % in Dry Plant Tissues	Biochemical Roles	Symptoms of Deficiency
Chlorine—Cl	0.01 to 1	Osmosis and ionic balance, splitting of water in photosynthesis	Rarely, if ever, deficient because of availability relative to requirements
Iron—Fe	0.01 to 0.03	Component of proteins involved in redox reactions	Chlorosis between leaf veins, especially in younger leaves, eventual loss of all color in leaves followed by death of tissues
Manganese—Mn	0.005 to 0.08	Required for activity of many enzymes, involved in splitting of water in photosynthesis, membrane stabilizer	Intervenous chlorosis and eventual necrosis
Boron—B	0.002 to 0.008	Component of cell walls and membranes	Black necrosis in younger tissues, including terminal buds
Zinc—Zn	0.002 to 0.01	Required for activity of many enzymes	Reduced growth of internodes
Sodium—Na	~0.001	Osmosis and ionic balance, involved in C_4 and CAM metabolism	Chlorosis and necrosis in C_4 and CAM species
Copper—Cu	0.0006 to 0.003	Activator or a component of many enzymes	Production of dark green leaves with necrotic spots at tips and progressing toward leaf base around margins
Molybdenum—Mo	0.00001 to 0.0005	Component or cofactor for enzymes involved in nitrogen metabolism	Chlorosis between veins and necrosis of older leaves, twisting and death of leaves without necrosis, other symptoms similar to nitrogen deficiency because of the role of Mo in nitrogen metabolism
Nickel—Ni	~0.00001	Component of urease enzyme	Chlorotic leaf tips from urea accumulation

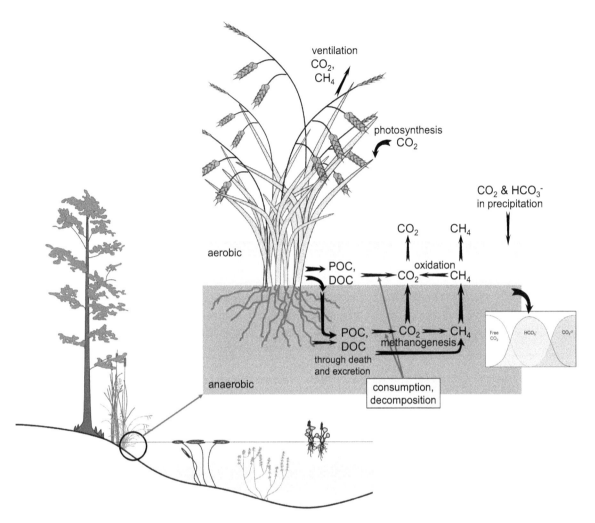

FIGURE 7.1 Simplified Overview of some of the Major Components of the Carbon Cycle in a Freshwater Wetland. Included are Uptake of Carbon Via Photosynthesis, Deposition of Particulate (Poc) and Dissolved Organic Carbon (Doc) Via Excretion from Live Plants and Death of Plant Tissues (Both Above- and Belowground), Release of Co_2 Via Degradation of Detrital Carbon (I.e., Dead Organic Carbon), Conversion of Co_2 and Organic Carbon into Methane (Ch_4) Via Methanogenesis, and the Diffusion of Co_2 and Ch_4 Out of the Wetland from the Water Column and Through the Plant's Aerenchyma System. Note that Oxidative Transformations Will Occur not Only in the Water Column (When Oxygen is Present), but also in Narrow Zones of Oxygenated Sediments at the Water Interface and within the Rhizosphere. Redrawn in Part from Mitsch and Gosselink (2007).

major electron acceptors had been largely exhausted within anoxic soils (Figure 5.25, Table 5.2). Recall that these electron acceptors included oxygen, which provides the greatest energy yield, followed by nitrogen, manganese, iron, and sulfur (Table 7.4).

7.3   METHANOGENESIS AND THE CARBON CYCLE

Once reduction-oxidation potential in the soil falls below about −200 mV (when reducible sulfur begins to disappear; Table 7.4), methanogenesis begins to dominate the sediment bacterial metabolism. Production of methane is carried out by an assortment of archaebacteria using one of about six types of carbon sources (Table 7.5). One of the highest-energy yielding reactions among the methanogenic pathways is that used by the autotrophic methanogens.

These autotrophs use CO_2, HCO_3^-, or CO_3^{2-} as their sole carbon source and obtain electrons from hydrogen gas (H_2). Because they use no organic carbon source for energy or reduction potential, they are considered to be **chemolithotrophs**. As mentioned in the previous chapter, CO_2 is abundant in wetland sediments because of organic matter decomposition, so it provides an abundant resource for species such as methanogens that are well adapted to life in that anaerobic environment.

Upon examination of the energy yields for methanogenesis versus other forms of microbial metabolism within wetland sediments (Tables 7.4 and 7.5), it becomes clearer why we do not see a dominance of methanogenesis until other electron acceptors are largely exhausted. Even if acetate, a low-energy-yielding carbon source, were the only carbon source available, microbes that can use other alternate electron acceptors, such as

TABLE 7.3

Abundances within the upper layers of the earth's continental crust, on a percentage mass basis, of the 16 soil nutrients given in Tables 7.1 and 7.2. Data from Yaroshevsky (2006).

Rank	Nutrient	Percent
1	Si	30.34800
2	Fe	3.08900
3	Ca	2.94500
4	K	2.86500
5	Na	2.56700
6	Mg	1.35100
7	S	0.09530
8	P	0.06650
9	Cl	0.06400
10	Mn	0.05270
11	N	0.00830
12	Zn	0.00520
13	Ni	0.00186
14	B	0.00170
15	Cu	0.00143
16	Mo	0.00014

nitrate or manganese dioxide (MnO_2), typically will outcompete most types of methane-producing Archaea. Nevertheless, methanogenesis is an important process energetically and ecologically in wetlands because of the extensive presence of anaerobic conditions within the soil and sediments.

In this context of the energetics of wetland microbial metabolism, another interesting form of methanogenic metabolism that yields approximately the same amount of energy for the methanogen as those processes given in Table 7.5 is a process referred to as **syntrophy** (Atlas and Bartha 1998; Reddy and DeLaune 2008). In syntrophic methanogenesis, the methanogen obtains reducing power in the form of H_2 from a neighboring anaerobic fermenter species and uses the hydrogen to reduce CO_2 to CH_4. This process yields approximately 143 kJ per mole of CO_2 metabolized; however, the preparatory process carried out by the fermenter consumes about 42 kJ per mole of organic substrate oxidized (Reddy and DeLaune 2008). The whole of this interspecific syntrophic metabolism thus nets only about 100 kJ of energy per mole of organic substrate consumed. This type of symbiotic metabolism is incredibly common among microbes, especially in wetland sediments where we see complex and heterogeneous patterns of redox potential across space and time. Along these lines, there

TABLE 7.4

Common succession of electron acceptors in flooded wetland soils or sediments. Data from Reddy and DeLaune (2008), Wetzel (2001), Mitsch and Gosselink (2007), Atlas and Bartha (1998).

Electron Acceptor	Reduced Forms	Approximate Redox Potential (mV)	Approximate Energy Yield (kJ mol^{-1}) from Oxidizing:		
			Glucose	Benzoate	Acetate
O_2	H_2O	> +300	2,879	3,175	–
NO_3^-	N_2, NH_4^+	100 to 300	2,721	2,977	907
MnO_2	Mn^{2+}	100 to 300	2,027	–	572
$Fe(OH)_3$	Fe^{2+}	−100 to +100	441	303	48
SO_4^{2-}	S^{2-}, H_2S	−200 to −100	381	185	48

TABLE 7.5

Common reactions employed by methanogenic archaebacteria and their approximate energy yields per mole of substrate utilized. Data and information from Atlas and Bartha (1998), Daniels et al. (1984), Gottschalk (1986), and Reddy and DeLaune (2008).

Carbon Source	Reactants	Products	Type of Metabolism	Approximate Energy Yield
Carbon dioxide	$CO_2 + H_2$	$CH_4 + H_2O$	Autotrophic—Chemolithotrophy	139 kJ per mol CO_2
Methanol	CH_3OH	$CH_4 + CO_2 + H_2O$	Heterotrophic	78 kJ per CH_3OH
Methylamine	$CH_3NH_3^+ + H_2O$	$CH_4 + CO_2 + NH_4^+$	Heterotrophic	56 kJ per $CH_3NH_3^+$
Carbon monoxide	$CO + H_2O$	$CH_4 + CO_2$	Heterotrophic	46 kJ per CO
Formate	$HCOOH$	$CO_2 + CH_4 + H_2O$	Heterotrophic	30 kJ per HCOOH
Acetate	CH_3COOH	$CH_4 + CO_2$	Heterotrophic	28 kJ per CH_3OH

is at least one methane-generating species of Archaea that can contribute significant amounts of methane to wetland carbon budgets in generally oxygenated soils, potentially because of the existence of oxygen-free microsites, even within oxygenated zones of the wetland soil profile (Angle et al. 2017).

A final note about methanogenesis, relative to its importance in the carbon cycle, pertains to the role of methane in global warming. Carbon dioxide and methane are the two most important greenhouse gases; that is, both are known to absorb and retain heat and they collectively make up $\geq 90\%$ of annual global greenhouse gas emissions (IPCC 2014). Although concentrations of methane in the atmosphere (~1.8 parts per million [ppm]) tend to be in the neighborhood of two orders of magnitude lower than for CO_2 (~400 ppm), methane has a much stronger heat absorbing capacity than does CO_2 (Ciais et al. 2013; IPCC 2014; Mitsch et al. 2013). Fortunately, individual molecules of methane have a shorter residence time in the atmosphere, because of oxidative chemical reactions that remove about 90% of annual methane emissions (Ciais et al. 2013). Accounting for the different heat absorption capacities and lifetimes in the atmosphere, the average molecule of methane has approximately 25–30 times as much influence on climatic warming than does CO_2, when averaged over a 100-year period (IPCC 2014; Mitsch et al. 2013).

What does this mean, then, for the role of wetlands in climate change? Wetlands are the largest natural source of methane emissions into the atmosphere, but they contribute only about one-third of total annual emissions globally. Anthropogenic sources such as livestock, rice cultivation, landfills, fossil fuels, and burning of biomass yield roughly twice as much methane as wetlands do (Ciais et al. 2013; Mitsch et al. 2013; Zhang et al. 2017). Mitsch et al. (2013) assembled data from 14 published studies and their own measurements of carbon emissions from another seven wetlands to determine whether wetlands could overcome their significant methane emissions to become net *sinks* (i.e., sites of permanent storage) for these greenhouse gases. They found that, despite the high emissions of methane (assuming methane contributes 25× the heat absorption of CO_2), wetlands across latitudes from tropical to temperate would be expected to overcome the methane emissions through a net storage of carbon, usually within a timeframe of 100 years. For this to happen, however, the wetlands must remain undisturbed to facilitate burial of organic matter, which locks away carbon indefinitely because of the very slow rates of decomposition in oxygen-free sediments. Unfortunately, ongoing warming from climate change is already impacting future potential for wetland carbon storage. This is especially true for the boreal peatlands that currently contain a third of the world's soil carbon. Those wetlands are increasingly becoming net sources of both CO_2 and CH_4 emissions, as permafrost thaws and sediments begin to undergo active decomposition (Zhang et al. 2017). We will discuss this topic further in Chapter 11.

7.4 SOIL MACRONUTRIENTS

Unlike carbon, hydrogen, and oxygen, which the plants may obtain directly as CO_2 or H_2O, the remaining essential nutrients in Tables 7.1 and 7.2 are obtained as dissolved materials within the water that is taken up by the plant. The majority of these materials enter the water via dissolution from the soil (including from decomposing soil organic matter) and geological materials through which the water passes. Some of these nutrients, however, may arrive within a wetland (or other ecosystem) in precipitation or dry particulate outfall, because of natural or anthropogenic processes that introduce particulate mineral materials into the atmosphere. Those processes include volcanic eruptions and windstorms, as well as agricultural and industrial activities, among others. Regardless of which of these routes brings the nutrients to the wetland, the information covered in Chapter 4 on wetland hydrology is directly relevant to the movement of dissolved nutrients into, within, and out of a wetland or other aquatic habitat.

Because of the influence wetlands have on dissolved nutrients in surface waters (which we have touched on previously and will see again in Chapter 11), most people who have an interest in wetland management or conservation also have an interest in managing nutrient loads entering or leaving the wetlands. Excluding carbon, hydrogen, and oxygen, there are 16 essential nutrients in Tables 7.1 and 7.2: seven macronutrients and nine micronutrients. If we are interested in managing the supply of nutrients for a natural wetland, as a way of influencing the amount of plant biomass produced in that wetland (or vice versa, managing plants to influence nutrient concentrations), 16 is a relatively large number of components to attempt to manage. This has been recognized as a problematic issue for quite some time, however, and we are fortunate that previous scientists have simplified this for us somewhat.

In the early 1800s, German scientists were tackling the issue of how to improve agricultural productivity (i.e., the yield of crops). Two scientists in particular, Carl Sprengel and Justus von Liebig, are credited with developing and disseminating the idea of nutrient limitation in agriculture (Barbour et al. 1999; Hessen et al. 2013). The general premise of their ideas, now known as the "law of the minimum," is that growth of plants (or other organisms, as the concept is now used) in some finite area will be restricted based on the availability of the resource that is in lowest supply—relative to the needs of the organism(s) in question. A great illustration of this idea (in my opinion) is that of a barrel built from staves of unequal lengths and then filled with water (Figure 7.2). The amount of water the barrel can hold is limited by the shortest stave, regardless of the height of all the others. In the same way, the amount of plant biomass that can be produced in some area (in a wetland, for example) will be limited by the resource that is at the lowest level relative to the needs of the plant species that are present. The amounts of the other resources will have no influence on productivity of the plants until the level of the limiting resource is increased.

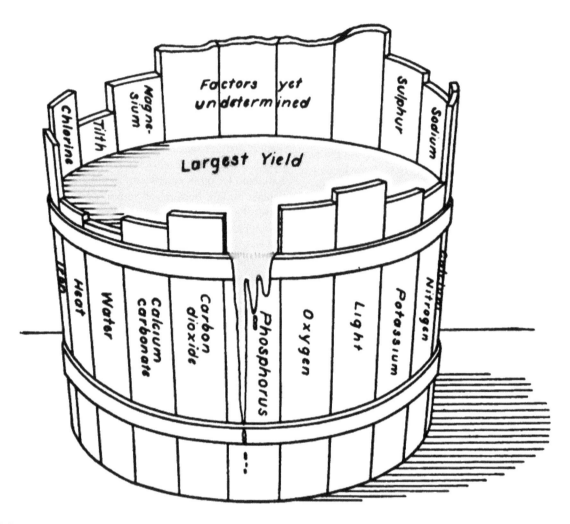

FIGURE 7.2 The Concept of Limiting Resources, as Conveyed in Liebig's "Law of the Minimum," Developed in the Field of Agronomy. Total Productivity of the Plants in an Area is Represented by the Water in the Barrel, Which is Limited to the Level of the Shortest Stave in the Barrel (Representing Phosphorus in this Illustration). To Increase Total Productivity, the Availability of that Limiting Nutrient must be Increased; Otherwise, Increasing the Availability of other Resources will not Affect Total Productivity. The Illustration here, taken from Whitson and Walster (1912), is an English-Language Revision of the Original, Published in German in 1903 by Hans Arnold Von Dobeneck.

In reality, resources often interact, and it may be that two or more resources interact with one another so closely that we need to focus on two or three co-limiting resources to most efficiently manage our system of interest, but the general idea of limiting resources nevertheless applies. For example, Olde Venterink, Van der Vliet, and Wassen (2001) examined nutrient limitation among six wet meadows in the Netherlands and compared the results of their work with those of another 45 published nutrient limitation studies to determine whether patterns in nutrient limitation were consistent among different types of wetland or in terms of the relationship between nutrient limitation and productivity. The experimental approach for determining the limiting nutrient is to experimentally fertilize small areas of vegetation, using fertilizer containing individual candidate nutrients, alone or in combination. The results from this then are evaluated to determine which nutrients result in increased growth when they are applied, in comparison with unfertilized plots. There are ways of increasing the complexity of these studies, but this very basic description will suffice for our present purposes.

In their study, Olde Venterink, Van der Vliet, and Wassen (2001) chose to add either nitrogen (N), phosphorus (P), potassium (K), or all three in combination, to their experimental wetland plots. When they analyzed the data and compared their results with the other 45 studies, they found that, across studies and ecosystem types (including wet meadows, fens, and bogs), most sites were limited by a single nutrient (34 of 51 sites), and nitrogen was the most commonly limiting nutrient, being the sole limiting nutrient at 23 of the sites. However, in the studies where the addition of more than one nutrient was required for increased plant growth, there was only one case where nitrogen was not one of the co-limiting nutrients. The point of this is to say that, even though I have included 16 essential soil nutrients in this chapter, if we are interested in managing growth of plants in a natural setting, it should be possible to focus on only a subset of these.

One question I hope you asked, regarding the prior example study by Olde Venterink, Van der Vliet, and Wassen (2001), is "Why did they choose those three nutrients?" A quick look at Table 7.1 will reveal that, among the soil macronutrients, nitrogen and potassium are present in the highest concentrations within plant tissues. Based on this, we might assume they would be the two most important to consider, but aside from calcium, the remaining macronutrients are present in plant tissues at roughly equivalent concentrations. Recall that the law of the minimum applies to the available supply relative to needs of the plant. So, we must consider not only the amount of resource needed to build plant tissues, but also the amount available to the plant. The approximate percent composition of the earth's continental crust, among our 16 soil nutrients, is given in Table 7.3. It turns out that phosphorus is often present in soil in forms that are relatively insoluble, and the soluble forms are often subject to **leaching** (i.e., being dissolved out of the soil root zone); this sometimes results in phosphorus being much less available to plants than the other remaining macronutrients (Hillel 2008). In addition, we see in Table 7.3 that phosphorus is, on average, about 20–40 times less abundant in the earth's crust than are calcium and magnesium. Thus, because of their individual combinations of requirements and availability, nitrogen, phosphorus, and potassium are often referred to as the three principal macronutrients (Hillel 2008), and these are the three that we usually see in commercial fertilizers, sometimes referred to as N-P-K fertilizer.

7.4.1 NITROGEN

Nitrogen is the soil macronutrient that is present in the highest concentration in plants, forming part of some critically important macromolecules within living tissues: nucleic acids (DNA, RNA) and proteins (within the "amino" part of amino acids). In addition to its occurrence in nucleic acids and proteins, nitrogen is also a key constituent of chlorophylls in the green plants, further emphasizing its importance for essentially all life on earth. In spite of this importance, limitation of plant productivity by nitrogen frequently occurs in nature, outside of human intervention (see, e.g., Vitousek et al. 1997). We saw this in the earlier example from the work of Olde Venterink, Van der Vliet, and Wassen (2001), for a variety of wetland types, but this is also the case for many other ecosystems. Vitousek and Howarth (1991) discussed the prevalence of nitrogen limitation in aquatic and terrestrial ecosystems worldwide and concluded that nitrogen limitation was quite common, although temperate freshwater lakes and tropical coastal marine ecosystems were some of the more prominent exceptions, tending to be limited more by phosphorus than nitrogen. In their review of nutrient limitation among wetland types, Sharitz and Pennings (2006) found that there is considerable variation among published estimates

on nutrient limitation. They also suggested that nutrient limitation might be driven to some degree by the **bioavailability** of nutrients within wetland sediments. For example, nitrogen contained within difficult-to-decompose organic materials or phosphorus that is chemically bound to detritus may be present in high concentrations but unavailable to contribute to plant growth.

7.4.1.1 Nitrogen Transformations in Wetland Soil

Nitrogen exists in a wide diversity of forms in nature (Figure 7.3), spanning chemical oxidation states from -3 to $+5$ (Barbour et al. 1999). Ammonia (NH_3), ammonium (NH_4^+), and the organic nitrogen in amino acids and proteins all represent the most reduced forms of nitrogen, with an oxidation state of -3. At the other end of the redox spectrum, we see that nitrate nitrogen (NO_3^-) has an oxidation state of $+5$. Most of the transformations shown in Figure 7.3 are carried out by some form of soil microorganisms, or more correctly, by one or more species of soil bacteria.

We will begin with nitrogen in its organic form, as incorporated into plant tissues (or other living cells). Once ammonia has been liberated from this organic material by consumption and excretion or decomposition, the ammonia may be converted to nitrite and then to nitrate through a process referred to as **nitrification** (Figure 7.3). This process is carried out stepwise by bacteria in the genus *Nitrosomonas* (converting ammonium to nitrite) and *Nitrobacter* (oxidizing nitrite to nitrate). At this point, nitrate is relatively easily moved about within the water column and/or soil water, and it is highly subject to being leached out of the root zone or carried downstream out of the wetland because of its negative charge. It is also this form that is predominantly taken up by plants, although many species are capable of taking up dissolved ammonium (or both nitrate and ammonium). Nevertheless, much of the nitrogen in a given soil may be lost because of leaching away of nitrate, which results from the incompatibility of the negative charge of the nitrate ion and the prevalence of negatively charged particles within the soil (Fitter and Hay 2002).

The other oxidative pathway illustrated here, termed **anammox** (for *an*aerobic *amm*onium *ox*idation; Figure 7.3), is a bit unusual because it takes place under reducing conditions. Anammox metabolism results in the reaction of ammonium with nitrite, which oxidizes the ammonium to dinitrogen gas (N_2). Anammox metabolism was only relatively recently discovered, in a laboratory setting, through experiments designed to follow up on the mysterious disappearance of ammonium from anaerobic biochemical reactors (Mulder et al. 1995). Later, Thamdrup and Dalsgaard (2002) were among the first to describe evidence of anammox metabolism from natural systems. They found that, under laboratory conditions, as much as 67% of N_2 gas emitted from anaerobic ocean sediments could be attributed to anammox bacterial metabolism. Kartal et al. (2007) mentioned that anammox bacteria have been studied in oceanic upwellings, Arctic marine habitats, the Black Sea, and estuaries and mangrove forests. Zhang et al. (2020) found

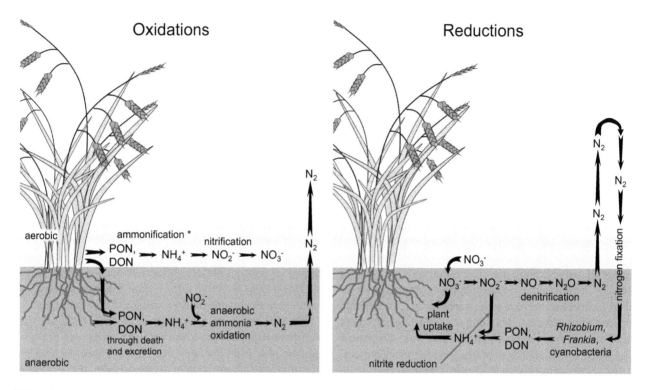

FIGURE 7.3   Some of the Major Pathways of Nitrogen Transformation in Aquatic and Wetland Environments. On the Left are Chemically Oxidative Pathways, Starting with Organic Nitrogen, in the Form of Plant Biomass. As the Plant Tissues Die and Decompose, they are Added to the Particulate (Pon) and Dissolved (Don) Organic Nitrogen Pools. During Decomposition of the Tissues, Organic-Bound Nitrogen is Released in the Form of Ammonium (NH_4^+), Which can be Oxidized to Nitrite (NO_2^-), then Nitrate (NO_3^-; Termed Nitrification). Within the Anaerobic Sediments, Ammonium Released Through Decomposition can be Combined with Nitrite to Yield Nitrogen Gas (N_2), Via a Microbially Mediated Process Referred to as Anammox Metabolism (*An*aerobic *Amm*onia *Ox*idation), Leading to Loss of Gaseous Nitrogen from the Wetland into the Atmosphere. On the Right are Chemically Reducing Processes. On the Left, We see Uptake of Nitrate by the Plants, Which Results in a Reduction of the Nitrate to Organic Amine Groups (such as those in Amino Acids). Plants also can take up Ammonium, but this Does not Lead to a Net Change in Oxidation State of the Nitrogen Atom. In Addition to being taken up by Plants, Nitrate can be Converted to Nitrogen Gas, Via a Process Known as Denitrification. another Important Biological Reduction of Nitrogen in Wetlands and Aquatic Habitats is the Uptake of Nitrogen Gas from the Atmosphere; This is a Microbially Mediated Process Referred to as Nitrogen Fixation. Nitrogen Fixation will be Discussed in more Detail Later in this Chapter. Based on Information in Atlas and Bartha (1998), Barbour et al. (1999), Fitter and Hay (2002), Kartal et al. (2007), and Mitsch and Gosselink (2007).

that, in mangroves of the Zhangjiang Estuary, in southeastern China, anammox could account for 12% or more of the nitrogen lost from deeper, anaerobic sediments. Similarly, Li et al. (2015) found anammox to account for 14%–34% of total nitrogen losses from intertidal wetlands in the Yangtze Estuary on the eastern cost of China, and Risgaard-Petersen et al. (2004) found it contributed 5%–24% of N_2 production in sediments from Danish estuaries. Extensive sampling across the freshwater Baiyangdian Lake in northern China showed similar results for the littoral wetland zones of this lake, with anammox metabolism contributing 11%–35% of total N_2 production from lake sediments (Zhu et al. 2013). In short, data that have been collected so far suggest this anaerobic nitrogen oxidation could have important implications for loss of nitrogen from a diversity of freshwater, brackish, and coastal marine wetlands.

One of the more important pathways contributing to loss of nitrogen from wetlands is the stepwise reduction of nitrate to nitrogen gas, referred to as denitrification

(Figure 7.3). Denitrification is carried out by bacteria in a number of genera, including *Paracoccus*, *Thiobacillus*, *Alcaligenes*, and *Pseudomonas* (Atlas and Bartha 1998), transforming nitrate to dinitrogen gas along the pathway:

$$NO_3^- \rightarrow NO_2^- \rightarrow NO \rightarrow N_2O \rightarrow N_2 \qquad (7.1)$$

The nitrogen gas resulting from this process then can escape from the wetland into the atmosphere. The nitrous oxide intermediate in this process (N_2O) also is gaseous in nature and can escape from wetlands, but it tends to represent a much smaller percentage of nitrogen losses than N_2 does (Mitsch and Gosselink 2007). However, nitrous oxide is recognized as a greenhouse gas, like CO_2 and CH_4 that were discussed earlier, so it is important to be aware of this process as a potential source of some greenhouse gases. Denitrification is functionally the opposite of nitrification, which has the effect of adding nitrate to the water or soil solution, whereas denitrification removes nitrate from the wetland. Thus, we have now seen that three of the major pathways for transformation of nitrogen in nature result in a form of nitrogen that is easily lost by leaching from the root zone or gaseous transfer to the atmosphere. This is one factor that contributes significantly to nitrogen limitation in many aquatic and terrestrial environments (Vitousek and Howarth 1991).

In contrast to denitrification, the other major pathway for nitrogen reduction that we will discuss involves the capture of nitrogen gas from the atmosphere (essentially reversing denitrification, which we just covered). This process is called **nitrogen fixation** and is carried out by quite a large number of free-living and symbiotic bacteria, including cyanobacteria. Some of the genera of cyanobacteria that can fix atmospheric nitrogen are *Anabaena*, *Aphanizomenon*, and *Nostoc*. Although cyanobacteria (once known as "blue-green algae") are photosynthetic, nitrogen fixation in most of these species takes place in cells called **heterocysts** (literally translated to "different cells"), which lack the water-splitting components of the photosynthetic pathway and, thus, do not produce oxygen (Atlas and Bartha 1998). In this way, cyanobacteria are able to maintain the redox conditions required for reduction of N_2 into organic nitrogen. In the cyanobacteria species that do not possess heterocysts, they employ behavioral mechanisms to create the chemically reducing conditions, such as temporal separation of photosynthesis and nitrogen fixation (Meeks 2009). There are other nitrogen-fixing bacteria, such as species in the genera *Rhizobium* and *Frankia*, that form tight symbioses with green plants, and these will be discussed later in the chapter.

Natural incorporation of atmospheric nitrogen into terrestrial and aquatic environments by nitrogen-fixing microbes accounts for roughly two-thirds of the global total of these transfers (Ciais et al. 2013). Lightning, another natural source of nitrogen fixation, contributes another 1% of total global atmospheric nitrogen fixation. The remaining third or so of nitrogen fixation is related to human activities, such as industrial fertilizer production and combustion of fossil fuels (Ciais et al. 2013). There are other significant anthropogenic contributions of nitrogen to natural environments, such as agricultural fertilizer runoff, runoff of nutrient-rich animal manure, and addition of effluent from waste treatment. These inputs of nitrogen have direct impacts on natural ecosystems, are increasing annually as crop production needs increase, and, as would be expected, tend to be highest in areas of high human activity (Glibert 2020; Figure 7.4).

Human activities have additional, indirect consequences on nutrient availability in natural systems. One of these is the acidification of terrestrial and aquatic ecosystems, caused by reaction of nitrogen-containing compounds with atmospheric water vapor to form nitric acid, which falls to the earth as acid precipitation. This acidification causes the leaching of cationic nutrients (e.g., Ca^{2+}, Mg^{2+}, K^+) from the soil, but also can impact nitrogen availability in somewhat complex ways. Increased acidification (i.e., the addition of protons, H^+), interferes with the process of nitrification, with

some research suggesting that nitrification ceases at pH values below about 5.6 (Rudd et al. 1988; Vitousek et al. 1997). At these pH levels, not only was nitrification impaired, but any additional nitrate that entered the system stimulated the process of denitrification, which uses nitrate as a reactant (Rudd et al. 1990). In this way, acidification through acid precipitation can have the effect of reducing access to important cations, but if the acidification results from nitrogen emissions to the atmosphere, it can compound the issue of nitrogen limitation in receiving ecosystems.

An additional means by which human activities can affect the availability of nitrogen in natural ecosystems is via climate change. As was mentioned briefly in Chapter 3 (and as we will see again in Chapter 11), one unfortunate consequence of climate change for northern peatland ecosystems is the increased prevalence of wildfire. Although fire can be beneficial with regard to availability of many soil nutrients, nitrogen can be volatilized into the atmosphere during the burning of vegetation, providing yet another means by which this nutrient can be lost from soils and vegetation (Vitousek and Howarth 1991). In wetlands, another effect of climate change is the periodic occurrence of increased drought intensity and duration (or shifts altogether in precipitation patterns), leading to drying of wetlands. Drying of wetland sediments exposes accumulated sediments to oxygen and results in accelerated decomposition, releasing organically bound nitrogen at much faster than background rates. In the end, the breadth of forms that nitrogen can assume in nature, in combination with the diverse ways that human activities are modifying the global environment, lead to frequent occurrence of nitrogen limitation in natural systems. At the same time that we may see loss of nitrogen from human-impacted terrestrial habitats or drained wetlands, the ever-increasing rates of human inputs of nitrogen from crops, livestock, and urban environments lead to an increased frequency of eutrophication (nutrient enrichment) of aquatic habitats.

7.4.2 Phosphorus

Unlike nitrogen, phosphorus takes relatively few forms in wetlands and aquatic habitats (Figure 7.5), and it most often occurs at a redox state of +5 in natural systems, occurring as one of several forms of phosphate (Reddy and DeLaune 2008). The specific phosphate ion encountered is influenced strongly by pH, with equilibrium transitions adding or removing a hydrogen at pH values of 2.2, 7.2, and 12.4, in these reactions (Reddy and DeLaune 2008):

$$H_3PO_4 \longleftrightarrow H_2PO_4^- \longleftrightarrow HPO_4^{-2} \longleftrightarrow PO_4^{-3} \quad (7.2)$$

Under extremely reduced conditions, as we see in wetlands with considerable organic matter accumulations under anoxic conditions (e.g., redox values below −500 mV; Reddy and DeLaune 2008), phosphates can serve as electron acceptors. In this process, the phosphates are converted to phosphine gas (PH_3), with phosphorus having an oxidation

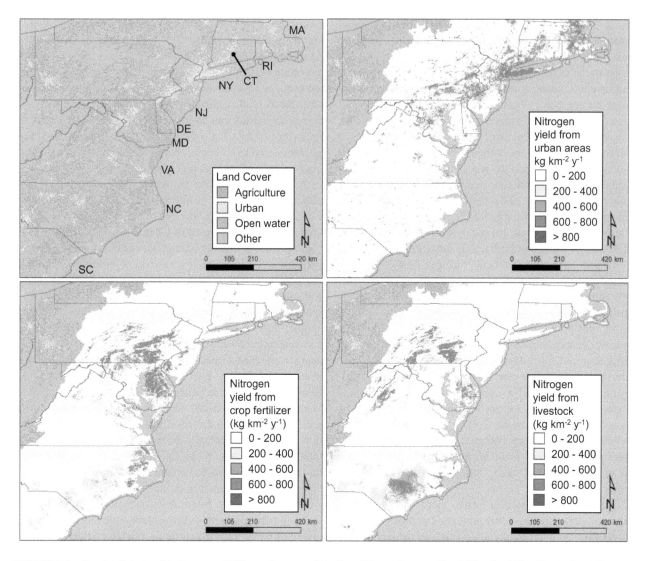

FIGURE 7.4   Spatial Relationship Between Different Sources of Surface Water Nitrogen (Total Dissolved Plus Particulate Nitrogen) and Land Use Along the Atlantic Coast of the United States. Land Cover (Upper Left) is Based on the 2016 Version of the Multi-Resolution Land Characteristics (Mrlc) Land Cover Data, Recategorized to Enhance Distribution of Urban and Agricultural Land Use. Surface Water Nitrogen Data are Based on Data from the US Geological Survey Spatially Referenced Regressions on Watershed (Sparrow) Database. Sparrow Data Accessed from www.usgs.gov/mission-areas/water-resources/science/sparrow-mappers in July 2020; Mrlc Landcover Data Accessed from www.mrlc.gov in july 2020.

state of −3. Phosphine is highly reactive with oxygen and usually ignites on contact with the atmosphere (Atlas and Bartha 1998; Reddy and DeLaune 2008). The resulting combustion and its associated green glow are sometimes referred to as "swamp gas." This phenomenon may even be accompanied by the ignition of methane gas that also is released from highly reduced wetland sediments.

Within wetlands, there are two primary processes influencing the mobility of phosphorus (Figure 7.5). One of these is the cyclical uptake of phosphorus by plants (or other organisms), followed by eventual release during decomposition of plant tissues after death (or after senescence, which is a form of programmed death of tissues in perennial species). The other process is the cyclical adsorption of phosphates to oxidized metals (iron, aluminum,

manganese, calcium, magnesium), followed by release of the phosphate once the phosphate-metal complex precipitates out of solution and into the chemically reduced sediments (Figure 7.6).

The first of these processes (uptake, death, and decomposition) is a relatively long process, dependent upon the life histories of the plant species involved. Some species may produce multiple sets (or cohorts) of leaves, shoots, and so on during a growing season, and these species would tend to put some of that phosphorus back into circulation at shorter intervals than species that produce only one set of tissues per growing season. We will see more about this pattern of growth in Chapter 9 when we talk about vegetative growth and asexual reproduction. The latter process (adsorption, precipitation, and re-release) takes place on

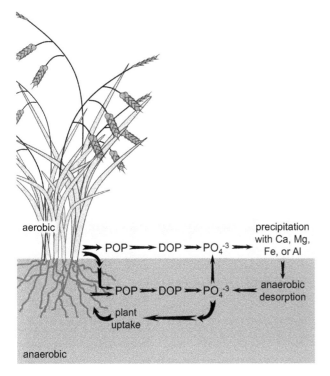

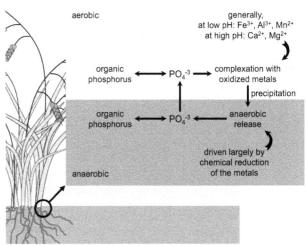

FIGURE 7.6 Interactions of Phosphorus with Water and Sediment Redox Processes. Once Phosphorus is Released from Organic Material, its Cycling within the Water and Sediments is Influenced Strongly by Interactions with Oxidized Cations within the Water Column or Soil/Sediment Water. Complexation of Phosphorus with Oxidized Cations Leads to Precipitation of Phosphorus Out of the Water Column and Deposition into the Sediments. If not taken up by Plants or Released by Sediment Disturbance, Phosphorus Will Tend to Remain within the Sediments, Owing to Cyclical Patterns of Oxygen Production and Release within the Water Column. Uptake by Plants and Eventual Release in Detritus or Plant Exudates are Mechanisms by Which Phosphorus may be Transported from the Sediment Phosphorus Pool Back into the Water Column, Making Plants an Important Link in Wetland Phosphorus Cycling (Figure 7.5). Based on Information in Atlas and Bartha (1998), Mitsch and Gosselink (2007), Reddy and Delaune (2008), and Wetzel (2001).

FIGURE 7.5 Some of the Major Pathways of Phosphorus Transformation in Aquatic and Wetland Environments. Decomposition of Plant Biomass Releases Particulate and Dissolved Organic Phosphorus (Pop, Dop) into the Aboveground and Belowground Phosphorus Pools. Mobility of the Dissolved Phosphorus then is Influenced Primarily by Oxidation State of the Sediments or Water Column and by Uptake into the Biota (Plants, Algae, Bacteria, Etc.). Oxidizing Microhabitats Lead to Precipitation of Phosphorus, often Bound to Oxidized Cations, Whereas Reducing Conditions Release Phosphorus, Enhancing its Mobility. Based on Information in Atlas and Bartha (1998), Mitsch and Gosselink (2007), and Reddy and Delaune (2008).

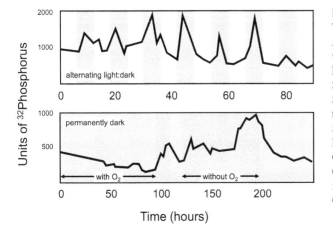

FIGURE 7.7 Release of Phosphorus from Sediments into the Water Column is Influenced Strongly by Oxygen Production by Alga Photosynthesis. In the Upper Panel, Sediment Surface Algae were Exposed to an Alternating Light/Dark Period. During Periods When Algae were Photosynthesizing (In the Light, Non-Shaded Parts of the Graph), Phosphorus was Retained within the Sediments. During Dark Periods, as Oxygen Concentrations in the Water Declined, Phosphorus was Gradually Released from the Sediments. In the Lower Panel, Gray-Shaded Periods Represent Periods during Which Oxygen Levels in the Experimental Chambers were Experimentally Reduced by Bubbling Nitrogen Gas through the Chambers. This Led to Release of Phosphorus from the Sediments during those Periods of Reduced Oxygen Concentration. Based on Data in Carlton and Wetzel (1988).

much shorter cycles, driven by cycles of oxygen availability within (or atop) the sediments. For example, Carlton and Wetzel (1988) found evidence of this in flow-through experimental chambers, driven by dissolved oxygen concentrations (Figure 7.7; from experiments that were discussed in Chapter 5). In chambers that were exposed to alternating

light/dark cycles, there were gradual declines of dissolved phosphorus during the light periods, presumably related to photosynthesis of algae atop the surface of the sediments. During the dark periods, however, concentrations of phosphorus increased in the water column. Similarly, in chambers that were kept dark but where dissolved oxygen

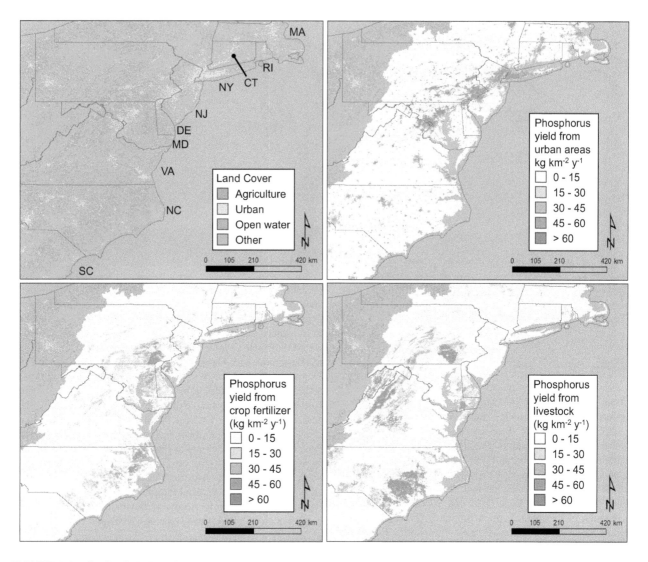

FIGURE 7.8 Spatial Relationship between Different Sources of Surface Water Phosphorus (Total Dissolved Plus Particulate Phosphorus) and Land use Along the Atlantic Coast of the United States. Land Cover (Upper Left) is Based on the 2016 Version of the Multi-Resolution Land Characteristics (Mrlc) Land Cover Data, Recategorized to Enhance Distribution of Urban and Agricultural Land Use. Surface Water Phosphorus Data are Based on Data from the US Geological Survey Spatially Referenced Regressions on Watershed (Sparrow) Database. Sparrow Data Accessed from www.usgs.gov/mission-areas/water-resources/science/sparrow-mappers in July 2020; Mrlc Landcover Data Accessed from www.mrlc.gov in july 2020.

concentrations were experimentally manipulated, concentrations of phosphorus in the water increased substantially during periods when oxygen was removed from the water and declined when oxygen was added.

As with nitrogen, we see close correspondence between human activities and phosphorus loads in surface waters (Figure 7.8). Similar to the maps shown here, Robertson, Saad, and Schwarz (2014), found that phosphorus transport in waters of the Mississippi River basin (US) was most strongly influenced by animal-based agriculture and density of high-volume wastewater treatment plants. This was in contrast with nitrogen loads in the same waters, which tended to be more strongly influenced by row crop agricultural land cover (see also Figure 7.4). Globally, Seitzinger et al. (2010) saw very similar patterns, with agricultural nutrient management being an important determinant of

surface water nitrogen loads, whereas sewage treatment and use of phosphate-based detergents seemed to have the greatest impact on phosphorus loads. Because nitrogen and phosphorus tend to be the most important plant (and algal) nutrients determining limitations on primary productivity in terrestrial and aquatic ecosystems, there is considerable interest in managing anthropogenic sources of these nutrients. Wetlands are one tool that is often used in this regard, and we will see more about this in Chapter 11.

7.4.3 Other Soil Macronutrients

Whereas nitrogen and phosphorus are incorporated into important biological **macromolecules** (nucleic acids and proteins), potassium is more freely circulated with plant tissues, serving roles in ionic balance, turgor pressure, and enzyme activity (Table 7.1). We also saw the involvement of potassium as a co-transported ion in the uptake of bicarbonate in Chapter 6 (Figure 6.32). This *lability* (freedom of movement) also makes potassium one of the nutrients that is most easily leached from soils and from detritus (dead organic matter) within both terrestrial and aquatic ecosystems (Barbour et al. 1999). Somewhat ironically, then, potassium makes up a significant fraction of tissues (after carbon, hydrogen, and oxygen) but is somewhat of a transient nutrient both within plant tissues and within the ecosystems they inhabit. In spite of its transient nature, potassium constitutes almost 3% of the mineral content of the earth's upper crust, making it the fourth most abundant of our 16 soil nutrients.

Calcium and magnesium constitute about 0.5% and 0.2% of plant tissues but are the third and sixth most abundant soil nutrients within the earth's upper crust (Tables 7.1 and 7.3). Calcium is a constituent of cell walls but also important in cell-to-cell signaling and membrane permeability. Magnesium serves an important role in plants as a component of chlorophyll, but it also serves as a cofactor for some enzymes (as does calcium). Both calcium and magnesium are divalent cations (i.e., carry a charge of +2) and thus are important in soil fertility, occupying cation exchange sites on the predominantly negatively charged soil particles. These two nutrients behave similarly in the environment, owing to their chemical similarities, with the exception that magnesium carbonates are more soluble in water (Atlas and Bartha 1998). Because they are positively charged ions, both calcium and magnesium are affected by pH, and both can be readily leached from soils under acidic conditions, such as in soils receiving inputs of acidic precipitation.

Sulfur is somewhat equivalently represented in plant tissues (sixth most abundant) and in the mineral component of the upper crust (seventh most abundant). Because of this relatively balanced availability, sulfur is infrequently encountered as a limiting nutrient in wetlands (Mitsch and Gosselink 2007). Sulfur is quite important biologically, as it is used in proteins and as a component of certain coenzymes and vitamins (Taiz et al. 2015). Sulfur is a bit more dynamic in nature than is phosphorus, occurring at oxidation states from −2 to +6 and regularly taking on a gaseous state in wetlands as hydrogen sulfide (H_2S) or dimethyl sulfide (DMS) gas, the latter of which is more common in marine habitats (Figure 7.9; Atlas and Bartha 1998; Mitsch and Gosselink 2015). As we discussed previously, reduction of sulfur under highly anaerobic conditions can result in the accumulation of H_2S within the sediments, which subsequently can impact plants via direct toxicity to roots or indirect impacts via chemical complexation with zinc, copper, or other trace metal nutrients (Mitsch and Gosselink 2015).

The final of the soil macronutrients, silicon (Si), is by far the most disproportionately abundant nutrient, relative to its contribution to plant tissues (30% of upper crust vs. 0.1% of plant tissues). Silicon serves as a component of cell walls, and its accumulation there varies widely among plant species. According to Taiz et al. (2015), species in

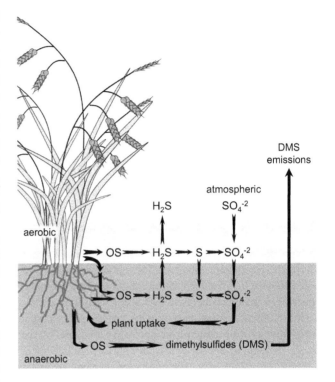

FIGURE 7.9 Some of the Major Pathways of Sulfur Transformation in Aquatic and Wetland Environments. As with other Mineral Nutrients, Decomposition of Plant Biomass Releases Particulate and Dissolved Organic Sulfur (Os) into the Aboveground and Belowground Sulfur Pools. Sulfur Cycling Involves Some Gaseous Exchange with the Atmosphere, with Hydrogen Sulfide and Dimethylsulfide Gases (From Reduction of Dimethylsulfates) being Released from the Wetland Sulfur Pools, as a Result of Anaerobic Sediment Microbe Metabolism. In Freshwater Systems, these Gaseous Releases are Dominated by H_2s. Otherwise, Sulfur is Dominated by Particulate and Dissolved Forms of Sulfur, with Rates of Deposition of Atmospheric Sulfur (Sulfates) being Determined to a Large Degree by Local and Regional Human Activities. Based on Information in Atlas and Bartha (1998) and Mitsch and Gosselink (2015).

the Equisetaceae (*Equisetum* species, or horsetails) are the only plants that require silicon to complete their life cycles. Other species will accumulate silicon in their cell walls to varying degrees, but *Equisetum* species accumulate so much silicon that one of their common names is "scouring rush." These species have been used by humans as a tool for cleaning cooking and eating ware, and even as a form of sandpaper, because of the silicon accumulations (Knowlton 2012; Taiz et al. 2015).

7.5 SOIL MICRONUTRIENTS

The soil micronutrients are so named because they are required in much smaller quantities than the macronutrients we have just covered. Percentages of these nutrients in plant tissues range from about 0.01% down to 0.00001% (or ten parts per million; Table 7.2). The most abundant of these in plant tissues is chlorine, but often this abundance results from chlorine accumulation in saline soils, such as in salt marshes, as one of many adaptations for water uptake and water conservation in these habitats (Batzer and Sharitz 2006). The other micronutrients make up a much smaller fraction of plant biomass but nonetheless are important for plant metabolism.

7.5.1 Influence of Redox Chemistry on Iron and Manganese

We learned a bit about the involvement of manganese and iron in the metabolism of sediment microbiota in Chapter 5 and again saw some of this in Table 7.4. Once sediment supplies of available oxygen and nitrogen are exhausted, these are typically the first two electron acceptors playing a role in hypoxic sediment metabolism. The relative availability of these nutrients is influenced by microbially driven chemical reduction of these elements, which itself usually proceeds in a hierarchical fashion, based on the energy each element yields via microbial metabolism. As we saw in Table 7.4, the specific amount of energy extracted by using each element as a terminal electron acceptor may vary among different carbon sources (that is, different sources of energy used by sediment microbes), but the relative energetic ranking among these four alternative electron acceptors is generally consistent within a single energy source. This consistency yields the sequential changes in availability of these nutrients that we discussed in Chapter 5 (Figure 5.25).

Chemically reduced iron and manganese can interfere with plant proteins by binding directly to root surface proteins, can interfere with transport across cell membranes, and can impact photosynthesis and cellular respiration, among other effects (Table 5.2). Oxidized forms of these elements also are involved in the cycling of phosphorus in wetlands (Figures 7.5, 7.6, and 7.10). Both iron and manganese, in their oxidized forms, are capable of pulling dissolved phosphorus out of solution and into the sediments, where chemically reducing conditions liberate the phosphorus, making it more mobile and accessible for plant uptake.

Related to this is the impact of radial oxygen loss (ROL) from roots on iron within the rhizosphere. As we saw in Chapter 6, ROL can protect plants to some degree from damage that may result from exposure to reduced metals, such as iron and manganese. At the same time, oxidized iron in the rhizosphere can complex with dissolved phosphate, immobilizing the phosphorus and again making it unavailable to the plants (Mitsch and Gosselink 2015). Oxidized iron from this interaction with the roots also is one of the diagnostic features used to determine whether soils in an area indicate wetland hydrology (Figure 7.11).

7.5.2 Other Soil Micronutrients

The remaining micronutrients are required in quite small quantities by plants, all being present in concentrations at or below 0.01% in plant tissues (Table 7.2). Most of these

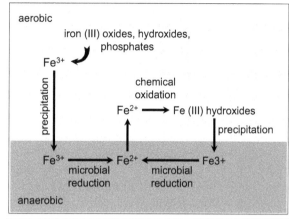

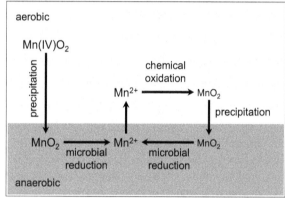

FIGURE 7.10 Cycling of Iron and Manganese in Wetlands are very Similar Processes. Both are Driven to a Large Degree by Redox Processes and Involve Precipitation of Oxidized Forms into the Sediments, often in Conjunction with Dissolved Phosphorus. Within the Sediments, Microbial Reduction of These Metals can Lead to their Diffusion Back into the Water Column, Where Chemical Oxidation Causes the Cycle to Repeat. Based on Information in Atlas and Bartha (1998), Mitsch and Gosselink (2007), Reddy and Delaune (2008), and Wetzel (2001).

nutrients are involved in enzyme activity. Two exceptions to this are sodium and boron. Sodium is involved in ionic balance and in the regeneration of phosphoenolpyruvate (PEP) in species using C_4 or CAM photosynthesis (Taiz et al. 2015). Recall from Chapter 6 that PEP is the compound to which carbon is attached, by PEP carboxylase, during carbon fixation in those species exhibiting C_4 or CAM photosynthesis. Boron has many roles in cellular metabolism and is involved in cell wall structure. Among suggested roles of boron in cellular metabolism are an involvement in nucleic acid synthesis and in cell cycle regulation (Taiz et al. 2015).

Zinc, copper, and molybdenum are all involved in one way or another with the activity of plant and/or microbial enzymes (Taiz et al. 2015). Zinc is important for the function of alcohol dehydrogenase, carbonic anhydrase, and other enzymes associated with forms of carbohydrate metabolism, in addition to having a role in chlorophyll synthesis in some species. Copper is associated with various enzymes involved in nitrogen metabolism and in the function of plastocyanin, which helps to shuttle around electrons

FIGURE 7.11   Evidence of Iron Redox Processes in the Soil and Rhizosphere of *Rhynchospora Corniculata*. The Gray-Colored Soil Matrix Indicates Reducing Conditions, whereas the Reddish Color Along the Roots themselves and on the Immediately Adjacent Soil Indicates Oxidation of Iron Induced By Oxygen Leakage from the Roots.

during the light-dependent reactions of photosynthesis. Molybdenum is a component of nitrate reductase and nitrogenase, both important enzymes in nitrogen metabolism; thus, molybdenum deficiency can have important implications for many other aspects of plant nutrition.

7.6   PLANT ADAPTATIONS FOR NUTRIENT ACCESS

As discussed throughout the preceding sections of this chapter, there are many scenarios in which aquatic and wetland plants may find themselves lacking in one or more of these critical nutrients. This may be the case chronically in some habitats, owing to sources of water inputs to the wetland, to rates of decomposition and nutrient release, to underlying geology, to the mix of potentially competing neighbor species, or to the degree of soil development, as a few examples. In such cases, we often find plant species may take advantage of one or more of a suite of biological adaptations, three of which are discussed in more detail next.

7.6.1   Nitrogen-Fixing Symbionts

We saw earlier that nitrogen fixation refers to the process of converting atmospheric dinitrogen gas (N_2) into reduced nitrogen, such as NH_3. Nitrogen-fixing prokaryotes account for roughly two-thirds of global nitrogen fixation, with the remaining third coming from lightning and industrial processes (Ciais et al. 2013). Biological nitrogen fixation is carried out by both free-living and symbiotic bacteria, with

symbiotic nitrogen fixation accounting for roughly one-fourth of biological fixation (Dresler-Nurmi et al. 2009). As mentioned in section 7.4.1.1, there are three categories to which the symbiotic nitrogen fixers belong: cyanobacteria, rhizobia, and bacteria of the genus *Frankia* (a genus of actinomycete bacteria; Dresler-Nurmi et al. 2009). In all of these, there is some degree of resource exchange between the partners in the **symbiosis**, with the plant host receiving a percentage of their nitrogen needs from the bacterial symbiont, and the bacteria benefitting from plant host–supplied carbohydrates.

7.6.1.1   Cyanobacteria

Cyanobacteria appeared somewhere in the range of 3.5 to 2.5 billion years ago, but the ancestors of those that form heterocysts (the differentiated cells that maintain hypoxic conditions for nitrogen fixation) are thought to have appeared more recently, at around 1.5 billion years ago (Meeks 2009). A group of about four genera of cyanobacteria (*Nostoc, Anabaena, Colothrix,* and *Chlorogloeopsis*) have been cultured from plant-cyanobacterial associations, but *Nostoc* species are the most commonly encountered nitrogen-fixing symbiont (Meeks 2009; Raven 2002). Cyanobacterial symbionts have been found in association with hornworts, liverworts, mosses (although this is considered to be rare), monilophytes (including *Azolla* species), cycads (the only group of extant gymnosperms known to form these associations), and one family of angiosperms (Gunneraceae; Meeks 2009; Raven 2002). All of these, except the cycads, are commonly encountered in wet habitats (wetlands, rainforests, etc.), and all of the approximately 50 examined species of

Gunnera (the only genus in the Gunneraceae, with about 60 total species) are known to form symbioses with *Nostoc* species (Stevens 2017; Raven 2002).

In the symbioses with *Gunnera*, the cyanobacterial symbionts are hosted within specialized "stem gland" cells near the leaf petioles (Johansson and Bergman 1992; Meeks 2009; Stevens 2017). The association is initiated via chemical communication between motile *Nostoc* filaments (termed hormogonia) and the host plant, followed by migration of the *Nostoc* cells into the *Gunnera* stem glands. The cyanobacterium enters into the plant tissues via intercellular channels (i.e., outside, but between, the plant's cells), eventually entering and colonizing cells within the gland tissues and establishing a symbiotic association with the plant. This is reported to be the only plant-cyanobacterial symbiosis in which the cyanobacterium takes up residence inside the plant cells (Meeks 2009). In this symbiosis, the *Nostoc* symbiont releases about 90% of the fixed nitrogen (in the form of ammonium, NH_4^+), which would then be available to the *Gunnera* host for its uptake and use.

In *Azolla* species (aquatic monilophyte ferns), the colonization process is a bit simpler, with the cyanobacterium being passed to the next generation in *Azolla* spores (Carrapico 2010; Eily, Pryer, and Li 2019; Meeks 2009). In this association, the cyanobacteria reside within a cavity in the *Azolla* fronds (the fern equivalent of leaves; Figure 7.12). The identity of the cyanobacterial symbiont appears to be inconclusively determined as either *Nostoc* or *Anabaena* species (or both); the difficulty in identification stems from problems in culturing the symbiont in laboratory studies. Data suggest wide variation in potential resource exchange between symbiotic cyanobacteria and their plant hosts. In *Azolla caroliniana*, it appears that the cyanobacterium

contributes 5% or less of the total carbon fixation for the symbiosis, whereas in the hornwort *Anthoceros punctatus*, this contribution may be as high as 30% of total photosynthetic carbon fixation (Meeks 2009). A slightly higher fraction of nitrogen fixed by the symbiont (~40%) is released as NH_4^+, which is then accessible to the *Azolla* host. Because of its combination of nitrogen-fixing ability and growth as a free-floating aquatic plant, *Azolla* has long been used in India and Asia as an organic fertilizer, especially in rice culture, where its use has been shown to be much more sustainable than the use of industrially produced fertilizers (Figure 7.13; Wagner 1997).

7.6.1.2 Rhizobia

Rhizobia are perhaps the best-known group of nitrogen-fixing symbionts, occurring in **legumes** (Fabaceae) including beans, peas, peanuts, and clover, among others. Some wetland taxa that host rhizobium symbionts include *Aeschynomene*, *Mimosa*, *Neptunia*, *Sesbania*, and *Vigna* (James et al. 2001), as well as *Apios*, which we saw in Figure 2.32. In contrast to the cyanobacterial symbioses earlier, rhizobial symbionts usually inhabit nodules within the root tissues of host plants, entering into these relationships through a variety of mechanisms, most of which involve infection of roots hairs of the host plant (Ibáñez, Wall, and Fabra 2017). There are at least 55 species of rhizobia, scattered among a diversity of evolutionary lineages that include many other bacterial life history types (Dresler-Nurmi et al. 2009). Although rhizobium symbiosis is quite common in the Fabaceae, it is neither universal within nor restricted to the family (Doyle and Luckow 2003). The Fabaceae can be divided into three major subgroups (Mimosoideae, Papilionoideae, and Caesalpinioideae), the first two of which exhibit rhizobium

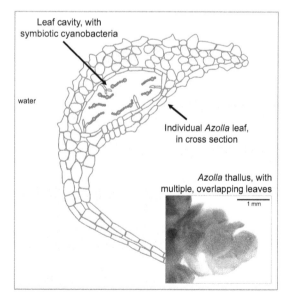

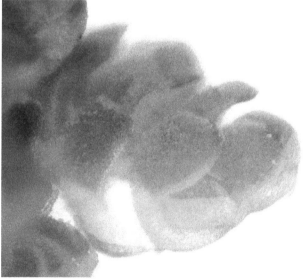

FIGURE 7.12 *Azolla* Leaf (Frond) Internal Anatomy, Showing General Location of Endosymbiotic, Nitrogen-Fixing *Nostoc* Colony Within an Individual Leaf (Redrawn from Carrapiço 2010). Inset Photo Shows the Imbricate (Overlapping) Nature of Individual Leaves Along a Frond of *Azolla*.

FIGURE 7.13   Demonstration Farm Displaying Different Species of the Aquatic Fern *Azolla* in use as an Organic Nitrogen Fertilizer at the Philippine Rice Research Institute in Nueva Ecija, Philippines. Top Photo is a Culture of *Azolla Caroliniana* Grown with Taro (*Colocasia Esculenta*); Sign in the Bottom Photo Describes the Use of *A. Caroliniana*, *A. Microphylla*, *A. Pinnata* Var. *Imbricata*, and *A. Pinnata* Var. *Pinnata* in Lowland Agriculture. By Judgefloro—Own Work, CC BY-SA 4.0, Top: https://commons.wikimedia.org/w/index.php?curid=42618342; Bottom: https://commons.wikimedia.org/w/index.php?curid=42618343.

nodulation in about 90% of their genera, in contrast to the Caesalpinioideae, where only about 5% of the genera display nodulation (Doyle and Luckow 2003). Given that there are roughly 20,000 species in the Fabaceae, nitrogen-fixing rhizobia are incredibly important components of almost every ecosystem type.

7.6.1.3 *Frankia*

There are 25 or so genera of angiosperms known to have species forming a symbiotic association with species of *Frankia*, a group of actinomycete bacteria. The plants involved in these symbioses, as a group, are often referred to as actinorhizal (compared with the term rhizobial, concerning the plants forming associations with rhizobium species, discussed earlier). Three actinorhizal genera are frequently encountered in wetlands: *Morella* and *Myrica*, in the Myricaceae (wax myrtle family), and *Alnus* in the Betulaceae (birch family; Figure 7.14).

The taxonomy of *Myrica* species was relatively recently revised, with only two species left as *Myrica* species and the remaining species assigned to the genus *Morella* (Herbert 2005a, 2005b). The genus *Myrica* comprises two species of temperate latitude shrubs (*M. gale* and *M. hartwegii*), both of which are known to occur in wetland and/or riparian habitats (Bornstein 1997). *Morella* is much more diverse, with more than 40 species occupying a diversity of habitats, such as sand dunes, volcanic deposits, and other nutrient-poor areas, in addition to riparian and wetland habitats (Herbert 2005a). The genus *Alnus* includes about 25 species of shrubs, mostly of northern temperate latitudes, although there are some species in high-altitude habitats of the Andes (Stevens 2017). *Alnus* species are frequently encountered in temperate riparian ecosystems, including in and around beaver impoundments.

Carbon and nitrogen fixation appear to be closely linked in these actinorhizal symbioses, with increases or declines in host photosynthesis rates being correlated with concomitant changes in symbiont nitrogen fixation rates (Persson and Huss-Danell 2009). Similarly, it is reported that density of root nodules and production of enzymes involved in bacterial nitrogen fixation are affected by host plant metabolism. As might be expected, based on earlier discussions of nutrient limitation, the access to nitrogen provided in these symbioses shifts the nutrient limitation from nitrogen to phosphorus for actinorhizal species. Experiments involving fertilization with phosphate fertilizers showed that fertilization of plots dominated by speckled alder (*Alnus incana* subspecies *rugosa*) resulted in stimulation of both plant growth and bacterial nitrogen fixation activity (Gökkaya, Hurd, and Raynal 2006). Research in south Alaskan

FIGURE 7.14   Left: *Frankia* Nodules on Roots of Hazel Alder (*Alnus Serrulata*); Right: Stem of *A. Serrulata* with Recently Formed Catkins for the Next Year's Flowering (Photo Taken in September in Mississippi, USA).

watersheds ranging in size from about 5 to 12 km² also has shown that stands of alder in stream watersheds appear to influence dissolved nitrogen in the streams themselves, even when alder coverage is as low as 2% of the watershed (Hiatt et al. 2017). Thus, these plant-microbe interactions can have far-reaching effects, especially in ecosystems that are somewhat more remote from direct human interference.

7.6.2 Mycorrhizae

Another common type of plant-microbe symbiosis involves association between plant roots and fungi. This category of symbiosis, referred to as **mycorrhizae** (myco = fungus; rhiz = root), involves the fungi associating themselves with plant roots, exchanging resources such as water and nutrients for carbohydrates from the plant and sometimes providing protection to the plant against pathogens or even potential competitors (Table 7.6). There are four major categories of mycorrhizal symbioses, categorized by morphology (ectomycorrhizae and arbuscular mycorrhizae; Figure 7.15) or by the taxa of plant with which they associate (ericoid and orchid mycorrhizae).

The oldest known fossil record of a mycorrhizal association comes from approximately 400 Mya (Brundrett and Tedersoo 2018). This particular fossil provided evidence of arbuscular mycorrhizae in symbiosis with what is potentially a very early species of vascular plant (*Aglaophyton major*), although the characteristics of the plant do not match perfectly with most other known vascular plant taxa (Wellman, Kerp, and Hass 2006). It is also interesting to note that some fossils of *Aglaophyton major* also show evidence of endosymbiotic cyanobacteria (Krings et al. 2009). Very well-preserved fossils of this plant show that the cyanobacteria entered the plant through stomatal pores, colonized the air spaces just inside the stomata, and then spread into other tissues, including tissues already well colonized by the arbuscular mycorrhizal fungi (illustrated with micrographs in Krings et al. 2009).

Because of the very long history of mycorrhizal associations, it is no surprise that mycorrhizae appear throughout the green plants, from liverworts and hornworts onwards (Table 7.6). About 72% of angiosperm species harbor arbuscular mycorrhizae, while only approximately 2% associate with ectomycorrhizae, 1.5% with ericoid mycorrhizae, and fewer than 5% with orchid mycorrhizae (based on the estimate of approximately 18,000 extant orchid species; Brundrett and Tedersoo 2018). Brundrett (2017) indicated that there are approximately 2,300 aquatic plant species that are not known to be mycorrhizal or are only occasionally associated with mycorrhizae. In the latter case, development of mycorrhizal associations appears to be dependent on the habitat the plants occupy, the environmental conditions within the habitat, and/or the neighboring species.

Although Brundrett (2017) listed at least 20 common aquatic and wetland plant families that were not known to include mycorrhizal species, many of those families have been demonstrated in other work to include members that form mycorrhizal symbioses. Among those families previously thought to be non-mycorrhizal that are now known to form mycorrhizal associations are the Brassicaceae and Typhaceae (Bohrer, Friese, and Amon 2004); Hydrocharitaceae and Ruppiaceae (Khan 2004); Droseraceae (Radhika and Rodrigues 2007); and Commelinaceae, Fabaceae, Juncaceae, and Polygonaceae (Fusconi and Mucciarelli 2018). In addition, a study of the microbiome (that is, all of the combined microbial genomes)

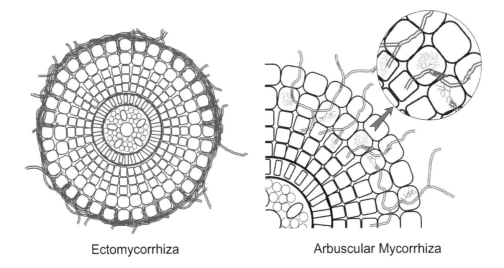

Ectomycorrhiza   Arbuscular Mycorrhiza

FIGURE 7.15 Ectomycorrhizae (Left) are Characterized Largely by the Extracellular Networks of Fungal Hyphae (Termed Hartig Net) that Surround Cells in the Outer Layers of the Roots of Infected Plants. The Roots are often Stunted in Morphology, Appearing Shorter and Thicker than Uninfected Roots. Arbuscular Mycorrhizae (Right), on the Other Hand, Grow More Deeply into the Plant Root and Penetrate the Cells Themselves, Usually Producing Highly Branched Structures Referred to as Arbuscules within Some of the Infected Cells. Based, in Part, on Drawings in Moore-Landecker (1996) and Photos in Brundrett and Tedersoo (2018) and Fitter and Hay (2002).

TABLE 7.6

Characteristics of the major groups of mycorrhizal fungi. Information from Brundrett and Tedersoo (2018), Fitter and Hay (2002), and Pressel et al. (2010).

	Ectomycorrhizae	Arbuscular Mycorrhizae	Ericoid Mycorrhizae	Orchid Mycorrhizae
Plant taxa involved	Relatively diverse; present in extant gymnosperms and angiosperms	Highly diverse; present in liverworts, hornworts, and throughout the vascular plants	Found in the order Ericales	Only in the Orchidaceae
Benefits to the plant	Mineral and organic nutrients	Mineral nutrients	Mineral and organic nutrients	Mineral and organic nutrients, carbon energy
Benefits to the fungus	Carbon energy, water	Carbon energy, water	Carbon energy, chemical protection from competitors	None?
Important morphological characteristics	Extracellular Hartig net in outer cell layers and hyphal sheath around roots	Intracellular hyphal penetration and formation of branched arbuscular hyphae within cells	Coils of fungal hyphae within cells, with cellular colonization from root surface only	Coils of fungal hyphae within cells, with cellular colonization from surface or adjacent root cells

associated with the seagrass *Zostera muelleri* showed genetic evidence for mycorrhizal fungi associated with non-photosynthetic portions of the plant (Hurtado-McCormick et al. 2019). Their study revealed that all *Z. muelleri* plants sampled from two coastal and two estuarine habitats, along an 86 km stretch of Australian coastline, yielded genetic evidence of this association. Another study of the seagrass *Posidonia oceanica* revealed a potential mycorrhizal symbiont isolated from 32 locations along more than 2,000 km of the northwestern coast of the Mediterranean Sea (Vohník et al. 2019). Thus, it seems that even in aquatic environments thought to be free of mycorrhizal symbioses, further exploration may reveal this relationship to be important and may even provide insight into new mechanisms for mycorrhizal symbioses (e.g., Borovec and Vohník 2018; Vohník et al. 2019).

Potential benefits to the plant in these symbioses include increased access to nutrients (especially phosphorus and nitrogen), increased access to soil water, and protection from pathogens, toxins, and drought/salinity stress (Table 7.6; Fitter and Hay 2002; Moore-Landecker 1996). The fungal symbiont essentially increases the radius of soil around the root system from which resources can be acquired by roughly sevenfold, adding the capability to degrade soil organic material to release otherwise inaccessible mineral nutrients (Fitter and Hay 2002). In exchange for providing these services to the plant, the fungus receives supplemental energy from the plant, in the form of carbohydrates.

Within wetlands, we see mostly arbuscular mycorrhizal associations, aside from some wetland tree species that may harbor ectomycorrhizae or ericoid mycorrhizae, and, of course, wetland orchid species, which host orchid mycorrhizae. This corresponds with the data from Brundrett and Tedersoo (2018), which showed that more than 70% of angiosperm species host arbuscular mycorrhizae. In studies examining the prevalence of mycorrhizae in wetlands, it seems that patterns of water depth, flooding, and soil

moisture availability are important determinants of the rate of root colonization (Ray and Inouye 2006; Wetzel and Van der Valk 1996). However, Bohrer, Friese, and Amon (2004) found that the season during which plants are examined might have more influence on colonization rate than wetland hydrology itself. Ramírez-Viga et al. (2018) conducted a detailed analysis of 48 published experiments examining benefits provided to wetland plants by mycorrhizal fungi. This analysis showed that experimental inoculation of plants with arbuscular fungi led to the following benefits for the plants:

- Reduced impacts of experimentally imposed salinity stress;
- Enhanced photosynthesis;
- Increased biomass; and
- Improved nutrient acquisition and assimilation, with stronger effects on uptake of phosphorus than for nitrogen.

A final note regarding microbial symbioses is that plants may harbor more than one type of microbial symbiont. As mentioned earlier regarding fossilized *Aglaophyton major*, plants may harbor both mycorrhizal symbionts and nitrogen-fixing bacterial symbionts, and *A. major* provides evidence for this from at least 400 million years ago. One example of an extant wetland taxon that hosts both mycorrhizae and nitrogen-fixing bacteria is the wetland shrub genus *Alnus* (alders, Figure 7.14). As seen in the earlier discussions of nitrogen-fixing symbioses, *Alnus* species host bacteria in the genus *Frankia* that help provide access to additional nitrogen. They also are capable of hosting both arbuscular mycorrhizae and ectomycorrhizae, with the ectomycorrhizae being more prevalent in adult trees and, thus, more prevalent in mature stands of alder (Tedersoo et al. 2009). Experimental work by Strukova, Vosátka, and Pokorný (1996) showed that there sometimes may be complex interactions among the plant, fungus, and bacteria, with the potential for competition between the two microbial symbionts for resources from the plant. Nevertheless, both of the symbioses have been shown to enhance growth of young *Alnus* seedlings under a variety of experimental conditions.

7.6.3 CARNIVOROUS PLANTS

The final biological adaptation we will discuss in regard to nutrient augmentation for wetland plants is the capture of invertebrates as an organic nutrient source, and these plants usually are called **carnivorous plants** (Table 7.7). Although these plants sometimes are referred to as insectivorous, perhaps, carnivorous is the better term in this regard, as the approaches used by these plants do not really discriminate between insects and other forms of invertebrate animals. I introduce the term insectivory here largely because this is the term used by Charles Darwin in the title of his book on this topic (*Insectivorous Plants*, Darwin 1875).

As are many of us, Darwin was enthralled with these plants, and he spent quite a lot of time exploring their behavior and cataloging his findings. He conducted numerous experiments with the sundew *Drosera rotundifolia* (Figure 7.16), examining the movements of leaves and the hairs (also known as **glandular trichomes**) that are used to capture prey, the process of prey digestion, and the influence of different mineral nutrients on movement of the glandular trichomes. He also conducted similar but less exhaustive studies on other species including *Dionaea muscipula* (Venus' flytrap), *Aldrovanda vesiculosa*, *Pinguicula vulgaris*, *Utricularia neglecta*, and a few others, including some that are not usually considered as carnivorous (e.g., *Saxifraga* species).

Arber (1920) also included a few examples of *Utricularia* species and *Aldrovanda*, but because she was focused almost exclusively on truly aquatic plants, these few species were the limit of her exploration of carnivory. A much more detailed recent treatment of carnivorous plants within a wetland context was provided by Cronk and Fennessy (2001), who dedicated a major portion of one chapter to discussing these plants' biological, taxonomic, and ecological diversity. Much of the information in Table 7.7 was gleaned from two tables included in Cronk and Fennessy (2001), with updates based on Stevens (2017) and internet resources provided via eFloras (2008). Incidentally, Stevens

TABLE 7.7

Taxonomic and geographic distribution of carnivorous plants. Based on information in Cronk and Fennessy (2001), eFloras (2008), Ellison and Gotelli (2001), Li et al. (2019), Lin et al. (2021), Plachno et al. (2006), Stevens (2017).

Family, Order (Age, Mya)	Genus	Number of Species	Distribution	Trap Type
Droseraceae, Caryophyllales (87)	*Aldrovanda*	1	Eastern hemisphere	Snap trap
	Dionaea	1	North and South Carolina, USA	Snap trap
	Drosera	200	Cosmopolitan	Active adhesive
Nepenthaceae, Caryophyllales (87)	*Nepenthes*	176	Eastern hemisphere tropics, except Africa	Pitfall (pitcher)
Sarraceniaceae, Ericales (87)	*Darlingtonia*	1	Northwestern North America	Pitfall (pitcher)
	Heliamphora	15	Northeastern South America	Pitfall (pitcher)
	Sarracenia	11	Eastern North America	Pitfall (pitcher)
Cephalotaceae, Oxalidales (79)	*Cephalotus*	1	Southwestern Australia	Pitfall (pitcher)
Tofieldiaceae, Alismatales (60–80)	*Triantha*	4	North America, Japan	Passive adhesive
Dioncophyllaceae, Caryophyllales (~50)	*Triphyophyllum*	1	Tropical western Africa	Passive adhesive
Plantaginaceae, Lamiales (<50)	*Philcoxia*	7	Brazil	Passive adhesive
Drosophyllaceae, Caryophyllales (~40)	*Drosophyllum*	1	Iberian Peninsula, Morocco	Passive adhesive
Byblidaceae, Lamiales (36)	*Byblis*	< 10	Southwestern and northern Australia, New Guinea	Passive adhesive
Lentibulariaceae, Lamiales (30)	*Genlisea*	30	Tropics of western Africa and eastern South America	Lobster pot
	Pinguicula	80	Throughout temperate northern hemisphere, a few species in South America	Active adhesive
	Utricularia	240	Cosmopolitan	Bladder
Roridulaceae, Ericales (10–20)	*Roridula*	2	Southern Africa	Passive adhesive
Bromeliaceae, Poales (14)	*Brocchinia*	20	Tropical South America	Pitfall (leaf rosette pitchers)
	Catopsis	30	Mexico, Caribbean, South America	Pitfall (leaf rosette pitchers)

FIGURE 7.16   Common Carnivorous Plants from Near the Gulf of Mexico Coast in Alabama and Mississippi (USA). The Sundews *Drosera Rotundifolia* (A) and *Drosera Tracyi* (B) Use Active Adhesive Traps to Capture Small Insect Prey on the Tips of Glandular Trichomes (Hairs) Along their Trap Leaves. Pitcher Traps such as *Sarracenia Alata* ((C–D), *Sarracenia Psittacina* (E), and *Sarracenia Leucophylla* (F) Use Passive Pitfall Traps to Capture Insect Prey. The Walls of the Pitchers (Which are Modified Leaves) are Lined with Inward-Pointing Hairs that Direct Prey Items into Liquid that Contains Plant Enzymes and a Diverse Decomposer Community, Where Prey are Digested Before Nutrient Absorption by the Plant. One Type of Habitat Where these Species are Commonly Encountered is the Longleaf Pine Savanna, a Fire-Maintained, Moist, and Nutrient-Poor Ecosystem of Coastal Plains Along the Gulf of Mexico in the US (D).

(2017) provides a great deal of detail on not only the taxonomy and systematics of these species, but also their ecology and biogeography.

As indicated in Table 7.7, there are a half dozen or so types of prey capture methods (i.e., traps) employed by carnivorous plants. All of these involve some modification of leaves or stems to produce active or passive prey capture structures, and they are roughly equally divided among active and passive trap types. Among the active types, we see active adhesive traps employed by *Drosera* species (Figure 7.16) as studied by Darwin, the bladders of *Utricularia* species that implement a "trapdoor" type of trapping mechanism (Figure 7.17), and *Dionaea muscipula*, that uses a snap trap capture mechanism (Figure 7.18). Passive traps include the pitchers of species in the Sarraceniaceae (Figure 7.16) and Nepenthaceae, passive (i.e., nonmoving) adhesive traps of *Byblis* species, and the "lobster pot" traps of *Genlisea* species (Figure 7.19).

Among the traps that use motion to capture prey, the speed with which the traps function varies quite widely. In his experiments with *Drosera rotundifolia*, Darwin (1875) reported that the leaves and/or glandular trichomes would bend towards the center of the leaf blade over periods of time ranging from roughly one hour to 24 hours, depending on the size and type of prey item and the frequency with which a given leaf had been stimulated during preceding days. Leaves then re-expanded within a period of as little as one day to as much as ten days, following digestion of the prey item. At the other end of the speed gradient, the trap doors on bladders of *Utricularia inflata* open within 3 milliseconds (0.003 seconds!) following mechanical stimulus of the trigger hairs (Vincent et al. 2011). Resetting of the traps takes approximately one hour, during which time water is actively pumped out of the bladder, creating a negative pressure inside. These negative pressures sometimes result in spontaneous opening of the traps, even without

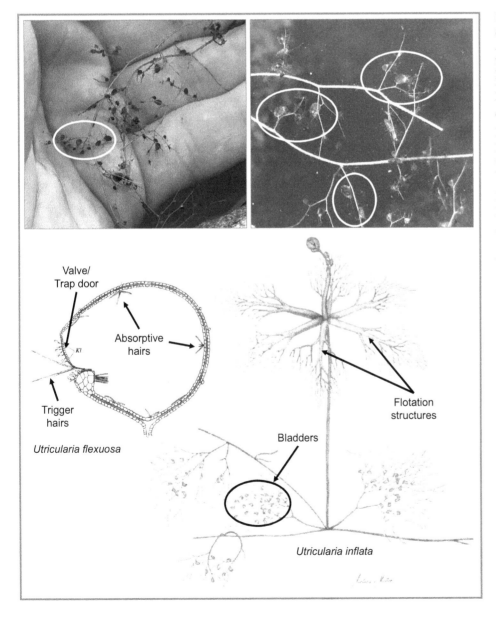

FIGURE 7.17 *Utricularia* Species use Bladders with Active Trap Door Mechanisms to Capture Aquatic Invertebrate Prey Items, Along with other Materials Contained Within Water Taken up When the Traps are Activated. Upper Photos Show Numerous Bladders of *Utricularia* with Captured Prey (Left) and Without (Right). The Drawing at Bottom Left Illustrates the Hairs that, When Contacted With Potential Prey, Trigger the Opening of the Trap Door Valve, Leading to Intake of Water Immediately Outside the Door. Digested and Decomposed Organisms and Organic Matter Taken in By the Traps Release Nutrients that are Then Absorbed by the Plant. Some *Utricularia* Species, such as *U. Inflata*, Use Flotation Structures to Maintain their Position Near the Water's Surface. Line Drawings of *Utricularia* Taken from Goebel (1893), with Labels Added here.

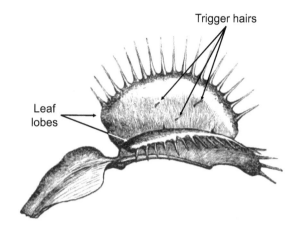

FIGURE 7.18 Active Trap Leaf of *Dionaea Muscipula*, Illustrating the Trio of Trigger Hairs on One Trap Lobe (Another Set of Three Trigger Hairs Can be Found on the Other Lobe). The Bristles Along the Margins of the Leaf Lobes Interlock When the Trap Closes to Aid in Securing the Prey. Digestive Enzymes Released within the Trap Aid in Degrading Prey to Release Nutrients for Absorption by the Plant. Taken from Darwin (1875).

an external mechanical stimulus. The speed of *Dionaea muscipula* snap traps falls between that of *Utricularia* and *Drosera*, closing to capture prey within about 0.3 seconds of receiving a mechanical stimulus (Volkov et al. 2008). *Dionaea* traps may remain closed for up to a week while digesting prey.

The traps of *Dionaea* function by detecting mechanical stimuli via three trigger hairs on each lobe of the distal half of the leaf (Figure 7.18). The stimulus induces ionic changes and corresponding electrical shifts within cells at the base of each trigger hair, followed by similar changes throughout the trap portion of the leaf, leading to changes in the volume and shape of cells in the leaf, closing the trap (Volkov et al. 2008). In contrast to the chemically induced trigger response in *Dionaea*, the opening of traps in *Utricularia* appears to be solely mechanical in nature. In addition to the negative pressure created inside the traps by pumping water out, the door is constructed of two layers of cells, one of which has cells arranged in such a way as to direct an inward buckling of the door once the trigger hairs are disturbed. The hairs are attached at the bottom, center of the door, near the edge that opens to allow water to rapidly enter when the trap is deployed. Because of their location and the degree of pressure that builds within the bladders, mechanical stimulation of the triggers causes an essentially immediate buckling response of the trap door. In *U. inflata*, whose bladders are approximately 2.0 mm in diameter, the traps take in water (along with any suspended materials) from a radius of approximately 0.5 mm around the trap orifice (Vincent et al. 2011).

Although quite a few species employ passive adhesive traps on leaves or stems to capture prey that are lured with nectar or scent attractants, another 200 or so capture prey by directing them towards a "collection area" through the use of inward-pointing hairs (Cronk and Fennessy 2001). In plants of *Genlisea*, the collection area is an enlarged, bulb-like structure near the base of plants, to which prey are directed by a series of rings of inward-pointing hairs along the neck of the trapping leaves (Figure 7.19). In plants using pitcher-type pitfall traps, the collection area in many species is an enlarged tubular leaf that also uses inward-pointing hairs to direct prey towards the fluids that accumulate within the trap, which is often referred to as a "pitcher" (Figures 7.16 and 7.20). The collection areas of these plants contain not only water, but also a milieu of plant-derived digestive enzymes and various invertebrates and microbes that act as decomposers, further aiding the process of degrading captured prey (Plachno et al. 2006). Similarly, carnivorous plants that collect their prey in basal leaf rosettes (i.e., carnivorous species of Bromeliaceae) also secrete digestive enzymes to assist in the breakdown of prey tissues (Plachno et al. 2006).

Among the enzymes produced by and/or contained within the traps of carnivorous species are phosphatases, proteases, chitinases, and glucosidases (Sirová, Adamec, and Vrba 2003; Matušíková et al. 2005). These, in order of mention in the previous sentence, would be responsible for such processes as phosphorus uptake, degradation of proteins (usually to access amino acids or nitrogen), degradation of insect exoskeletons, and breakdown of sugars. In many cases, it is clear that the enzymes are produced by the plant itself, as in the case of *Drosera* species, where enzymes are contained with the sticky secretions of the glandular trichomes, or *Dionaea*, which releases the enzymes within the trap leaves after prey capture. In other cases, the source of the enzymes is somewhat less clear, as in the bladders of *Utricularia* species, where some of the enzymes are potentially produced by organisms in the water taken up when traps activate (Sirová, Adamec, and Vrba 2003; Sirová et al. 2009).

We might assume that the production and activation of traps and the production and release of digestive enzymes would incur a cost to the plant, in terms of nutrient or energy allocation that might have otherwise gone towards other plant functions. Cronk and Fennessy (2001) mentioned this in their overview of carnivorous plants, giving several examples of studies that had shown, in some carnivorous species, that a large fraction of plant nitrogen was derived from captured prey. In examinations of recently discovered carnivory in *Triantha occidentalis*, a monocot species occupying bogs of northwestern North America, Lin et al. (2021) evaluated enzyme secretions of glandular hairs on the scapes of inflorescences as well as nutrient uptake from trapped prey. They found that the hairs secreted phosphatases on their surfaces, in patterns similar to those of *Drosera rotundifolia* (which they also tested, as a reference species). They also found that 0.1% of nitrogen incorporated into leaves of *Triantha occidentalis* during their experiments was derived directly from insect prey, while 7% of nitrogen incorporated into the stems used to capture prey came from entrapped insects.

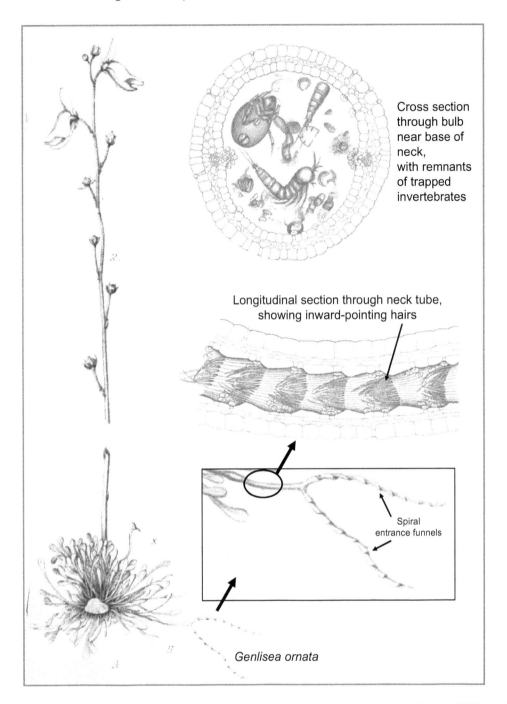

Cross section
through bulb
near base of
neck,
with remnants
of trapped
invertebrates

Longitudinal section through neck tube,
showing inward-pointing hairs

Spiral
entrance funnels

Genlisea ornata

FIGURE 7.19   Illustrations of *Genlisea Ornata* (Now Renamed *Genlisea Aurea*) Taken from Goebel (1893; Labels Added), Highlighting the Passive Trapping Structures Employed by this Species. Referred to as a "Lobster Pot" Mechanism, the Trap Uses a Series of Inward-Pointing Hairs to Direct Prey Towards the Bulb-Like Digestion Chamber. As with the Funnels in Actual Lobster Traps, the Hairs Serve a Second Role in Preventing Prey from Escaping Once they Enter the Neck of the Trap.

It has often been assumed that, because of the relatively high nitrogen content of the prey items captured by these species, that nitrogen is the limiting nutrient targeted for uptake by carnivory. Many studies have supported this by showing that as much as 50% to 75% of plant nitrogen is obtained from captured prey (Cronk and Fennessy 2001; Matušíková et al. 2005), but other studies have shown that other nutrients also are absorbed from prey items. These include phosphorus, potassium, sulfur, calcium, magnesium, and even organic carbon (Adamec 1997). Furthermore, in *Utricularia* species, Kibriya and Jones (2007) and Sirová, Adamec, and Vrba (2003) concluded that the pattern of enzyme production and release within the bladders was more in line with an emphasis on enhanced phosphorus acquisition (vs. nitrogen, for example), as the plants produced relatively high concentrations of

phosphatase enzymes within the bladders and their investment in carnivory declined with phosphorus fertilization.
Probably, the relative importance of different nutrients
varies among species, among habitats, and possibly even
among different suites of neighboring species.

Knight (1992) examined the photosynthetic production
of leaves and bladders of *Utricularia macrorhiza* to evaluate whether the bladders provide enough photosynthate to
compensate for their production. She found that production of bladders costs the plant around 15% to 80% of the
potential production of plants with only leaves (no bladders). Similarly, Adamec (2006) showed that bladders in
several *Utricularia* species have higher rates of respiration
than of photosynthesis, suggesting they are a net consump-
tive sink for photosynthetic carbon, and that their primary
benefit is in the acquisition of phosphorus or nitrogen.
Under some conditions, however, such as when light or carbon dioxide are at low levels, uptake of carbon from prey
items may help offset the cost of producing the bladders
(Adamec 1997, 2006). Furthermore, *Aldrovanda vesiculosa*
and some species of *Drosera* have been shown to assimi-
late organic carbon from digested prey and, in the case of
Drosera erythrorhiza, sometimes incorporate that carbon
into belowground storage organs (Adamec 1997).

Other work, using the pitcher plant *Sarracenia purpurea*
(Figure 7.20), has shown that the plants can adjust the relative allocation of leaf tissue to prey capture versus photosynthesis in response to mineral nutrient availability. Ellison
and Gotelli (2002) experimentally added nitrogen and/or
phosphorus to *S. purpurea* plants in a Massachusetts (USA)
bog and subsequently measured rates of photosynthesis and

relative proportions of leaf allocated to the pitcher versus
a leafy keel that was deemed more useful in photosynthesis than prey capture. They monitored the same set of 90
experimental plants over a three-year period and found that
as the amount of nitrogen added into the pitchers increased,
the plants shifted their biomass allocation more toward
production of the photosynthetic keel portion of the pitcher
versus the prey-capturing pitcher itself (Figure 7.21). This
shift was observed irrespective of the amount of phosphorus added to the pitchers and in the absence of prey capture by the plants (each pitcher was blocked to prevent prey
capture). This study thus not only demonstrated that there
appear to be costs to the production of the trap itself, but
over time, the plants are capable of shifting resource allocation to compensate for those costs.

Finally, Adamec (1997) suggested, in his review of
mineral nutrient acquisition by carnivorous plants, that
carnivory in some cases appears specifically to be an adaptation for growth in saturated soils where conditions are suboptimal for root performance. Brewer et al. (2011) expanded
upon this hypothesis to evaluate whether carnivory might
be an adaptation that specifically reduces the reliance on
roots for mineral nutrient uptake. They hypothesized that,
if this were true, carnivorous plants would be competitively
superior to non-carnivorous, but also non-aerenchymous,
plants under saturated soil conditions. Using a variety of
experimental and observational approaches, they evaluated
eight carnivorous plants and 48 non-carnivorous plants
from pine savannas of southern Mississippi (USA; see,
e.g., Figure 7.16). They found that only six of their eight
carnivorous plants produced roots, and, among those, the

FIGURE 7.20 A Close Relative of *Sarracenia Purpurea*, *Sarracenia Rosea*, at Elgin Air Force Base, Florida (USA). *Sarracenia Rosea*
Previously was Considered a Subspecies of *Sarracenia Purpurea*, and the Two are Morphologically Very Similar. Note the Variation
in Size, Shape, and Color of the Keels on the Pitchers. Photo Courtesy of Matthew Abbott.

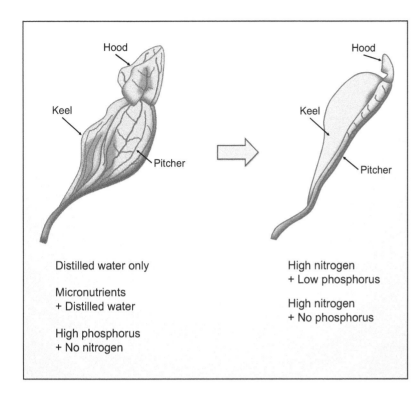

Hood

Keel

Pitcher

Distilled water only

Micronutrients
+ Distilled water

High phosphorus
+ No nitrogen

Hood

Keel

Pitcher

High nitrogen
+ Low phosphorus

High nitrogen
+ No phosphorus

FIGURE 7.21 Pitchers of *Sarracenia Purpurea* Modified their Morphology in Response to Three Years of Experimental Nutrient Addition. Nutrient Additions were Applied Multiple Times Each Year, Over a Three-Year Period, to Pitchers of Randomly Selected Plants from an Ombrotrophic Bog in Massachusetts (USA). Pitchers that Received No Nitrogen Additions (Addition of Only Water, Water Plus Micronutrients, or Water, Micronutrients, and Phosphorus) Displayed the Typical Morphology of Large Pitchers and a Relatively Small Photosynthetic Keel. In Contrast, Plants Whose Pitchers Received High Levels of Nitrogen Addition Modified their Leaves during the Experiment to Eventually Produce Leaves with Minimal Allocation to the Pitcher itself and Greater Allocation to a Larger Photosynthetic Keel Structure. Based on Data and Illustrations in Ellison and Gotelli (2002).

root porosity was 0.0% in all six. This included two species each of *Drosera*, *Pinguicula*, *Sarracenia*, and *Utricularia* (the last of which produced no roots). In contrast, mean root porosity of the non-carnivorous plants was 22%, with a range of 0% to 75%. Observational and experimental evaluation of the role of soil moisture on species distributions in these pine savanna habitats suggested that soil moisture did, in fact, influence distribution of both the carnivorous and the aerenchymous plant species, with these species occupying different habitats than the non-aerenchymous species. Brewer et al. (2011) concluded that their results support the hypothesis that carnivory might be yet another means by which wetland plants reduce the physiological costs associated with living in, or on, hypoxic wetland soils. They further stated that the rareness of carnivorous plants in nutrient rich wetlands is further evidence of this, as the soils and sediments in eutrophic wetlands might be expected to have relatively high concentrations of reduced minerals (iron, manganese, sulfur), to which carnivorous plants would be highly susceptible, given their lack of root aerenchyma.

7.7   IN SUMMARY

In this last chapter covering abiotic factors within the wetland environment, we have taken a relatively in-depth look at the mineral nutrient requirements of aquatic and wetland plants. Although there are some 16 nutrients discussed in this chapter, we typically find wetland plants to be limited in their growth primarily by nitrogen or phosphorus. This is due to the particular combination of plant requirements versus natural availability of the other 14 macro- and

micronutrients covered in this chapter. However, as the human population continues to grow, we often see anthropogenic deposition of these nutrients, particularly nitrogen and phosphorus, that may change the usual patterns of nutrient availability and limitation for wetland plants. To better understand the dynamics of mineral nutrients in wetlands, as a component of understanding the biology of aquatic and wetland plants, it is important to understand the different forms in which the major mineral nutrients occur and the role that plants play in transitions among those forms.

Cycling of carbon in wetland and aquatic habitats is dominated largely by uptake of carbon through photosynthesis and re-release via respiration, either by living organisms or by those organisms responsible for their eventual decomposition. One important component of carbon cycling in wetlands, from the perspective of environmental drivers of climate change, is the production of methane (methanogenesis) in anoxic wetland sediments. Methanogenesis is carried out by a variety of sediment microbes, using an assortment of small organic molecules derived from decomposition of sediment detritus (dead organic matter). A key determinant of whether wetlands serve as net sources of this greenhouse gas over long time periods is the degree of human disturbance the wetland experiences, with less disturbed wetlands becoming net carbon sinks much more quickly than those exposed to frequent disturbances.

Nitrogen cycling in wetlands is a relatively complex process, involving cycles of nitrogen oxidation and reduction driven by plant uptake and release, as well as metabolism of sediment microorganisms. Because nitrogen is frequently the primary limiting nutrient for freshwater wetland plants, it is important to understand the role that wetland plants play in transforming and/or sequestering nitrogen as it passes through these ecosystems. In addition to direct uptake and retention of nitrogen, plants strongly influence nitrogen processing through their deposition of detritus into wetland sediments, which, in turn, drives the availability of oxygen for microbially mediated nitrogen transformations.

Unlike nitrogen cycling, phosphorus cycling in wetlands is a relatively simple process, driven by plant uptake, death and decomposition of organic matter, and cycles of oxygen availability within the water column and the sediments. Phosphorus interacts strongly with various metal cations when they are in their oxidized forms, leading to deposition of phosphorus-metal complexes into wetland sediments.

As discussed in the previous two chapters, iron, manganese, and sulfur can form compounds that are toxic to plants when they are chemically reduced by sediment microbes. The oxygen that is released from roots as a by-product of transport through aerenchyma networks helps to mitigate this toxicity within the thin rhizosphere region that surrounds metabolically active roots. This, however, can have drawbacks, as the oxidized metals sometimes interfere with uptake of phosphorus, owing to the reactions mentioned in the previous paragraph.

As with the deficiencies in carbon, light, and oxygen discussed in Chapters 5 and 6, plants have accumulated a number of biological adaptations for mitigating deficiencies in major limiting nutrients. With respect to nitrogen limitation, aquatic and wetland plants have evolved several forms of symbiosis with nitrogen-fixing bacteria that reduce their reliance on soil (or water-borne) nitrogen. We saw this in the aquatic fern *Azolla*, which harbor endosymbiotic cyanobacteria, in a host of species, such as those in the Fabaceae, that form symbioses with rhizobia, and in alders (*Alnus* species) that form similar relationships with bacteria in the genus *Frankia*.

The less nutrient-specific adaptations we covered included the mycorrhizal relationships formed between wetland plants and mycorrhizal fungi, as well as the adaptation of carnivorous plants for acquiring supplemental nutrients from invertebrate prey captured through a variety of modified leaf traps. Mycorrhizal relationships enhance plant access to nitrogen, phosphorus, water, and other resources by increasing the volume of soil from which those resources are acquired. Many of the fungi involved in these relationships also have the ability to degrade soil organic matter further increasing plant access to soil nutrients. Similarly, carnivory provides access to a diversity of mineral nutrients by actively or passively capturing invertebrate prey whose decomposition and digestion allows direct absorption of liberated nutrients by the plant. Carnivory has, in fact, been postulated by some to serve as yet another adaptation not only for enhancing access to mineral nutrients, but also for minimizing negative impacts experienced by plant roots struggling to survive in anoxic wetland soils.

7.8   FOR REVIEW

1. How do microbes influence the availability of mineral nutrients for plants? Consider processes that reduce nutrient availability as well as those that increase availability.
2. What is an essential nutrient? What is the difference between a macronutrient and a micronutrient?
3. How do carbon, hydrogen, and oxygen differ from the other essential nutrients discussed in this chapter?
4. How and why is methanogenesis important in the context of wetland biology and management?
5. What is the central idea of the "law of the minimum?"
6. Why is the law of the minimum important from a wetland management perspective?
7. What are the most commonly limiting nutrients for plants? Why is this?
8. In what ways are plants involved in the cycling of nitrogen in wetlands?
9. How is the availability of phosphorus influenced by water and sediment redox processes?
10. What are some similarities and differences among the nitrogen-fixing symbioses covered in this chapter?
11. How are mycorrhizal symbioses different from nitrogen-fixing symbioses?
12. What are some of the costs and benefits of carnivory among plants?

7.9   REFERENCES

Adamec, L. 1997. "Mineral Nutrition of Carnivorous Plants: A Review." *The Botanical Review* 63: 273–99.
———. 2006. "Respiration and Photosynthesis of Bladders and Leaves of Aquatic *Utricularia* Species." *Plant Biology* 8: 765–69.
Angle, J. C., T. H. Morin, L. M. Solden, A. B. Narrowe, G. J. Smith, M. A. Borton, C. Rey-Sanchez, R. A. Daly, G. Mirfenderesgi, D. W. Hoyt, W. J. Riley, C. S. Miller, G. Bohrer, and K. C. Wrighton. 2017. "Methanogenesis in Oxygenated Soils is a Substantial Fraction of Wetland Methane Emissions." *Nature Communications* 8: 1–9.
Arber, A. R. 1920. *Water Plants: A Study of Aquatic Angiosperms.* Cambridge, UK: Cambridge University Press.
Atlas, R. M., and R. Bartha. 1998. *Microbial Ecology: Fundamentals and Applications.* 4th ed. Menlo Park, CA: Benjamin Cummings Science Publishing.
Barbour, M. G., J. H. Burk, W. D. Pitts, F. S. Gilliam, and M. W. Schwartz. 1999. *Terrestrial Plant Ecology.* 3rd ed. Menlo Park, CA: Benjamin/Cummings.
Batzer, D. P., and R. R. Sharitz. 2006. *Ecology of Freshwater and Estuarine Wetlands.* Berkeley, CA: University of California Press.
Bohrer, K. E., C. F. Friese, and J. P. Amon. 2004. "Seasonal Dynamics of Arbuscular Mycorrhizal Fungi in Differing Wetland Habitats." *Mycorrhiza* 14: 329–37.

Bornstein, A. J. 1997. "Myricaceae." www.efloras.org/florataxon. aspx?flora_id=1&taxon_id=10594.

Borovec, O., and M. Vohník. 2018. "Ontogenetic Transition from Specialized Root Hairs to Specific Root-fungus Symbiosis in the Dominant Mediterranean Seagrass *Posidonia oceanica*." *Scientific Reports* 8: 1–11.

Brewer, J. S., D. J. Baker, A. S. Nero, A. L. Patterson, R. S. Roberts, and L. M. Turner. 2011. "Carnivory in Plants as a Beneficial Trait in Wetlands." *Aquatic Botany* 94: 62–70.

Brundrett, M. C. 2017. "Global Diversity and Importance of Mycorrhizal and Nonmycorrhizal Plants." In *Biogeography of Mycorrhizal Symbiosis*, edited by L. Tedersoo, 533–56. Cham, Switzerland: Springer International Publishing.

Brundrett, M. C., and L. Tedersoo. 2018. "Evolutionary History of Mycorrhizal Symbioses and Global Host Plant Diversity." *New Phytologist* 220: 1108–15.

Carlton, R. G., and R. G. Wetzel. 1988. "Phosphorus Flux from Lake Sediments: Effect of Epipelic Algal Oxygen Production." *Limnology and Oceanography* 33: 562–70.

Carrapiço, F. 2010. "*Azolla* as a Superorganism. Its Implication in Symbiotic Studies." In *Symbioses and Stress. Cellular Origin, Life in Extreme Habitats and Astrobiology*, edited by J. Seckbach and M. Grube, Vol. 17, 227–41. Dordrecht: Springer.

Ciais, P., C. Sabine, G. Bala, L. Bopp, V. Brovkin, J. Canadell, A. Chhabra, R. DeFries, J. Galloway, M. Heimann, C. Jones, C. Le Quéré, R. B. Myneni, S. Piao, and P. Thornton. 2013. "Carbon and Other Biogeochemical Cycles." In *Climate Change 2013 the Physical Science Basis: Working Group I Contribution to the Fifth Assessment Report of the Intergovernmental Panel on Climate Change*, edited by T. F. Stocker, D. Qin, G.-K. Plattner, M. Tignor, S. K. Allen, J. Boschung, A. Nauels, Y. Xia, V. Bex, and P. M. Midgley, 465–570. Cambridge, UK: Cambridge University Press.

Cronk, J. K., and M. S. Fennessy. 2001. *Wetland Plants: Biology and Ecology*. Boca Raton, FL: CRC Press.

Daniels, L., R. Sparling, and G. D. Sprott. 1984. "The Bioenergetics of Methanogenesis." *Biochimica et Biophysica Actaphysica acta* 768: 113–63.

Darwin, C. 1875. *Insectivorous Plants*. London: Appleton.

Dismukes, G. C., V. V. Klimov, S. V. Baranov, Y. N. Kozlov, J. DasGupta, and A. Tyryshkin. 2001. "The Origin of Atmospheric Oxygen on Earth: The Innovation of Oxygenic Photosynthesis." *Proceedings of the National Academy of Sciences of the United States of America* 98: 2170–75.

Doyle, J. J., and M. A. Luckow. 2003. "The Rest of the Iceberg. Legume Diversity and Evolution in a Phylogenetic Context." *Plant Physiology* 131: 900–10.

Dresler-Nurmi, A., D. P. Fewer, L. A. Räsänen, and K. Lindström. 2009. "The Diversity and Evolution of *Rhizobia*." *Microbiology Monographs* 8: 3–41.

eFloras. 2008. "eFloras." eFloras.org Home

Eily, A. N., K. M. Pryer, and F. W. Li. 2019. "A First Glimpse at Genes Important to the *Azolla–Nostoc* Symbiosis." *Symbiosis* 78: 149–62.

Ellison, A. M., and N. J. Gotelli. 2001. "Evolutionary Ecology of Carnivorous Plants." *Trends in Ecology and Evolution* 16: 623–29.

———. 2002. "Nitrogen Availability Alters the Expression of Carnivory in the Northern Pitcher Plant, Sarracenia purpurea." *PNAS* 99: 4409–12.

Fitter, A. H., and R. K. M. Hay. 2002. *Environmental Physiology of Plants*. 3rd ed. San Diego: Academic Press, Inc.

Fusconi, A., and M. Mucciarelli. 2018. "How Important Is Arbuscular Mycorrhizal Colonization in Wetland and Aquatic Habitats?" *Environmental and Experimental Botany* 155: 128–41.

Glibert, P. M. 2020. "From Hogs to HABs: Impacts of Industrial Farming in the US on Nitrogen and Phosphorus and Greenhouse Gas Pollution." *Biogeochemistry* 150: 139–80.

Goebel, K. 1893. *Pflanzenbiologische Schilderungen*. Germany: N.G. Elwert.

Gökkaya, K., T. M. Hurd, and D. J. Raynal. 2006. "Symbiont Nitrogenase, Alder Growth, and Soil Nitrate Response to Phosphorus Addition in Alder (*Alnus incana* ssp. *rugosa*) Wetlands of the Adirondack Mountains, New York State, USA." *Environmental and Experimental Botany* 55: 97–109.

Gottschalk, Gerhard. 1986. *Bacterial Metabolism*, 2nd ed. New York: Springer-Verlag.

Herbert, J. 2005a. "Systematics and Biogeography of Myricaceae." PhD thesis, University of Saint Andrews.

———. 2005b. "New Combinations and a New Species in *Morella* (Myricaceae)." *Novon* 15: 293–95.

Hessen, D. O., J. J. Elser, R. W. Sterner, and J. Urabe. 2013. "Ecological Stoichiometry: An Elementary Approach Using Basic Principles." *Limnology and Oceanography* 58: 2219–36.

Hiatt, D. L., C. J. Robbins, J. A. Back, P. K. Kostka, R. D. Doyle, C. M. Walker, M. C. Rains, D. F. Whigham, and R. S. King. 2017. "Catchment-scale Alder Cover Controls Nitrogen Fixation in Boreal Headwater Streams." *Freshwater Science* 36: 523–32.

Hillel, D. 2008. *Soil in the Environment*. San Diego: Academic Press.

Hurtado-McCormick, V., T. Kahlke, K. Petrou, T. Jeffries, P. J. Ralph, and J. R. Seymour. 2019. "Regional and Microenvironmental Scale Characterization of the *Zostera muelleri* Seagrass Microbiome." *Frontiers in Microbiology* 10: 1–22.

Ibáñez, F., L. Wall, and A. Fabra. 2017. "Starting Points in Plant-Bacteria Nitrogen-fixing Symbioses: Intercellular Invasion of the Roots." *Journal of Experimental Botany* 68: 1905–18.

IPCC. 2014. *Climate Change 2014: Synthesis Report: Contribution of Working Groups I, II and III to the Fifth Assessment Report of the Intergovernmental Panel on Climate Change*. Geneva, Switzerland: IPCC.

James, E. K., M. De Fatima Loureiro, A. Pott, V. J. Pott, C. M. Martins, A. A. Franco, and J. I. Sprent. 2001. "Flooding-tolerant Legume Symbioses from the Brazilian Pantanal." *New Phytologist* 150: 723–38.

Johansson, C., and B. Bergman. 1992. "Early Events during the Establishment of the *Gunnera/Nostoc* Symbiosis." *Planta* 188: 403–13.

Kartal, B., M. M. M. Kuypers, G. Lavik, J. Schalk, H. J. M. Op Den Camp, M. S. M. Jetten, and M. Strous. 2007. "Anammox Bacteria Disguised as Denitrifiers: Nitrate Reduction to Dinitrogen Gas via Nitrite and Ammonium." *Environmental Microbiology* 9: 635–42.

Khan, A. G. 2004. "Mycotrophy and Its Significance in Wetland Ecology and Wetland Management." In *Developments in Ecosystems*, edited by M. H. Wong, 95–114. Amsterdam, The Netherlands: Elsevier B.V.

Kibriya, S., and J. I. Jones. 2007. "Nutrient Availability and the Carnivorous Habit in *Utricularia vulgaris*." *Freshwater Biology* 52: 500–9.

Knight, S. E. 1992. "Costs of Carnivory in the Common Bladderwort, *Utricularia macrorhiza*." *Oecologia* 89: 348–55.

Knowlton, A. 2012. "Equisetum." *Current Biology* 22: R388–R390.

Krings, M., H. Hass, H. Kerp, T. N. Taylor, R. Agerer, and N. Dotzler. 2009. "Endophytic Cyanobacteria in a 400-million-yr-old Land Plant: A Scenario for the Origin of a Symbiosis?" *Review of Palaeobotany and Palynology* 153: 62–69.

Li, H. T., T. Shuang Yi, L. Ming Gao, P. Fei Ma, T. Zhang, J. Bo Yang, Matthew A. Gitzendanner, et al. 2019. "Origin of Angiosperms and the Puzzle of the Jurassic Gap." *Nature Plants* 5 (5): 461–70. https://doi.org/10.1038/s41477-019-0421-0.

Li, X., L. Hou, M. Liu, Y. Zheng, G. Yin, X. Lin, L. Cheng, Y. Li, and X. Hu. 2015. "Evidence of Nitrogen Loss from Anaerobic Ammonium Oxidation Coupled with Ferric Iron Reduction in an Intertidal Wetland." *Environmental Science and Technology* 49: 11560–68.

Lin, Q., C. Ané, T. J. Givnish, and S. W. Graham. 2021. "A New Carnivorous Plant Lineage (*Triantha*) with a Unique Sticky-inflorescence Trap." *Proceedings of the National Academy of Sciences* 118: 1–6.

Matušíková, I., J. Salaj, J. Moravčíková, L. Mlynárová, J. P. Nap, and J. Libantová. 2005. "Tentacles of In vitro-grown Round-leaf Sundew (*Drosera rotundifolia* L.) Show Induction of Chitinase Activity Upon Mimicking the Presence of Prey." *Planta* 222: 1020–27.

Meeks, J. C. 2009. "Physiological Adaptations in Nitrogen-fixing *Nostoc*–Plant Symbiotic Associations." *Microbiology Monographs* 8: 181–205.

Mitsch, W. J., B. Bernal, A. M. Nahlik, Ü. Mander, L. Zhang, C. J. Anderson, S. E. Jørgensen, and H. Brix. 2013. "Wetlands, Carbon, and Climate Change." *Landscape Ecology* 28: 583–97.

Mitsch, W. J., and J. G. Gosselink. 2007. *Wetlands*. 4th ed. Hoboken, NJ: John Wiley & Sons, Inc.

———. 2015. *Wetlands*. 5th ed. Hoboken, NJ: John Wiley & Sons, Inc.

Moore-Landecker, Elizabeth. 1996. *Fundamentals of the Fungi*. Upper Saddle River, NJ: Prentice-Hall.

Mulder, A., A. A. van de Graaf, L. A. Robertson, and J. G. Kuenen. 1995. "Anaerobic Ammonium Oxidation Discovered in a Denitrifying Fluidized Bed Reactor." *FEMS Microbiology Ecology* 16: 177–84.

Olde Venterink, H., R. E. Van der Vliet, and M. J. Wassen. 2001. "Nutrient Limitation Along a Productivity Gradient in Wet Meadows." *Plant and Soil* 234: 171–79.

Persson, T., and K. Huss-Danell. 2009. "Physiology of Actinorhizal Nodules." In *Prokaryotic Symbionts in Plants*, edited by K. Pawlowski, 155–78. Berlin: Springer-Verlag.

Plachno, B. J., L. Adamec, I. K. Lichtscheidl, M. Peroutka, W. Adlassnig, and J. Vrba. 2006. "Fluorescence Labelling of Phosphatase Activity in Digestive Glands of Carnivorous Plants." *Plant Biology* 8: 813–20.

Pressel, Silvia, Martin I. Bidartondo, Roberto Ligrone, and Jeffrey G. Duckett. 2010. "Fungal Symbioses in Bryophytes: New Insights in the Twenty First Century." *Phytotaxa* 9: 238–53.

Radhika, K. P., and B. F. Rodrigues. 2007. "Arbuscular Mycorrhizae in Association with Aquatic and Marshy Plant Species in Goa, India." *Aquatic Botany* 86: 291–94.

Ramírez-Viga, T. K., R. Aguilar, S. Castillo-Argüero, X. Chiappa-Carrara, P. Guadarrama, and J. Ramos-Zapata. 2018. "Wetland Plant Species Improve Performance When Inoculated with Arbuscular Mycorrhizal Fungi: A Meta-analysis of Experimental Pot Studies." *Mycorrhiza* 28: 477–93.

Raven, J. A. 2002. "The Evolution of Cyanobacterial Symbioses." *Biology and Environment* 102: 3–6.

Ray, A. M., and R. S. Inouye. 2006. "Effects of Water-Level Fluctuations on the Arbuscular Mycorrhizal Colonization of *Typha latifolia* L." *Aquatic Botany* 84: 210–16.

Reddy, K. R., and R. D. DeLaune. 2008. *Biogeochemistry of Wetlands: Science and Applications*. Boca Raton, FL: CRC Press.

Risgaard-Petersen, N., R. L. Meyer, M. Schmid, M. S. M. Jetten, A. Enrich-Prast, S. Rysgaard, and N. P. Revsbech. 2004. "Anaerobic Ammonium Oxidation in an Estuarine Sediment." *Aquatic Microbial Ecology* 36: 293–304.

Robertson, D. M., D. A. Saad, and G. E. Schwarz. 2014. "Spatial Variability in Nutrient Transport by HUC8, State, and Subbasin Based on Mississippi/Atchafalaya River Basin SPARROW Models." *Journal of the American Water Resources Association* 50: 988–1009.

Rudd, J. W. M., C. A. Kelly, D. W. Schindler, and M. A. Turner. 1988. "Disruption of the Nitrogen Cycle in Acidified Lakes." *Science* 240: 1515–17.

———. 1990. "A Comparison of the Acidification Efficiencies of Nitric and Sulfuric Acids by Two Whole-lake Addition Experiments." *Limnology and Oceanography* 35: 663–79.

Seitzinger, S. P., E. Mayorga, A. F. Bouwman, C. Kroeze, A. H. W. Beusen, G. Billen, G. Van Drecht, E. Dumont, B. M. Fekete, J. Garnier, and J. A. Harrison. 2010. "Global River Nutrient Export: A Scenario Analysis of Past and Future Trends." *Global Biogeochemical Cycles* 24: 1–16.

Sharitz, R. R., and S. C. Pennings. 2006. "Development of Wetland Plant Communities." In *Ecology of Freshwater and Estuarine Wetlands*, edited by D. P. Batzer and R. R. Sharitz, 177–241. Berkeley, CA: University of California Press.

Sirová, D., L. Adamec, and J. Vrba. 2003. "Enzymatic Activities in Traps of Four Aquatic Species of the Carnivorous Genus *Utricularia*." *New Phytologist* 159: 669–75.

Sirová, D., J. Borovec, B. Černá, E. Rejmánková, L. Adamec, and J. Vrba. 2009. "Microbial Community Development in the Traps of Aquatic *Utricularia* Species." *Aquatic Botany* 90: 129–36.

Stevens, P. F. 2017. "Angiosperm Phylogeny Website." Version 14, July 2017 [and more or less continuously updated since]. www.mobot.org/MOBOT/research/APweb/.

Struková, Silvie, Miroslav Vosátka, and Jan Pokorný. 1996. "Root Symbioses of *Alnus glutinosa* (L.) Gaertn. and Their Possible Role in Alder Decline: A Preliminary Study." *Folia Geobotanica & Phytotaxonomica* 31: 153–62.

Taiz, L., E. Zeiger, I. M. Moller, and A. Murphy. 2015. *Plant Physiology and Development*. 6th ed. Sunderland, MA: Sinauer Associates, Inc.

Tedersoo, L., T. Suvi, T. Jairus, I. Ostonen, and S. Põlme. 2009. "Revisiting Ectomycorrhizal Fungi of the Genus *Alnus*: Differential Host Specificity, Diversity and Determinants of the Fungal Community." *New Phytologist* 182: 727–35.

Thamdrup, B., and T. Dalsgaard. 2002. "Production of N$_2$ through Anaerobic Ammonium Oxidation Coupled to Nitrate Reduction in Marine Sediments." *Applied and Environmental Microbiology* 68: 1312–18.

Vincent, O., C. Weißkopf, S. Poppinga, T. Masselter, T. Speck, M. Joyeux, C. Quilliet, and P. Marmottant. 2011. "Ultra-fast Underwater Suction Traps." *Proceedings of the Royal Society B: Biological Sciences* 278: 2909–14.

Vitousek, P. M., J. D. Aber, R. W. Howarth, G. E. Likens, P. A. Matson, D. W. Schindler, W. H. Schlesinger, and D. G. Tilman. 1997. "Human Alteration of the Global Nitrogen Cycle: Sources and Consequences." *Ecological Applications* 7: 737–50.

Vitousek, P. M., and R. W. Howarth. 1991. "Nitrogen Limitation on Land and in the Sea: How Can It Occur?" *Biogeochemistry* 13: 87–115.

Vohník, M., O. Borovec, Z. Kolaříková, R. Sudová, and M. Réblová. 2019. "Extensive Sampling and High-throughput Sequencing Reveal *Posidoniomyces atricolor* gen. Et sp. Nov. (Aigialaceae, Pleosporales) as the Dominant Root Mycobiont of the Dominant Mediterranean Seagrass *Posidonia oceanica*." *MycoKeys* 55: 59–86.

Volkov, A. G., T. Adesina, V. S. Markin, and E. Jovanov. 2008. "Kinetics and Mechanism of *Dionaea* Muscipula Trap Closing." *Plant Physiology* 146: 694–702.

Wagner, G. M. 1997. "*Azolla*: A Review of Its Biology and Utilization." *Botanical Review* 63: 1–26.

Wellman, C. H., H. Kerp, and H. Hass. 2006. "Spores of the Rhynie Chert Plant *Aglaophyton* (*Rhynia*) *Major* (Kidston and Lang) D.S. Edwards, 1986." *Review of Palaeobotany and Palynology* 142: 229–50.

Wetzel, P. R., and A. G. van der Valk. 1996. "Vesicular—Arbuscular Mycorrhizae in Prairie Pothole Wetland Vegetation in Iowa and North Dakota." *Canadian Journal of Botany* 74: 883–90.

Wetzel, Robert G. 2001. *Limnology: Lake and River Ecosystems*. San Diego: Academic Press.

Whitson, A. R., and H. L. Walster. 1912. *Soils and Soil Fertility*. St. Paul, MN: Webb Publishing Co.

Yaroshevsky, A. A. 2006. "Abundances of Chemical Elements in the Earth's Crust." *Geochemistry International* 44: 48–55.

Zhang, M., P. Dai, X. Lin, L. Lin, B. Hetharua, Y. Zhang, and Y. Tian. 2020. "Nitrogen Loss by Anaerobic Ammonium Oxidation in a Mangrove Wetland of the Zhangjiang Estuary, China." *Science of the Total Environment* 698: 134291.

Zhang, Z., N. E. Zimmermann, A. Stenke, X. Li, E. L. Hodson, G. Zhu, C. Huang, and B. Poulter. 2017. "Emerging Role of Wetland Methane Emissions in Driving 21st Century Climate Change." *Proceedings of the National Academy of Sciences of the United States of America* 114: 9647–52.

Zhu, G., S. Wang, W. Wang, Y. Wang, L. Zhou, B. Jiang, H. J. M. Op Den Camp, N. Risgaard-Petersen, L. Schwark, Y. Peng, M. M. Hefting, M. S. M. Jetten, and C. Yin. 2013. "Hotspots of Anaerobic Ammonium Oxidation at Land-freshwater Interfaces." *Nature Geoscience* 6: 103–7.

8 Reproduction

8.1 PLANT REPRODUCTION IN THE CONTEXT OF LIFE HISTORY

In Chapter 1, I introduced an assortment of anatomical terms, most of which are related in some way with sexual reproduction or asexual propagation in plants, depicted again here in Figure 8.1. These illustrations are a great starting point for our discussion in this chapter of aquatic and wetland plant reproduction, and we will carry parts of this with us into the chapters that follow. One concept included in Figure 8.1 that is important here is that of plant **life history**. Life history is a term that may be used to refer to the typical patterns of birth, growth, reproduction, and death in a species.

There are numerous ways in which one might consider a species' life history. In this chapter, we will look at life history from the combined perspective of longevity and reproductive frequency. At one extreme, we have plants that exhibit an **annual life history**, such as that in Figure 8.1A and B (and the first panel of Figure 8.2). Individuals of these species are born from (or born as) a seed, and they live until they complete their only season of sexual reproduction, and then they die. The time required for this in annuals is, by definition, less than one year. The actual time required for some annual species to go from seed to seed (i.e., from one generation of seed to the next) may be as little as six weeks, and some plant species can produce flowers within only six days after the seeds are exposed to suitable germination conditions (Harper 1977).

At the other extreme, we have plants with a **perennial life history** that live multiple years and undergo multiple seasons of reproduction. These plants exhibit longevity up to hundreds of years, with some individuals having been dated at almost 12,000 years of age (Vasek 1980), and they have the potential to undergo sexual reproduction during many, if not most, of those years. Many perennial species propagate both sexually and asexually, the latter resulting in interconnected modules (ramets) of the original plant that can live, grow, and reproduce independently if they become separated from the rest of the individual. We see this illustrated in Figure 8.1 (C–E), where individual ramets are interconnected via stolons and rhizomes as a single large genet. A genet comprises all the individual ramets that arise from the germination of a single seed. The collection of all connected ramets of an individual are sometimes referred to as a **clone**, and the process of numeric increase in ramets is referred to as **clonal propagation**. Because these species can continue, year after year, to produce ramets, each of which has its own apical **meristems**, or growing points, it is unknown whether perennial species ever simply die what we might term a "natural death" at the whole-genet level (Munné-Bosch 2008).

In addition to annual and perennial species, we also have an intermediate group sometimes referred to as **biennials** (Figure 8.2). These species live longer than annuals, many living for decades, but they characteristically undergo only a single season of sexual reproduction before producing seed and dying. Other terms sometimes used to describe both annuals and biennials are **monocarpic** and **semelparous** species, both of which refer to their single reproductive season (e.g., mono- = one; -carpic, referencing the carpel, which is the reproductive structure producing the ovaries and eggs). Corresponding terms for perennial species are **polycarpic** and **iteroparous**. The terms semelparous ("one birth") and iteroparous ("repeated births") are perhaps more commonly used in reference to animals than plants, so we will use the other terms in this text. In the context of this terminology, biennials often are referred to as *monocarpic perennials*, and "perennials" are referred to as *polycarpic perennials*.

In the present chapter, this distinction among life histories is useful in its influence on the relative importance of sexual versus asexual propagation among species. Annual species reproduce exclusively via sexual reproduction, and vegetative or asexual propagation typically plays a minor role in monocarpic perennials (i.e., biennials). However, in those species traditionally referred to as perennials (i.e., polycarpic perennials), asexual propagation can be very important. In fact, in many aquatic and wetland plants, especially those species that have been introduced outside their native range, asexual propagation is often more important than sexual reproduction. Thus, the section of this chapter dealing with asexual propagation (section 8.3) deals almost exclusively with polycarpic perennial species.

8.2 SEXUAL REPRODUCTION IN PLANTS

This book deals predominantly with angiosperms, and, consequently, our discussion of sexual reproduction will focus on flowers, fruit, and seeds. Gymnosperms, such as the conifers of boreal wetland systems, do not produce flowers or fruit, but they do produce seeds; as such, much of this chapter's discussion of seeds will apply to those species as well.

Our starting point for examining sexual reproduction is the flower (Figure 8.3). Flowers are specialized branches consisting of a series of **whorls** (concentric rings) of highly modified leaves (Judd et al. 2008). The whorls, of which there are four in a complete flower, emerge from a **receptacle**, which is the tip of the modified branch bearing the flower, or cluster of flowers in some species. If we begin with the outermost whorl and work inward, the four whorls are the sepals, petals, **stamens**, and the **carpels** or **pistils**.

DOI: 10.1201/9781315156835-8

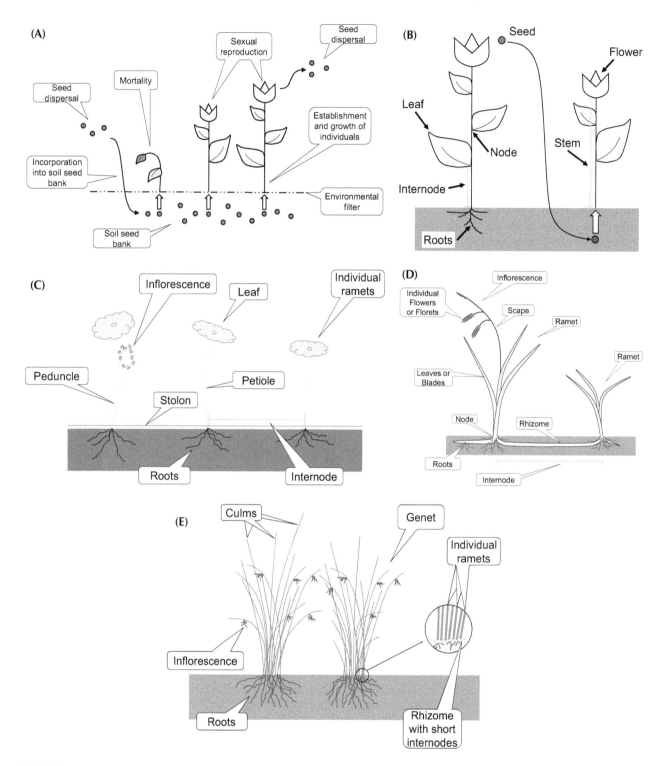

FIGURE 8.1 (A) Critical Stages in a Generalized Plant Life History. (B) Annual Plant Species Follow the Pattern in (A), Beginning Each Generation from a Seed and Ending with the Production (or not) of the Next Generation of Seeds. ((C–D) Perennial Species, in Contrast, Exist for Multiple Years as Repeating, Genetically Identical Modules (Ramets), Each of Which has the Potential to Live Separately from the Parent Plant. Along the Way, these Plants May Reproduce Sexually One or Many Times. Vegetatively, Perennial Species can be Generally thought of as Developing Along Main Stems that often will be either (C) Aboveground Stolons or (D–E) Belowground Rhizomes. The Length of Internodes also Varies Among Species (E.g., D Vs. E), with this Influencing the Relative Performance of a Given Species under Different Wetland Conditions. As we will see Later in this Chapter, there are Numerous other Anatomical Structures that Function in Asexual Reproduction, in Addition to the Stolons and Rhizomes Illustrated here.

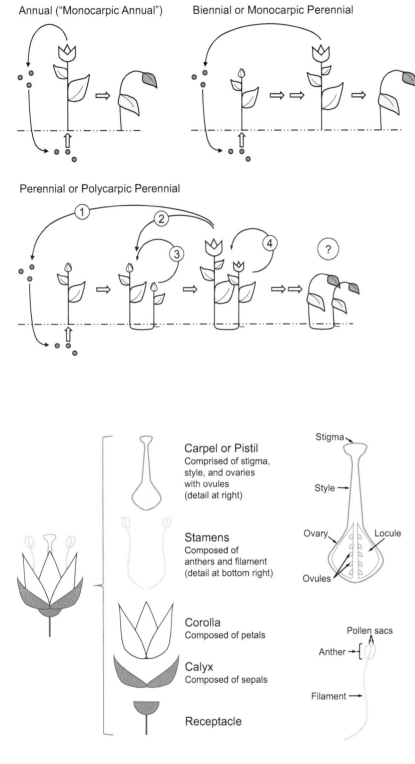

Annual ("Monocarpic Annual")

Biennial or Monocarpic Perennial

Perennial or Polycarpic Perennial

Carpel or Pistil
Comprised of stigma, style, and ovaries with ovules (detail at right)

Stamens
Composed of anthers and filament (detail at bottom right)

Corolla
Composed of petals

Calyx
Composed of sepals

Receptacle

Stigma
Style
Ovary
Locule
Ovules

Pollen sacs
Anther
Filament

FIGURE 8.2 Three Common Plant Life Histories. Annuals (Upper Left), Strictly Speaking, Germinate, Grow, Reproduce Sexually, Set Seed, and Die in a Single Growing Season (Less than One Calendar Year). Perennial Species, Broadly Speaking, are of Two General Life Histories. One (Upper Right) is Characterized by an Individual Plant that Lives for Multiple Growing Seasons, during Which Time it may or may not Produce Vegetative Clonal Offshoots (Daughter Ramets). After some Species-Specific Period of Growth, the Individual Will Undergo Sexual Reproduction, Set Seed, and Then Die. Polycarpic Perennials, in Contrast (Lower), Live Multiple Growing Seasons, Undergoing Sexual Reproduction in many of those Years (but not Necessarily All), during Which Time they may Produce Many Daughter Ramets, Each of which also may undergo Sexual Reproduction during Multiple Growing Seasons and may Produce their Own Daughter Ramets. during any Given Year after Establishment, a Polycarpic Perennial Might (1) Undergo Sexual Reproduction, (2) Revert to or (3) Remain in a Nonreproductive, Vegetative Condition, or (4) Reiterate Through Sexual Reproduction. It is not Fully Understood Whether Polycarpic Perennial Plants ever Truly Die at the Whole-Plant Level as a Regular Programmed Aspect of their Life Histories (Munné-Bosch 2008). Because Annual Species Only Exhibit a Monocarpic Life Cycle, It is not Necessary to Specify that they are Monocarpic.

FIGURE 8.3 General Anatomy of a Flower. Central Column of Illustrations Show the General Component Whorls of Structures Forming this Flower, With the Innermost Whorl (The Carpel) at the Top. The Carpel(S) Will be Surrounded by the Stamens in Bisexual Flowers, and these Two Whorls Will be Surrounded Successively by the Petals and then Sepals, in Flowers with all Four Whorls of Floral Structures. On the Right, we See the Generalized Internal Anatomy of a Carpel (Upper Right) and of a Stamen (Lower Right).

The sepals collectively are referred to as the **calyx**, and the combined whorl of petals are the **corolla**. In some cases, it may be difficult to distinguish between the sepals and petals, in which case individual units of both whorls are referred to as **tepals**. We also see cases where undifferentiated sepals and petals are collectively referred to as the **perianth**, although this term can also be used to reference the combination of the calyx and corolla even if the two are clearly differentiable. To complicate matters further, some plant families or genera may exhibit modifications of one or more floral parts, show intergradation between adjacent whorls, or produce flower parts in a spiral arrangement with no distinct whorls (Figure 8.4).

Although sepals and petals vary widely among taxa in their morphology and importance for sexual reproduction, the stamens and carpels have direct roles. At the tips of the stamens are the *anthers*, where pollen production occurs. Within the carpel is the *ovary*, where the *ovules*

FIGURE 8.4   Dogwoods (*Cornus* Species) and Water Lilies (*Nymphaea* Species) Provide Some Examples of Floral Part Modifications That Contrast with the Idealized Flower Shown in Figure 8.3. Top Row: Flowering Dogwood (*Cornus Florida*; Pictured here as a Horticultural Variety) is Known for its Showy Flowers, but the Showy "Petals" are, in Fact, Bracts That Surround Clusters of Tiny Individual Flowers. The Individual Flowers Themselves, with Small, Greenish-Yellow Petals, can be Seen in the Center of the Top Right Photo. Middle Row and Bottom: *Nymphaea Odorata* Flowers Show a Gradual Transition from an Outer Whorl of Pale Green Sepals (Middle Left and Bottom Left) to Green and White Petals to White Petals to Petaloid Stamens Surrounding the Gynoecium.

are produced, each containing an egg cell. Not all flowers will host anthers *and* carpels. Those that produce both are referred to as **perfect** flowers, also called **bisexual**, or **hermaphroditic**, flowers; those that produce only one or the other are termed **imperfect** or **unisexual** flowers. Imperfect flowers that produce only stamens are called **staminate** flowers, and those that produce only carpels (pistils) are referred to as *carpellate* or **pistillate** flowers. In species that produce imperfect flowers, we sometimes see species that produce both staminate and pistillate flowers on the same plant. We call these **monoecious** species (mono- = one; -ecious = house or home) because both flower types live together on the same plant. That is, they have the same home, together on an individual plant. Species where staminate and pistillate flowers are produced on separate plants are called **dioecious** species (di- = two) because they live in two separate homes. There are other variations to this as well, such as species that produce staminate, pistillate, and perfect flowers all on the same plants (**polygamous**

species). **Dioecy** (i.e., being dioecious) is one means by which plants enhance chances of outcrossing during sexual reproduction, as we will see later.

This discussion of sexual reproduction is a good place to revisit the phenomenon of alternating generations in plants (Figure 2.9). We saw in Chapter 2 that, among land plants, the bryophyte life cycle is dominated by the gametophyte generation, but that the tracheophytes are dominated by the sporophyte generation (refer to Figure 2.9). Thus, in the angiosperms, the plants that we most often see are the sporophyte generation (Figure 8.5). This dominant sporophyte generation is **diploid**, meaning it has two complementary sets of chromosomes. However, the diploid state is not always an accurate description of sporophytes because plants have an amazing ability to duplicate their **genomes** (their entire complement of genes, or DNA), resulting in plants that may contain four, six, or even more sets of chromosomes within each somatic cell. The more subordinate generation in tracheophytes is the gametophyte generation,

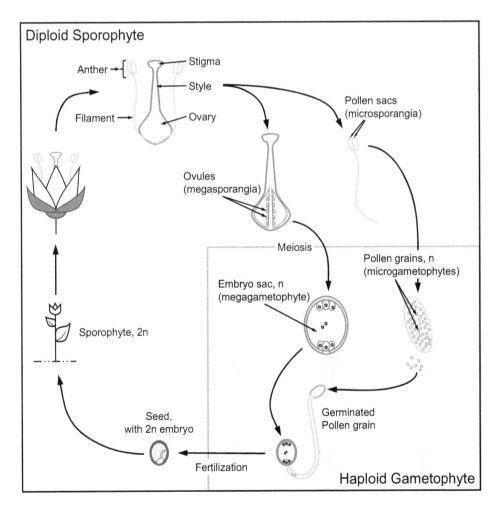

FIGURE 8.5 General Pattern of Alternation of Generations in Angiosperms. The Life Cycle is Dominated by the Diploid (2n) Sporophyte Generation, Which begins after Union of Haploid Nuclei from the Male and Female Gametophytes during Fertilization (Bottom Center). Meiosis within the Diploid Sporangia Produces Haploid (N) Embryo Sacs within the Ovules (Female Gametophytes) and Haploid Pollen within the Pollen Sacs (Male Gametophytes; Right Center). Note that Neither of the Gametophytes is Truly Free-Living, although Pollen usually is Dispersed from the Anthers to Initiate Pollen Dispersal and Pollination. Ploidy (I.e., the Number of Sets of Chromosomes) is Indicated in Shorthand By a Lowercase "N," Where N = Haploid, 2n = Diploid, 3n = Triploid, Etc.

and it is **haploid**, having only one set of chromosomes, except in cases where **genome duplication** has occurred.

In angiosperms, the sporophyte generations produce two types of **sporangia** (spore-producing organs; Figure 8.5). The ovules house the *megasporangia*, and the pollen sacs function as the *microsporangia* (Judd et al. 2008). The female gametophytes, or **megagametophytes**, are the embryo sacs produced within the ovules. Similarly, the male gametophytes (**microgametophytes**) are the pollen grains produced within pollen sacs of the anthers. An important note here is that the embryo sac and the pollen grains (the gametophyte generation) are the only parts of angiosperms that are correctly referred to as female or male. Flowers, technically speaking, are not male or female because they produce spores rather than gametes (egg and sperm). As mentioned previously, the flowers that produce the sporangia are referred to as being staminate, pistillate, or perfect (containing both stamens and pistils).

Once the sexual reproductive parts are mature, anthers usually will release pollen grains, which may then be transported by various vectors from the anther to potentially receptive stigmas on the same flower, other flowers on the same plant, or flowers on other plants of the same or a different species (Figure 8.6). Mechanisms by which pollen is transported (referred to as **pollination**) include wind, water, and insects (Table 8.1). These pollination mechanisms offer varying degrees of specificity for the species involved, and we will examine this a bit later. The degree of specificity, however, is critically important in determining the likelihood of pollen grains finding compatible and receptive stigmas to ensure successful fertilization. **Compatibility** here refers to pollen grains landing on stigmas of plants with which successful fertilization could occur if the receiving flower is receptive to fertilization at that time. **Receptivity** refers to a biological state of the stigma in which the processes of fertilization will occur once the pollen reaches a compatible stigma.

Upon arrival of the pollen grains on a compatible and receptive stigma (Figure 8.6), a series of physiological changes in both the stigma and the pollen take place that

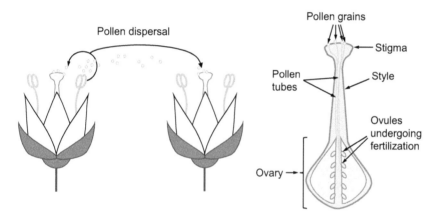

FIGURE 8.6 Once Released from the Anthers, Pollen Grains may be Transported by Various Means (Wind, Water, Animals) to Stigmas on the same Flower or other Flowers. After Reaching the Stigma of a Receptive, Compatible Flower, a Pollen Grain Will Germinate and Send Out a Pollen Tube that Works its Way Down through the Tissues of the Style until it Reaches the Opening of an Ovule, Called the Micropyle. Once the Tube Reaches the Micropyle, it Will Penetrate the Embryo Sac and Fertilization Will Occur (Figure 8.7).

TABLE 8.1
Spectrum of pollination mechanisms found among aquatic and wetland angiosperms.

General Mechanism	Terminology/ Specific Forms	Description
Self-pollination	Autogamy	Pollination within individual open flowers; requires that flowers be hermaphroditic (i.e., perfect)
	Cleistogamy	Pollination occurring within individual closed flowers; requires hermaphroditic flowers
	Geitonogamy	Pollination between different flowers on the same plant; usually found in monoecious species but can occur among separate flowers in hermaphroditic species
Animal (insect) pollination	Entomophily	Sticky pollen attaches to insects and is carried to stigmas of pistillate flowers during insect movement from flower to flower
Wind pollination	Anemophily	Copious dry pollen released by dehiscing anthers and transported by air to stigmas of pistillate flowers
Water pollination	Above the water surface	Pollen usually transferred by intact staminate flowers floating atop the water, thus keeping anthers and pollen dry until staminate and pistillate flowers collide; also referred to as "dry epihydrophily"
	Below the water surface	Pollen released from staminate flowers below the water surface, solitarily or in mucilaginous clusters, sometimes precociously germinating; pollination occurs once neutrally buoyant or sinking pollen reaches stigmas of pistillate flowers; also referred to as "hypohydrophily"
	On the water surface	Pollen released from staminate flowers and floating atop the water, solitarily or in chains or mucilaginous clusters, until encountering pistillate flowers; also referred to as "wet epihydrophily"

lead to hydration and subsequent germination of the pollen grain, growth of the pollen tube through the tissues of the stigma (with the aid of digestive enzymes), and eventual delivery of two sperm nuclei to the embryo sac (Edlund, Swanson, and Preuss 2004; Figure 8.7). The initial step here is of special significance for aquatic and wetland plants, owing to the onset of tube germination after hydration of the pollen grain. For most species, premature hydration of pollen can result in germination prior to the pollen reaching a receptive stigma and, thus, will result in failed fertilization for that pollen grain. This presents a problem for species living in wetlands and aquatic habitats, and later we will address adaptations for mitigating this challenge.

Fertilization in angiosperms is distinct from other plants in that it involves not one but two fertilizations, initiated by delivery of two sperm nuclei to the ovule via the growing pollen tube (Linkies et al. 2010). One fertilization involves fusion of a sperm nucleus with the egg cell inside the ovule; the other involves fusion of the second sperm nucleus with the embryo sac's central cell (Figure 8.7). Within aquatic and wetland angiosperms, we see two slight variations on this double fertilization process. The first, seen in the Nymphaeales (*Nuphar*, *Nymphaea*, *Cabomba*, etc.), is the production of a diploid endosperm from the fusion of the second sperm nucleus and a mononucleate central cell (Figure 8.7). In taxa from the Magnoliidae onward (see Figure 2.17), the second sperm nucleus fuses with a dinucleate central cell to form a triploid endosperm.

The **endosperm** tissue formed from fertilization of the central cell has multiple functions (Linkies et al. 2010). Perhaps the best-known function of endosperm is that it provides a source of nutrition for the developing embryo and for the seedling after germination. It is also believed the endosperm plays a role in developmental signaling for the embryo within the seed. Finally, endosperm tissue in the region of the **micropyle** (entry point of pollen tube to the ovule and exit for the growing seedling at germination; Figure 8.7) has been demonstrated to influence germination

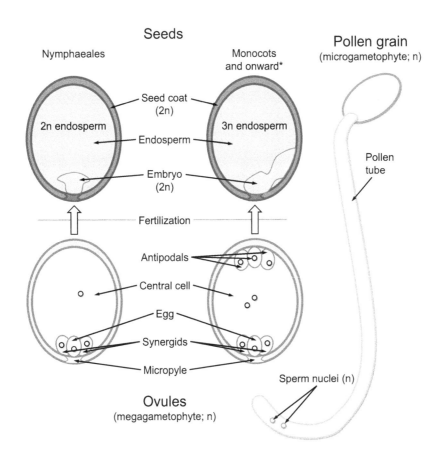

FIGURE 8.7 Fertilization in Aquatic and Wetland Angiosperms Takes Two Different Forms (Linkies et al. 2010). In Both Forms, Fertilization Involves Two Separate, but Concomitant, Fertilization Events. In Both Cases, One of these Fertilizations Involves a Union of One Sperm Nucleus from the Pollen Grain with the Egg Cell inside the Embryo Sac. The Difference between the Two Forms Involves the Second Fertilization. In the Nymphaeales (Left), the Central Cell of the Embryo Sac Possesses Only One Nucleus, and Union with a Second Sperm Nucleus Produces a Diploid (2n) Endosperm. In Taxa from the Monocots Onward (Figure 2.17), the Central Cell of the Embryo Sac Possesses Two Nuclei, and Fertilization Produces a Triploid (3n) Endosperm. The Synergids Appear to be Involved in Interactions with the Pollen Tube that Facilitate Fertilization (Leydon et al. 2015), while the Function of Antipodal Cells is Somewhat Unclear and may Vary Among Plant Species (Song, Yuan, and Sundaresan 2014). *In some Aquatic Monocot Families (Cymodoceaceae, Hydrocharitaceae, Najadaceae, Posidoniaceae, Potamogetonaceae, Zannichelliaceae, and Zosteraceae), the Mature Seed Contains No Endosperm. In those Families, the Nutritive Reserves are Stored Within the Embryo itself (Baskin and Baskin 2014).

by weakening just prior to the onset of germination in many angiosperm species (Linkies et al. 2010).

8.2.1 Points of Potential Modification

At numerous points earlier, I noted that the structures and behaviors described were the general, typical, or usual features that we observe among angiosperms. There are many exceptions to these general rules, driven by selective pressures serving to optimize **fitness** among individuals within species of plants living in aquatic and wetland habitats. By fitness, I mean the ability of individuals of a species to produce viable offspring, or offspring that are fertile and that can survive under a given set of environmental conditions.

Before getting into specific examples of how those various plant reproductive characteristics have been modified, I want to look generally at some of the key steps in reproduction whose modifications are typically associated with life in aquatic and wetland ecosystems (Figure 8.8). Perhaps surprisingly, the first floral trait given in Figure 8.8 is

simply whether the flower opens at all. Non-opening flowers might seem disadvantageous, considering that pollination is a prerequisite for production of seeds; however, there are some aquatic species in which **cleistogamy**, or reproduction via closed flowers, is very common (Hutchinson 1975; Sculthorpe 1967). In most species, wherein the flowers open for reproduction (**chasmogamy**), the flowers may open beneath the water, on the water's surface, or at some elevated position above the water. The position of the flower, then, will play a major role in how pollen is transferred from one flower to another, and even some of the species using water to transport pollen have adaptations to avoid wetting the pollen during transport. Pollination by water, or hydrophily, is observed in surprisingly few species of aquatic plants, while most hydrophytes will be pollinated by either wind (**anemophily**) or animals, with the animal pollinators usually being insects (**entomophily**; Table 8.1).

Assuming pollen grains survive their journey and find flowers that are both compatible and receptive to reproduction, successful pollination will result in the production of

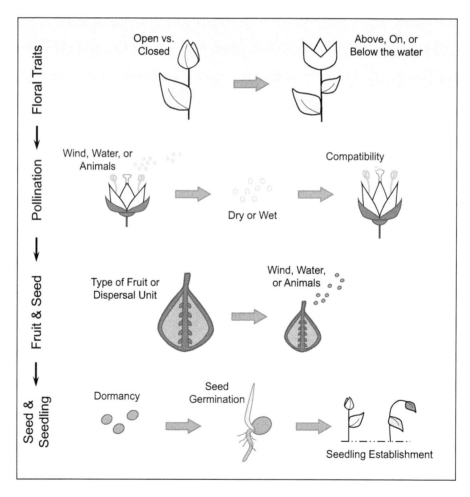

FIGURE 8.8 Stages in Sexual Reproduction Where we see Significant Adaptations Evidenced Among Aquatic and Wetland Plant Species. Some of the More Important Environmental Features for Which these Species have Adapted Include the Potential Negative Impacts of Prematurely Wetted Pollen, the Generally Dynamic Nature of Wetland Ecosystems (E.g., Interannual Variation in Hydroperiod), and the Frequently Intense Levels of Intra- and Interspecific Competition. Detailed Examples of Specific Adaptations are Discussed in the Accompanying Text.

seeds enclosed within some type of fruit. Across the angio-sperms, there is considerable variation in the form taken by the combination of seed and fruit. **Dehiscent** fruits will open by some means, releasing seeds, and this dehiscence may take place while the fruit remains attached to the parent plant or after it has been dispersed by wind, water, or animal transport. **Indehiscent** fruits, on the other hand, do not split open, and it frequently is difficult to distinguish, visually, fruit from seed in these species. Because of this, indehiscent fruit, with seed enclosed, often function as the dispersal unit in those species. Mature seeds, at the time of dispersal, may be **dormant** or **nondormant**. Dormant seeds are incapable of immediate germination and require some ecophysiologi-cal mechanism to remove their dormancy before germi-nation can take place. In the meantime, while the embryo within the seed awaits suitable conditions for germination, that seed will join the other dormant seeds as part of the soil **seed bank** (Figure 8.1), from which future plant species assemblages will emerge to occupy the wetland. Those spe-cies with nondormant seeds can germinate right away, but this means that the timing of seed dispersal will have been fine-tuned through evolution so that it coincides with favor-able conditions for newly germinated seedlings to establish in the environment. Finally, regardless of whether seeds ger-minate right away or after the removal of dormancy, the new seedling typically has a much narrower set of environmental conditions under which it can successfully establish than is the case for growth and reproduction in mature plants.

8.2.2 Pollination within Closed Flowers—Cleistogamy

In species that are obligately cleistogamous, self-pollination is the only option available for carrying out sexual reproduc-tion. Sculthorpe (1967) listed several species of submersed and floating-leafed species that are known to be cleistog-amous, from 11 genera, across eight families; one genus, *Ranunculus*, was stated as having multiple cleistogamous species, but the number was not given. Philbrick and Les (1996) cite another four aquatic genera, in four additional families, as having cleistogamous representatives, without giving the number of known species. Further investiga-tion of the four genera given by Philbrick and Les (1996) revealed that *Myriophyllum farwellii* was described as hav-ing flowers whose arrangement "suggested cleistogamy," although cleistogamy was not directly observed (Aiken 1981); no species of *Potamogeton* were reported to be cleistogamous (Philbrick and Anderson 1987); and *Ruppia maritima* was reported to exhibit a facultative cleistogamy (discussed later; Richardson 1983). The last of the four gen-era they discussed was the genus *Utricularia*, which does include at least one wetland species that regularly produces cleistogamous, along with chasmogamous (reproduction via open flowers), flowers (Kondo 1972). Thus, strict cleis-togamy, in which flowers never open, seems to be relatively uncommon (known from only about a dozen genera) but still present among aquatic angiosperms.

In *Utricularia juncea*, a wetland species from the east-ern coasts of tropical to temperate Americas, from 65% to 80% of the flowers among four North Carolina (USA) pop-ulations were observed to be cleistogamous (Kondo 1972). In this species, cleistogamous and chasmogamous flowers are sometimes found on the same scape (flowering stalk), but more typically, it seems that scapes produce either all cleistogamous or all chasmogamous flowers. In those cases where scapes produce only one type of flower, scapes with cleistogamous flowers were shorter (generally less than 15 cm in height), and the flowers were approximately half the size or smaller than the chasmogamous flowers. Kondo (1972) speculated that, in cases where both flower types were produced on individual scapes, this was caused by changes in environmental conditions (e.g., moisture or tem-perature) as the individual scapes developed.

Similarly, *Ottelia ovalifolia*, in the Hydrocharitaceae, is known to shift flower production throughout the year, from small, submersed, cleistogamous flowers during spring to normal aerial chasmogamous flowers during summer and then small, submersed, cleistogamous flower production again in autumn (Sculthorpe 1967). However, during sum-mer, if the plants were crowded, they also would produce the small, submersed, cleistogamous flowers. Those flow-ers were observed to be fertile, and pollination would occur within the unopened floral perianth, resulting in small fruit with fewer seeds than were produced in normal fruit.

Cleistogamy has also been observed in *Ruppia mari-tima*, in shallow, highly evaporative estuarine areas of New Hampshire (USA; Richardson 1983). Approximately one-third of inflorescences in those shallow wetlands were observed to carry out some degree of cleistogamous pol-lination, and aquarium studies confirmed a high incidence of successful fertilization from cleistogamous pollination in this species. Richardson (1983) speculated that cleistogamy in *Ruppia maritima* may have been triggered by exposure to reduced water levels or to environmental factors that changed along with reduced water level, such as increased salinity or temperature. Richardson further suggested the possibility that avoidance of floral opening in this species serves as an adaptation to tolerate abiotic stress in wetlands experiencing significant evaporative losses, by helping to keep the flowers moist and allowing pollination to take place. It seems then, based on these examples from *Ruppia*, *Ottelia*, and *Utricularia*, that expression of cleistogamy sometimes may be strongly influenced by the environmen-tal conditions experienced during sexual reproduction. This might not be terribly surprising, given the numerous exam-ples of environmental influence on phenotype that we saw in the discussion of heterophylly in Chapter 6.

8.2.3 Pollination of Open Flowers

Although there are some species that are capable of suc-cessful cleistogamous pollination, most pollination occurs using open flowers (chasmogamy). The opening of flowers, however, does not preclude self-pollination, either within individual hermaphroditic flowers or among flowers on the same plant. It is possible that the somewhat random and unpredictable nature of where individual plants can and do establish has selected for, in some species, the ability to self-pollinate as a means of producing seeds even in low-density, pioneer populations, to allow persistence of a population during conditions generally unfavorable for plant growth and reproduction (e.g., winter or drought). Given the importance of sexual reproduction for promotion of genetic diversity and evolutionary novelty, we ought to expect low incidences of self-pollination (Cheptou 2019). However, a glance through a list of hydrophilous genera of aquatic plants shows that monoecy and **hermaphroditism** are very common (Table 8.2), setting the stage for self-pollination. This is not unique to aquatic species, as most angiosperms produce hermaphroditic flowers, but rates of self-pollination and inbreeding can vary quite widely among populations, within a given pollination mode or even within a single species (Whitehead et al. 2018). Angiosperms thus must balance maintenance of small populations (through self-pollination when outcrossing is not possible) against longer-term negative impacts of inbreeding, including reduced genetic diversity, accumulation of deleterious **alleles** (versions of a given gene), and loss of beneficial alleles (Cheptou 2019).

8.2.3.1 Self-Pollination within Open Flowers—Autogamy

As with cleistogamy, **autogamy** requires that chasmogamous flowers be hermaphroditic, as it involves, by definition, pollen transfer from an anther to a stigma within the same flower (Table 8.1). An examination of our list of hydrophilous families (Table 8.2) shows that there are four genera with species that regularly produce bisexual, or hermaphroditic, flowers: *Groenlandia* and *Potamogeton*, in the Potamogetonaceae, *Posidonia*, in the Posidoniaceae, and *Ruppia*, in the Ruppiaceae. Some authors include a fifth genus, *Stuckenia* (Potamogetonaceae), but Stevens (2017) groups *Stuckenia* into the genus *Potamogeton*. Because I have relied on Stevens' Angiosperm Phylogeny website for much of the phylogenetic information in earlier chapters, I am including *Stuckenia* within *Potamogeton* here. Be aware that other treatments may separate these genera; thus, *Stuckenia* would be another example of potentially autogamous aquatic plants. Among the four genera given earlier, *Posidonia* includes at least one species that employs a strategy known as **protandry** to avoid autogamous self-pollination. In *Posidonia australis*, the anthers mature in advance of the pistils (protandry = the anthers mature first; "pro" = first, "andry" = male), and the pistil matures after the anthers have dehisced to release their pollen. This and similar strategies to avoid self-pollination and inbreeding will be covered in more detail later in this chapter.

Guo and Cook (1990) examined the pollination biology of *Groenlandia densa* in detail, under controlled conditions in aquaria and experimental tanks at the University of Zürich Botanic Garden. They found that although the usual

condition seems to be for *Groenlandia* to produce flowers just above the water surface, flowers sometimes were also produced on shorter, submersed shoots. When inflorescences were held below water, bubbles formed inside the anthers, and pollen was observed to collect at the surface of the bubble, both on the inside and outside (Figure 8.9). When bubbles eventually escaped from the flowers, some of the pollen was deposited on the adjacent stigmas of that same flower. From a group of 15 inflorescences kept permanently submerged in that study, all but one produced fruit. In natural populations, however, the inflorescences usually protrude as much as 5 mm above the water surface, so it is unknown how important bubble autogamy is in natural pollination of this species.

Two better known examples of autogamy in submersed flowers occur in *Potamogeton pusillus* and *Ruppia maritima*. As seen earlier in *Groenlandia*, these species demonstrate bubble-mediated autogamous pollen transfer below the water surface. In *Ruppia maritima*, Richardson (1983) found that bubble autogamy was more likely to occur in estuarine sites than in coastal sites. In the coastal populations, *Ruppia* was more prone to disperse pollen as a film on the water surface, where it was available for self- or cross-pollination. Detailed anatomical studies of *R. maritima* by Richardson indicated lacunar aerenchyma led to the point where anther sacs connected to the filaments of the stamen, and it was determined that the air bubbles functioning in pollination resulted from lacunar transport of gases during active photosynthesis. Similarly, Philbrick (1988) discussed the role of the internal lacuna system in *Potamogeton pusillus* as it related to bubble autogamy, and also suggested that bubble autogamy may have been an intermediate form of pollination between aerial pollination and forms of hydrophily that utilize wettable pollen (e.g., non-autogamous pollination below the water surface). This is an interesting hypothesis, given current information on the phylogenetic relationships of hydrophilous taxa (Li et al. 2019), wherein the five families exhibiting high frequencies of both bubble autogamy and submersed hydrophily also form the five most recently diverged branches of the Alismatales (Table 2.3). Note, however, that in the genus *Potamogeton*, some of the basal (earlier derived) North American species (e.g., *P. spirillus* and *P. diversifolius*) produce both aerial and submersed flowers and that much of the more recent diversification within that genus has exclusively yielded species with aerial, rather than submersed, flowers (Ito et al. 2016; Philbrick and Anderson 1987). Thus, it seems that submersed hydrophily and bubble autogamy may have been possible in some of the earliest derived species in this genus, with that trait having been abandoned in favor of aerial pollination alone in later emerging species. This leaves the origin of submersed flowers in *Potamogeton* a bit of a mystery for the time being.

Zannichellia was indicated by Les (1988) to be another genus that can undergo autogamous pollination, based on information given by Aston (1973), who indicated that bisexual plants of *Zannichellia palustris* "can occur,"

TABLE 8.2

Aquatic plant families exhibiting hydrophily, modified and updated from Cronk and Fennessy (2001). The first six families (Hydrocharitaceae–Ruppiaceae) are listed from oldest to youngest, as in Table 2.3. The families Zosteraceae, Posidoniaceae, and Cymodoceaceae comprise exclusively marine species. Lists of genera in each family and approximate numbers of species per genus were obtained from Flora of North America (2021), Global Biodiversity Information Facility (2021), and Stevens (2017). Note that numbers of species refer to total within each genus, not solely those that exhibit hydrophily. Additional information here was obtained from Cook (1988), Cook and Urmi-König (1984, 1985), Cox (1988), Efremov et al. (2019), Guo and Cook (1990), Les (1988), Philbrick (1991), and Rashmi, Krishnakumar, and Les (2017).

Family	Noteworthy Genera	Approx. Species	Sexual Condition	Life History	Type of Hydrophily	Pollen Apparatus
Hydrocharitaceae	Appertiella	1	Dioecious	Annual	Above surface	Staminate flower
	Blyxa	9	Dioecious, rarely Monoecious	Annual & Perennial	*Entomophily*	Pollen grains
	Egeria	2	Dioecious	Perennial	*Entomophily*	Pollen grains
	Elodea	5	Di- & Monoecious	Annual & Perennial	On surface	Pollen grains
	Enhalus	1	Dioecious	Perennial	Above surface	Staminate flower
	Halophila	10	Di- & Monoecious	Annual & Perennial	On surface	Mucilaginous chains and germinated pollen
	Hydrilla	1	Di- & Monoecious	Annual & Perennial	On surface	Pollen grains or staminate flower
	Hydrocharis	3	Monoecious	Perennial	*Entomophily*	Pollen grains
	Lagarosiphon	10	Di- & Monoecious	Perennial	Above surface	Staminate flower
	Limnobium	2	Monoecious	Perennial	*Anemophily*	Pollen grains
	Maidenia	1	Di- & Monoecious	Annual & Perennial	Above surface	Staminate flower
	Najas	40	Di- & Monoecious	Annual & Perennial	Below surface	Pollen grains & pollen tubes
	Nechamandra	1	Dioecious	Annual & Perennial	Above surface	Staminate flower
	Otelia	21	Di- & Monoecious	Annual & Perennial	*Entomophily*	Pollen grains
	Stratiotes	1	Di- & Monoecious	Perennial	*Entomophily*	Pollen grains
	Thalassia	2	Dioecious	Perennial	Below surface	Mucilaginous chains and germinated pollen
	Vallisneria	26	Dioecious	Annual & Perennial	Above surface	Staminate flower
Potamogetonaceae	Althenia	10	Di- & Monoecious	Annual & Perennial	On surface or autogamy	Pollen grains
	Groenlandia	1	Hermaphroditic	Perennial	Bubble autogamy, but also *anemophilous*	Pollen grains
	Potamogeton	>100	Hermaphroditic	Annual & Perennial	A few species below surface, but *most are anemophilous*	Pollen grains
	Zannichellia	5	Di- & Monoecious	Annual & Perennial	Below surface, primarily geitonogamy	Pollen grains & pollen tubes in mucilage
Zosteraceae	Phyllospadix	5	Dioecious	Perennial	On and below surface	Pollen grains
	Zostera	12	Monoecious	Annual & Perennial	On and below surface	Chains of filamentous pollen grains
Posidoniaceae	Posidonia	9	Hermaphroditic	Perennial	Below surface	Filiform pollen grains
Cymodoceaceae	Amphibolis	2	Dioecious	Perennial	On and below surface	Chains and filiform pollen
	Cymodocea	4	Dioecious	Perennial	On surface	Pollen grains
	Halodule	6	Dioecious	Perennial	On surface	Filiform pollen grains
	Syringodium	2	Dioecious	Perennial	Below surface	Pollen grains
	Thalassodendron	3	Dioecious	Perennial	Below surface	Pollen grains and chains of pollen
Ruppiaceae	Ruppia	10	Hermaphroditic	Annual & Perennial	On or below surface, also bubble autogamy	Chains of V-shaped pollen
Ceratophyllaceae	Ceratophyllum	6	Monoecious	Perennial	Below surface	Detached anthers or germinated pollen grains
Plantaginaceae	Callitriche	75–91	Monoecious	Annual & Perennial	A few species below surface, but *most are anemophilous*	Pollen grains; at least 7 species with internal geitonogamy

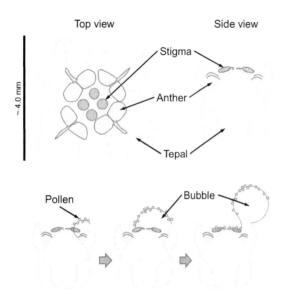

Top view   Side view

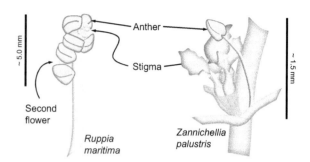

FIGURE 8.10 Arrangements of Flowers in *Ruppia Maritima* (Left) and *Zannichellia Palustris* (Right) are Highly Conducive to Self-Fertilization. In *Ruppia*, Stigmas are Surrounded by the Anthers in Each Flower, and in *Zannichellia*, the Anther Sits Just Above the Funnel-Shaped Stigmas, and Pollen is Released within a Mucilaginous Material that Causes Pollen to Sink within the Water Column. Redrawn from Correll and Correll (1972).

FIGURE 8.9 Bubble Autogamy in *Groenlandia Densa*. Upper Diagrams Illustrate General Anatomy of the Hermaphroditic Flowers of *G. Densa* (Upper Left, Top-Down View; Upper Right, Lateral View). In this Species, the Stamens are Attached to the Inner Surfaces of the Tepals and are Situated Adjacent to a Pair of Carpels, Atop Which the Stigmas are Located. During Pollination, the Anthers Produce Air Bubbles that Gradually Expand to Cover the Surfaces of One or More Stigmas, as Depicted in the Bottom Sequence of Illustrations. As the Air Bubble Finally Escapes the Flower (Bottom Right), Pollen Grains are Deposited on the Stigmatic Surfaces. Based on Observations and Illustrations in Guo and Cook (1990).

In *Zannichellia*, the pistillate and staminate flowers are produced in very close proximity to one another, within the axils of the leaves (Figure 8.10). As the anthers mature, the filaments grow, extending the anther directly above the funnel-shaped styles of the adjacent pistillate flowers (Arber 1920). The pollen grains are released along with starch grains that produce a gelatinous mucilage upon contacting the water. This mixture of mucilage and pollen is slightly denser than the water, resulting in pollen being carried directly downward to the stigmas below, leading to a predominance of geitonogamous self-pollination between the adjacent staminate and pistillate flowers (Arber 1920; Les 1988).

Another genus of aquatic and wetland plants that produces tiny, axillary, unisexual flowers is *Callitriche*, in the Plantaginaceae (Figure 8.11). At least seven species of *Callitriche* are known to undergo an unusual form of geitonogamous self-fertilization (Philbrick and Les 2000). In these species, rather than dropping pollen directly onto the stigma as in *Zannichellia*, the pollen grains germinate while within the anthers, and pollen tubes grow down the filament, through vegetative tissues, and up through the base of pistillate flowers into the ovary (Philbrick 1984). Microscopic examination of flowers showed that some pollen tubes would grow from the staminate flowers into ovaries of immediately adjacent pistillate flowers (within 0.1 mm of the stamens), while others would grow more than one millimeter across the stem to a nearby axillary pistillate flower, and still others would grow several millimeters down the stem to more distant flowers. In *Callitriche heterophylla* and *C. palustris*, this phenomenon, termed internal geitonogamy, was found to occur most commonly in underdeveloped flowers produced on plants that appeared to have experienced unfavorable conditions (e.g., complete flooding or growing completely exposed out of the water). Underdeveloped staminate flowers on those stressed plants tended to produce approximately 0.5% to 1.5% as many pollen grains as fully developed flowers on unstressed plants (Philbrick 1984). Internal geitonogamous fertilization was

although the frequency of this was not stated. The putatively bisexual flowers were described as carpels and an individual stamen all within a single cup-like perianth. This species produces axillary, unisexual flowers, with the pistillate flowers encircled by a membranous, cup-like perianth and the staminate flower originating within the same leaf **axil**, but with no perianth (Haynes and Hellquist 2020; Figure 8.10). It is possible, then, that some specimens either appeared to have had a perianth that surrounded both flowers (when, in fact, it did not) or that occasional plants undergo some form of developmental aberration in which the perianth of the pistillate flowers encircles both the staminate and the pistillate flowers, owing to their proximity to one another. It is known that dioecy sometimes occurs in *Zannichellia* populations (Les 1988), so other forms of atypical development are conceivable. Regardless, the literature on this species suggests that monoecy (i.e., separate pistillate and staminate flowers) is the dominant sexual condition in *Zannichellia palustris*; thus, autogamy would be expected to occur relatively infrequently.

8.2.3.2 Self-Pollination between Open Flowers—Geitonogamy

Because *Zannichellia palustris* predominantly produces separate but adjacent pistillate and staminate flowers, with both flower types on the same plant, self-pollination in this species is most likely to occur via **geitonogamy** (Table 8.1).

FIGURE 8.11 *Callitriche Heterophylla*, Exhibiting its Characteristic Submersed, Linear Leaves and Floating Rosettes of Spatulate Leaves. Spatulate Leaves within the Floating Rosettes are Approximately 3–4 Mm Wide. This is One of Seven Species Known to Exhibit Internal Geitonogamy, Wherein Pollen Tubes Grow Through Vegetative Tissues from the Anther of Staminate Flowers into Ovaries of Pistillate Flowers Across Distances of Less than One Millimeter to Several Millimeters.

observed in as many as 85% of fully developed flowers in *Callitriche heterophylla* and *C. palustris*, but 100% of the underdeveloped flowers that were examined in these species showed evidence of internal geitonogamy. Philbrick (1984) hypothesized that this strange form of self-fertilization, at the time only described in one other angiosperm family (the Malpighiaceae), may be a means of ensuring successful seed production during unfavorable environmental conditions that might prevent successful outcrossing. In some ways, this is similar to the observations of Richardson (1983) in studies of cleistogamy in *Ruppia maritima*, wherein cleistogamy was most often observed in plants that appeared to have experienced environmental stress during declining water levels.

8.2.3.3 Pollination by Animals

Approximately 60%–70% of angiosperm taxa are pollinated by some form of animal (Cook 1988; Cronk and Fennessy 2001), and the vast majority of animal pollinators are insects (entomophily; Figure 8.12) among all angiosperms as well as within aquatic and wetland taxa. Owing to their critical importance in successful reproduction, which has led to multiple evolutionary radiations among angiosperms (Soltis et al. 2018), the relationships between plants and their animal pollinators typically exhibit suites of characteristics that are diagnostic of a particular type of pollinator (Table 8.3). These relationships between plants and insect pollinators are so tightly interwoven that much of the early diversification of angiosperms has been shown to correlate with major diversifications among the Hymenoptera (bees, wasps), Diptera (flies), Lepidoptera (moths, butterflies), and Coleoptera (beetles) roughly 150 million years ago (Soltis et al. 2018).

In addition to the more easily observable characteristics of animal-pollinated species given in Table 8.3, there are a few features of the pollen grains themselves that tend to favor dispersal by insects, in particular (Cook 1988). Among these are:

- Pollen grains with spiny surfaces;
- Pollen grains having rough surfaces, sometimes caused by indentations or apertures in the **exine** (outer covering);

- Sticky pollen, often forming clumps of multiple grains;
- Production of relatively few grains per anther; and
- Small-diameter grains, usually less than 20 µm in diameter.

Among aquatic and wetland plant species, insect pollinators often are either bees or flies; consequently, we see many white, yellow, and purple flowers among wetland angiosperms, as these are the colors preferred by those pollinators (Figure 8.12). Examples of insect pollination in common wetland plant genera include species of *Nuphar*, *Nymphoides*, *Cabomba*, *Utricularia*, *Pontederia*, and several species in the Hydrocharitaceae (Table 8.2 and examples in Cronk and Fennessy 2001). Some of these insect pollinators can be highly specialized. Sculthorpe (1967) gives an example of a species of bee (*Dufourea novae-angliae*) that was known to emerge only at the time when flowering occurred in *Pontederia cordata*, and that was the only plant species known to be pollinated by this bee.

Although there are certain suites of characteristics that strongly suggest that insects may be the typical pollinator of a given plant species (Table 8.3), there are other times when the pollinator of a given species may be unclear by a cursory inspection alone. Cook (1988) provides information for species in several families of aquatic plants that appear, based on floral structure, scent, or other characteristics, to be wind pollinated, but they are actually pollinated by insects. Floral characters such as small or absent perianths, small flowers overall, or unisexual flowers with long slender filaments supporting the anthers tend to suggest pollination by wind (anemophily). However, in some cases, there are finer details of the pollen grains themselves or direct observations of entomophily that suggest insect pollination. One of the more interesting groups covered by Cook (1988) was the duckweeds, including species of the tiny *Lemna* and *Wolffiella*, that were suggested to be pollinated by insects.

There also may be species that are predominantly pollinated by one means, such as water movements (i.e., hydrophilous species), but in which other pollination vectors may sometimes operate. A study by Van Tussenbroek et al. (2016b) found that numerous types of marine invertebrates are capable of successfully pollinating the seagrass *Thalassia testudinum*, which is typically considered to be a water-pollinated species (Table 8.2). They found that several species of polychaete worms and crustacean larvae visited staminate and pistillate flowers of *Thalassia* (a dioecious genus) during their foraging activities and that those organisms often carried viable pollen grains on their bodies. When pollination was allowed to occur in experimental tanks without water circulation, there were frequently more than 100 germinated pollen grains found on individual stigmas of pistillate plants in tanks with invertebrates (30% of tanks), in contrast with a predominance of stigmas with no germinated pollen grains at all in tanks lacking invertebrates (~70% of tanks). It was hypothesized that invertebrates possibly cue in on the release of pollen in *Thalassia* because of the thick carbohydrate mucilage in which the pollen grains are released. This potentially provides an energy-rich food source for the animals, while having the additional benefit of allowing pollen grains to travel with the animals as they move about within seagrass beds. It is quite possible, then, that this process also occurs in other aquatic plants that release their pollen in a matrix of carbohydrates, such as *Halophila* and *Zannichellia* species, although most pollen transfer in those species may be mediated by water movements.

Finally, it is important to note that vertebrate animals also have been shown to pollinate wetland plants. In Western Australia, there is a species of *Utricularia* in which pollination has been observed to be carried out by a bird known commonly as the western spinebill (Figure 8.13). Plachno et al. (2019) carried out detailed micromorphological studies on flowers of *Utricularia menziesii* to better understand how the flowers of this species differ from insect-pollinated *Utricularia* species. In most *Utricularia* species, there are pigmented nectar guides that assist insect pollinators in locating nectar rewards as part of the co-evolved relationship between the plant and the insect species (Plachno et al. 2019). This includes *Utricularia* species pollinated by flies, bees, butterflies, and moths. In *Utricularia menziesii*, there are no nectar guides. Instead,

TABLE 8.3

General characteristics of aquatic and wetland angiosperms using animals as pollination vectors (modified from Judd et al. 2008).

	Typical Flower Colors	Scent	Shape of Corolla	Flowering Time
Bees	Blue, yellow, purple	Strong	Bilaterally symmetrical with landing platform	Day
Butterflies	Red or other bright colors	Weak	Landing platform, sometimes with nectar spurs	Day
Flies, with reward	Light colors	Faint	Radially symmetrical	Day
Birds	Bright, often red	None	Tubular, pendant	Day
Beetles	Green or white	Strong	Various	Day or night
Flies, carrion-type	Brownish or purplish	Rotten, strong	Various	Day or night
Moths	White or pale colors	Sweet, strong	Sometimes with nectar spurs	Night or dusk

FIGURE 8.12 A Few Examples of Insect Pollination of Wetland Plants. Note the Yellows, Whites, and Purplish Colors that are Prominent in these Bee and Butterfly Pollinated Plants. Clockwise, from Top Left: *Butomus Umbellatus*, *Nymphaea Odorata*, *Cephalanthus Occidentalis*, and *Ludwigia Grandiflora*.

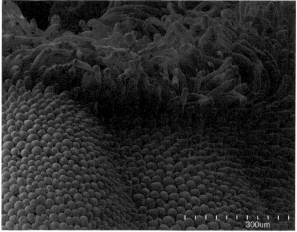

FIGURE 8.13   The Red Flowers of *Utricularia Menziesii* in Western Australia (Left) are Pollinated by the Western Spinebill (*Acanthorhynchus Superciliosus*; Top Right). It is Believed that the Yellow Area at the Top of the Red Flower, Termed the Palate, Serves as a Tactile Guide for the Birds as they Probe the Flowers for Nectar During Pollination. Lower-Right Photo is an Electron Micrograph of Papillae that make up the Basal Lining of the Palate and the Entryway of the Floral Throat. A Scale Bar (300 Microns) is Shown in the Lower Left of the Micrograph. Photos of the Flower and Bird Pollinator © Jean and Fred Hort; Attribution 2.0 Generic (CC BY 2.0); https://creativecommons.org/licenses/by/2.0/; Micrograph Provided Courtesy of Dr. Bartosz Jan Plachno. The Flower in the Left Photo is Approximately 2 Cm Long.

the flowers are primarily a bright red color, with a large yellow palate at the opening of the floral tube, which leads to a nectar reward at the base of a long floral "spur" (the elongated bottom of the corolla in Figure 8.13). That spur approximates the shape of the western spinebill beak, and the birds have been observed hopping along the ground from flower to flower, with pollen grains accumulated at the base of their beaks (Plachno et al. 2019). Plachno et al. (2019) mentioned that other species of *Utricularia* are described as being pollinated by hummingbirds, as have a few species of *Pinguicula* (also in the Lentibulariaceae), but *Utricularia menziesii* appears to have the strongest body of evidence supporting bird pollination, including direct observation of pollination taking place.

8.2.3.4   Pollination by Wind

Wind-pollinated, or anemophilous, species, in general, show many contrasts to the characteristics discussed earlier

for animal-pollinated species (Cook 1988). For example, they typically produce unisexual flowers (i.e., separate pistillate and staminate flowers) that are separated spatially or temporally, in the case of many anemophilous tree species, from leaves and that have small (or absent) perianths. Separation of unisexual flowers from the leaves and production of small perianth segments helps avoid the problem of such structures obstructing wind dispersal of the pollen, once it is released from the anthers. The stigmas of pistillate flowers in wind-pollinated species usually will have morphological adaptations that enhance the efficiency of pollen capture. Additionally, wind-dispersed pollen also shows contrasts with animal-dispersed pollen, including:

- Pollen grains with smooth surfaces;
- Non-sticky, powdery pollen that disperses in clouds of individual grains (rather than in clumps or clusters);

- Production of large numbers of grains per anther, usually with relatively large anthers; and
- Pollen grains usually in the range of 20–40 μm in diameter.

According to Cook (1988), more than one-third of the genera of angiosperms that contain aquatic species are exclusively wind pollinated. Some of the more common families of wetland plants that are anemophilous are the Cyperaceae, Poaceae, Typhaceae, and Juncaceae. In 1988, that amounted to 119 genera of plants, 100 of which appeared to have descended from wind-pollinated terrestrial ancestors. Many of the 19 genera that lacked known terrestrial ancestors were predominantly from families that were thought to be exclusively wind pollinated. That group included families such as the Haloragaceae and Typhaceae, but also the Potamogetonaceae and Callitrichaceae (now part of the Plantaginaceae), which are now known to include at least a few water-pollinated species. There also are some genera, including *Brasenia* and *Limnobium*, that appear to have evolved to wind pollination from insect-pollinated ancestors (Cook 1988).

8.2.3.5 Pollination by Water

Based on most available estimates, there are fewer than 200 species of angiosperms that are evolved to use water as the primary means of transporting pollen during sexual reproduction (Table 8.2). Those species are found in eight families, within genera that, in most cases, include ten or fewer water-pollinated species. Those eight families also have a median molecular age of about 50 million years (Table 2.3). Hydrophily thus appears to be a relatively new advent for most of these species, with relatively little diversification having occurred within the groups that have evolved to take full advantage of the aquatic environment (Philbrick 1991).

Les (1988) discussed at length potential factors that have led to, or at least are associated with, the relatively low diversity of hydrophilous taxa. One conclusion was that the predominance of asexual propagation among aquatic species may have contributed to low rates of diversification, despite hydrophilous species having a higher than usual frequency of **dicliny** (see Table 8.4 for descriptions of this and related terms), compared with other angiosperms. For example, among terrestrial angiosperms, as many as 75% of species within a region may be hermaphroditic (Gurevitch, Scheiner, and Fox 2006), in contrast with approximately 15% of the hydrophilous species represented in Table 8.2. However, even in comparison with the younger wind-pollinated families of the Cyperaceae (> 5,000 species), Poaceae (> 11,000 species), and Juncaceae (> 400 species), which are wind pollinated and exhibit considerable tendency towards asexual propagation, hydrophilous species diversity is astonishingly low. Likely, it is the interaction of the reliance on water for pollination in combination with other constraints of life in the aquatic environment that have led to this low diversity among hydrophilous species. As Les (1988) put it: "the complex interactions of hydrophily

and other aspects of their reproductive biology are likely to have profoundly influenced the patterns of their present diversity."

As discussed in the previous section, most hydrophilous taxa appear to have evolved within or out of predominantly wind-pollinated clades, and there are numerous similarities between anemophily and hydrophily (Philbrick 1991). Anemophily and hydrophily both rely on nonspecific abiotic pollen transport, and in most cases, distribute pollen through a three-dimensional fluid medium (Philbrick and Anderson 1987). A correlate of the reliance upon nonspecific pollen dispersal through space is that the species with greater frequency of outcrossing during pollination often produce large quantities of dry, spherical pollen, but we will see later that not all water-pollinated species fit this description.

Among the water-pollinated taxa, we see three general pathways of pollen transport (Table 8.1): above the water surface, on the water surface, and below the water surface. Some authors use the terms **epihydrophily** and **hypohydrophily** to distinguish between pollination above (epi-) or below (hypo-) the water surface. However, the former term, epihydrophily, also is sometimes modified with the terms "dry" and "wet" (e.g., dry epihydrophily) to distinguish between pollen above or on the water surface. This creates a somewhat inaccurate category of "wet epihydrophily" for species whose pollen is transported on the water surface. Although the pollen in these species is transported on the surface of the water (a place that we might assume to be "wet"), the pollen in many of these species has characteristics that prevent the individual grains from becoming wet (Cook 1988; Haynes and Holm-Nielsen 2001). This "unwettable" nature of the pollen grains provides at least two important contributions to pollination in these species. First, the avoidance of wetting prevents premature germination of the pollen grains, which would lead to a higher incidence of failed pollination for those pollen grains. Second, the unwettable surface of this pollen allows the grains to float atop the water surface, increasing the efficiency of pollination, versus the case where wettable pollen grains might adhere to the water surface itself and thus exhibit restricted movement across the water.

8.2.3.6 Pollination above the Water Surface

There are two general types of water-mediated pollination that occur above the water surface. In the more common of these, staminate flowers are released from the plant, and they float atop the water until contacting pistillate flowers. The pistillate flowers have unwettable surfaces that help to create depressions into which the staminate flowers are drawn (Figure 8.14). Pollination then occurs when pollen from anthers of the staminate flower encounters the stigma of the recipient pistillate flower. Genera exhibiting this form of hydrophily include *Vallisneria*, *Lagarosiphon*, *Maidenia*, and *Appertiella*, among others. Common characteristics among species that use this form of pollination are (Cox 1988):

TABLE 8.4

Mechanisms associated with minimization of self-pollination in angiosperms. Information gleaned from Barrett (2002), Gurevitch, Scheiner, and Fox (2006), and Les (1988).

Category	Mechanism	Description
Primary categories	**Cosexuality**	State where both stamens and pistils can be found on individual plants; includes both hermaphroditic and monoecious plants (and others described later)
	Dioecy	Staminate and pistillate flowers produced on separate individual plants
	Hermaphroditism	Plants produce flowers having both stamens and pistils (bisexual flowers) or produce both staminate and pistillate flowers on the same individual plant
	Monoecy	Plants produce both staminate and pistillate flowers on the same individual
Variations on cosexuality	**Androdioecy**	Within a population, some individual plants produce only staminate flowers, while others produce bisexual or both staminate and pistillate flowers
	Andromonoecy	Individual plants produce both staminate flowers and bisexual flowers; more common than **gynomonoecy**
	Cryptic dioecy	Situation in a monoecious or hermaphroditic population where some flowers produce nonviable pollen and others produce nonviable ovules
	Gynodioecy	Within a population, some individual plants produce only pistillate flowers, while others produce bisexual or both staminate and pistillate flowers; more common than androdioecy
	Gynomonoecy	Individual plants produce both pistillate flowers and bisexual flowers
	Sequential hermaphroditism	Individual plants initially produce either staminate or pistillate flowers only (usually staminate), then switch to the other flower type; the switch sometimes is gradual, with both staminate and pistillate co-occurring briefly during the transition
Segregation of sexes	Dichogamy	Separation of sexes in time, as in sequential hermaphroditism
	Dicliny	Separation of stamens and pistils into separate flowers; standard monoecy and dioecy are examples of dicliny
	Herkogamy	Separation of sexes in space; may be found in conjunction with dichogamy, where staminate flowers, produced in one part of an inflorescence, mature prior to pistillate flowers, which are produced in another part of the inflorescence
Morphological Variations	**Distyly**	Production of two different floral morphs, e.g., long and short styles or filaments
	Enantiostyly	Production of mirror-image floral morphs, such as flowers with styles that bend either left or right of the flower's centerline
	Heterostyly	Production of flowers differing in lengths of styles or filaments, leading to "morphs" that encourage outcrossing between, vs. within, morphs
	Tristyly	Production of three different floral morphs

- Dry, spherical pollen grains;
- Staminate flowers released below the water then floating atop the water, transported by water movements;
- Staminate flowers with structures that function similarly to sails, to aid water-borne transport of pollen within or upon the anthers; and
- Floating pistillate flowers with unwettable perianths and attached to plant by long, flexible peduncles that allow movement with the water.

The second form of above-water hydrophily is demonstrated by *Hydrilla verticillata*. In this species, the staminate flowers are produced underwater, as earlier, and they are released from the plant to float to the water surface. However, the major difference in *Hydrilla* is that, upon reaching the water surface, the staminate flowers open forcefully, propelling the pollen into the air (Figure 8.15).

According to Cook (1988), however, pollen of *Hydrilla* lacks characteristics typically associated with wind-pollinated species and, similarly, the stigmas are not characteristic of anemophilous species. For optimal effectiveness in pollination in *Hydrilla*, the pollen must drop almost vertically into the pistillate flowers, rather than being transported horizontally by air currents. The pollen grains are large (> 90 microns diameter) and covered with a dense pattern of papillae that aid in adhesion of the pollen to the relatively small stigmas once contact is made. Because of these floral and pollen characteristics, the staminate flower must open in close proximity to a pistillate flower for successful pollination; otherwise, all of the pollen will simply fall to the water surface and be lost (Haynes and Holm-Nielsen 2001).

8.2.3.7 Pollination on the Water Surface

As with pollination above the surface, there are two general mechanisms by which pollination on the water

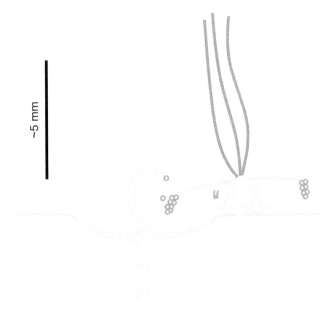

surface takes place. In the first, unwettable pollen grains are released from staminate flowers onto the surface of the water. In *Elodea* species, for example (Figure 8.16), staminate flowers are produced below the water but are released from the plant and allowed to float to the surface (Haynes 1988; Haynes and Holm-Nielsen 2001). Upon reaching the air, the staminate flowers open, with the perianth forming a raft upon which the flower can move about the surface of the water. After these flowers open, the anthers dehisce and release unwettable pollen, which also floats atop the water until it encounters pistillate flowers. The pistillate flowers form a slight indentation in the surface of the water, aided by hydrophobic properties of the perianth parts, but the styles often will directly contact the water, aiding in capture of pollen.

In the second general mechanism of surface pollination, the pollen grains are released from anthers within a polysaccharide-based mucilage or as masses of filamentous pollen grains (Figure 8.17; Cox 1988; Haynes and Holm-Nielsen 2001). These aggregations of pollen serve to increase the overall surface area by which pollen can "search" for stigmas of the pistillate flowers for pollination. This mechanism is common among marine seagrasses, which use the movement of tides to assist in transport of the pollen rafts from anther to stigma. The mucilage in which the pollen is embedded not only serves as a means of adhesion among pollen grains, but also may aid in preventing the wetting of the individual pollen grains until contact is made with stigmas (Cox 1988; Cronk and Fennessy 2001).

Common characteristics among species that use this form of pollination are (Cox 1988):

FIGURE 8.14 Pollination above the Water Surface in *Lagarosiphon Muscoides*. Pistillate Flower (Left) is Receiving Pollen Directly from the Staminate Flower (Right), Which has Blown across the Water Surface with the Assistance of a "Sail" Formed by Three Sterile Stamens (Haynes 1988). An Indentation Formed in the Water Surface By Unwettable Perianth Parts of the Pistillate Flower Aids in Forcing Contact between Staminate and Pistillate Flowers. Upon Contact between the Flowers, the Outstretched Fertile Stamens Transfer Pollen from Anther to Stigma. Note that the Pollen Does not Contact the Water During Pollination in this Species and those with a Similar Pollination Mechanism. Redrawn from Cook (1988).

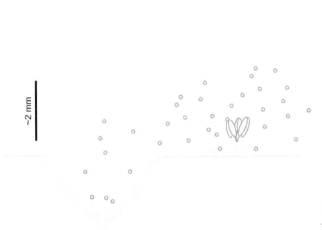

FIGURE 8.16 Pollination on the Water Surface in *Elodea Nuttallii* and *Elodea Canadensis*. As in *Hydrilla* and *Lagarosiphon* (And Others), the Staminate Flower is Released from the Parent Plant, After Which it can Serve as a Search Vehicle to Aid in Pollen Transport to the Pistillate Flowers. In this Species, Unwettable Pollen Grains Fall to the Water Surface and are Aided in Collision with Pistillate Flowers by an Indentation in the Water Surface around Each Pistillate Flower. Unwettable Surfaces of the Pistillate Flower Perianth Aid in Formation of an Indentation of the Water Surface, Facilitating Interaction between the Staminate and Pistillate Flowers. Redrawn from Cook (1988).

FIGURE 8.15 Pollination above the Water Surface in *Hydrilla Verticillata*. The Staminate Flower (Right) Opens Upon Emerging from Below the Water, Resulting in Explosive Release of Pollen into the Air. Some of this Pollen May Contact the Stigma of the Pistillate Flower (Left), but any that Falls upon the Water itself Will Become Unavailable for Pollination. Redrawn from Cook (1988).

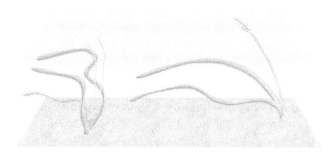

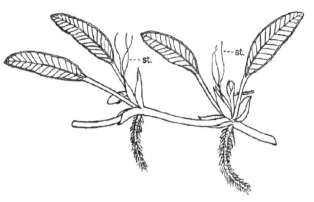

FIGURE 8.17 Pollination on the Water Surface in *Halodule Pinifolia*. Anthers Appears above the Water Surface at Low Tide, at Which Point, they Dehisce and Release Pollen (Plant on Right Side of Figure). Rafts of Filiform Pollen then are Transported Across the Water Surface by Tidal Movements, Where Some Will Contact Filamentous Stigmas of Pistillate Flowers, also Floating Atop the Water Surface (Plant on Left). Redrawn from Cox (1988).

FIGURE 8.18 Pistillate Flowers of the Hydrophilous Marine Plant *Halophila Ovalis*, Each with Three Filamentous Stigmas (St.), Emanating from a Single Style on Each Flower. Pollination Takes Place When Rafts of Long, Mucilaginous Strands of Pollen Encounter the Floating Stigmas, Assisted by Tidal Movements (Haynes and Holm-Nielsen 2001). Illustration Taken from Arber (1920).

- Low-density pollen with hydrophobic (i.e., unwettable) surfaces;
- Filamentous pollen or pollen embedded in mucilage (in most species), facilitating formation of rafts that function in search for stigmas;
- Floating stigmas that are long and thread-like (marine species) or that form depressions in the water surface (freshwater species); and
- Marine species whose flowering is timed to coincide with tidal movements.

Some of the genera that use this mode of pollination include *Cymodocea*, *Halodule*, *Amphibolis*, and *Halophila* (Figure 8.18).

8.2.3.8 Pollination below the Water Surface

Pollination below the water surface is a difficult phenomenon to observe in nature. As a result, many of the published studies on submersed hydrophily have taken place in controlled laboratory settings, where such factors as currents and tidal movements are either absent or artificially recreated. Recall the studies described earlier of pollination in *Groenlandia densa* where flowers that usually emerge from the water, or just at the surface, were forcibly held below water to observe the process of bubble autogamy (Figure 8.9). Another consequence of the difficulties associated with observing submersed hydrophily in nature is that we lack a broad, general understanding of how hydrophily evolved and why it is so much more common among monocot taxa (Philbrick 1991). Nevertheless, we do have detailed information about submersed pollination in a few species, and we have already seen a few examples of submersed pollination in preceding sections on autogamy and geitonogamy (Figures 8.9 and 8.10).

In *Ceratophyllum* species (Figure 8.19), for example, stamens are released from submersed axillary staminate flowers, and pollen is released upon arrival of the stamen at the water surface (Cox 1988). The pollen is negatively buoyant

(i.e., slightly denser than water), and the grains descend back through the water column where some may encounter stigmas of the axillary pistillate flowers (Figure 8.19). Other species, such as *Althenia bilocularis*, use a similar approach, but the pollen is released within a mucilaginous cloud and captured on highly dissected, feathery stigmas (Cox 1988). In *Najas* species, pollen germinates prior to release from the anthers, and pollination is aided in these species by the germinated pollen tube intersecting with stigmas upon descent of the grains through the water or by pollen tubes growing towards the stigmas from grains that fell upon leaves or other parts of the recipient plants (Huang et al. 2001). Some species also employ a combination of submersed and surface pollination, wherein buoyant pollen is released from submersed anthers, but it may be intercepted by receptive stigmas as it passes upward through the water column (Cronk and Fennessy 2001). Species in the Zosteraceae (*Phyllospadix* and *Zostera* species) are some of the more common examples of this.

Although we need more research on mechanisms of underwater pollination *in situ*, there are some general trends that have been observed, including (Cox 1988):

- Pollen grains denser than water and filamentous or spherical;
- Pollen often released in mucilaginous strands or clouds;
- Pollen released in large quantities in freshwater species; and
- Stigmas highly variable but with morphological features that aid in three-dimensional pollen capture (e.g., stiff with sticky papillae or feathery with large surface area).

One genus of aquatic plants in which observations of submersed pollination are lacking is *Callitriche*, in the Plantaginaceae. This genus was previously thought to be exclusively wind pollinated, except for cases of autogamy or internal geitonogamy (Cook 1988). However, there are species of *Callitriche* that produce only submersed inflorescences, suggesting that they could be pollinated underwater. For one of these species, *Callitriche hermaphroditica*, Philbrick (1993) conducted DNA-based paternity exclusion analyses to determine whether it is possible that this species experiences genetic **outcrossing** (successful interbreeding between genetically distinct individuals). His analyses revealed multiple genetic markers that indicated multiple seedlings of the sampled mother plants had received DNA from a second parent plant; in other words, his work demonstrated that outcrossing had taken place in the sampled population of *C. hermaphroditica*.

In addition to direct genetic evidence of outcrossing in *C. hermaphroditica*, Philbrick and Osborn (1994) found that the pollen structure in this species differed substantially between it and species that produce aerial flowers.

The authors were unsure whether there was a functional explanation for the differences, wherein the outer layer of the pollen grains (the exine) was very thin or absent in *C. hermaphroditica* but well defined in the aerial flowering species examined. They mentioned that the pollination mechanism in another species, *C. truncata*, is relatively poorly understood but that it also produces submersed flowers. Ito et al. (2017) found that, phylogenetically, *C. hermaphroditica* and *C. truncata* formed a monophyletic clade with only one other species, *C. lusitanica*, that appears to be somewhat poorly studied as well (Lansdown et al. 2017). Cooper, Osborn, and Philbrick (2000) compared the pollen structure of *C. truncata* and *C. lusitanica* with several other *Callitriche* species, and they found that, as in *C. hermaphroditica*, the pollen lacked an exine layer. Thus, is it possible that this three-species clade represents a cluster of *Callitriche* species that make use of submersed hydrophily?

FIGURE 8.19   *Ceratophyllum Demersum* is an Example of a Submersed Hydrophilous Species that is Pollinated below the Water Surface. Upper Right Illustrates the Axillary Flowers, While the Lower Photos Highlight Flowers (Left) and Fruit (Right). Illustrations (Upper Right) Taken from Arber (1920).

8.2.4 Compatibility, Inbreeding, and Inbreeding Avoidance

Flowering plants have evolved an array of mechanisms that regulate the frequency with which a given species may engage in self-fertilization (Table 8.4). As we have seen in previous sections, some species are not only capable of self-fertilization, but this seems to be their primary means of sexual reproduction (e.g., species that carry out bubble autogamy, cleistogamy, or internal geitonogamy). Species that produce hermaphroditic flowers or that produce some combination of staminate and pistillate flowers on individual plants (see cosexuality in Table 8.4) are much more likely to self-pollinate than are species that segregate sexes among individual plants. However, even when sexes are segregated spatially among individual plants, **inbreeding** (breeding among genetically closely related individuals) is still a possibility.

As further insurance against inbreeding, some species exhibit not only morphological and anatomical adaptations to discourage pollination of themselves or close relatives, but they also may exhibit physiological and biochemical **incompatibility** mechanisms. There are at least two broad recognition systems that are employed among angiosperms to recognize pollen from themselves or close relatives, and thus, to distinguish between compatible and incompatible pollen (Figure 8.20; Silva and Goring 2001).

These compatibility systems rely on protein-based recognition mechanisms that allow the style of the pistillate, or recipient, flower to recognize pollen as self or non-self.

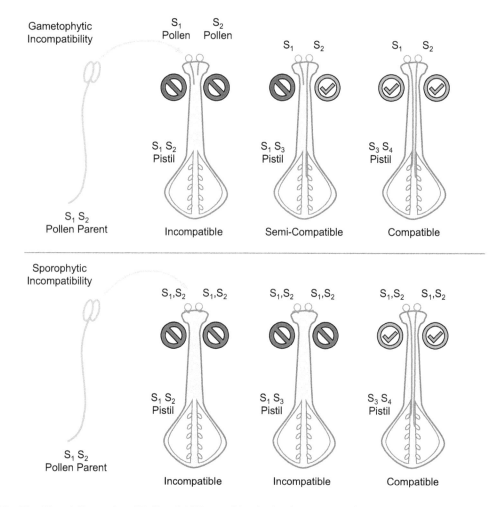

FIGURE 8.20 Two Broad Categories of Pollen Self-Recognition in Angiosperms. In Gametophytic Incompatibility (Upper Panel), It is the Haploid Nature of the Microgametophyte (Pollen Grain) that Determines whether the Stigma will Recognize Pollen as Self and, thus, as Incompatible. As a Result, if the Compatibility Allele Carried by the Pollen Grain Matches either Allele Carried by the Pistillate Sporophyte Tissue, the Pollen Grain Might Germinate, but Pollen Tube Growth will be Arrested in the Upper Portion of the Style. Non-Match between the Pollen Grain and Pistil Results in Compatibility and Successful Pollination (Top, Right). In Sporophytic Incompatibility (Lower Panel), it is the Diploid **Genotype** of the Staminate Pollen Parent that Determines Incompatibility. Here, If Either Incompatibility Allele of the Pollen Parent Matches Either Allele of the Pistillate Parent Tissue, Pollen Tube Growth will not Occur, and Pollination will Fail. Compatibility Occurs When None of the Parental Compatibility Alleles Match (Bottom, Right). Redrawn from Silva and Goring (2001).

Recognition of incompatible self pollen initiates one of a suite of biochemical responses in the recipient plant; the specific process depends on the species involved. Some of the best studied self-incompatibility systems are those of species in the poppy family (Papaveraceae), the tomato family (Solanaceae), and the mustard family (Brassicaceae). You may notice that only one of these has appeared regularly among the aquatic and wetland plant subjects of this book (the Brassicaceae). Indeed, it is the agricultural and economic importance of those three families that likely has led to them having been the focal subjects of these compatibility studies.

In the Papaveraceae and the Solanaceae, the mechanisms by which self-pollination is prevented rely on arresting growth of the germinated pollen tube as it makes its way down through the style of the recipient flower. This is accomplished either by shutting down gene expression (Solanaceae) or by interfering with cytoskeletal proteins (Papaveraceae). In the Brassicaceae, pollination is arrested much sooner, by preventing hydration of the pollen, preventing pollen germination, or excluding germinated pollen tubes from the surface of the stigma (Silva and Goring 2001). It appears at present that these incompatibility mechanisms have received little attention among the aquatic and wetland plants.

Up to this point, we have seen that plants exhibit a wide variety of physical and physiological methods by which they can minimize inbreeding. Why is this the case? Why do we see such a diversity of mechanisms to ensure outcrossing, when some species appear (perhaps) to suffer no negative impacts from a high incidence of self-pollination?

One often cited rationale for an emphasis on outcrossing, or **xenogamy**, during sexual reproduction is the greater degree of genetic variability that is provided by intermixing genes from different individuals, populations, or even species. Recall that wetlands are highly dynamic ecosystems, in part because of the potential for substantial year-to-year variation in hydrology. That variation in hydrology leads to potential for large swings in essentially all components of wetland ecosystems, including soils, soil microbes, plant species composition, potential pollinators, herbivores, and so on. The genetic admixture that is provided through outcrossing during pollination may provide a buffer within the local population against negative effects of those interannual environmental changes.

Another important effect of outcrossing is that it helps to minimize the potential negative impacts of inbreeding, which can lead to an accumulation of unfavorable alleles within a population. This process, or the outcome of this process, is referred to as **inbreeding depression**, wherein one or more phenotypic traits resulting from inbreeding leads to a decline in fitness for inbred individuals.

Reusch (2001) examined populations of the marine seagrass *Zostera marina* (eelgrass; similar morphologically to *Halodule*, illustrated in Figure 8.17) in the Baltic Sea and found that isolated patches tended to be highly inbred. In those isolated patches, individual *Zostera* ramets sometimes were found to produce no more than two viable seeds total. Inbreeding and inbreeding depression, which was based on rates of genetic **homozygosity** (presence of two identical alleles at a sampled chromosome location), were higher in isolated patches than in larger, continuous eelgrass meadows. In terms of rates of outcrossing, Reusch (2001) saw that, as genetic diversity within patches of *Zostera* increased, there was a concomitant increase in outcrossing rates of viable offspring, along with greater successful seed production.

An interesting aspect of inbreeding in *Zostera* is that, although it is monoecious with staminate and pistillate flowers both present within individual inflorescences, it is capable of dichogamous (Table 8.4) floral maturation (Reusch 2000), with staminate flowers maturing prior to maturation of the pistillate flowers in some populations (protandry), pistillate flowers first in others (**protogyny**), and some populations having both flowers maturing simultaneously (Reusch 2000; Ruckelshaus 1995). Although **dichogamy** has been demonstrated to reduce the likelihood of self-pollination from within a given inflorescence in *Zostera marina* (Ruckelshaus 1995), it does not preclude pollination by close relatives. *Zostera marina* grows as a clonal, polycarpic perennial (Figure 8.2), which provides opportunities for self-fertilization among different ramets within a given clone, or genet. This, in combination with the presence of self-compatibility in this species, provides ample opportunities for inbreeding within a clone or between nearby related clones, as was observed by (Reusch 2001).

Studies similar to the previous were conducted in another perennial seagrass species, *Posidonia australis*, by Sinclair et al. (2014). *Posidonia australis* produces bisexual flowers, which potentially provides greater opportunities for inbreeding; however, *Posidonia* is protandrous. In this species, the filaments of the stamens block access of pollen to the stigmas until after anthers have dehisced to release their pollen. At that point, the style begins to grow as the pistil matures. This process reduces chances for autogamous pollination, but, because flowers within an inflorescence mature at different times, geitonogamy is still possible between different flowers of an inflorescence (Sinclair et al. 2014). In addition, the pollen is released into the water column where it can remain suspended for several weeks, extending opportunities for pollination.

In studies of *Posidonia australis* pollination in southwestern Australia, Sinclair et al. (2014) found that, within populations of *P. australis*, there was a moderate level of genetic diversity, but considerably higher genetic diversity among offspring. In fact, they found that essentially all the offspring they sampled (421 in total) were outcrossed, based on the set of seven genetic markers that were used for their analyses. They also estimated that pollen dispersal distances in their sampled populations (~30 m) averaged two to three times the size of the average clone, or genet.

The maximum pollen dispersal recorded in their study was almost 180 meters, while the mean diameter of individual genets was approximately 13 meters. Thus, *Posidonia australis* appears to rely on the combination of dichogamy and long-distance pollen dispersal to ensure outcrossing and appears to do so quite effectively.

In contrast to the relatively long-distance pollen dispersal in *Posidonia*, *Vallisneria americana* in Chesapeake Bay (US) was found to have relatively short pollen dispersal (Lloyd, Tumas, and Neel 2018). As a consequence, this species was found to exhibit frequent inbreeding, with both maternal and paternal relatives contributing to inbreeding in as much as 17% of offspring in some sites, even though this is a dioecious species. Genetic diversity within patches, however, was believed to be maintained via long-distance dispersal of seeds from other populations within the region. This case of seed dispersal exceeding pollen dispersal is relatively rare, however, and is likely aided by adaptations for water-assisted dispersal of the seeds.

In addition to mating systems (e.g., monoecy vs. dioecy), phenological traits (dichogamy), and biochemical or physiological compatibility, variations in flower morphology can enhance likelihood of outcrossing during pollination. One common example of this is **heterostyly** (Table 8.4), in

which different floral morphs encourage directional pollen delivery among, rather than within, individual plants (Figure 8.21). This is commonly seen in insect-pollinated species, where small spatial separation of anthers and stigma can be very effective at preventing a particular pollinator from depositing pollen on stigmas in the flower where the pollen was produced. Typically, in self-incompatible tristylous species, successful pollination will occur when pollen has been taken from stamens that are of the same length as the style being pollinated. For example, flowers of the morph shown in Figure 8.21 should be most successfully pollinated with pollen from a flower having medium-length stamens. Note that, as is illustrated in Figure 8.21, we typically do not see stamens and styles of the same length co-occurring in flowers of heterostylous species.

Eichhornia crassipes exhibits a tristylous system but is self-compatible (Haynes 1988). Experimental pollination trials showed high rates of seed production in both self-pollinated and outcrossed plants, but outcrossed plants tended to yield higher fruit production. Another *Eichhornia* species, *E. azurea*, is tristylous but only partially self-compatible. In *Eichhornia azurea*, self-pollination in flowers with long styles was more successful when using pollen from the medium-length stamens (94% success) than from

FIGURE 8.21 *Eichhornia Crassipes* Illustrates Tristyly, One Morphological Means of Encouraging Outcrossing During Entomophilous Pollination. Tristyly is Expressed in this Species with Short, Medium, and Long Styles and Filaments. In the Photos here, we see Short and Long Filaments with Medium-Length Styles. This Species also is Fully Self-Compatible (Pollen from any Length Stamen can Fertilize any Length Style), unlike its Congener *Eichhornia Azurea*, Which is Tristylous but Partially Self-Incompatible (Haynes 1988).

the short stamens (12% success; Barrett 1978). Thus, as a system of ensuring outcrossing, heterostyly breaks down somewhat in species that are self-compatible if the pollinator isn't completely directed away from the style during its visit to a given flower.

8.2.5 Fruit and Seed Development and Dispersal

Recall that one distinction between the gymnosperms and the angiosperms is that the angiosperms, as the true flowering plants, produce fruit, which is derived from the ovary and in which the fertilized ovules develop into seeds. The gymnosperms, in contrast, produce "naked seeds" (gymno- = naked), even though those seeds may be encased in structures, such as cones, that may appear somewhat fruitlike. For example, compare the cones of the coniferous Pinaceae with the conelike fruit of alders (Figure 8.22). In the conifers, the cones comprise a central axis that produces modified shoots that become scales that bear the ovules (Judd et al. 2008). After fertilization, those scales become the wing that we see attached to conifer seeds in many species and which aids, to some degree, in wind dispersal of the seeds (Figure 8.22). Each of these scales is subtended by a bract that is closely attached to the ovule-bearing scales in some species. Those bracts may have a woody appearance, as in the Pinaceae, or it may be leathery, as in the Cupressaceae (e.g., *Taxodium*). In other species (within *Juniperus*, for example), the bract is fleshy, and the cones appear berrylike.

In angiosperms, pollination of flowers and fertilization of ovules initiates a cascade of signals within the developing seed and fruit that begin the processes of fruit and seed development. These signal cascades involve at least three of the plant hormones we saw in Chapter 6: auxins, ethylene, and gibberellins (Table 6.1; Seymour et al. 2013; Taiz and Zeiger 2006). Auxins are produced by pollen, as well as in the seed that begins to form after fertilization of the ovule. Auxin produced by the seed stimulates production of gibberellins that, in turn, stimulate growth of the developing fruit. Gibberellins also appear to be involved in such processes as the development of specific tissues within the maturing fruits, such as zones of dehiscence, in fruit that open to release seeds at maturity (Seymour et al. 2013). Later in fruit development, ethylene begins to play a role in ripening, along with certain auxins. We learned previously that ethylene is a gaseous plant hormone, and, as such, it can act not only on the individual fruit in which it is produced, but also on nearby fruits. This aspect of fruit ripening is often exploited in commercial agriculture to control rates and timing of fruit ripening (Taiz and Zeiger 2006).

Simultaneous with the development of the fruit, the seeds (one or multiple) also develop as each zygote (fertilized ovule) matures, forming the embryo, the endosperm tissues that provide nourishment for that developing embryo, and the outer seed coat that protects the seed contents (Figure 8.7). The seed coat may be modified in many ways, from species to species, to further protect the seed or

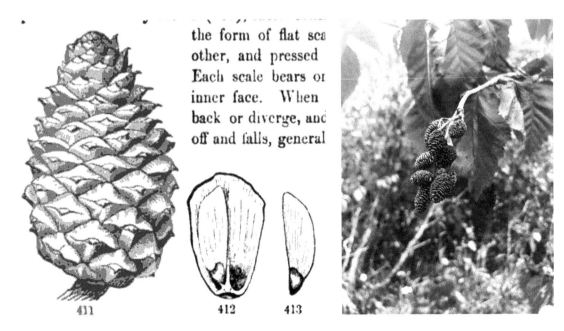

FIGURE 8.22 Top: Cone and Seeds from "A Common Pitch Pine," as Given By Gray (1887). The Cone is Shown at Left (Labeled "411"), with One Scale and an Attached Seed ("412"), and a Single Removed Seed with its Attached Wing ("413"). Bottom: Fruit of *Alnus* (Alder). Much like the Pinecone, the Alder **Infructescence** (Cluster of Fruit) Comprises a Central Axis Surrounded by Multiple Scales within Which the Fruit are Produced. The Fruit of Alder, Coincidentally, are Single-Seeded, Winged **Samaras** that are Produced in Pairs Atop Each Scale, very much Like Pine Seeds.

to aid in dispersal, in species where the seed disperses separate from the mature fruit (Judd et al. 2008). For example, the outer coat may be brightly colored to encourage bird dispersers, or it may produce energy-rich outgrowths (**arils**) that encourage invertebrate dispersal.

The structure of the fruit that is produced by a given species will depend on the complex genetics and biochemistry involved in fruit development but also on the physical arrangement of the flowers and their parts (e.g., Figure 8.23). We typically see fruits grouped into at least three general categories: dry, indehiscent fruits; dry, dehiscent fruits; and fleshy fruits (Table 8.5). In indehiscent fruits, the seed is fused, to varying degrees, to the surrounding ovary tissues, and these will disperse as one unit. In many cases, the tissues that are attached to the seed function to aid in seed dispersal, such as wings that aid in wind dispersal of the samara-like mericarps of maples (Figure 8.24). Dehiscent fruits, on the other hand, open at maturity to release seeds that are dispersed independently of the fruit. **Capsules**, such as those shown at the bottom of Figure 8.25, are one example of dehiscent fruits, each of which contains many small seeds. Three common types of fleshy fruits are the **berry**, **pome**, and **drupe** (Figure 8.26). Note that the *Rubus spectabilis* example in Figure 8.26 is an **aggregate**

fruit, formed from many individual, unfused carpels of a single flower. While aggregate fruits are formed from multiple carpels of one flower, another fruit type, the **multiple fruit**, is formed from the carpels of many clustered flowers. *Platanus* species (known as sycamores in the US), produce multiple fruits formed from many clustered achenes (Judd et al. 2008).

In addition to the wind- and animal-dispersed fruits with which most of us are familiar, we see that many aquatic species have adapted to use water as a means of fruit and seed dispersal. Dispersal by water is often referred to as **hydrochory**, in contrast with **anemochory** (wind dispersal) and **zoochory** (animal dispersal). There are a variety of mechanisms that permit hydrochory, including (Cronk and Fennessy 2001):

- Production of tiny, lightweight seeds;
- Production of fruits with structures that provide temporary buoyancy to seeds that sink when released;
- Production of fruit or seeds with outer corky tissues or tissues with internal air spaces; and
- Use of buoyant accessory plant parts to allow release of seeds during or after dispersal.

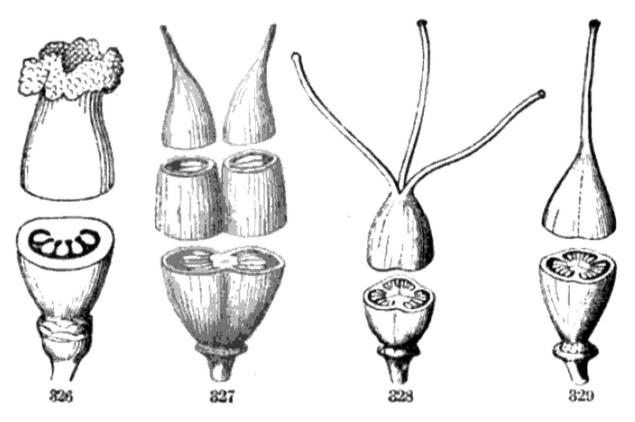

FIGURE 8.23 Some Examples of Pistil Structure and Carpel Arrangement. The Physical Arrangement and Degree of Unification of Floral Parts Plays a Role in Determining the Structure of the Mature Fruits. Left to Right: Simple Pistil with One Carpel, Showing Attachment of Ovules to the Placenta Within; Pistil Formed from Two Carpels United at their Bases Only; Compound Pistil Formed from the Union of Three Carpels, with the Styles Remaining Separate; another Individual of the Previous Species with a Three-Carpellate Compound Pistil, but with Carpels and Styles United. Illustrations from Gray (1887).

TABLE 8.5

Some of the more common fruit types produced among aquatic and wetland angiosperms. Information taken from Cronk and Fennessy (2001), Godfrey and Wooten (1979, 1981), Judd et al. (2008).

Fruit types	Families (*Genera*)	Description
		Dry, indehiscent fruits
Achene	Asteraceae, Cyperaceae, Ceratophyllaceae, Cymodoceaceae, Platanaceae, Polygonaceae	Hard, one-seeded fruit formed from a single ovary, with thin outer wall (*pericarp*) surrounding the seed; generally, relatively small, in contrast to nuts; includes the fruit of many species in the Asteraceae (where the fruit is sometimes called a *cypsela*), in which the sepals are modified to form a hairy *pappus* that assists in wind dispersal
Caryopsis	Poaceae	Single-seeded fruit with pericarp fused to the seed coat, in contrast to achenes and nuts, where the pericarp simply surrounds the seed
Loment	Fabaceae (*Aeschynomene*)	Fruit formed from a single, multi-seeded, carpel that splits transversely into multiple, one-seeded segments at maturity
Nut	Fagaceae (*Quercus*), Nelumbonaceae	Relatively large, one-seeded fruit formed from a compound ovary in which a single carpel is functional; thick pericarp surrounds the seed
Nutlet	Alismataceae, Lamiaceae, Potamogetonaceae	One-seeded fruit formed from separate (i.e., unfused) carpels
Samara	Oleaceae (*Fraxinus*), Ulmaceae	One- or, rarely, two-seeded fruit in which an outgrowth of the ovary wall forms a wing that functions in wind dispersal of the fruit
Schizocarp	Apiaceae, Callitrichaceae, Sapindaceae (*Acer*)	Fruit derived from an ovary with two or more carpels that splits into multiple, usually one-seeded, segments (*mericarps*); the mericarps themselves often structurally resemble achenes, samaras, or other single-seeded fruits
Utricle	Araceae (*Lemna*)	Single-seeded fruit with a thin pericarp that sits loosely around the seed
		Dry, dehiscent fruits
Capsule	Altingiaceae (*Liquidambar*), Hydrocharitaceae, Iridaceae, Juncaceae, Lentibulariaceae, Lythraceae, Onagraceae, Primulaceae, Saxifragaceae	Dehiscent fruit formed from an ovary comprising multiple carpels that splits open at maturity (often along sutures between carpels) to release seeds
Follicle	Aponogetonaceae, Butomaceae, Cabombaceae (*Cabomba*), Magnoliaceae	Dehiscent fruit formed from a single carpel that opens along one longitudinal suture
Legume	Fabaceae	Dehiscent fruit formed from a single carpel that opens along two longitudinal sutures
Silique	Brassicaceae	Dehiscent fruit formed from an ovary having two carpels; the two halves of the fruit open outwards, away from a central partition between the carpels
		Fleshy fruits
Berry	Araceae, Ericaceae (*Vaccinium*), Nymphaeaceae	Fleshy fruit formed from a compound ovary that may produce one or a few, but usually many, seeds; the seeds may have a fleshy covering or a fleshy appendage referred to as an aril
Drupe	Caprifoliaceae (*Sambucus*), Cornaceae, Nyssaceae, Rosaceae (*Rubus*)	Fleshy fruit containing a hard pit that encloses, usually, one seed
Pome	Rosaceae (*Amelanchier*, *Aronia*, *Crataegus*)	Fleshy fruit with soft outer tissue surrounding papery to cartilaginous structures that enclose the seeds

8.2.6 Dormancy and Germination

Recall that, when the stages of sexual reproduction were introduced with Figure 8.8, it was mentioned that mature seeds may be capable of germination at the time they are released from the plant, or they may be dormant. Dormant seeds are incapable of immediate germination and require removal of dormancy before germination can take place (Table 8.6). Typically, the mechanism for removal of dormancy is related to processes or conditions that occur just

prior to, or coincident with, favorable conditions for seedling establishment. As was mentioned in that earlier section, the lack of dormancy also will typically be observed in species whose seeds (and fruit) mature during times of favorable conditions for seedling establishment.

A major consideration in determining the type of dormancy present in seeds of a given species is whether the embryo itself is *differentiated* (i.e., embryonic organs, such as cotyledons and **radicle**, or embryonic root, are present) and/or fully developed (that is, it has reached its final

FIGURE 8.24 Red Maple (*Acer Rubrum*) Produces Small, Wind-Pollinated Flowers during Late Winter–Very Early Spring in Mississippi (Upper Photo). Over the Next Few Weeks, Fertilized Flowers Develop into Bright Red Samara-Like **Schizocarps**, Each Comprising Two Wind-Dispersed Mericarps (Lower Photo).

FIGURE 8.25 An Assortment of Examples of Dry Fruits Produced by Common Wetland Species in the Southeastern United States. Note, Especially, the Diversity of Achenes Illustrated Among the First Seven Species, As Well As the Visual Similarities Between many of the Achenes and the Caryopses Produced by Grasses.

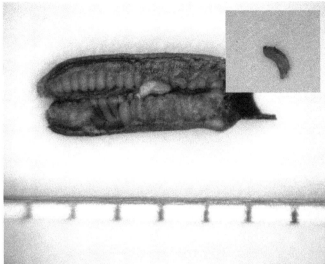

FIGURE 8.25 (Continued)

FIGURE 8.26 Examples of Fleshy Fruit Types Included in Table 8.5. *Rubus* (Bottom Right) is an Example of a "Fruit" that is Composed of many Tightly Aggregated Individual Fruits, Each Formed from a Separate (Unfused) Carpel and, in this Case, Each Containing a Single Seed. The Individual Styles Remain Visible in this Photo of *Rubus Spectabilis* (Salmonberry).

FIGURE 8.26   (Continued)

embryonic size within the seed). Those species whose embryos do not complete differentiation or development prior to dispersal of the seeds are said to exhibit **morphological dormancy** (Baskin and Baskin 2014; Table 8.6). Note that morphological dormancy also may be combined with physiological dormancy (described later), in which case the seeds are said to exhibit **morphophysiological dormancy** (Table 8.6). In species with morphological (or morphophysiological) dormancy, the embryo must complete its development before the seed can germinate. This results in a delay of germination until environmental conditions are suitable for embryonic growth to resume. Factors that have been shown to influence completion of embryonic development in species with morphological dormancy include soil moisture, ambient temperatures, and photoperiod (Baskin and Baskin 2014).

We see two general categories of dormancy in seeds whose embryos are fully developed at the time of dispersal: physical and physiological dormancy (Table 8.6). **Physical dormancy** occurs in species whose seeds are incapable of absorbing, or imbibing, water from their surroundings. Impermeability of seed or fruit coats to water results from one of more layers of cells whose walls are enriched with hydrophobic materials such as cutin, lignin, suberin, and wax (Baskin and Baskin 2014). Germination in species with physical dormancy requires that the impermeable layer be disrupted to allow movement of water into the

seed. Disruption of that layer may occur through physical or environmental weathering that degrades the hydrophobic substances or leads to formation of cracks within it, or it may occur through mechanical or chemical **scarification** (abrasion of the outer layers of the coat) of the seed.

In cases where the embryo is fully developed and the seed is capable of imbibing water but the embryo possesses some physiological inhibitor, the seed is said to exhibit **physiological dormancy** (Table 8.6). Physiological dormancy is expressed at varying levels, depending on the strength of the inhibiting mechanism, and this influences the length of stratification required to break dormancy. **Stratification** is the process of storing seeds in or on a moist substrate at temperatures similar to those encountered in nature after dispersal but prior to the typical time of germination (Baskin and Baskin 2014). Stratification thus is one of the laboratory techniques used to simulate the conditions to which seeds are exposed after natural dispersal and, thus, under which patterns of germination would have evolved.

For seeds with physiologically dormant embryos, the period of stratification allows time for degradation or leaching out of the inhibiting mechanism(s). Seeds expressing low levels of physiological dormancy (termed "nondeep" physiological dormancy; Baskin and Baskin 2014) exhibit stratification requirements from as little as a few days to as long as three months. Species with deep physiological dormancy, on the other hand, may require two or more years of

TABLE 8.6

Major seed dormancy types, along with diagnostic characteristics, causes, and removal mechanisms of each. Based on information provided by Baskin and Baskin (2014).

Dormancy Type	Diagnostic Characteristics	Causes	Removal Mechanism
Morphological	Embryo undifferentiated or underdeveloped. Following dispersal and imbibition, the embryo completes development, and seeds germinate within a month.	Underdeveloped embryo	Conditions that permit embryo growth and subsequent germination
Morphophysio-logical	Embryo undifferentiated or underdeveloped. Following dispersal and imbibition, the embryo completes development, but seeds fail to germinate within a month.	Underdeveloped embryo accompanied by physiological inhibiting mechanisms	Stratification (warm and/or cold, depending upon the species)
Nondormant	Seed, with differentiated and fully developed embryo, **imbibes** water. Root and shoot emergence usually occur within a few days.	-NA-	-NA-
Physical	Seed contains differentiated and fully developed embryo, but the seed does not imbibe water. Scarified seeds imbibe water and usually germinate within a few days.	Seed or fruit coat is impermeable to water	Opening of seed or fruit or scarification of the seed or fruit coat
Physiological	Seed, with differentiated and fully developed embryo, imbibes water, but emergence of root or shoot usually delayed by at least a month.	One of many physiological inhibiting mechanisms	Stratification (warm and/or cold, depending upon the species)
Physical + Physiological	Seed contains differentiated and fully developed embryo, but the seed does not imbibe water. Scarified seeds imbibe water, but emergence of root or shoot usually delayed by at least a month.	A combination of the previous two causes	Opening of seed or fruit or scarification of the seed or fruit coat, followed by warm and/or cold stratification

stratification prior to germination. Some of the more common factors that inhibit germination in seeds with physiological dormancy are (Baskin and Baskin 2014):

- Seed layers that inhibit diffusion of oxygen into the seed which, in turn, can prevent growth of the embryo and/or prevent oxidative degradation of inhibitors within the seed (a combination of physical and physiological dormancy);
- Presence of inhibitory concentrations of the plant hormone abscisic acid within the seed (see Table 6.1); and
- Unfavorable red:far red light ratios (e.g., low red:far red ratios, indicative of light filtered through a canopy of green leaves, often inhibit germination).

Mechanisms of dormancy removal given in Table 8.6 can be linked to ecological factors to which seeds are exposed during the interval between dispersal and the time of year during which they typically germinate. For example, seeds with physiological dormancy that are dispersed during autumn in temperate latitudes will usually require a period of cold stratification, as would be experienced while overwintering in moist soil or sediments, for removal of germination inhibitors. Species with physical dormancy often must pass through the gut of some animal prior to germination, as a means of removing or damaging the outer physical covering (scarification) that prevents absorption of water by the seed. As a result, a better understanding of the natural history of a species of interest should grant one a better understanding of the germination requirements for seeds of that species. Answering questions such as "Is the species dispersed by waterfowl?", "Do seeds mature in spring or in autumn?", or "What is the dispersal unit?" can reveal much about the germination biology of the species.

Baskin and Baskin (2014) summarized data on seed germination trials for more than 370 species of aquatic and wetland plants. They found that more than 80% of those species exhibited physiological dormancy, which was most often removed by a period of cold stratification. Lengths of stratification ran from five days to nine months; however, those authors pointed out that few ecosystems experience more than about 200 days of temperatures that would equate to cold stratification (0–10 °C). Thus, they concluded that many seed germination studies likely have employed excessive periods of stratification for the species they investigated.

If we consider that much of the scientific literature has been published by investigators in the temperate latitudes, there may be seeds of many species of aquatic and wetland plants yet to be investigated that require warm stratification for dormancy removal. Another similar issue noted by Baskin and Baskin was that roughly 75% of the studies they summarized had been carried out on species with an emergent growth form (species that are rooted in the soil but whose leaves extend beyond the water's surface). They further indicated that the specific mechanism responsible for removal of physiological dormancy in nature was very poorly studied. They did give one example of a species with physical dormancy (*Neptunia oleracea*) in which the dormancy was removed by alternating warm and cold periods, which may have resulted in weathering of the seeds. They went on to suggest that those responses could indicate that the typical wetland hydroperiod (i.e., cycles of flooding and drawdown) could provide both a cycle of alternating temperatures as well as a period of stratification that would permit germination to occur shortly after wetland sediments become exposed during periods of low water.

Given the role of oxygen in breaking dormancy, it would seem appropriate that seeds in most species would be adapted to germinate during low water periods in wetlands, when the sediments would be more uniformly oxygenated. This is the case for most species, but there are some species, from multiple growth forms and taxonomic lineages whose seeds germinate at higher percentages under flooded conditions than non-flooded. The studies summarized by Baskin and Baskin (2014) included around 30 such species. Many of the species' seeds that germinate under flooded conditions still require light for germination, which is another common pattern among aquatic and wetland species in general. In other cases, there are species whose seeds germinate well under hypoxic or anoxic conditions, but seedlings of those species, once germinated, fail to develop normally. Some of those species, such as *Echinochloa crus-galli*, appear to fuel the embryonic growth that occurs during anoxic germination by anaerobic metabolism (ethanol fermentation; Figure 6.1; Baskin and Baskin 2014). Seeds of another species, *Nelumbo nucifera*, were shown to germinate under oxygen-free experimental conditions, but germination appeared to have been facilitated by oxygen in the gases held within the seeds themselves.

In contrast to the generally high frequency of physiological dormancy among most aquatic and wetland plants, seeds of mangrove species tend to be nondormant; approximately 70%–80% of mangrove species produce nondormant seeds (Baskin and Baskin 2014). In some species, the seeds germinate, and the seedlings begin to grow while still attached to the parent plant (Spalding, Kainuma, and Collins 2010; Figure 8.27). In the Rhizophoraceae, a family with widespread vivipary, seedlings undergo significant development while attached to the parent. In other groups, such as the genus *Avicennia*, seeds germinate while attached to the parent plant, but they disperse prior to any significant growth of the seedling. This phenomenon is referred to as **cryptovivipary**, and the dispersal units are often referred to simply as **propagules** because of the diversity in developmental stages that exist among mangrove species (Baskin and Baskin 2014; Spalding, Kainuma, and Collins 2010; Figure 8.27). Another important feature of the propagules in many mangrove species is that they are buoyant, allowing them the potential to migrate to distant sites for establishment. Propagules of many species are capable of floating for three months or longer, providing considerable time for waterborne dispersal.

The other major group of marine macrophytes, the seagrasses (aquatic marine species from Cymodoceaceae, Hydrocharitaceae, Posidoniaceae, Ruppiaceae,

Potamogetonaceae, and Zosteraceae), also include a large proportion (~40%) of species with nondormant seeds. The remaining seagrass species are believed to exhibit nondeep physiological dormancy, usually with short periods of stratification required to break dormancy. The type of stratification, as one might expect, appears to correlate with conditions under which the seeds typically germinate. Even within a single species, the type of stratification required to remove dormancy can vary among populations, reflective of local conditions prior to seed germination. *Zostera marina*, for example, responded to different stratification temperatures if collected in The Netherlands, where cold stratification removed dormancy, than if collected in Mexico, where warm stratification was used (Baskin and Baskin 2014). Thus, responses of germination to environmental conditions appear to be much more closely tied to the local environment in habitats to which the species is adapted than to the species' phylogenetic history.

8.2.7 REGENERATION NICHE

A species' **niche** can be referred to as the sum of the physical and biological properties of the environment to which the species is physiologically and behaviorally adapted and under which the species can maintain non-negative population growth over time (Chase and Leibold 2003; Grinnell 1917; Hutchinson 1957). Within the context of a species' niche, those conditions specifically associated with successful replacement of individuals within the population, especially factors that influence seed production, dispersal, germination, and seedling establishment, are often collectively referred to as the **regeneration niche** (Grubb 1977; Gurevitch, Scheiner, and Fox 2006).

In his discussion of the importance of the regeneration niche in plant **autecology** (the ecology of an individual species), community ecology, and plant evolutionary biology, Grubb (1977) discussed many factors that influence the four components of regeneration given earlier. He noted that many species exhibit patterns in seed production, wherein some may experience moderate levels of seed production in all years or may demonstrate cycles of high and low seed production, but other species may show what appears to be irregular patterns of seed production among years. Variation in environmental conditions is likely responsible for year-to-year variation in irregularly productive species. For example, occasional drought years may result in declines in seed production, and we have seen some potential biological explanations for this in the preceding sections, such as switches in the mating system for aquatic species during drought conditions.

Regarding spatial dispersal of seeds, we have discussed some of the adaptations of seeds for dispersal by wind, water, and animals. Another component of dispersal, dispersal through time, is regulated by the patterns of dormancy and germination discussed in the immediately preceding section. The ecological importance of seed dormancy as a means of dispersing plants through time is an

FIGURE 8.27 Examples of Dispersal Propagules in Mangrove Species. Upper Photos Show Germinated Seeds (Propagules) Still Attached to the Parent Plant in *Rhizophora Mangle* (Left, a Viviparous Species) and *Avicennia Germinans* (Right, a Cryptoviviparous Species). Lower Photos Show Propagules that have Dropped from the Parent, some of Which have Taken Root and are Beginning to Grow.

important factor that often is inadequately incorporated into seed ecology. In wetlands especially, the interannual variations in hydrology make dispersal through time incredibly important for ensuring persistence of plant populations. As we've seen earlier, cycles of oxygen availability that are driven by hydroperiod provide key signals for dormant seeds within aquatic and wetland seed banks as to the potential suitability for seedling establishment, based upon a particular species' requirements for germination as well as establishment.

Given the variety of constraints that determine a species' regeneration niche, each species might be expected to have a fairly unique combination of characters that allow

production of viable seed, determine the type of dormancy, allow or encourage seed germination, facilitate dispersal to suitable sites and times, and, finally, establish seedlings that can repeat the process, ensuring population persistence over time (Figure 8.8). Warwick and Brock (2003) investigated regeneration patterns among species present in the seed banks of shallow wetlands in a hydrologically variable region of New South Wales, Australia. They used eight different hydrologic regimes to study emergence, establishment, and reproductive output of 100 or so species present in the seed banks of those wetlands. They found that the species establishing from seed banks of those wetlands responded not only to the patterns of hydrology but also to the season in which the experiment was conducted, in terms of the species present as well as their respective reproductive outputs. A similar study was carried out by Nicol and Ganf (2000), using nine different combinations of water depth and water-level drawdown to determine regeneration niches for three wetland species: *Typha domingensis*, *Triglochin procerum*, and *Melaleuca halmaturorum*. Considerable variation was found among these species' responses to the hydrologic manipulations. The *Typha* species was found to germinate and establish in all nine treatments, while the *Melaleuca* species only established in treatments that ended with exposed soil. *Triglochin procerum* could avoid unfavorable conditions altogether by producing buoyant seedlings that floated atop the water in all treatments that included soil inundation (flooding).

The practical implication of all this diversity in species responses to environmental conditions to which they are exposed, across their lifetimes, is that we cannot ignore any aspect of a species' life history, if we want to truly understand its biology. Because an understanding of the biology of plant species is an essential element in managing a single species or an entire wetland plant assemblage, those of us interested in managing wetlands and/or wetland plants should always seek opportunities to learn more about the life histories of the species with which we work.

8.3   CLONAL PROPAGATION IN AQUATIC AND WETLAND PLANTS

The objective of reproduction, from a biological standpoint, is the passing of an individual's genes to the next generation, with the goal that the next generation will do the same. From this perspective, it might seem that it does not matter whether those genes are packaged in a sexual or an asexual dispersal unit (fruit, seed, vegetative fragment, etc.). However, there are some important distinctions between the two modes of propagation.

Perhaps the clearest distinction between sexual and asexual reproduction is that sexual reproduction (except in cases of self-pollination) combines genes from a maternal and a paternal parent, resulting in offspring with **recombined genomes** (i.e., genomes comprising different mixtures of genetic information than was found in either parent). As discussed earlier, this reduces negative consequences of inbreeding, such as reduced genetic diversity, accumulation of deleterious alleles within local populations, and potential loss of beneficial alleles through death of individuals with reduced vigor (Cheptou 2019). In contrast with the genetically diverse offspring produced through sexual reproduction, individuals arising from asexual reproduction are, with minor exceptions, genetic equivalents (i.e., clones) of the original plant.

Harper (1977) emphasized a somewhat more subtle difference between asexual reproduction via vegetative structures and sexual reproduction by pointing out that the process we recognize as sexual reproduction involves formation of individuals of the subsequent generation via

production of a zygote. Asexual reproduction via vegetative structures (e.g., meristematic regions of stems), on the other hand, does not produce a new individual; instead, it produces a clonal ramet (Figure 8.1) via growth and differentiation of pre-existing organs. "Individuals" formed in this way thus do not represent a subsequent generation but a continuation of the same generation via growth or clonal propagation, even though the clonal offspring may separate, physically, from the producing plant for dispersal. With this in mind, clonal propagation does not fulfill the objective of passing an individual's genes to the subsequent generation; rather, clonal propagation extends the time during which a genetic individual has opportunities for passing along its own genes via sexual reproduction (Silvertown 2008).

Recall that those species that live and reproduce through multiple growing seasons are termed polycarpic perennial species; these are the species usually referred to as "perennials" (Figure 8.2). Clonal propagation is much more prevalent among species with polycarpic perennial life histories than among monocarpic species, whether annual or biennial. In large part, this results from the need for perennial species to possess modified vegetative structures to protect meristems from unfavorable environmental conditions (e.g., excessively cold or excessively dry conditions) between growing seasons.

8.3.1   MULTIPLE FUNCTIONALITY OF ASEXUAL PROPAGULES

Most plant cell types are **totipotent**; that is, under appropriate conditions, they have the potential to dedifferentiate and reinitiate differentiation into other cell and tissue types (Taiz and Zeiger 2006). Because of this, essentially any plant part may play a role in clonal numeric increase, although the capacity for a particular plant part to do so varies among species.

Grace (1993) looked at a variety of issues surrounding clonal propagation among aquatic and wetland plants, specifically addressing the topic from an evolutionary perspective. This approach to discussing clonal propagation brings an inherent focus on ecological and evolutionary trade-offs between sexual reproduction and asexual propagation, as well as among different types of **clonal propagules** (dispersal units generated through asexual propagation). The term trade-off here refers to the relative allocation of limited energy or resources among some set number of critical functions that a species must carry out to survive, grow, and reproduce.

One classic example of a trade-off in sexual reproduction is the "decision" to either produce a large number of small seeds or a small number of large seeds. If all resources (including available space) were unlimited, the most advantageous choice for a plant would be to produce a large number of large, well-provisioned seeds that were equipped to disperse to any of the infinitely available locations for germination and establishment. However, this is not the case. Resources are limited, and there is a finite number of places matching the regeneration niche of any given species. This brings about the unavoidable circumstance in which all species exist, wherein selective forces determine whether individuals have arrived at the appropriate combination of resource allocation among all the functions they must accomplish to ensure survival, growth, and production of viable offspring that will carry their genes forward into subsequent generations. I placed "decision" in quotes earlier because the decisions, such as they are, really are made by the process of selection acting on the outcomes of resource allocation among an individual's necessary functions. If the individual allocates resources appropriately, in the context of existing selective pressures, then it will succeed in establishing, growing, and producing viable offspring; otherwise, it fails to pass along its genes. We will revisit these ideas in the next two chapters.

As noted previously, clonal propagules can arise from different parts of the plant. Grace (1993) applied the prior concepts of ecological and evolutionary trade-offs to clonal propagules, in a discussion of why we might see certain combinations of clonal increase being used by plants that live under a given set of environmental conditions. A critical component of his discussion was an overview of some of the potential trade-offs that exist among key functions of the propagules themselves. As we see in Figure 8.28, clonal propagules arising from different plant parts have contrasting abilities to perform functions complementary to the role of propagation itself.

Grace (1993) selected six functions of clonal propagules around which to base his discussion of the importance and utility of clonal propagation in aquatic plants (Table 8.7). Four of these—resource acquisition, resource storage, protection of resources and meristems, and anchorage—are functions found in vegetative structures regardless of their potential role in propagation. This is an important aspect of what Grace was trying to convey in his paper; because clonal propagules arise from pre-existing vegetative structures, they all provide one or more functions that are basic to plant growth and survival. This contrasts with fruit or seeds, most of which generally serve only in dispersal and numeric increase of individuals that can pass along the parents' genes. Along these lines, it is no accident that the other two of the six functions in Figure 8.28 and Table 8.7 are dispersal and numeric increase.

If we look, for example, at clonal propagules that arise from modified buds (Figures 8.28–8.30), we see that these structures appear to be well equipped to carry out photosynthesis, given that they all readily produce viable foliage upon establishment elsewhere. Turions, which are highly condensed shoot apices surrounding a dormant bud or apical meristem (Figure 8.29), may even remain photosynthetically active while dormant (Adamec 2011, 2018). Adamec (2011) further found that turions of many species carry sufficient mineral resources to initiate growth and reach photosynthetic rates comparable to fully established plants in a relatively short time. Although numbers of these structures that may be formed per plant vary by propagule type (e.g.,

turions vs. pseudoviviparous buds), they can be an effective means of increasing the number of distinct clones. Furthermore, once detached from the original plant, they also are highly effective at dispersal, especially those capable of entering dormancy, because dormancy helps to protect them from physiological stresses such as cold temperatures and desiccation. In contrast, this category of propagules provides essentially no anchorage of the regenerating clone and relatively little capacity for resource storage (aside from the mineral nutrients mentioned earlier), and some types of modified buds have low ability to protect the meristem during dispersal (Figure 8.28). Because of this suite of characteristics, modified buds tend to be advantageous for species typical of habitats that experience frequent disturbance and in which suitable sites for establishment and growth may be widely separated within or among habitats (Table 8.7).

Upon examining the combinations of functions provided by different propagules (Figure 8.28), we begin to see patterns of correlation emerge among the six functions illustrated here (Figure 8.31). As we saw earlier with turions, they tend to have high capacity for dispersal, which makes them relatively poor at anchoring the new clone into the soil or sediments; conversely, because they provide little anchorage, they are excellent at dispersal. Similarly, we tend to see structures that have high capacity for numeric increase also quite well adapted for dispersal, but relatively poor at storing accumulated resources, as this would increase the mass per propagule and reduce their effectiveness at dispersal.

Another pattern that appears among propagule types is a general difference in the functions performed by aboveground versus belowground structures (Figure 8.28). While we see that aboveground structures tend to be highly adapted for numeric increase and dispersal, clonal structures formed from belowground organs tend to be adapted for anchorage and the protection and storage of accumulated resources. This contrast is especially distinct when we look at propagules arising from modified shoots (middle row of Figure 8.28). Layers, runners, and stolons, which are aboveground structures, generally function well in numeric increase and resource acquisition, whereas rhizomes and stem tubers tend towards providing anchorage and storage and protection of accumulated resources. Again, these differences are attributable largely to the different functions performed by the vegetative organs from which these propagules arise, reminding us of the advantages that are provided in co-opting pre-existing vegetative organs to aid in numeric increase and dispersal of clonal individuals.

8.3.2 Types of Asexual Propagules

As discussed in Chapter 1, plants are modular organisms. Perennial species, existing as a network of interconnected ramets of differing ages, are excellent examples of this modularity. Furthermore, the potential for individual ramets to exist independently from the original plant gives some species an even greater modular character. It has long been recognized that there is a high frequency of asexual propagation

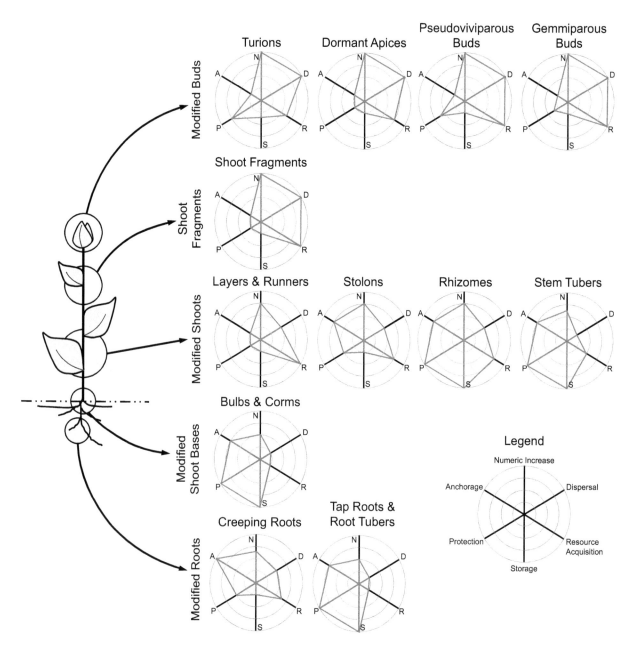

FIGURE 8.28 Clonal Propagules Potentially can Arise from any Part of a Plant, Owing to the Totipotency of most Plant Cell Types. Here we see Examples of Combinations of Adaptive Features and their Relative Importance Among many of the most Common Vegetative Structures that Function in Clonal Propagation. The Greater the Distance the Shaded Area Extends from the Center of Each Diagram, the Greater Degree to Which that Structure Serves in a Particular Function. For Example, Turions are much more Important for Numeric Increase and Dispersal than for Anchorage or Storage of Resources. Based on Grace (1993), in Combination with Cronk and Fennessy (2001), Personal Communication with Grace, and my Own Interpretations of Roles that these Structures Play in Clonal Species Ecology. Refer to Subsequent Figures and Section 8.3.2 for Descriptions of these Organs.

in aquatic plants (Grace 1993; Hutchinson 1975; Philbrick and Les 1996), and many species are well adapted to use the modular perennial life history to their advantage in maintenance of populations in temporally dynamic aquatic and wetland ecosystems. Sculthorpe (1967) suggested that one factor contributing to the prevalence of asexual propagation among aquatic plants is the difficulty of carrying out sexual reproduction in water, as we saw in previous sections of this chapter. Santamaría (2002) further suggested that some of

the conditions experienced by aquatic plants, especially submersed species, such as low light and nutrient availability and cooler temperatures, may select for species with a greater ability to take advantage of clonal propagation. The abilities of clonal propagules to carry stored resources or to access resources from the environment during dispersal may provide an advantage over dispersal by seeds and fruit under conditions commonly encountered in aquatic and wetland habitats.

TABLE 8.7

Six key functions carried out by vegetative structures that commonly play a role in clonal propagation. "Selective habitat characteristics" refers to characteristics that are expected to select for species having a high potential to carry out the function. Based on information in Grace (1993).

Functions	Description	Selective Habitat Characteristics
Anchorage	Anchoring the plant in place, while also sometimes providing additional structural support	Frequently subjected to erosive forces of water or wind
Dispersal	Physical movement of propagules away from the parent plant	Spatially variable habitats
Numeric increase	Increase in the number of physiologically independent, spatially discrete ramets	Frequently disturbed
Protection	Protection of the propagule, including enclosed resources and meristems, from injury resulting from biotic or abiotic sources; includes perenniation, overwintering, and drought avoidance	Frequent or long periods of unfavorable conditions such as cold, drought, fire, or herbivory
Resource acquisition	Capture of energy via photosynthesis or water and nutrients via absorption	Low resource availability or brief growing season
Storage of resources	Storage of carbohydrates, lipids, proteins, mineral nutrients, or water within specialized storage tissues and organs, usually belowground in aquatic and wetland plants	High density of neighbors or brief growing season

 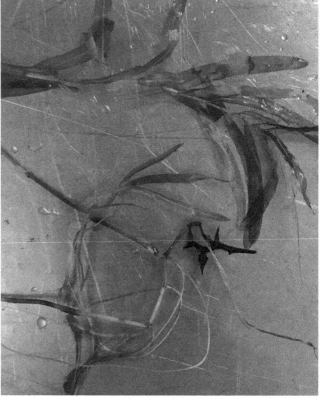

FIGURE 8.29   Turions in *Potamogeton* Species. Left: Turion Developing within a Shoot Tip, with Rigid Upper Leaves and Hardened Leaf Bases. Right: Plant that Established from a Turion Formed the Previous Year. The Dark Structure in the Right Center of this Photo is What Remains of the Turion itself.

Considerable variation exists among species in the abilities of different plant parts to play a role in clonal propagation, and those abilities generally are thought to correlate with characteristics of the habitats in which the species most commonly occur (Table 8.7). However, we often see many different types of clonal propagules represented within any individual wetland, and this likely is a further reflection of the spatial and temporal heterogeneity of wetland ecosystems. For example, within the deeper, aquatic zones of a wetland, we might see floating-leafed species with large

FIGURE 8.30 Left: Pseudoviviparous Buds Formed Within an Inflorescence of *Butomus Umbellatus*. Center: Normal Inflorescence of *B. Umbellatus*. Right: Gemmiparous Buds/Plantlets on Leaf Margins of "Alligator Plant" (*Kalanchoe Daigremontiana*, not a Wetland Plant).

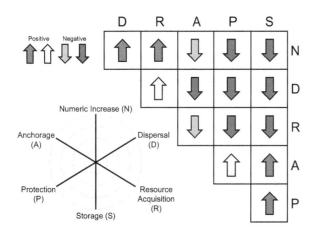

FIGURE 8.31 Generalized Relationships Among Six Functions Provided by Clonal Propagules. Owing to Trade-Offs in Resource Allocation Patterns and the Mechanisms by Which Each Function is Carried Out, there are Inherent Trade-Offs Among Functions. For Example, a Structure that Provides Substantial Anchorage is Unlikely to Provide Dispersal Capabilities, thus these Two Functions (A and D) are Strongly Negatively Correlated. However, that same Structure may Provide Considerable Storage Capacity for Accumulated Resources, Yielding a Strong Positive Correlation between Anchorage and Storage Functions (A and S). Intensity of Shading of the Arrows is Meant to Represent Strength of Relationship.

rhizomes for resource storage (e.g., *Nuphar*, *Nymphaea*, or *Nelumbo*) growing amidst submersed populations of *Ceratophyllum*, *Hydrilla*, or *Najas*, which are known to heavily utilize shoot fragmentation as a means for dispersal and numeric increase. With this in mind, it is important to be aware of the roles of different propagule types in persistence of plant populations, whether one is interested in

conserving populations of rare species or reducing populations of invasive species.

8.3.2.1 Modified Buds

Buds at the tips of a plant's shoots are structures that protect the shoot meristems from unfavorable conditions. This could mean they are responsible for protecting the meristem from cold temperatures and desiccation during winter, from excessive heat and desiccation during a warm dry season, or from desiccation during dry periods in an intermittently or seasonally flooded wetland. In the context of a modular perennial plant, each bud represents the potential to generate a new modular shoot, as part of the clone. In the context of clonal propagation, each bud also represents the potential to generate a propagule that may be dispersed to a new location, diversifying that individual's opportunities for survival and eventual sexual reproduction.

Turions, which we discussed earlier, perhaps are the best-known form of modified bud among aquatic species (Figure 8.29). Turions may be represented as condensed apical shoot sections or hardened apical or axillary buds that serve to protect shoot meristems during unfavorable conditions. Other names for turion include winter buds and hibernacula, and they may be produced on aboveground shoots or on rhizomes (Cronk and Fennessy 2001; Hutchinson 1975). Turions are known to be formed in species of 14 genera from at least nine families spread across the angiosperms, from basal genera (e.g., *Brasenia*) to eudicots including *Aldrovanda* and *Utricularia* (Adamec 2018).

Hutchinson (1975) gives examples of five different types of turion that form in *Potamogeton* species, some from rhizomes and others from aboveground shoots. These include a group of at least seven species of narrow-leaved

Potamogeton that produce turions from aboveground shoots, after which the entire plant decomposes, leading Hutchinson to describe these species as essentially exhibiting an asexual annual life history.

Both Cronk and Fennessy (2001) and Adamec (2018) cite numerous environmental cues that seem to be correlated with turion formation, such as temperature, photoperiod, nutrient availability, presence of competitors, and certain plant hormones. Adamec (2018), however, indicates that in most cases, the specific factor(s) responsible for turion formation are not precisely known for turion-forming species. As with the expression of heterophylly, many of these factors are correlated with one another in nature, and detailed experimental studies to tease apart the potential signals have been undertaken in very few species.

Other modified bud types that function as clonal propagules include pseudoviviparous buds that may form in inflorescences of species such as *Butomus umbellatus* and *Myriophyllum verticillatum* (Cronk and Fennessy 2001) and gemmiparous plantlets or gemmiparous buds (Figure 8.30). The latter are perhaps best known from *Ceratopteris* and *Armoracia* species in aquatic ecosystems. In *Ceratopteris* and *Armoracia*, gemmiparous plantlets can form along leaf margins, and they disperse once the leaf decomposes sufficiently to allow the plantlets to move freely in the water column.

8.3.2.2 Shoot Fragments

As mentioned by Cronk and Fennessy (2001), shoots of plants, especially those of aquatic plants, are prone to fragmentation resulting from physical damage (Figure 8.32). The resulting shoot fragments are, essentially, nodal stem sections containing at least one axillary meristem. Aquatic plants, in part because they can rely on water for at least a portion of their support, tend to have less rigid shoots than terrestrial plants, or even plants of wetland margins. While this is an advantage from the standpoint of managing limited resources (i.e., the plants can allocate resources that would have been spent on structure support towards other functions), it does make the plants more susceptible to fragmentation from storms, wind, wave action, boats, or other sources of physical damage. Fortunately, however, many of these species can take advantage of the fragmentation by relying on periodic disturbances to produce clonal propagules that can be disseminated for colonization of distant areas by the new clones.

Some species, such as *Myriophyllum spicatum*, have the ability not only to regenerate from fragments created by physical disturbance, but also to generate their own shoot fragments through a process called autofragmentation (Kimbel 1982; Smith et al. 2002). After flowering, *Myriophyllum spicatum* naturally produces autofragments from shoot tissues, each of which develops adventitious roots before abscission and dispersal from the clone (Kimbel 1982). Furthermore, in climates where *Myriophyllum spicatum* can undergo multiple bouts of flowering, it may undergo autofragmentation after each round of flowering. Kimbel (1982) found that fragments produced through autofragmentation had up to 15% of their mass in the form of nonstructural carbohydrate content, thereby providing an internal energy source for regeneration after separation from the plant. It further was found that experimentally generated fragments having lower nonstructural carbohydrate content experienced substantial mortality, indicating the importance of autofragments being provisioned with these carbohydrates prior to dispersal.

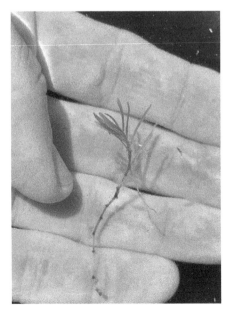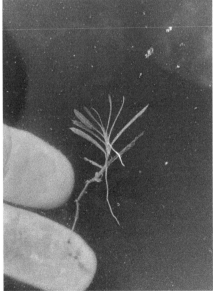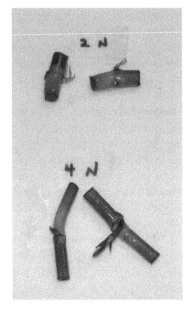

FIGURE 8.32   Left and Center: Shoot Fragment of a *Potamogeton* Species that has Begun to Produce Roots. Right: Excised Shoot Fragments (Nodal Sections) of *Alternanthera Philoxeroides* Producing New Growth During an Experiment. The Upper Fragments Measured 2 Cm in Length; the Lower Fragments were 4 cm Long.

Smith et al. (2002) examined the effect of sediment nitrogen concentration on autofragmentation and found that *Myriophyllum spicatum* produced more than twice as many autofragments under low sediment nitrogen (0.03 mg NH_3–N per g sediment) as under elevated nitrogen treatments (0.54 mg NH_3–N per g sediment). In the face of low nitrogen availability, plants produced an average of 8.5 autofragments per plant, in contrast with an average of 3.5 autofragments per plant in the high nitrogen treatments. In this study and another conducted by Cao, Wang, and Zhu (2012), autofragments accounted for 10% to 15% of the total plant biomass, across a broad range of sediment nitrogen availability. Thus, while the potential for dispersal and numeric increase by autofragmentation appears to be much higher under low resource availability (Smith et al. 2002), *Myriophyllum spicatum* appears to have a narrow range of biomass allocation available for producing the fragments. The increased number of fragments produced under low nitrogen conditions likely is an adaptation for dispersing clonal propagules away from low-resource areas in favor of the potential to colonize better habitats elsewhere (Smith et al. 2002).

8.3.2.3 Modified Shoots

There are five common types of modified shoots that can function in clonal propagation. Four of these (layers, runners, stolons, and rhizomes) are quite similar to one another, in that they simply represent a progression from a slightly modified aerial stem towards a strictly belowground stem (Cronk and Fennessy 2001). The first three of these are aboveground stems, while rhizomes are strictly belowground structures.

Layers, as seen in some species of *Alternanthera* and *Ludwigia* (Figure 8.33), might be thought of as opportunistic in their role as clonal propagules. That is, stems in these species often grow upright but are quite prone to grow horizontally, depending on the specific habitat conditions and population densities. When growing horizontally, layers often root adventitiously at nodes that contact the water or moist soil. This layering ability is a characteristic feature of species included in the creeping emergent growth form discussed by Rejmánková (1992). Runner species, in contrast, usually exhibit the runner morphology as a regular part of their life history, often because they produce top-heavy shoots. *Eleocharis rostellata*, for example, produces inflorescences near the tips of shoots, and the weight of those inflorescences causes individual shoots to bend towards the sediment. After contacting the sediment, tips of the shoots root adventitiously, producing a new clonal ramet (Cronk and Fennessy 2001; Grace 1993). *Echinodorus cordifolius* (Figure 8.34) is another species that commonly exhibits a running morphology, wherein plantlets develop at the tips of inflorescences and take root once the tip contacts moist soil or sediments.

Like layers and runners, stolons and rhizomes are functionally very similar to one another. The key distinction between these types of modified stem is that stolons are

FIGURE 8.33 Layering in Two Species of Creeping Emergent Plants, *Alternanthera Philoxeroides* (Top) and *Ludwigia Peploides* (Bottom). Species with this Growth Form Regularly Root at Nodes that Contact the Water or Moist Sediments, Allowing Radial Expansion of the Clone Across the Sediments or the Water's Surface.

produced aboveground, whereas rhizomes are produced belowground (Figure 8.35–8.36). Perhaps because of the locations in which they are produced, stolons generally have a lower capacity for resource storage than do rhizomes (Figure 8.28). Rhizomes, because they are produced belowground, also have a greater capacity for protecting stored resources than do stolons. Perhaps because of this increased capacity for storage and protection of accumulated resources, we often see rhizomes playing a major role in the clonal propagation of competitively dominant species in wetlands (Figure 8.36).

Within the rhizomatous species, we see two distinct types of plant growth form, although a gradient exists between these two extremes (Figure 8.36). At one extreme, we see species with relatively long rhizome internodes that are capable of relatively rapid horizontal clonal expansion (e.g., *Typha* species). This form is sometimes referred to as a diffuse, or guerilla, morphology. The contrast to this is a compact, or phalanx, growth form, in which the internodes are relatively short and individual clones produce high densities of shoots within a small area of wetland (Figure 8.36, *Juncus effusus*). Species that use the compact morphology

FIGURE 8.34 *Echinodorus Cordifolius* Frequently Produces New Ramets Via Runners, Where Tips of the Inflorescences Take Root after Contacting Moist Soil or Sediments. An Inflorescence that became Rooted and Anchored in Soil is Shown in the Lower Photo. Note the New Leaf that has been Produced at the Base of the Cluster of Flowers/Fruit in the Lower Photo.

tend to be able to capture and hold small areas of habitat with individual clones, whereas those with a diffuse form tend to spread out and sample greater areas of a wetland.

Some species, such as *Phragmites australis*, take advantage of multiple types of shoot modifications for clonal propagation (Packer et al. 2017). *Phragmites australis*, like many other competitively dominant wetland monocots, employs rhizomes in clonal expansion, but is also capable of expansion via aboveground stoloniferous growth or layering, where aboveground stems root adventitiously after contacting the soil. Packer et al. (2017) mention occasions where *P. australis* has produced stolons as long as 10 m, with 70 or more upright stems growing along an individual stolon.

The final type of shoot modification we will consider here is the stem tuber (Figure 8.37). As indicated in Figure 8.28, tubers are much less suited to numeric increase and dispersal than are other types of stem modifications. Tubers are, however, particularly suited to accumulating and protecting resources, as evidenced by one of my favorite stem tubers, the potato (*Solanum tuberosum*). Tubers essentially serve as a storage reserve between growing seasons, from which rhizomes sprout the following growing season, allowing further clonal expansion (Cronk and Fennessy 2001). Many species of wetland plants produce stem tubers for perenniation, while using rhizomes or stolons for clonal expansion, including species of *Nelumbo*, *Scirpus*, *Cyperus*, *Potamogeton*, and *Sagittaria* (Cronk and Fennessy 2001; Sculthorpe 1967).

8.3.2.4 Modified Shoot Bases

Bulbs and corms (Figure 8.37) are subterranean structures specialized for resource storage and perenniation. Bubs are short underground clusters of axillary buds surrounded by leaves that usually are modified for resource storage, while corms are compact underground stems, similarly modified for storage (Cronk and Fennessy 2001; Graham, Graham, and Wilcox 2006). The very compact structure of these organs, combined with the fact that they are produced belowground, severely limits their use in dispersal (Grace 1993). Furthermore, because they are produced at the base of stems, or, in the case of corms, are stem bases themselves, they also are limited in the degree to which they can contribute to numeric increase of clonal ramets. In contrast, bulbs and corms are excellent at resource storage and protection of both resources and meristems and serve as another example of trade-offs in functional capacity of clonal propagules.

8.3.2.5 Modified Roots

Modified roots are slightly more heterogeneous than modified stems, in terms of the functions they provide to the plant, falling into two general groups. Creeping roots, as the name suggests, spread horizontally, providing opportunities for dispersal similar to rhizomes and stolons (Grace 1993). In addition to aiding in horizontal subterranean spread, creeping roots are capable of water and nutrient uptake and contribute towards anchorage of the genet within the soil or sediments. Furthermore, if the horizontal connections between ramets are severed via physical damage, erosion, or other processes, each of the resulting fragments contributes to numeric increase of clones within the wetland. Creeping roots have been documented in both herbaceous (*Rorippa* species) and woody (*Morella* species) wetland plants (Cronk and Fennessy 2001; Grace 1993).

FIGURE 8.35 *Pistia Stratiotes* Produces Stolons that Give Rise to Vegetative Offshoots. The Individual Ramets may Remain Connected, or they can Become Separated, Aiding in Dispersal of the Clone.

The other group of modified roots includes tap roots and root tubers, which overlap little with creeping roots in their functions (Figure 8.28). Although they provide storage and protection for acquired resources, they generally contribute little to dispersal and numeric increase of ramets. Root tubers of *Nymphoides* species (Figure 8.38) might be an exception to this, by virtue of their being attached to floating leaves. This feature allows much greater dispersal than would be found in species producing tubers attached to subterranean root systems, such as in some *Eleocharis* or *Nymphaea* species (Grace 1993; Sculthorpe 1967).

8.4 BALANCING SEXUAL VERSUS ASEXUAL REPRODUCTION

In this chapter, we have seen a great diversity of mechanisms plants use for persisting in aquatic and wetland habitats and for passing along their genes through sexual reproduction. We saw that obstacles to pollination in aquatic habitats have led to a host of adaptations for ensuring the delivery of viable pollen to receptive stigmas. We also learned that many species also have retained mechanisms for clonal propagation to help ensure multiple opportunities for success at sexual reproduction. Given the reliance on asexual clonal propagation in ensuring species persistence in habitats where sexual reproduction is so difficult, an important question, and one that has often been asked in the context of evolutionary biology, is why those species would bother with sexual reproduction at all.

Clonal propagation is a relatively simple extension of a plant's vegetative growth, and all the structures used for production of new clones provide multiple services to the plant (Figure 8.28). In addition to the advantages of clonal propagules, there are some key ecological and evolutionary drawbacks of sexual reproduction. Two major drawbacks in this regard are the ecological costs of producing sexual reproductive structures and the evolutionary costs to maternal parents of sharing genes with paternal parents (Antonovics and Ellstrand 1984; Harper 1977; Jaenike 1978; Lloyd 1982; Silvertown 2008; Williams 1975).

Regarding the costs associated with flower and fruit production, Harper (1977) provided numerous examples of

FIGURE 8.36 Many Wetland Species Realize Clonal Expansion Via Rhizomes. The Top Two Photos here Illustrate Diffuse Branching of Rhizomes in *Typha* Species. Photos in the Bottom Left and Bottom Center Show the Characteristically Large Rhizomes Produced By *Nuphar Advena*, while the Bottom Right Photo Shows the Starkly Contrasting Short Internodes on Slender Rhizomes of *Juncus Effusus*.

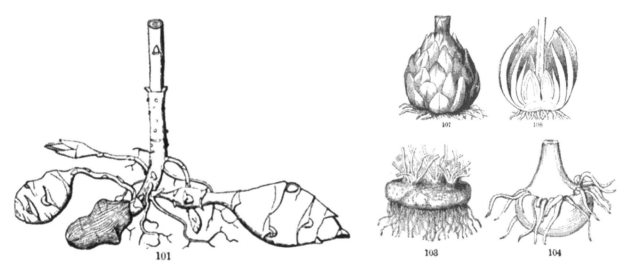

FIGURE 8.37 Left: Belowground Tubers Produced by the Aptly Named *Helianthus* Tuberosus. Top Right: External Structure and Internal Section of the Bulb of a Wild Lily. Bottom Right: Corms of Cyclamen and Arisaema; Illustrations from Gray (1887).

FIGURE 8.38   Root Tubers Produced by *Nymphoides Cristata*. Because the Shape of these Tubers Somewhat Resembles that of a Banana (Cronk and Fennessy 2001), some Species of *Nymphoides* are Called by the Common Name "Banana Plant."

reduced growth in plants after highly productive years of sexual production. For example, 36 years of data on seed production and annual growth increments in *Fagus sylvatica* (European beech) showed substantial declines in annual growth in years with high seed production. A doubling of seed production in those trees typically resulted in an approximately 50% reduction in the width of annual growth rings. Harper (1977) further mentioned that this growth reduction was equivalent to reductions observed after a year of heavy herbivory by defoliating caterpillars. In some spruce species (*Picea* species) where seeds are not produced every year, seed production was correlated with an approximately 40% reduction in growth, with a 20% reduction observed in the year following reproduction (Harper 1977). Thus, there is considerable cost associated with producing fruit versus continuing with production of only vegetative structures. These costs in terms of resource allocation and reduced growth may also be compounded

by costs associated with attracting pollinators (provision of nectar reward, production of showy flowers, etc.), as well as the potential for failed mating (Lloyd 1980), which is a very real risk in aquatic habitats.

Because of the costs associated with fruit and seed production, which are borne by maternal, or pistillate, parents, evolutionary biologists have often discussed costs of sexual reproduction in terms of maternal genetic costs. As was the case with clonal propagules, we can think about trade-offs that exist among costs and benefits of different propagation systems to help us think about why particular traits may be more common than others under certain circumstances. For example, in Figure 8.39, hypothetical dispersal "shadows" are given for maternal (seed, fruit, or clonal propagules) and paternal (pollen) parents of plants of three different propagation systems. If we consider a focal maternal parent (the darker, stippled central circle in the three propagation systems) and the dispersal of that parent's genes versus dispersal of paternal genes (the large, cross-hatched circles), we see that in both the cosexual and dioecious mating systems, the longer dispersal distances of pollen grant a much greater **dispersal shadow** to the paternal parents' genes. The benefit of cosexual mating systems, however, is that the increased dispersal of the genes carried by pollen contributes to dispersal of the central maternal parent's genes because the pollen was produced by that same plant. Of course, there is a corresponding cost to the four neighboring plants in any seeds that are fertilized by the focal plant's pollen, but that is inconsequential, or even advantageous, for the focal plant. This could be one reason that cosexuality tends to be more common than dioecy among angiosperms (Lloyd 1982).

The material costs of seed and fruit production and dispersal cost of seed (versus pollen) are two important costs incurred by maternal parents during sexual reproduction. Both cost categories can impact the competitive abilities of the plant or its offspring. Production of seed and fruit consumes resources that would have been available for other functions, such as root production for soil resource acquisition, stem and leaf production for increased light capture, or production of perenniation structures. Reduced dispersal distances of offspring also can result in increased density of offspring, leading to greater likelihood of competition among siblings.

Another cost of sexual reproduction for maternal parents is the genetic and evolutionary cost of sharing genes with paternal parents (Antonovics and Ellstrand 1984; Williams 1975). Because sexual offspring result from the union of a sperm and an egg from two different parents (except in cases of self-fertilization), the maternal parent, who

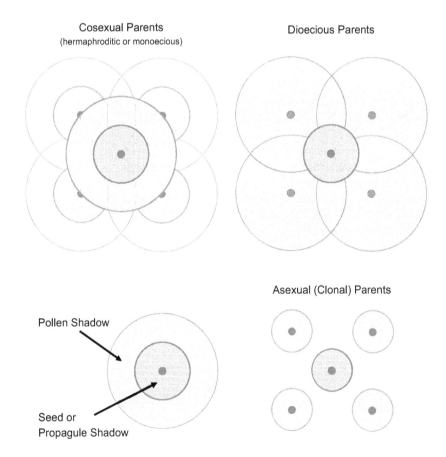

FIGURE 8.39 Conceptual Comparison of the Relative Genetic Neighborhoods Covered by Maternal (Dark/Stippled Circles) Vs. Paternal Parents (Cross-Hatched Circles) under Different Mating and Propagation Systems: Sexual Reproduction with Cosexual Parents, Sexual Reproduction with Dioecious Parents, and Asexual Propagation. Reduced Dispersal of the Maternal Parent's Genes in Sexual Systems, Via Seed Dispersal, Compounds the Ecological and Evolutionary Costs of Sexual Reproduction, As Discussed in the Text. However, a Trade-Off of this is that Pollen and Seeds Generally are Better Adapted for Dispersal than are Clonal Propagules, Which are Shown with a Smaller Dispersal Distance in the Lower Right. This Latter Aspect of Sexual Reproduction Results in Increased Average Rates of Spatial Spread and in Reduced Competition Among Sexually Produced Offspring. Diagrams Based on those of Lloyd (1982). Pollen Dispersal Distance is Illustrated as Roughly Twice that of Seed Dispersal, Based on Published Examples (E.g., Lloyd 1982; Ruckelshaus 1996; Van Tussenbroek et al. 2016a).

allocates resources toward production of seed and fruit, sacrifices half of the genetic content of the offspring in sexual reproduction. Such a large sacrifice, compounded by the resource and dispersal costs mentioned earlier, certainly makes one wonder why plants retain sexual reproduction, when we have seen that there are many means available for asexual propagation. This is where trade-offs become important considerations.

Williams (1975) provides a brief list of trade-offs in some key attributes of sexual versus asexual dispersal units (Table 8.8). The first two of these, size of propagule and proximity to parents at which the offspring develop, are illustrated by the asexual propagation system given in Figure 8.39, where we see a restricted dispersal shadow for asexual propagation, in comparison with dispersal of seeds or fruit. Larger propagules generally will have shorter dispersal distances, which potentially will result in even higher densities of clonal offspring than we might expect to see with sexual offspring. Thus, in sexual reproduction there exists a trade-off for the maternal parent between sacrificing half the genetic content of sexually produced offspring that may disperse greater distances or sacrificing dispersal distance to retain 100% of the genetic contribution by producing clonal offspring. Other advantages of clonal offspring include potentially continuous production of clonal propagules that may develop directly to sexual maturity, bypassing the seedling stage. The seedling stage is known to be more sensitive to environmental conditions, thus developmentally bypassing that stage reduces mortality rates among clonal offspring (Table 8.8).

The remaining trade-offs in Table 8.8 are tied to environmental variability, which we know to be high in wetlands. This is an area where sexual dispersal begins to provide a return on the sacrifices made by maternal parents. By mixing genes from a maternal and a paternal plant, along with recombination of alleles by crossing-over among chromosomes during meiosis, sexual reproduction yields significant genetic diversification among offspring. This genetic diversification is an important source of adaptation to changing environmental conditions, and there are expected to be at least three scenarios under which genetic variation can offset the costs of sex for maternal parents (Antonovics and Ellstrand 1984; Ellstrand and Antonovics 1985).

First, if environmental conditions vary spatially and temporally, added genetic variation from sexual recombination can improve fitness of individuals within a given population by increasing the diversity of phenotypes for environmentally relevant traits among offspring (Jaenike 1978). This is especially relevant for wetlands, where environmental conditions can vary greatly in space across a wetland, as well as seasonally and among years. A second scenario given by Antonovics and Ellstrand (1984) under which genetic variation arising through sexual reproduction can be beneficial is when conditions within an environment are heterogeneous *and* when different genotypes have adaptations to ensure they arrive at the most suitable microenvironments (or regeneration niches). This aspect of beneficial variation contributes even further to long-term fitness increases if there is competition among genotypes within a microenvironment that results in the most fit genotype succeeding within each microenvironment. Dispersal vectors often contribute to successful arrival in suitable microenvironments, such as when water birds transport seeds from wetland to wetland, when tidal movement carries seeds or fruit to appropriate positions along a shoreline, or when wind carries seeds or fruit from one stretch of lake shoreline to another.

Finally, Antonovics and Ellstrand (1984) and Ellstrand and Antonovics (1985) discussed the potential benefits to offspring whose genotypes differ from that of the majority of individuals within a population. Thus, as posited by Jaenike (1978), the fitness advantages afforded by sexual recombination may lie not in the novelty of the recombined genotype itself but in the fact that the genotype is rare, while threats such as predators, pathogens, and competitors generally are adapted to the majority genotype. Antonovics

---

TABLE 8.8

Selected contrasts between sexual vs. asexual propagation. Taken from Williams (1975).

Trait	Sexual Offspring	Clonal Propagules
Initial size	Small	Large
Proximity to parents	Widely dispersed	Frequently develop near parents
Timing of production	Seasonally	Continuously
Development	Often progress though distinct stages to sexual maturity	May develop directly to sexually mature stage
Optimal genotype for local environment	Offspring genotype may differ from parental, impacting optimal environmental conditions	Optimum environment generally predictable from parental genotype × environment combination
Mortality rate	Generally high	Generally low
Strength of selection	Natural selection can be strong, thus the high mortality rates	Natural selection generally weaker, owing to close match between genotype and environment and genetic homogeneity of propagules

and Ellstrand (1984) and Ellstrand and Antonovics (1985) demonstrated the fitness advantage afforded to minority genotypes through field experiments with the grass *Anthoxanthum odoratum*. In their study, they manipulated the density of different genotypes of the grass and found higher growth, higher survival, and higher reproductive output among the minority genotypes in two separate experiments.

Ironically, however, even though there are known advantages to sexual reproduction, and those advantages have been demonstrated to offset costs of sex under various selective pressures, clonal propagation is widespread among aquatic and wetland plants (Cronk and Fennessy 2001; Grace 1993; Hutchinson 1975; Philbrick and Les 1996). In fact, Silvertown (2008) compiled data from 248 published studies of clonal plants and found that aquatic plants tended to have the highest frequency of clonal propagation from among a dataset of 218 species across 74 families. He specifically cited the frequent production of readily dispersible clonal propagules and the ease with which they are dispersed by water as factors contributing to this tendency. Others, such as Philbrick and Les (1996), have emphasized the difficulties associated with pollination in aquatic habitats as major drivers of the tendency towards clonal propagation in aquatic species. As with most traits that are subject to diverse pressures, it is likely that each species, or even each population of a given species, has been fine-tuned by the combination of selective forces in their local environments to exhibit a phenotype that optimizes trade-offs to maximize fitness in those environments.

8.5   IN SUMMARY

We have covered a great deal of information regarding species reproductive biology in this chapter, beginning with a general focus on plant life histories based on plant longevity (annuals vs. perennials) and reproductive periodicity (monocarpic vs. polycarpic species). In spite of the inherent differences among those life history backgrounds, all angiosperms share one basic similarity in the way in which sexual reproduction takes place: the flower. From that very basic starting point, in this chapter we have covered quite a diversity of mechanisms that individual species use for sexual reproduction, ensuring highly specific mating and facilitating the evolution of the thousands of species that occupy wetlands and aquatic habitats.

From plants that carry out pollination without ever opening their flowers to plants that have adapted specifically to use water as a means of conveying pollen from flower to flower, we see suites of plant and floral traits that are finely tuned to the environments within which the species typically occur. Additionally, we see cases where there seem to be relatively tight constraints within phylogenetic groups, such as the Hydrocharitaceae, with 18 genera and more than 130 species, almost three-fourths of which are hydrophilous. At the same time, there are other groups, such as the genus *Callitriche*, that exhibit many different mating

systems and means of pollination, including internal pollination through vegetative tissues.

Within species, consistency in floral structure and pollination mechanisms combines with genetically controlled compatibility systems to help ensure fertilization takes place among individuals of the same species, while minimizing pollination among closely related individuals. Incompatibility mechanisms help to avoid the problem of inbreeding depression, which can compromise the sustainability of a population, while also encouraging maintenance of genetic diversity to ensure persistence through the environmental variation often exhibited by wetland ecosystems.

After successful pollination and fertilization, the seeds and their enclosing fruits begin to develop and then mature into vehicles for seed dispersal. The structure of the mature fruit strongly influences the dispersal mechanism, including whether seeds are dispersed with, or separate from, the fruit. In some cases, the seed and fruit are so intimately connected that they can scarcely be distinguished from one another; examples of the latter are the caryopses produced by grasses (Poaceae) and the achenes produces by sedges (Cyperaceae). As with pollination, we see in some truly aquatic species the development of seed and fruit structures that enable dispersal by the water itself, or hydrochory.

Once the seeds are released from the parent plant and dispersed spatially to their eventual resting place, the timing of germination will be determined by the type of dormancy present in the seed, or the lack thereof. Seeds that are dormant at maturity are further dispersed in time until local conditions match those required for germination and, usually, establishment of the new seedling. There is a diversity of dormancy types, requiring different mechanisms for removal of dormancy and the possibility for germination. Those removal mechanisms correlate generally with local environmental conditions between the time of release from the parent plant and the season, or other time, during which seeds of that species typically germinate. All the environmental conditions encompassed by the production of viable seed, the dispersal of those seeds in space and time (via dormancy, if present), and eventual germination and seedling establishment are conceptually included in the idea of a species' regeneration niche.

In contrast with sexual reproduction, clonal propagation makes use of structures that often are already produced by the plant during growth or for surviving seasonally unfavorable growth conditions. In doing so, clonal propagules contribute not only to dispersal and numeric increase of clones, but also to such processes as resource acquisition, anchorage, and storage and protection of accumulated resources. Depending on the species, clonal dispersal units may be derived from essentially any part of the plant, and the functions provided by the propagules often are correlated with the structure from which they are derived. For example, modified buds and shoot fragments generally contribute resource acquisition via photosynthesis, while modified roots might contribute to storage and protection of resources while also helping to anchor the plant in place.

The multi-functionality of clonal propagules is important for aquatic plants in particular because of the difficulties associated with carrying out successful pollination in aquatic environments. The use of vegetative structures to prolong the longevity and spatial spread of a clone (thereby increasing opportunities for eventual sexual reproduction) reduces the long-term ecological and evolutionary costs of producing the propagules, which, in turn, helps to offset other costs associated with sexual reproduction. Those costs, such as the ecological costs of producing flowers and fruit or the evolutionary costs of sharing genes, also may be offset through increased success of genetically diverse offspring in spatially and temporally dynamic habitats. The trade-offs discussed here in the context of reproduction and clonal propagation are examples of some of the features in natural populations upon which selection acts and which we will examine further in the following chapter.

8.6 FOR REVIEW

1. What are the relationships between the terms annual, biennial, and perennial and the terms monocarpic and polycarpic?
2. Why is it technically incorrect to use the terms "male/female flower" or "male/female plant?"
3. In angiosperms, what happens during alternation of generations? What are the "generations?"
4. Essentially all stages of sexual reproduction have the potential to be modified via evolution. Which stages seem to have been most affected among truly aquatic plants, what are some of the modifications to those stages, and why were those modifications important for reproductive success in aquatic plants?
5. What are some mechanisms of self-pollination among aquatic plants and what plant attributes are correlated with syndromes of self-pollination?
6. What are some potential problems with self-pollination? Given those problems, why do we see some groups of plants in which self-pollination seems to be the dominant form of pollination?
7. What are some similarities between and differences among animal-, wind-, and water-pollinated plants? How do the differences function in facilitating their respective forms of pollination?
8. What are some of the more common means by which plants avoid inbreeding? With which form(s) of pollination does each of these mechanisms most closely function?
9. What are the two most common means of removing dormancy in seeds? What processes in nature are associated with each of those dormancy removal mechanisms?
10. What is the primary function of dormancy in plants?

11. What are some of the key differences between sexual and asexual propagules?
12. How do clonal propagules derived from aboveground structures differ from those arising from belowground structures?
13. What are some ways in which the multiple functions provided by clonal propagules benefit the plants that produce those structures?
14. Clonal propagation provides little opportunity for genetic diversification. What are some disadvantages resulting from this? How do plants compensate for or avoid those disadvantages?

8.7 REFERENCES

Adamec, Lubomír. 2011. "Dark Respiration and Photosynthesis of Dormant and Sprouting Turions of Aquatic Plants." *Fundamental and Applied Limnology* 179 (2): 151–58. https://doi.org/10.1127/1863-9135/2011/0179-0151.
———. 2018. "Ecophysiological Characteristics of Turions of Aquatic Plants: A Review." *Aquatic Botany* 148 (April): 64–77. https://doi.org/10.1016/j.aquabot.2018.04.011.
Aiken, Susan G. 1981. "A Conspectus of *Myriophyllum* (Haloragaceae) in North America." *Brittonia* 33: 57–69.
Antonovics, Janis, and Norman C. Ellstrand. 1984. "Experimental Studies of the Evolutionary Significance of Sexual Reproduction I. A Test of the Frequency-dependent Selection Hypothesis." *Evolution* 38: 103–15.
Arber, Agnes R. 1920. *Water Plants: A Study of Aquatic Angiosperms*. Cambridge, UK: Cambridge University Press.
Aston, Helen I. 1973. *Aquatic Plants of Australia*. Melbourne, Australia: Melbourne University Press.
Barrett, Spencer C. H. 1978. "Floral Biology of *Eichhornia azurea* (Swartz) Kunth (Pontederiaceae)." *Aquatic Botany* 5: 217–28.
———. 2002. "The Evolution of Plant Sexual Diversity." *Nature Reviews Genetics* 3 (4): 274–84. https://doi.org/10.1038/nrg776.
Baskin, Carol C., and Jerry M. Baskin. 2014. *Seeds: Ecology, Biogeography, and Evolution of Dormancy and Germination*. 2nd ed. New York: Academic Press.
Cao, Jia Jie, Ying Wang, and Zun Ling Zhu. 2012. "Growth Response of the Submerged Macrophyte *Myriophyllum spicatum* to Sediment Nutrient Levels and Water-Level Fluctuations." *Aquatic Biology* 17 (3): 295–303. https://doi.org/10.3354/ab00484.
Chase, J. M., and M. A. Leibold. 2003. *Ecological Niches: Linking Classical and Contemporary Approaches*. Chicago, IL: The University of Chicago Press.
Cheptou, P. O. 2019. "Does the Evolution of Self-Fertilization Rescue Populations or Increase the Risk of Extinction?" *Annals of Botany* 123 (2): 337–45. https://doi.org/10.1093/aob/mcy144.
Cook, Christopher D. K. 1988. "Wind Pollination in Aquatic Angiosperms." *Annals of the Missouri Botanical Garden* 75 (3): 768–77.
Cook, Christopher D. K., and Katharina Urmi-Keonig. 1984. "A Revision of the Genus *Egeria* (Hydrocharitaceae)." *Aquatic Botany* 19: 73–96.
———. 1985. "A Revision of the Genus *Elodea* (Hydrocharitaceae)." *America* 21: 111–56.

Cooper, Ranessa L., Jeffrey M. Osborn, and C. Thomas Philbrick. 2000. "Comparative Pollen Morphology and Ultrastructure of the Callitrichaceae." *American Journal of Botany* 87 (2): 161–75. https://doi.org/10.2307/2656902.

Correll, Donovan S., and Helen B. Correll. 1972. *Aquatic and Wetland Plants of Southwestern United States*. Washington, DC: U. S. Environmental Protection Agency.

Cox, Paul Alan. 1988. "Hydrophilous Pollination." *Annual Review of Ecology and Systematics* 19: 261–80.

Cronk, Julie K., and M. S. Fennessy. 2001. *Wetland Plants: Biology and Ecology*. Boca Raton, FL: CRC Press.

Edlund, Anna F., Robert Swanson, and Daphne Preuss. 2004. "Pollen and Stigma Structure and Function: The Role of Diversity in Pollination." *Plant Cell* 16 (Suppl): 84–98. https://doi.org/10.1105/tpc.015800.

Efremov, Andrey N., Boris F. Sviridenko, Cezary Toma, Attila Mesterházy, and Yury A. Murashko. 2019. "Ecology of *Stratiotes aloides* L. (Hydrocharitaceae) in Eurasia." *Flora: Morphology, Distribution, Functional Ecology of Plants* 253 (January): 116–26. https://doi.org/10.1016/j.flora.2019.03.009.

Ellstrand, Norman C., and Janis Antonovics. 1985. "Experimental Studies of the Evolutionary Significance of Sexual Reproduction II. A Test of the Density-Dependent Selection Hypothesis." *Evolution* 39: 657–66.

Flora of North America Editorial Committee, eds. 1993+. 2021. "Flora of North America North of Mexico [Online]. 22+ Vols." http://floranorthamerica.org/.

Global Biodiversity Information Facility. 2021. "GBIF: The Global Biodiversity Information Facility." www.gbif.org/.

Godfrey, Robert K., and Jean W. Wooten. 1979. *Aquatic and Wetland Plants of Southeastern United States: Monocotyledons*. Athens, GA: University of Georgia Press.

———. 1981. *Aquatic and Wetland Plants of Southeastern United States: Dicotyledons*. Athens, GA: University of Georgia Press.

Grace, J B. 1993. "The Adaptive Significance of Clonal Reproduction in Angiosperms: An Aquatic Perspective." *Aquatic Botany* 44: 159–80.

Graham, Linda E., James M. Graham, and Lee W. Wilcox. 2006. *Plant Biology*. 2nd ed. Upper Saddle River, NJ: Pearson Education, Inc.

Gray, Asa. 1887. *The Elements of Botany for Beginners and for Schools*. New York: American Book Company.

Grinnell, J. 1917. "The Niche-Relationships of the California Thrasher." *The Auk* 34: 427–33.

Grubb, P. J. 1977. "The Maintenance of Species-Richness in Plant Communities: The Importance of the Regeneration Niche." *Biological Reviews* 52: 107–45.

Guo, You Hao, and Christopher D. K. Cook. 1990. "The Floral Biology of *Groenlandia densa* (L.) Fourreau (Potamogetonaceae)." *Aquatic Botany* 38 (2–3): 283–88. https://doi.org/10.1016/0304-3770(90)90011-9.

Gurevitch, Jessica, Samuel M. Scheiner, and Gordon A. Fox. 2006. *The Ecology of Plants*. 2nd ed. Sunderland, MA: Sinauer Associates, Inc.

Harper, John L. 1977. *Population Biology of Plants*. New York: Academic Press, Inc.

Haynes, Robert R. 1988. "Reproductive Biology of Selected Aquatic Plants." *Annals of the Missouri Botanical Garden* 75 (3): 805–10.

Haynes, Robert R., and C. Barre Hellquist. 2020. "Zannichellia." *Flora of North America*. http://floranorthamerica.org/Zannichellia.

Haynes, Robert R., and L. B. Holm-Nielsen. 2001. "The Genera of Hydrocharitaceae in the Southeastern United States." *Harvard Papers in Botany* 5 (2): 201–75.

Huang, Shuang-Quan, You-Hao Guo, Gituru W. Robert, Yao-Hua Shi, and Kun Sun. 2001. "Mechanism of Underwater Pollination in *Najas marina* (Najadaceae)." *Aquatic Botany* 70 (1): 67–78. https://doi.org/10.1016/S0304-3770(00)00141-8.

Hutchinson, G. Evelyn. 1957. "Concluding Remarks." *Population Studies: Animal Ecology and Demography. Cold Spring Harbor Symposia on Quantitative Biology* 22: 415–427.

Hutchinson, G. Evelyn. 1975. *A Treatise on Limnology: Volume III, Limnological Botany*. New York: John Wiley & Sons, Inc.

Ito, Yu, Gerardo Robledo, Laura Iharlegui, and Norio Tanaka. 2016. "Phylogeny of *Potamogeton* (Potamogetonaceae) Revisited: Implications for Hybridization and Introgression in Argentina." *Bulletin of the National Museum of Nature and Science. Series B, Botany* 42 (4): 131–41.

Ito, Yu, Norio Tanaka, Anders S. Barfod, Robert B. Kaul, A. Muthama Muasya, Pablo Garcia-Murillo, Natasha De Vere, Brigitta E. E. Duyfjes, and Dirk C. Albach. 2017. "From Terrestrial to Aquatic Habitats and Back Again: Molecular Insights into the Evolution and Phylogeny of *Callitriche* (Plantaginaceae)." *Botanical Journal of the Linnean Society* 184 (1): 46–58. https://doi.org/10.1093/botlinnean/box012.

Jaenike, John. 1978. "A Hypothesis to Account for the Maintenance of Sex within Populations." *Evolutionary Theory* 3: 191–94.

Judd, Walter S., Christopher S. Campbell, Elizabeth A. Kellogg, Peter F. Stevens, and Michael J. Donoghue. 2008. *Plant Systematics: A Phylogenetic Perspective*. 3rd ed. Sunderland, MA: Sinauer Associates, Inc.

Kimbel, Jeffrey C. 1982. "Factors Influencing Potential Intralake Colonization by *Myriophyllum spicatum* L." *Aquatic Botany* 14: 295–307.

Kondo, Katsuhiko. 1972. "A Comparison of Variability in *Utricularia cornuta* and *Utricularia juncea*." *American Journal of Botany* 59 (1): 23–37. https://doi.org/10.2307/2441227.

Lansdown, Richard V., Ioannis Bazos, Maria Carmela Caria, Angelo Troia, and Jan J. Wieringa. 2017. "New Distribution and Taxonomic Information on *Callitriche* (Plantaginaceae) in the Mediterranean Region." *Phytotaxa* 313 (1): 91–104. https://doi.org/10.11646/phytotaxa.313.1.6.

Les, Donald H. 1988. "Breeding Systems, Population Structure, and Evolution in Hydrophilous Angiosperms." *Annals of the Missouri Botanical Garden* 75 (3): 819–35.

Leydon, Alexander R., Tatsuya Tsukamoto, Damayanthi Dunatunga, Yuan Qin, Mark A. Johnson, and Ravishankar Palanivelu. 2015. "Pollen Tube Discharge Completes the Process of Synergid Degeneration That Is Initiated by Pollen Tube-Synergid Interaction in *Arabidopsis*." *Plant Physiology* 169 (1): 485–96. https://doi.org/10.1104/pp.15.00528.

Li, Hong Tao, Ting Shuang Yi, Lian Ming Gao, Peng Fei Ma, Ting Zhang, Jun Bo Yang, Matthew A. Gitzendanner, et al. 2019. "Origin of Angiosperms and the Puzzle of the Jurassic Gap." *Nature Plants* 5 (5): 461–70. https://doi.org/10.1038/s41477-019-0421-0.

Linkies, Ada, Kai Graeber, Charles Knight, and Gerhard Leubner-Metzger. 2010. "The Evolution of Seeds." *New Phytologist* 186 (4): 817–31. https://doi.org/10.1111/j.1469-8137.2010.03249.x.

Lloyd, David G. 1980. "Benefits and Handicaps of Sexual Reproduction." In *Evolutionary Biology, Volume 13*, edited by Max K. Hecht, William C. Steere, and Bruce Wallace, 301. New York: Plenum Press.

———. 1982. "Selection of Combined Versus Separate Sexes in Seed Plants." *The American Naturalist* 120 (5): 571–85.

Lloyd, Michael W., Hayley R. Tumas, and Maile C. Neel. 2018. "Limited Pollen Dispersal, Small Genetic Neighborhoods, and Biparental Inbreeding in *Vallisneria americana*." *American Journal of Botany* 105 (2): 227–40. https://doi.org/10.1002/ajb2.1031.

Munné-Bosch, Sergi. 2008. "Do Perennials Really Senesce?" *Trends in Plant Science* 13 (5): 216–20. https://doi.org/10.1016/j.tplants.2008.02.002.

Nicol, Jason M., and George G. Ganf. 2000. "Water Regimes, Seedling Recruitment and Establishment in Three Wetland Plant Species." *Marine & Freshwater Resource* 51: 305–9.

Packer, Jasmin G., Laura A. Meyerson, Hana Skálová, Petr Pyšek, and Christoph Kueffer. 2017. "Biological Flora of the British Isles: *Phragmites australis*." *Journal of Ecology* 105: 1123–62.

Philbrick, C. Thomas. 1984. "Pollen Tube Growth within Vegetative Tissues of *Callitriche* (Callitrichaceae)." *American Journal of Botany* 71 (6): 882–86.

———. 1988. "Evolution of Underwater Outcrossing from Aerial Pollination Systems: A Hypothesis." *Annals of the Missouri Botanical Garden* 75 (3): 836–41.

———. 1991. "Hydrophily: Phylogenetic and Evolutionary Considerations." *Rhodora* 93 (873): 36–50.

———. 1993. "Underwater Cross-Pollination in *Callitriche hermaphroditica* (Callitrichaceae): Evidence from Random Amplified Polymorphic DNA Markers." *American Journal of Botany* 80 (4): 391–94. https://doi.org/10.2307/2445385.

Philbrick, C. Thomas, and Gregory J. Anderson. 1987. "Implications of Pollen/Ovule Ratios and Pollen Size for the Reproductive Biology of *Potamogeton* and Autogamy in Aquatic Angiosperms." *Systematic Botany* 12 (1): 98–105.

Philbrick, C. Thomas, and Donald H. Les. 1996. "Evolution of Aquatic Angiosperm Reproductive Systems." *BioScience* 46: 813–26.

———. 2000. "Phylogenetic Studies in *Callitriche*: Implications for Interpretation of Ecological, Karyological and Pollination System Evolution." *Aquatic Botany* 68 (2): 123–41. https://doi.org/10.1016/S0304-3770(00)00114-5.

Philbrick, C. Thomas, and Jeffrey M. Osborn. 1994. "Exine Reduction in Underwater Flowering *Callitriche* (Callitrichaceae): Implications for the Evolution of Hydrophily." *Rhodora* 96 (888): 370–81.

Plachno, Bartosz J., Małgorzata Stpiczynska, Piotr Swiatek, Hans Lambers, Vitor F. O. Miranda, Francis J. Nge, Piotr Stolarczyk, et al. 2019. "Floral Micromorphology of the Bird-Pollinated Carnivorous Plant Species *Utricularia menziesii* r.Br. (Lentibulariaceae)." *Annals of Botany* 123 (1): 213–20. https://doi.org/10.1093/aob/mcy163.

Rashmi, K., G. Krishnakumar, and Donald H. Les. 2017. "*Blyxa mangalensis*, a New Species of Hydrocharitaceae from India." *Kew Bulletin* 72 (1): 1–8. https://doi.org/10.1007/s12225-016-9663-4.

Rejmánková, Eliška. 1992. "Ecology of Creeping Macrophytes with Special Reference to *Ludwigia peploides* (H.B.K.) Raven." *Aquatic Botany* 43 (3): 283–99. https://doi.org/10.1016/0304-3770(92)90073-R.

Reusch, T. B. H. 2000. "Pollination in the Marine Realm: Microsatellites Reveal High Outcrossing Rates and Multiple Paternity in Eelgrass *Zostera marina*." *Heredity* 85 (5): 459–64. https://doi.org/10.1046/j.1365-2540.2000.00783.x.

———. 2001. "Fitness-Consequences of Geitonogamous Selfing in a Clonal Marine Angiosperm (*Zostera marina*)." *Journal of Evolutionary Biology* 14 (1): 129–38. https://doi.org/10.1046/j.1420-9101.2001.00257.x.

Richardson, Frank David. 1983. "Variation, Adaptation and Reproductive Biology in *Ruppia maritima* L. Populations from New Hampshire Coastal and Estuarine Tidal Marshes." Durham, NH: University of New Hampshire.

Ruckelshaus, Mary H. 1995. "Estimates of Outcrossing Rates and of Inbreeding Depression in a Population of the Marine Angiosperm *Zostera marina*." *Marine Biology* 123: 583–93.

———. 1996. "Estimation of Genetic Neighborhood Parameters Form Pollen and Seed Dispersal in the Marine Angiosperm *Zostera marina* L." *Evolution* 50: 856–64.

Santamaría, Luis. 2002. "Why Are Most Aquatic Plants Widely Distributed? Dispersal, Clonal Growth and Small-Scale Heterogeneity in a Stressful Environment." *Acta Oecologica* 23 (3): 137–54. https://doi.org/10.1016/S1146-609X(02)01146-3.

Sculthorpe, C D. 1967. *Biology of Aquatic Vascular Plants*. New York: St. Martin's Press.

Seymour, Graham B., Lars Ostergaard, Natalie H. Chapman, Sandra Knapp, and Cathie Martin. 2013. "Fruit Development and Ripening." *Annual Review of Plant Biology* 64: 219–41. https://doi.org/10.1146/annurev-arplant-050312-120057.

Silva, N. F., and D. R. Goring. 2001. "Mechanisms of Self-Incompatibility in Flowering Plants." *Cellular and Molecular Life Sciences* 58: 1988–2007.

Silvertown, Jonathan. 2008. "The Evolutionary Maintenance of Sexual Reproduction: Evidence from the Ecological Distribution of Asexual Reproduction in Clonal Plants." *International Journal of Plant Sciences* 169 (1): 157–68. https://doi.org/10.1086/523357.

Sinclair, Elizabeth A., Ilena Gecan, Siegfried L. Krauss, and Gary A. Kendrick. 2014. "Against the Odds: Complete Outcrossing in a Monoecious Clonal Seagrass *Posidonia australis* (Posidoniaceae)." *Annals of Botany* 113 (7): 1185–96. https://doi.org/10.1093/aob/mcu048.

Smith, Dian H., John D. Madsen, Kenneth L. Dickson, and Thomas L. Beitinger. 2002. "Nutrient Effects on Autofragmentation of *Myriophyllum spicatum*." *Aquatic Botany* 74 (1): 1–17. https://doi.org/10.1016/S0304-3770(02)00023-2.

Soltis, Douglas, Pamela Soltis, Peter Endress, Mark W. Chase, Steven Manchester, Walter Judd, Lucas Majure, and Evgeny Mavrodiev. 2018. *Phylogeny and Evolution of the Angiosperms*. Chicago: University of Chicago Press.

Song, Xiaoya, Li Yuan, and Venkatesan Sundaresan. 2014. "Antipodal Cells Persist through Fertilization in the Female Gametophyte of *Arabidopsis*." *Plant Reproduction* 27 (4): 197–203. https://doi.org/10.1007/s00497-014-0251-1.

Spalding, Mark, Mami Kainuma, and Lorna Collins. 2010. *World Atlas of Mangroves*. London: Routledge.

Stevens, P. F. 2017. "Angiosperm Phylogeny Website." Version 14, July 2017 [and More or Less Continuously Updated Since]. www.mobot.org/MOBOT/research/APweb/.

Taiz, Lincoln, and Eduardo Zeiger. 2006. *Plant Physiology*. 4th ed. Sunderland, MA: Sinauer Associates, Inc.

Van Tussenbroek, Brigitta I., L. M. Soissons, T. J. Bouma, R. Asmus, I. Auby, F. G. Brun, P. G. Cardoso, et al. 2016a. "Pollen Limitation May Be a Common Allee Effect in Marine Hydrophilous Plants: Implications for Decline and Recovery in Seagrasses." *Oecologia* 182 (2). https://doi.org/10.1007/s00442-016-3665-7.

Van Tussenbroek, Brigitta I., Nora Villamil, Judith Márquez-Guzmán, Ricardo Wong, L. Verónica Monroy-Velázquez, and Vivianne Solis-Weiss. 2016b. "Experimental Evidence of Pollination in Marine Flowers by Invertebrate Fauna." *Nature Communications* 7: 1–6. https://doi.org/10.1038/ncomms12980.

Vasek, Frank C. 1980. "Creosote Bush: Long-Lived Clones in the Mojave Desert." *American Journal of Botany* 67 (2): 246. https://doi.org/10.2307/2442649.

Warwick, Nigel W. M., and Margaret A. Brock. 2003. "Plant Reproduction in Temporary Wetlands: The Effects of Seasonal Timing, Depth, and Duration of Flooding." *Aquatic Botany* 77 (2): 153–67. https://doi.org/10.1016/S0304-3770(03)00102-5.

Whitehead, Michael R., Robert Lanfear, Randall J. Mitchell, and Jeffrey D. Karron. 2018. "Plant Mating Systems Often Vary Widely among Populations." *Frontiers in Ecology and Evolution* 6 (APR): 1–9. https://doi.org/10.3389/fevo.2018.00038.

Williams, George C. 1975. *Sex and Evolution*. Princeton, NJ: Princeton University Press.

9 Population Biology and Evolutionary Ecology

9.1 POPULATION BIOLOGY IN THE CONTEXT OF EVOLUTION

In the previous chapter, we saw many examples of important differences in attributes of sexual reproduction among species and, in some cases, populations within a species. For example, we saw that about half the genera of aquatic plants (Table 8.2) included species with annual as well as with perennial life histories. In that same table, approximately one-third of the genera (11 of 32 genera) included both dioecious and monoecious species. We also saw that individual species can sometimes exhibit considerable variation among populations. In the submersed species *Ruppia maritima*, for example, it was observed that self-fertilization by bubble autogamy was more likely to occur in estuarine populations than in those of coastal marine habitats (Richardson 1983). Even greater variation was observed among populations of the seagrass *Zostera marina*, with some populations exhibiting protandry (staminate flowers matured before pistillate), some exhibiting protogyny (pistillate flowers matured first), and some populations having plants where both flower types matured simultaneously (Ruckelshaus 1995; Reusch 2000). This species also was shown to produce seeds that responded differently to stratification temperatures, dependent upon the location at which they were collected (Baskin and Baskin 2014).

From these examples, especially those illustrating differences among populations of *Zostera marina*, we can begin to see that the central question from Chapter 1—"Why does that plant grow there?"—could have different answers in different locations, even within a single species. Furthermore, the environmental and genetic factors responsible for the variation that we sometimes see among populations can lead to similar variation within populations. This latter process requires a certain amount of environmental and genetic heterogeneity among microhabitats and among individuals across the space that the population occupies. Interactions among individuals within a population and between the individuals and their environment drive changes in genetic composition over time, through such processes as selection and gene flow. Outcomes of those interactions drive both short- and longer-term evolution of populations, and thus collectively, the species, giving rise to the diversity of species and environmental adaptations that we have seen in the first eight chapters of this book.

In this chapter, we will learn about approaches used to examine population-level dynamics of plant species. We begin in the following section with an overview of population biology and progress ultimately to ecological and evolutionary processes responsible for the patterns of aquatic and wetland plant diversity that we see in nature. Before we discuss population biology, however, we should take a moment to define what is meant by the term **population**. A population consists of all individuals of a given species that co-occur within some determined space. For example, all the individuals of *Zostera marina* occurring in a particular estuary might be considered a population of this species. At first glance, this appears quite straightforward, but we will see later in the chapter that the boundaries of populations are not always so easily defined. In fact, this is the case not only for boundaries of populations, but of species as well, because of the ways in which individuals interact with one another across the true spatial extents of populations. For now, however, we will use the prior simple definition as we begin to look at **population biology**, which is the process of analyzing the roles of reproduction, survival, and life history on changes in populations over time (Gurevitch, Scheiner, and Fox 2006).

9.2 POPULATION BIOLOGY

According to John Harper (1977, p. 1), population biology aims to "answer questions about the differences in the numbers of organisms from place to place and from time to time." Harper went on to specify that an investigation into population biology should include such things as species' life cycles, responses of species to the physical environment, stresses caused by increases in population density, and other factors that influence evolutionary change in populations. Before we begin to incorporate all those ideas into a formal description of population change, however, we will begin with a very simple equation to describe changes in population density over time.

$$N_{t+1} = N_t + \text{Births} - \text{Deaths} + \text{Immigrants} - \text{Emigrants} \qquad (9.1)$$

In this formula, **population density** (the number of individuals in the population) at a future point in time (N_{t+1}) is calculated as the current population (N_t) plus births and immigrants during the time interval, reduced by the numbers of deaths and emigrants during that period. Immigration and emigration often are treated as negligible in natural plant populations, owing to the very small percentage of seeds and pollen that travel "long" distances (i.e., outside of the spatial limits of the population), although the actual percentages in nature vary among modes of dispersal

DOI: 10.1201/9781315156835-9

(Harper 1977; Jordano 2017; Van Tussenbroek et al. 2016). If dispersal into and out of the population is sufficiently low, then population dynamics will be dominated by births and deaths within the population, and we can focus on those two components of the population in studying and forecasting changes in the population over time. Eliminating dispersal from Equation 9.1 thus yields Equation 9.2:

$$N_{t+1} = N_t + \text{Births} - \text{Deaths} \qquad (9.2)$$

We will build on Equation 9.2 in the following section as we explore simple models of unconstrained, or density-independent, population growth. Then, in the subsequent section, we will look at expanding this model a bit further by incorporating density effects. This allows the model to represent natural population dynamics more realistically, where resource availability will limit the number of individuals that can be supported in a population. From there, we will examine stage-based approaches to modeling populations. Stage-based approaches allow one to account for the effects of a plant species' stage-based life history on changes in population density. For example, we know that, under ideal conditions, a plant transitions from seed to seedling to an established pre-reproductive stage and, eventually, to reproductive status. Stage-based models allow one to evaluate specific effects of and threats to each life history stage on changes in a population over time. Let's first take a closer look at the use of Equation 9.2 to model population growth independent of resource or other density-based constraints.

9.2.1 DENSITY-INDEPENDENT POPULATION GROWTH

Harper (1977) emphasized, with support from Malthus, Darwin, and Wallace, the conceptual importance of **density-independent growth**, or population growth that is unconstrained by the per capita use of available resources. This form of population growth, in more mathematical terms, proceeds exponentially, determined solely by the per capita reproductive capacity of the individuals in the population. Thomas Malthus (1798) is widely known for his essay on the unsustainability of density-independent growth of populations (specifically the human population), and both Charles Darwin and Alfred Russel Wallace recognized that this aspect of population biology is, to a large degree, responsible for selective pressures on individuals and traits that lead to evolutionary change in populations (Harper 1977). Specifically, the presence of excess individuals in a population, beyond what resources can support, provides opportunities for selective pressures to remove the least fit genotypes from the population. Thus, before we can fully appreciate the evolutionary ecology of aquatic and wetland plants, we should have a basic understanding of factors influencing population growth.

Equation 9.2 represents population growth influenced only by the per capita capacity for reproduction and susceptibility to mortality. In this representation of density-independent population growth, the rates of reproduction

and of death are simply proportional to the number of individuals in the population (Hastings 1997), although this is not directly indicated in the formula. However, we can modify Equation 9.2 so that it does directly incorporate the proportionality of birth and death in population growth. This can be accomplished in one of two ways, with respect to the manner in which we consider the temporal nature of population change. First, we could consider generations to be discrete in time; that is, at discrete time points we would see the addition of individuals to the population through births. This is a relatively accurate depiction of sexual reproduction in plants, for example, where we typically see one flowering phase per year within a population, or of seed germination, which often is closely connected to seasonal changes in habitat, such as exposure of mudflats during the dry season. A **discrete time model** of density-independent population growth might look like:

$$N_{t+1} = R * N_t \qquad (9.3)$$

Here, we see the net population change over time resulting from births and deaths represented as R, the net reproductive rate of the population (Hastings 1997). Note that, when R > 1, the population will grow exponentially, and when R < 1, the population will decline. Thus, when R > 1, births exceed deaths, and when R < 1, deaths exceed births. Illustrations of simulated density-independent growth for increasing, as well as decreasing, populations are given in Figure 9.1. If we want to extend Equation 9.3 beyond a single generation, to some number t generations, the equation becomes:

$$N_t = R^t * N_0 \qquad (9.4)$$

Here, N_0 represents the initial population size, N_t represents the population size after t generations, and R^t is the net reproductive rate multiplied by itself t times. Although a discrete time model represents well such processes as annual cycles of flowering or seasonal patterns of seedling emergence, other aspects of plant life history are not represented so well by this type of model. For example, in clonal plant species, vegetative shoots, or ramets, typically appear in waves called **cohorts** throughout the year (Figures 9.2 and 9.3). Similarly, some annual herbaceous species display similar patterns in the production of leaves throughout the growing season. The data on *Juncus effusus* culm production given in Figure 9.3 illustrate very well the difficulty one would encounter in applying discrete time models to populations of ramets within clonal plant species, as Wetzel and Howe (1999) encountered monthly production of hundreds to thousands of culms in each tussock of *Juncus effusus* included in their study. If we are interested in modeling processes such as these, which lack distinct generations fitting Equations 9.3 or 9.4, we can use a continuous (i.e., nondiscrete) time model, such as:

$$N_t = e^{rt} * N_0 \qquad (9.5)$$

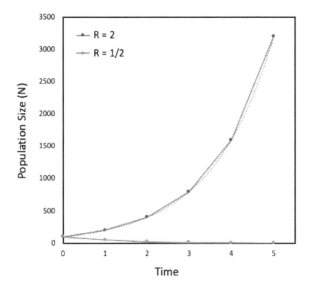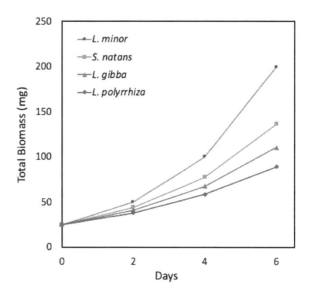

FIGURE 9.1   Density-Independent Growth Curves. On the Left, we see Generic Examples of a Growing Population, with R = 2, and a Declining Population Where R = 1/2. The Dotted Line Along the R = 2 Curve Illustrates Growth at the Equivalent Rate Modeled with the Continuous Time Growth Equation (Equation 9.5 With R = Ln(2)). On the Right are Experimental Data for Density-Independent Growth of *Lemna Minor*, *L. Gibba*, *L. Polyrrhiza*, and *Salvinia Natans*, taken from Clatworthy and Harper (1962).

In this **continuous time model**, t represents the amount of time passing between two population samples, r is the intrinsic rate of population increase, and *e* is the base of the natural logarithm (Hastings 1997). An example of the contrast between discrete and continuous population growth is given for the increasing population growth curve in Figure 9.1. Analogous to the net reproductive rate, R, the intrinsic rate of population increase can inform us about the general state of the population. If r > 0, the population is growing, with births outnumbering deaths; when r < 0, deaths outnumber births, and the population will decline. If r = 0, then the term e^{rt} will equal 1, and the population size remains constant over time. For more direct comparison between discrete and continuous growth rates, we can compute the value of r from that of R as follows:

$$r = \ln(R) \qquad (9.6)$$

Clatworthy and Harper (1962) conducted a series of experiments with four species of floating aquatic plants (*Lemna minor*, *L. gibba*, *L. polyrrhiza*, and *Salvinia natans*) to examine the relationship between unconstrained growth and the relative competitive abilities of these species. To determine baseline growth and develop hypotheses for their competition experiments, they grew each species in laboratory culture with ample light, water, and mineral nutrients. In those monospecific laboratory cultures, Clatworthy and Harper also replenished the nutrient solution and removed fronds of the experimental plants once every two days, ensuring ideal conditions for unconstrained, density-independent population growth. Data from the first six days of these experiments are shown in the right-hand panel of

Figure 9.1 as an example of density-independent growth in four species of aquatic plants.

The examples provided in Figures 9.1 through 9.3 also illustrate another important aspect of plant population biology: the modular nature of the plant individual is conducive to studying population biology at diverse biological scales. That is, we can think of an individual plant as a population of modules. These modules may be leaves, buds, or flowers on a distinct individual plant, or they may be ramets (or portions thereof) that form clonal populations. Thus, we can study populations of whole plants, of ramets that make up whole but indistinct individual plants, or of plant parts, as with the flax example in Figure 9.2. The appropriate approach would depend on the nature and objective of the study being undertaken.

9.2.2   Density-Dependent Population Growth

Although many species can exhibit density-independent growth under certain circumstances, such as during early colonization of a barren or resource-rich environment, this form of population growth typically is short-lived. Note, for example, that the *Lemna* and *Salvinia* examples given in Figure 9.1 span less than a week in time. Even under the ideal conditions of that study, after the first week, each of the plant species switched from exponential growth to linear growth, adding an average of 9 to 19 mg of new live tissue per day during weeks two through six (Clatworthy and Harper 1962). More often, what we see in nature is that over longer periods of time, various factors will begin to limit, or restrict, growth of a population. Often this is through reductions in birth rates resulting from resource limitation or via factors that cause increased mortality,

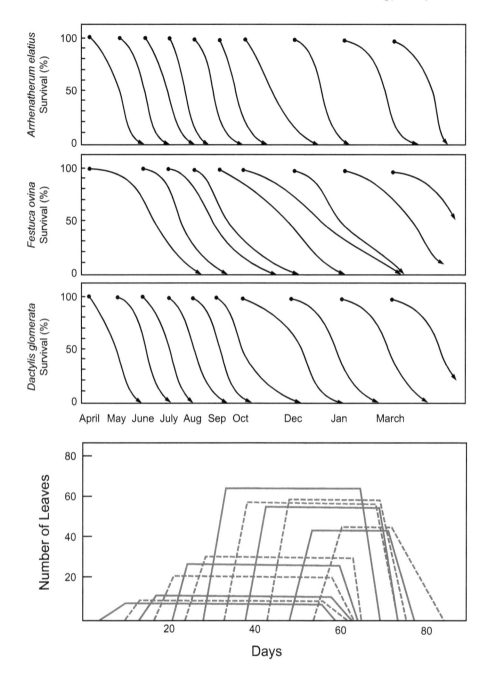

FIGURE 9.2 Patterns of Leaf Cohort Emergence and Survival in Three Perennial Grass Species (Upper Panels) and the Annual Herb *Linum Usitatissimum* (Flax, Bottom Panel). These Examples Illustrate not Only the More or Less Continuous Production of New Cohorts of Plant Parts, but also the Overlapping Nature of the Cohorts through Time. Plant Ecologists often are Interested in Observing and Modeling Processes such as these as they Relate to Clonal Expansion of Plant Populations or Interactions between Plants and Populations of Herbivores or Pathogens. In the Upper Panel, Each Curve Represents a New Cohort of Leaves, or Blades; Each Begins at 100% and Decreases as Leaves Die Over Time. Redrawn from Harper (1977).

such as predation or parasitism. To better reflect these realities, we may wish to use a population growth model that incorporates factors limiting population growth, such as those resulting from interactions among individuals within the population as density increases. We term these models **density-dependent growth** models.

To accomplish this, we can add a density-dependent term to reduce the intrinsic rate of population increase, based on

how close the population density is to the density that can be supported at that location. When we do this, the change in population density over some period of time, written as dN/dt ("change in N with respect to some change in time"), becomes (Hastings 1997):

$$\frac{dN}{dt} = r * N * \left(1 - \frac{N}{K}\right) \qquad (9.7)$$

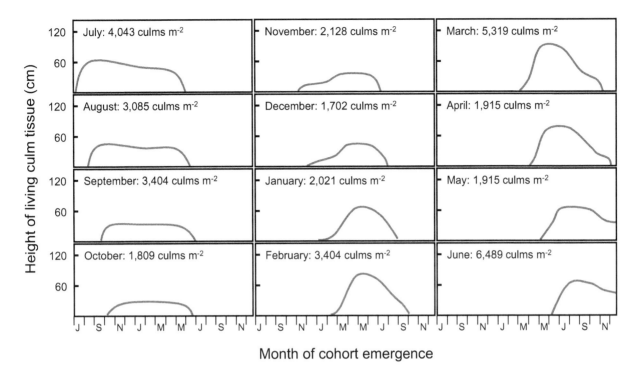

Month of cohort emergence

FIGURE 9.3   Seasonality in Culm Production of *Juncus Effusus* in the Talladega Wetland Ecosystem (Alabama, USA). Each Curve Approximates the Median Pattern in Culm Emergence, Growth, and Senescence for Each Monthly Cohort of Culms from July 1993 Through June 1994. Density of Culms (Culms Per M²) is Given for Each Month; Data from Wetzel and Howe (1999).

Here, we see that unconstrained population growth (r * N, analogous to Equation 9.4) is multiplied by a density-dependent constraining factor (1–N/K) that reduces the number of individuals added to the population as the population size, N, nears the value of the newly introduced parameter, K. This term K represents the **carrying capacity** of the population, or the number of individuals that can be supported by available resources at the location in which the population occurs. If, for example, the population size is very small, perhaps almost zero, the term N/K will approximate zero, and Equation 9.7 can be approximated as r * N * 1, and the population will experience exponential growth (see Equations 9.4 and 9.5). On the other hand, as the population density, N, grows very near K, the value of the constraining factor (1–N/K) will decrease to approximately zero (because 1–K/K = 1–1 = 0), and the population growth will slow, approaching zero itself. The shape of this type of constrained, or density-dependent, growth is shown in Figure 9.4. The equivalent equation for forecasting future population size under density-dependent growth, obtained from Equations 9.5 and 9.7 with the help of calculus, is (Hastings 1997):

$$N_t = \frac{e^{rt} * N_0}{1 + N_0 * (e^{rt} - 1)/K} \qquad (9.8)$$

Costa et al. (2003) provided data on growth of *Eichhornia crassipes* in wastewater treatment wetlands, where the plant was found to grow more or less exponentially at low population densities; however, growth slowed significantly at higher densities (Figure 9.4). They found that the capacity for populations of the plant to improve water quality also declined with the decrease in population growth. Similarly, Frighetto et al. (2019) used these same *Eichhornia crassipes* growth data to explore approaches for managing population densities to obtain optimal water quality improvement in wastewater treatment systems. They proposed that maintenance of the populations near 50% of carrying capacity would allow optimal uptake of contaminants by preventing the plants from entering the phase of declining population growth that occurs as density approaches carrying capacity. These studies thus provide an example of a practical application of population biology to solve an issue that is becoming more common with increasing use of aquatic and wetland plants in water quality improvement.

9.2.3   AGE- OR STAGE-BASED APPROACHES

All the aforementioned population growth models treat plant populations as essentially homogeneous with respect to the plants' age, size, maturity, reproductive status, and other characteristics that we know to vary among individuals within a population, as well as over time. More sophisticated approaches exist, however, for studying population dynamics while accounting for differences among seeds, seedlings, juvenile plants, and mature plants of various

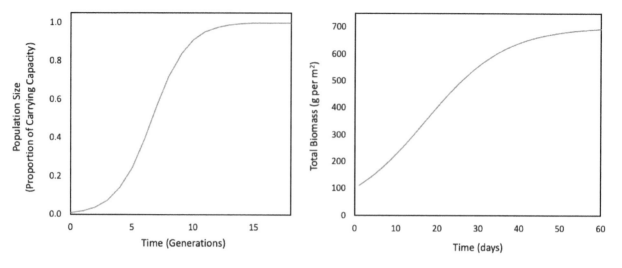

FIGURE 9.4   Density-Dependent (also Referred to as Logistic) Growth Curves. On the Left of these Curves, When the Population Size is Small, Density-Dependent Factors are Negligible, and the Population Will Grow Exponentially, as in Figure 9.1. However, as the Population Size Approaches the Habitat's Carrying Capacity, Population Growth Slows to Near Zero, Plateauing to an Asymptote at (Or Around) Carrying Capacity. The Figure on the Right Shows Data Obtained from Growth of *Eichhornia Crassipes* in Wetlands Designed to Treat Domestic Wastewater (Data from Costa et al. 2003; Frighetto et al. 2019).

reproductive states. These models are generically termed age-based models, but because of the way that plants experience age versus growth, we often call these **stage-based models** when dealing with plant populations (Gurevitch, Scheiner, and Fox 2006). For example, it is relatively easy to distinguish among a seed, a seedling, a pre-reproductive established plant, and a flowering individual of a plant species. In contrast, it is very difficult to determine the chronological age of a grass, sedge, or a rush; thus, stages, rather than ages, are typically used when studying plant populations.

A life cycle graph representing a stage-based population growth model for *Sarracenia purpurea* is given in Figure 9.5. This graph and the accompanying data in Table 9.1 were taken from a study conducted by Gotelli and Ellison (2002) aimed at understanding the influence of atmospheric nitrogen deposition on population dynamics and extinction risk in this pitcher plant species. A life cycle graph illustrates transitions among life cycle stages in a population proportional to the number of individuals known to be present within a stage at some time, t. For example, in Figure 9.5, the arrow labeled "P_{33}" represents the proportion of nonflowering adults present in one year that remained nonflowering adults the following year. The proportion of nonflowering adults in one year that flowered the following year is represented by P_{43}, and the variable P_{23} represents nonflowering adults from one year that were found to have decreased in size the following year. The values associated with these transitions are called **transition probabilities** because they represent, generally, the probability that an individual from one stage will transition into some other stage in the time between censuses. If there were no mortality among the nonflowering adults between

the first and second year, the three transition probabilities given earlier (P_{23}, P_{33}, and P_{43}) would sum to 1.0. However, if we sum the values in the "Nonflowering Adult" columns for each population in Table 9.1, those values sum to 0.97 in Hawley Bog and 0.99 in Molly Bog. This indicates that, on average, from 1% to 3% of the nonflowering adults will die between one year and the next in these two bogs.

Inspection of the transition probabilities in these two populations of *S. purpurea* (Table 9.1) reveals a few important details about the biology of these pitcher plant populations. First, new seedlings experienced very high mortality, with only 10% of the plants surviving to become established juvenile plants. In contrast, the other three life stages have very high survival, with at least 97% of individuals expected to survive from one year to the next. Another interesting value in Table 9.1 is the seemingly very large "transition" between flowering adults and seedlings, referred to as recruits in this study. Rather than a transition of adults to recruits, this value ($P_{14} = 4.0$) represents the average number of new plants in the population each year, per flowering adult. Thus, it is expected that each flowering plant will contribute four seedlings to the population each year. If we do a bit of math on the transitions in Table 9.1, we see that an average adult in Hawley Bog will contribute four seedlings, only 10% of those will become established juveniles, 4% of the established juveniles will grow to nonflowering adult size, and 18% of those adults will be expected to flower the following year. Thus, flowering adults in Hawley Bog in one year will contribute approximately 0.003 flowering adults to the population three years later; this value is approximately 0.005 for Molly Bog. Although this value is greater than zero, 0.003 and 0.005 seem to be relatively slow rates of per capita population increase.

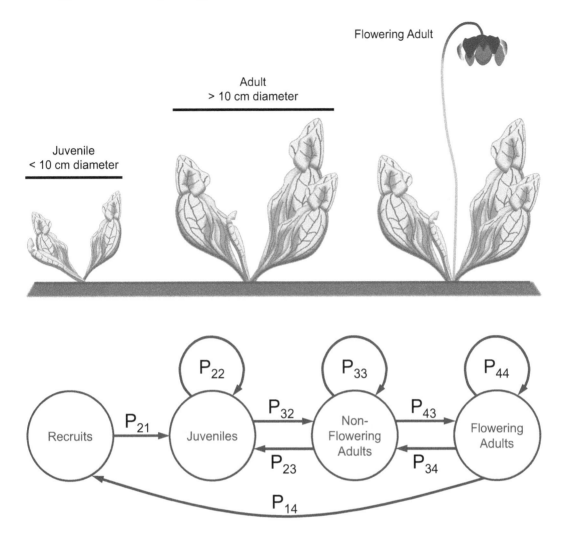

FIGURE 9.5 A Life Cycle Graph of *Sarracenia Purpurea* Based on the Work of Gotelli and Ellison (2002). In their Study, they Separated Established Plants into Juvenile and Adult Plants Based on the Diameter of the Plant (Upper Diagrams). Recruits were New Plants that Appeared in the Population after the First Year of the Study. Each Arrow in this Diagram Represents a Transition from One Stage to another. Note that it is Possible for Plants to become Smaller Between Census Years, Indicated by the Two Left-Pointing Horizontal Arrows. The Circular Arrows Indicate Plants that Remain within a Stage between Census Years.

TABLE 9.1

Transition matrices for two populations of *Sarracenia purpurea* in Massachusetts and Vermont (USA). Data from Gotelli and Ellison (2002).

	Stage at Time *t*			
Stage at Time *t+1*	**Recruit**	**Juvenile**	**Nonflowering Adult**	**Flowering Adult**
Hawley Bog, MA				
Recruit	0	0	0	4.00
Juvenile	0.10	0.95	0.09	0
Nonflowering Adult	0	0.04	0.70	0.84
Flowering Adult	0	0	0.18	0.16
Molly Bog, VT				
Recruit	0	0	0	4.00
Juvenile	0.10	0.85	0.18	0
Nonflowering Adult	0	0.13	0.71	0.67
Flowering Adult	0	0	0.10	0.31

Beyond simply providing transition probabilities among life cycle stages, studies that make use of stage-based population models can be used to forecast future population increase, considering the contributions that each stage makes to future populations (Figure 9.6). Although they look much more complex than the population growth models we have seen previously, these stage-based approaches are modifications of Equation 9.3 or 9.4, where:

$$N_{t+1} = A * N_t \quad (9.9) \text{ or}$$

$$N_t = A^t * N_0 \quad (9.10)$$

In these equations, the only new parameter we see is that A is substituted for R, the net reproductive rate. Here, A represents the transition matrix (Table 9.1), which itself represents the contributions of all the stage transitions to population change over time. In other words, the transition matrix, A, is simply a more complex representation of R. At each time interval in forecasting future population size using Equation 9.9 or 9.10, matrix algebra is used to compute the population components in the future time step, as illustrated in Figure 9.6.

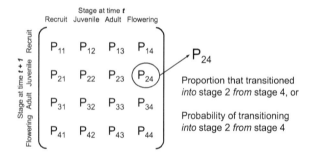

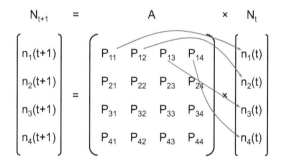

FIGURE 9.6 Generalized Diagram of a Transition Matrix (Upper Diagram) and the Process of Multiplying a Transition Matrix (A) by a Vector of Stage-Based Data from a Given Census Period (N_t). The Transition Matrix is Parameterized from Field Data Collected in the Population(S) of Interest, Based on the Observed Proportion of Plants Transitioning Among the Possible Stages between Each Pair of Censuses. In Iterating the Transition Matrix Across Each Period of Census Data, each Element in Each Row of Data in A is Multiplied by its Corresponding Element in the Vector N_t. In the Next Iteration, the Solution from the First Round of Multiplication (N_{t+1}) Becomes N_t for the Next Iteration, and so on.

Gotelli and Ellison (2002) used the data from Table 9.1 to project population changes in the Hawley and Molly bog ecosystems. They found expected population doubling times of 152 years for Hawley Bog and 125 years for Molly Bog; thus, the apparently slow rates of per capita increase that were calculated earlier do contribute to relatively slow rates of population change over time. These models not only can be used to predict future population size, but they also can help evaluate which stages are most susceptible to environmental threats or other factors that may influence population growth or decline over time (Fox and Gurevitch 2000; Gurevitch, Scheiner, and Fox 2006).

Gotelli and Ellison (2002) also conducted nutrient addition experiments to determine the effects of increased nitrogen deposition on population dynamics of these pitcher plant populations. They found that when they increased rates of nitrogen deposition to values representing high deposition rates for the region in which the studies were conducted, there were significant declines in intrinsic rate of population increase (r) for this *Sarracenia* species. Those declines in intrinsic rate of population increase were found to predict an approximately 38% risk of population extinction for *S. purpurea* populations experiencing a 1% per year increase in nitrogen deposition, with populations expected to take from 200 to 300 years to disappear at those rates of increased nitrogen deposition.

This example illustrates numerous ways in which a stage-based approach to population biology can provide important insight into population dynamics that may be overlooked with the approaches discussed in the previous sections of this chapter. Similarly, Smith, Caswell, and Mettler-Cherry (2005) used stage-base models to examine factors associated with population declines of a federally threatened plant (*Boltonia decurrens*, Asteraceae) in the floodplain of the Illinois River (USA). Their modeling efforts indicated that changes in the management of water levels on the impounded Illinois River (increased frequency of summer flooding) had led to shifts in the life history of this species, from an annual life history to biennial and perennial life histories. These shifts in the timing of sexual reproduction of the plants had, in turn, resulted in reduced sexual reproductive rates and population decline through the Illinois River floodplain. Others have used stage-based models to determine appropriate timing of control efforts for invasive aquatic plants (Erwin et al. 2013), to assess the effect of fire regimes on overstory trees in savannas of Kakadu National Park (Werner and Peacock 2019), and to study the impacts of herbivory on populations of sexual and vegetative buds of arctic willows (*Salix arctica*; Tolvanen, Schroderus, and Henry 2002).

9.3 METAPOPULATIONS: POPULATIONS OF POPULATIONS

In the introductory section of this chapter, I said that a population consists of all individuals of a given species that co-occur within some space. I then suggested as an example that all the individuals of *Zostera marina* occurring in a particular estuary might be considered a population, but that boundaries of populations are not always so easily defined. Consider this example of an estuarine population of *Z. marina*. What biological mechanism separates individuals in that estuary from those outside the estuary along the coastline, or from those in a second nearby estuary, and so on? What if, instead of a submersed marine species that disperses pollen and fruit via the water, we were interested in an emergent species that uses the wind for pollen and seed dispersal? Would we expect the true geographical extents of populations in these two contrasting species to be similar? Would we expect to be able to determine the true geographical extents of these populations at all?

What if, instead of a species that lives in habitats with poorly defined physical boundaries, as between estuaries and coastal marine ecosystems, we were to consider a species that lives in habitats with clearer physical boundaries? Take, for example, the alpine lakes and ponds in Figure 9.7. There are many aquatic plant species that live in lake ecosystems such as these (e.g., species of *Callitriche*, *Eleocharis*, *Potamogeton*, and *Ranunculus*), many of which are dispersed by water birds (Figuerola et al. 2005; Santamaría 2002; van Leeuwen 2018). Birds could easily disperse fruit and seeds of these species from lake to lake, but would all species be able to survive and reproduce in all lakes in all years? The spatial distribution of suitable and unsuitable lakes for each species would not only influence the likelihood of establishment within lakes but also the potential for plants in one lake to interact with plants in other lakes via seed or pollen transfer. With an example such as this, the spatial boundaries of populations are quite unclear, despite the individual lakes themselves being very clearly defined.

One area of ecology that addresses issues such as these is the study of metapopulations. A **metapopulation**, in simplest terms, is a population of populations. More specifically, a metapopulation is a network of populations or habitat patches that are linked via migration of individuals or genes among the patches (Hanski 1998). Levins (1969) introduced the concept of metapopulations in a paper aimed at expanding the strategy of pest control from a focus on managing local populations to an approach that accounted for recolonization of local habitat patches from nearby occupied patches within a metapopulation network. He built on basic population biology to consider colonization and extirpation of local subpopulations as analogous to births and deaths of individuals within a population over time (Figure 9.8).

In the construct presented by Levins (1969), successful colonization of vacant sites by migrating individuals occurs at a rate proportional to availability of vacant sites (Figure 9.8). As modeled, the rate at which sites are colonized will increase from low to intermediate levels of site occupancy, after which it will decrease as fewer and fewer vacant sites are available (Figure 9.9). This is similar to the pattern we saw in density-dependent population

growth, where the rate of change in population density (i.e., the slope of the population growth curve) increases from low to intermediate population density and then declines as density-dependent ecological factors become more important (Figure 9.4). Extirpation of local populations, termed local extinction by Levins (1969), on the other hand, is density independent, being driven by a different suite of ecological factors. Extinction is thus modeled as a linear rather than curvilinear process (Figure 9.9). Taken together, these patterns suggested that changes in local extinction (i.e., eradication of local subpopulations) would be expected to exert the greatest per-unit effect on equilibrium levels of site occupancy at intermediate levels of occupancy, where rates of successful migration among sites were highest (Figure 9.9). From a pest management perspective, this suggests that pests ought to be more easily managed at intermediate to low levels of site occupancy across a metapopulation. Similarly, if the focal species were rare or threatened, we would expect that species to encounter greater difficulty persisting on the landscape if relatively few suitable habitat patches are occupied.

Hanski (1985) noted that Levins' approach requires an assumption that local population density is uncorrelated with the proportion of suitable sites that are occupied. This is often not the case; high levels of site occupancy across a region typically are accompanied by high population density at sites within a metapopulation network (Hanski 1985). When this is true, probability of local extinction from occupied sites will vary with the proportion of sites occupied because of the tendency for common species to also be abundant at sites they occupy, and vice versa. Hanski (1985) thus argued that local population density must also be considered in metapopulation studies, along with the spatial distribution of suitable sites and the proportion of sites occupied at a given point in time. Variation in local population density will be influenced by this natural correlation between commonness and abundance, as well as by ecological factors influencing local population densities within a site, and sites may fluctuate over time from supporting large populations to supporting small populations to lying vacant (Figure 9.10).

It was further emphasized that human activities that lead to habitat degradation and landscape fragmentation are expected to lead to declines in the number of sites suitable for colonization, and this process will disrupt metapopulation dynamics across a landscape (Hanski 1985). For wetland species, activities that lead to drainage and/or filling in of wetland habitats will impact local populations in two ways (Figure 9.11). First, degradation and increased isolation of suitable sites will increase the difficulty with which a species can successfully colonize sites within a metapopulation network; that is, isolation of suitable sites decreases their biological connectedness to other parts of a metapopulation. Second, alteration of the ecological conditions within and around suitable sites will facilitate the colonization of a different assemblage of plant species, frequently resulting in establishment of invasive species that may have

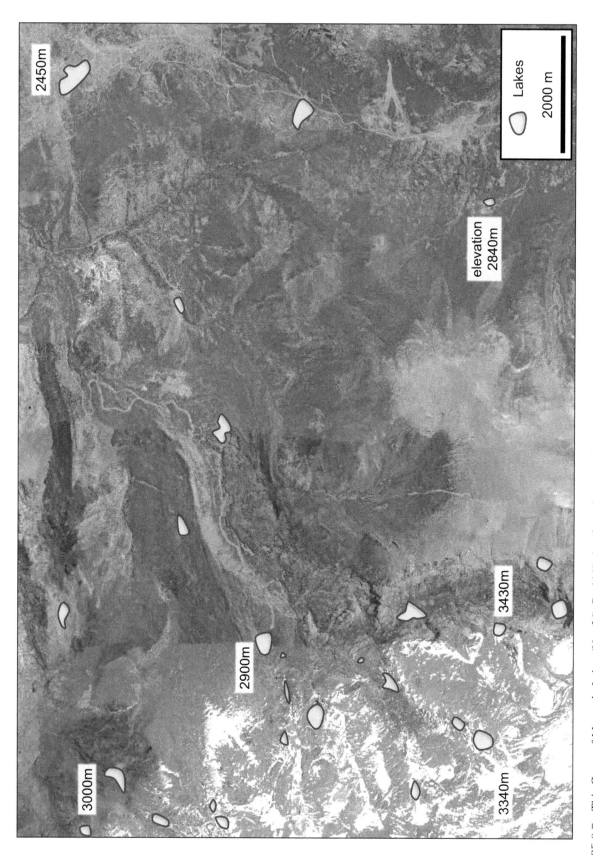

FIGURE 9.7 This Group of Mountain Lakes (N = 26) Could Harbor from One to 26 Populations of Plants. Factors such as the Spatial Density of Lakes, Stream Connections Among Lakes, Local Environmental Conditions, Pollen and Seed Dispersers, and Obstacles to Dispersal all Influence the Degree to Which Plants in any One of the Lakes Interact with Plants in the Others. Subsets of these Lakes in Which Plants Exchange Genetic Material Through Pollen or Seed Dispersal would Form Metapopulations of those Species. This Area Lies Immediately Southwest of Estes Park, Colorado (USA). Aerial Imagery from Esri (2022) "World Imagery" www.arcgis.com/home/item.html?id=10df2279f9684e4a9f6a7f08febac2a9.

Site colonization:	$mN\left(1-\dfrac{N}{T}\right)$	Modeled after Equation 9.7: $rN\left(1-\dfrac{N}{K}\right)$
		m = migration rate; N = number of local populations or sites occupied; T = total suitable sites
Local extinctions:	EN	Rate at which local populations disappear
Metapopulation change:	$\dfrac{dN}{dt}=mN\left(1-\dfrac{N}{T}\right)-EN$	Analog of Equation 9.2, where site colonization represents "births" and local extinctions represent "deaths"
Equilibrium:	$mN\left(1-\dfrac{N}{T}\right)=EN$	Metapopulation equilibrium, the point when $\dfrac{dN}{dt}=0$, is achieved when migration rate equals extinction rate.

FIGURE 9.8 Equations Used in Describing Metapopulation Dynamics by Levins (1969). Site Colonization is Represented as a Density-Dependent Process, Similar to that Included in the Logistic Growth Curves Obtained from Equation 9.7. Only Empty Sites can be Colonized, and the Probability of Encountering an Unoccupied Site at Random from a Randomly Selected Occupied Site Will Reach its Maximum at Intermediate Levels of Occupancy (Figure 9.9). Local Extinction, or the Likelihood of a Site Becoming Unoccupied, is Modeled here as a Density-Independent Process, Hypothesized to be Independent of the Number or Proportion of Suitable Locations that are Occupied. Finally, the Change in Number of Occupied Sites in the Metapopulation Over Time is Simply the Number that are Colonized Minus the Number that become Unoccupied during an Interval of Time. Equilibrium is Attained When Successful Colonization Equals Local Extinction.

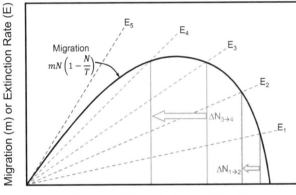

FIGURE 9.9 Similar Degrees of Change in Extinction Likelihood have much Greater Effects on Metapopulation Dynamics at Intermediate Levels of Site Occupancy than at High Levels of Occupancy. Because Equilibrium Levels of Site Occupancy are Achieved When Migration Equals Extinction, the Points in this Figure Where Extinction Vectors Intersect the Migration Curve Represent Metapopulation Equilibrium Points. If a Large Proportion of Suitable Sites are Occupied, the Likelihood of Reoccupation of a Vacated Site is Relatively High Because there are many Sites from Which it may be Colonized. However, as Extinction Rate Increases, E.g., from E_1 to E_2 or E_3 to E_4, Each Unit Increase will have a Greater Proportional Impact on Equilibrium Levels of Site Occupancy Until Extinction Exceeds Migration and Capacity for Site Recolonization within the Metapopulation (E_5). Redrawn from Levins (1969).

further negative impacts on the naturally occurring species (Hanski 1985; Lázaro-lobo and Ervin 2021).

Dynamics of species colonization among suitable sites varying in size and spatial isolation also were examined in the context of the theory of insular zoogeography, also known as the *theory of island biogeography*, put forth by

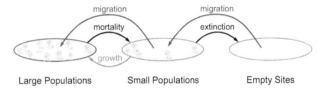

FIGURE 9.10 Conceptual Model of Within-Site Dynamics at One Node (Location) Within a Metapopulation. Instead of Simply Considering Whether a Site is Occupied or Vacant, This Model Incorporates Some Information About the Status of the Population at That Location. Here, for Example, the Local Population May Be Large (High Density or Large Spatial Extent) or Small (Low Density or Small Spatial Extent). Local Ecological Factors, in Combination With Dispersal Among Sites, Will Determine Whether a Local Population Grows, Declines, or Goes Extinct. Migration to Empty Sites Will Result Initially in Small Populations at those Sites. Small Populations Can Become Large Populations Through Additional Immigration or Growth of the Resident Population. Redrawn from Hanski (1985).

MacArthur and Wilson (1963, 1967). This theory posited that (1) the likelihood of a site or habitat patch being successfully colonized by a species is higher when the distance that must be crossed to reach the site is shorter, and (2) the likelihood of a species disappearing from a habitat patch is greater in smaller patches, because of the lower carrying capacity of small habitat patches (Figure 9.12). Although the theory was developed specifically for community ecology of island faunas, it can be generalized to any group of taxa dispersing among any type of variously isolated islands of habitat, including plants in aquatic and wetland ecosystems. Because extinction represents the loss of individuals within a colonizing species, we also could generally apply this theory to spatial dynamics of individual species just

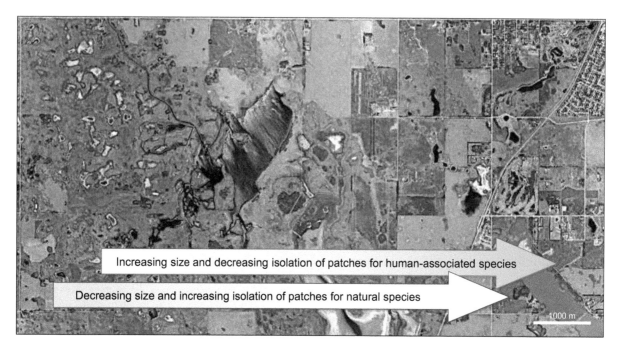

FIGURE 9.11 Illustration of a Landscape Exhibiting a Gradient in Human Modification, from a Somewhat Minimally Impacted Prairie Pothole Wetland Landscape on the Left to a Highly Altered Anthropogenic Landscape on the Right. Effects on Aquatic and Wetland Plant Species Across this Gradient Will Depend on the Species' Affinity for Human-Caused Ecosystem Disturbance. The Town of Yorkton, Saskatchewan, Canada, Sits in the Upper-Right Corner. Aerial Imagery from Esri (2022) "World Imagery" www.arcgis.com/home/item.html?id=10df2279f9684e4a9f6a7f08febac2a9.

as easily as we apply it to species assemblages. The mountain lakes in Figure 9.7 and pothole wetlands of Figure 9.11 are examples of habitat islands or potential archipelagos, depending on the scale at which we view them, that could be studied in this framework. Furthermore, the concept of habitat patch carrying capacity could be considered on the basis of resource availability or on other factors that influence site suitability for colonizing species. That is, one could replace island size with habitat quality in considering the potential size of populations that could be supported at a site. In this way, we might consider that poor-quality habitat patches would behave similarly to small patches with respect to species and population dynamics. Both of these aspects of habitat carrying capacity came into play in a recent study by Dahlgren and Ehrlén (2005), where they found that the larger of the 51 lakes they studied tended to be more nutrient-rich than the smaller lakes. In correspondence with expectations from island biogeography and metapopulation dynamics discussed earlier, they found that larger lakes tended to harbor more species of aquatic plants and that proximity of lakes influenced species occurrences. The presence of at least 50% of the plant species they encountered was positively correlated with geographic proximity to nearby lakes as well as the surface area of upstream lakes within a given lake's watershed.

Taken together, these ideas of metapopulation dynamics and habitat island biogeography can provide a framework for better understanding distribution and persistence

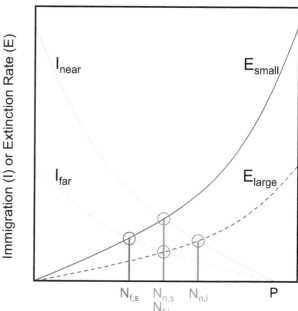

FIGURE 9.12 Interrelationship Among Species Occupancy, the Size of a Habitat Patch, and Degree of Patch Isolation According to Island Biogeography Theory. Habitat Patches (Sites) that are Larger and Nearer a Source of Potential Colonists are Expected to have Higher Species Richness ($N_{n,l}$) at Equilibrium than Smaller Sites that are Farther from a Colonization Source ($N_{f,S}$). I_{near} = Immigration Rate for a Site Near a Source of Immigrants; I_{far} = Immigration Rate for a Site Far from a Source of Immigrants; E_{small} = Extinction Rate for a Small Habitat Patch; E_{large} = Extinction Rate for a Large Habitat Patch; P = Total Number Species in the Available Species Pool. After Macarthur and Wilson (1963) and Simberloff (1974).

of aquatic plant populations across time and space (Hanski 1998). Across a metapopulation, all these factors will determine the degree to which neighboring populations (and neighboring metapopulations) interact, and thus will influence the potential for continued genetic exchange or genetic isolation among populations. Factors that lead to isolation and eventual genetic divergence among subpopulations within a metapopulation or between neighboring populations can ultimately lead to the formation of distinct varieties, subspecies, or even species, as we will explore further in the following section.

9.4   METAPOPULATIONS, LOCAL ADAPTATION, AND SPECIATION

In the 1930s and 1940s, a team of plant ecologists carried out a series of observational and experimental studies of plants in the species *Achillea lanulosa*, in the Asteraceae (Figure 9.13). These studies were carried out across an approximately 300 km transect from California's Pacific coast across the Sierra Nevada mountains. This study was aimed at clarifying relationships "of the individual to the local population, of the local population to the climatic race, of the climatic race to the species and the species complex" and to generally improve understanding of evolutionary processes giving rise to these units of biological organization (Clausen, Keck, and Hiesey 1948). In this context, the phrase climatic race was another term for what had been called **ecotypes** in previous research. Ecotypes are discrete populations or subpopulations of species that have genetically controlled, usually **adaptive**, differences from other populations (Gurevitch, Scheiner, and Fox 2006). Adaptive differences are phenotypic differences in characters that can be associated with aspects of the environment that influence growth, survival, or reproduction.

Prior to the discovery of the genetic code and methods for determining DNA sequence similarities and differences among individuals and populations, there were a variety of nonmolecular approaches that could be used to infer genetic control of individuals' phenotypes, and many of these approaches remain in use today. Two of these approaches used by plant ecologists are the **common garden experiment** and the **reciprocal transplant experiment**. In these experiments, plants are grown at one or more common locations from seeds that have been collected in natural populations of interest; the reciprocal transplant involves

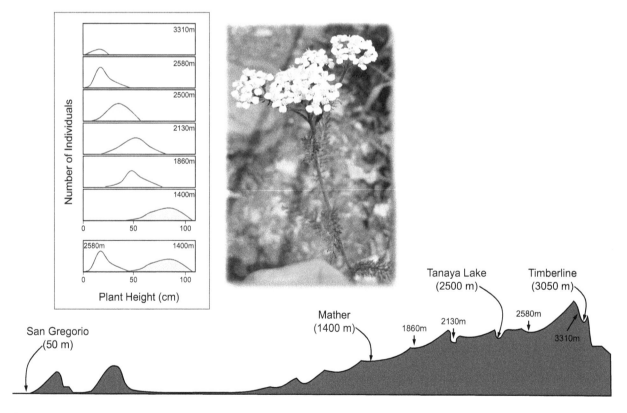

FIGURE 9.13   Distribution of Plant Height in a Common Garden Experiment Investigating Ecotypic Variation in *Achillea Lanulosa* (Asteraceae) from Populations Across the Sierra Nevada Mountain Range in California (USA). Elevations Given are those from Which Seeds were Collected; the Experiment was Conducted in a Common Garden at 30 M Elevation (Stanford University, Ca, USA). Comparative Plant Height Data in the Lowermost Graph Emphasizes the Difference in Heights of Plants Grown from Seed Collected at 1,400 M Elevation Versus those Taken from 2,580 M. The Entire Transect, Illustrated in Profile here, Encompassed Approximately 300 Km. Plant Height and Elevation Data Taken from Clausen, Keck, and Hiesey (1948). Photo: *Achillea Millefolium*, the Parent Taxon for What is now Treated as *Achillea Millefolium* Variety *Lanulosa*.

plants from all locations being grown in common gardens at all locations to further explore environmental effects on variation. Each location is referred to as a common garden, and the act of growing the plants together at that location removes the effect of environmental variation on the realized phenotype of the plants. Thus, the variation observed among individual plants, or populations of plants, can be assumed to have resulted from genetic factors controlling the phenotype.

For example, Clausen, Keck, and Hiesey (1948) collected seeds of *Achillea lanulosa* from more than a dozen locations extending from the Pacific coast up through the Sierra Nevada and into the Great Basin, at elevations from sea level to more than 3,000 m above sea level (Figure 9.13), and they grew plants from those seeds in a common garden near Stanford University (CA, USA). Resulting distributions of plant height from six of those populations are illustrated in Figure 9.13. Those data reveal two important patterns. First, there is considerable variation in plant height within each population of seed-grown plants, but the majority of individuals cluster around some central plant size, characteristic of each elevation. Second, despite having been grown in a common garden, plants exhibited decreasing size corresponding with their parent populations along this section of the elevation gradient. Furthermore, the distribution of heights of plants grown from seeds collected at 2,580 m and 3,310 m showed virtually no overlap with heights of the plants whose seeds were collected at 1,400 m elevation (Figure 9.13).

An additional component of the work of Clausen, Keck, and Hiesey (1948) was a reciprocal transplant experiment where plants were taken from three locations along the climate gradient and grown together in each of three locations representing climates of the three parent populations (Figure 9.14; Table 9.2). In this experiment, 30 *Achillea lanulosa* plants from different populations were separated into three vegetative clones each, and those clones planted in each of three locations: Stanford, Mather, and Timberline common gardens. Tracking the growth of individual clones across three parts of the climate gradient further strengthened insight on both the genetic and the environmental controls present in the experiment. Results of the reciprocal transplant study indicated that, generally speaking, plants performed best in climates most similar to those in which their parents had lived.

Other ecologists have since conducted additional studies on these *Achillea lanulosa* populations to evaluate potential mechanisms by which the plants are adapted to their local climates. Gurevitch (1992b) found differences in leaf morphology between the Mather and Timberline populations that appeared to be genetically regulated and related to climatic differences between the two locations (Table 9.2). Leaves from the warmer, wetter, lower elevation Mather location were larger, with a more open architecture of their leaf dissections, whereas those from Timberline exhibited a much more compact branching pattern and were smaller in outline (Gurevitch 1992b). These differences could result in

greater control over water loss for plants in the Timberline population in comparison to those at Mather (Gurevitch 1992a). In a separate study, Gurevitch (1992a) found that plants from the higher altitude Timberline population had significantly greater photosynthetic carbon capture efficiency than did plants from the Mather location. Gurevitch speculated that the shorter growing seasons or lower atmospheric CO_2 concentrations at the higher elevation site may have selected for greater photosynthetic efficiency in that population. More recently, Ramsey, Robertson, and Husband (2008) conducted a large-scale study of population genetics in the *Achillea millefolium* complex (to which *Achillea lanulosa* in the aforementioned studies belongs) across the Pacific coast of the United States and Canada. They found that some genetic **haplotypes** (groups of individuals exhibiting the same DNA sequences) were very widespread, occurring from Alaska to Arizona, but others were highly restricted to habitats exhibiting narrow ecological conditions. They suggested that the genetic patterns they encountered across populations of the species contained within this taxonomic complex represented a relatively rapid ecological (and population genetic) differentiation following migration of this group into North America sometime within the past few hundred thousand years. Thus, the patterns that were observed in the classical studies of Clausen, Keck, and Hiesey (1948) appear to be related to ongoing diversification of members of this group of terrestrial plants in response to the broad range of ecological conditions they have encountered on their migration from Eurasia into and across North America.

One of the goals stated by Clausen, Keck, and Hiesey (1948) was to help improve understanding of evolutionary processes giving rise to populations, ecotypes, and species. One very important concept illustrated by their work is that of adaptation to local environmental conditions, something we have been examining throughout the preceding eight chapters of this book. As we have seen, local adaptation is central to the ability of individuals to colonize and survive in new habitats, for the persistence of populations over time, and for differentiation among populations. To illustrate this further, we will consider the hypothetical plant populations in Figures 9.15 and 9.16. We can think about a metapopulation of a plant species (the orange-flowered plant in Figures 9.15 and 9.16) that at some time in the past (T1) was distributed across ten wetlands. Over time, genetic mutations arose that gave some individuals in some of the subpopulations different flower colors and leaf morphologies. By time T9, we see that habitat conditions in two of the ten wetlands have become unsuitable for any of the novel genotypes, and a total of four new genotypes have arisen. Perhaps the red-flowered plants are pollinated differently than the white- or yellow-flowered individuals, each of which may make use of different pollinators themselves (see Table 8.3). These different pollinators would potentially lead to genetic isolation among individuals with differing phenotypes, while differences in leaf morphology could allow individuals to

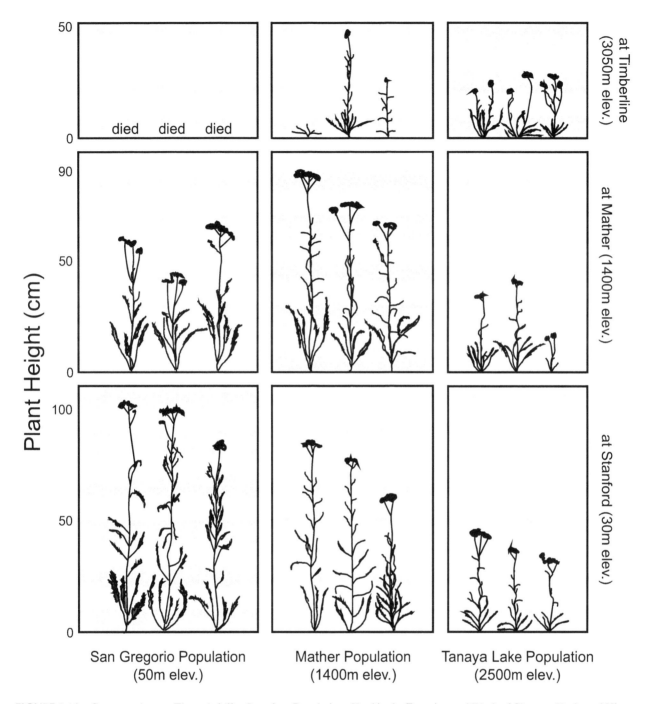

FIGURE 9.14 Contrasts Among Three *Achillea Lanulosa* Populations Used in the Experimental Work of Clausen, Keck, and Hiesey (1948). Clones from Multiple Seedlings Representing Each of Three Populations were Grown at Each of Three Common Garden Locations. Each Column of Graphs Represents a Single Population, and Each Row Represents Plants Grown at One of the Three Common Gardens. Within Each Column of Graphs, Clones are Stacked Vertically Across Rows; E.g., the Leftmost Plants Illustrated for the San Gregorio Population all Came from the Same Genet. Green Shaded Panels Along the Upward Left to Right Diagonal Indicate Plants Grown at the Elevation Closest to that for their Parental Population. Locations of Source Populations and Common Gardens, Except Stanford, are Indicated in Figure 9.13. Redrawn from Clausen, Keck, and Hiesey (1948).

take advantage of different abiotic habitat conditions. For instance, the white-flowered plants with heterophyllous leaf morphologies (leftmost, white-flowered individuals in Figures 9.15 and 9.16) may be able to survive and reproduce in wetlands with widely fluctuating water levels while the other three novel genotypes may perform better in deep water with little fluctuation. All these differences in physical morphology, pollinator syndromes, and habitat affinities lead to varying degrees (and mechanisms) of isolation among populations of the new genotypes, potentially placing them on four independent evolutionary trajectories (Figure 9.15).

TABLE 9.2

Summary characteristics of the three experimental common garden locations used by Clausen, Keck, and Hiesey (1948).

	Stanford	Mather	Timberline
Elevation (m)	30	1,400	3,050
Community type	Oak savanna	Coniferous forest	Alpine
Average growing season (days)	283	145	67
Winter snowfall	None	Moderate	Heavy
Active period for *Achillea lanulosa*	December to July	May to July	August
Average January temperature (C)	8	2	10
Average January temperature (C)	19	21	013
Average annual precipitation (cm)	39	103	78

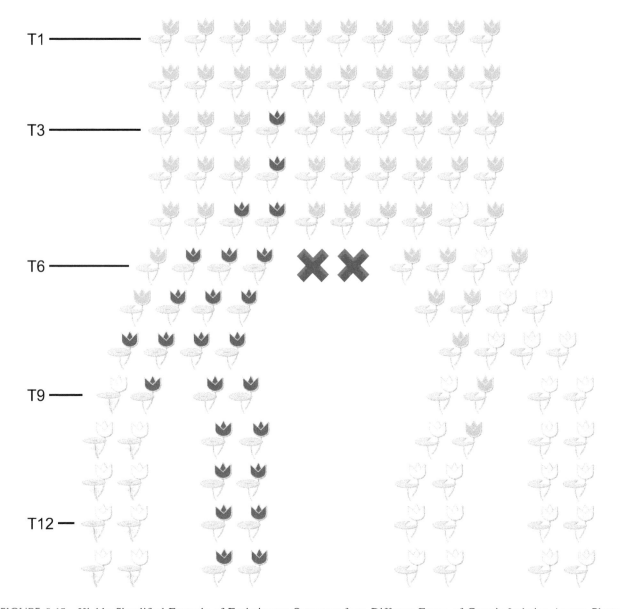

FIGURE 9.15 Highly Simplified Example of Evolutionary Outcomes from Different Forms of Genetic Isolation Among Plants. Beginning with a Genetically Intermixed Population or Metapopulation of Plants at Time T1, a Novel Phenotype (Red Flowers and Peltate Leaves) Appears at T3 that Eventually becomes the Dominant Phenotype across Part of the Original Distribution. At Time T6 a Portion of the Original Spatial Distribution becomes Uninhabitable (Red Crosses), Isolating Subpopulations on the Left from those on the Right and Setting the Stage for the Four New Phenotypes to Replace the Original. See Text and Figure 9.16 for Further Discussion. Modeled after Judd et al. (2008).

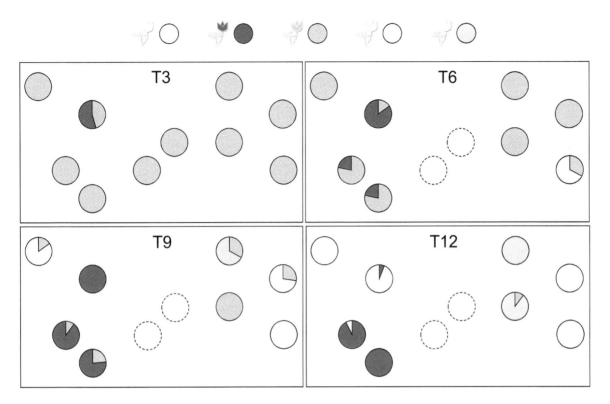

FIGURE 9.16 Hypothetical Metapopulation of Aquatic Plants Depicted in Figure 9.15. Here, Shading within Each Circle Represents the Relative Dominance of Each Genetic Lineage (Species, Subspecies, Ecotype, Etc.) within Each of the Ten Original Wetlands. For Simplification, Only Time Points 3, 6, 9, and 12 from Figure 9.15 are Represented. As Illustrated here, we can see that Changes in Genetic Structure of a Metapopulation Over Time are much more Complicated than Suggested in Figure 9.15. Note also the Contribution that Loss of the Two Central Habitats Made to Isolation of the Two Novel Phenotypes/Genotypes on the Left from those on the Right.

9.4.1 MECHANISMS OF DIVERGENCE

Isolation and local adaptation as discussed earlier are two elements in the process of divergence and speciation. At this scale of population biology, we tend to consider five mechanisms that influence the ability of populations to maintain their cohesion as a species or to diverge genetically from one another. These are **mutation**, **genetic drift**, migration or **gene flow**, and **selection**, with the latter of these taking the form of **natural selection** or **sexual selection** (Lomolino, Riddle, and Brown 2006). Mutation is simply a change in one or more nucleotides within an individual's DNA. When a mutation occurs in a region of DNA that codes for a protein, an important regulatory element such as an RNA, or a regulator binding region of the DNA, the mutation can have significant impacts on the fitness of individuals carrying that mutation. If the mutation is beneficial under environmental conditions experienced by an individual, that mutation may increase in prevalence in the population via natural selection. However, if the mutation is detrimental under existing environmental conditions, natural selection will tend to reduce its prevalence, such as via reduced fitness of offspring or death of individuals carrying the mutation. In the hypothetical metapopulation in Figure 9.16, for example, we saw that individuals that carried genes coding for the four novel phenotypes increased in prevalence over time, both within and among subpopulations. Migration of pollen or seeds among wetlands allowed the new versions of those genes to move among the wetlands, and selection allowed them to increase in prevalence. In contrast, the loss of suitable habitat at the two wetlands in the center of this metapopulation prevented gene flow between the subpopulations on the left and those on the right, allowing the two groups to diverge from one another.

Genetic drift is perhaps the least intuitive of the mechanisms influencing genetic cohesion and divergence. Genetic drift is a process in which the frequencies of alleles (different versions of a gene) change randomly over time due to random chance (Figure 9.17). Specifically, genetic drift results from alleles being lost from a population when those alleles are not passed along to offspring. As a result, genetic drift is most effective at influencing the genetic composition of small, randomly mating populations. Larger populations will tend to have more copies of even the rarest alleles than will small populations because there are more individuals to carry those alleles. In small populations, however, low abundance of alleles translates to a lower likelihood that those alleles will be retained across successive generations through sexual reproduction in a randomly mating population.

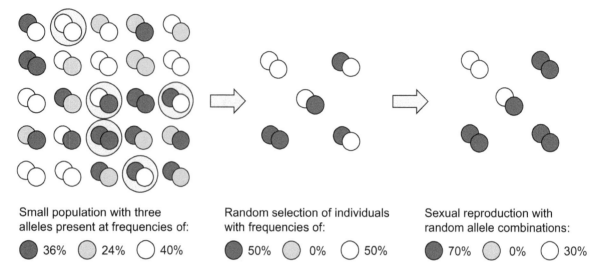

FIGURE 9.17 Genetic Drift can Substantially Impact Allele Frequencies in Small Populations. At Left, we have a Population of 25 Diploid Individuals, Each of Which Carries Two of the Possible Three Alleles for a Trait, Randomly Assigned Across Individuals. After Randomly Selecting 10% of the Population to Reproduce and then Randomly Selecting Combinations of Alleles from those Parents for Five Offspring, the Orange Allele was Eliminated, and the Red Allele Almost Doubled in Frequency.

Finally, natural selection and sexual selection can be strong forces influencing the persistence of alleles in a population. Both are represented in the hypothetical metapopulation of Figure 9.16. Novel flower color can influence pollination success, and when pollinators selectively visit flowers of a particular color, they will tend to maintain alleles coding for that color within the population. Strong preference for one flower phenotype over others will result in that phenotype increasing in prevalence over time, at the cost of other phenotypes present. Likewise, when individuals express phenotypes that make them better suited to local environmental conditions, such as heterophylly in a wetland with fluctuating water levels, those individuals will be more likely to pass along their genetic information to offspring who, in turn, will be well-suited to those same environmental conditions.

Because of the roles of population size, local adaptation, and gene flow in determining exchange of genetic information among populations, metapopulation theory is critical in understanding species and **speciation**, or the formation of new species. Levin (1995) argued that metapopulations are the perfect arenas for speciation based on three characteristics of metapopulations. First, metapopulations tend to be longer-lived than local populations. This longevity results from the ability of local populations within the metapopulation to be rescued from extirpation via migration from nearby occupied habitat patches. This allows metapopulations to survive long enough for evolutionary change to take place while also facilitating the spread of novel genotypes throughout the metapopulation. Second, because metapopulations comprise subpopulations across a broad geography, they typically represent a much greater heterogeneity in local environmental conditions than can be found in a single local population. This increased environmental

heterogeneity provides a broader range of potentially selective features that may drive local adaptation, increasing genotypic and phenotypic diversity across the metapopulation. Last, the structure of metapopulations, existing as a network of smaller populations connected by varying degrees of gene flow, can facilitate genetic drift in local populations. Localized genetic drift, especially towards locally adaptive genotypes, can result in the rapid **fixation** of those novel genotypes in the local populations (fixation is the exclusivity of those alleles or genotypes). Although this process results in lower genetic diversity within individual populations, it simultaneously leads to genetic diversification among subpopulations within a previously genetically intermixed metapopulation.

9.4.2 EXAMPLES FROM AQUATIC AND WETLAND ECOSYSTEMS

I introduced the discussion on metapopulations earlier with an example from Chapter 8 of the submersed *Zostera marina* exhibiting different patterns of dichogamy and seed germination across the range of the species. I then contrasted population dynamics of this marine species, whose populations may have indistinct boundaries, with a hypothetical wetland species that may experience more definite population boundaries. Nies and Reusch (2005) carried out a similar contrast, focusing on a single aquatic species, *Potamogeton pectinatus*, in northern Germany. They compared the population genetics of 12 freshwater lake populations of this plant with 14 coastal populations in the Baltic Sea. They found that gene flow among the marine populations was approximately 2.5 times greater than among the freshwater populations and concluded that the marine populations likely were highly connected via the essentially unrestricted potential for movement of water, carrying pollen and seed, among the marine habitats. In contrast, the freshwater lakes were completely isolated from one another, from the perspective of water movement. Nies and Reusch (2005) further found that there was essentially no evidence of gene flow between the lake metapopulation and that of the marine habitats, despite relatively close proximity (10 to 20 km) between some of the lakes and coastal populations.

At a slightly smaller spatial scale, Wei, Meng, and Jiang (2013) measured patterns of genetic connectivity in a riparian tree species (*Euptelea pleiospermum*; Eupteleaceae) within and among watersheds on headwater tributaries of the Yangtze River in the Shennongjia National Nature Reserve in China (Figure 9.18). These researchers sampled multiple individual trees at each of six to 12 populations along each of four neighboring rivers, where the uppermost populations on each river (at the highest elevation on each) were situated approximately 6 to 8 km apart. Despite the proximity of these sites, the intervening mountains and ridges served as effective barriers to gene flow, isolating trees along the rivers into four genetically distinct metapopulations. This species is wind pollinated and produces wind- and water-dispersed fruits (samaras) that are often carried downstream by air currents and streamflow. This serves to maintain upstream-to-downstream connectivity among populations along each individual river. Wei, Meng, and Jiang (2013) also found evidence for movement of pollen both upstream and downstream within each river valley;

nevertheless, the distance and physical barriers between rivers prevented similar genetic exchange between adjacent watersheds.

In Japan, Kato et al. (2011) investigated what appeared to be incipient speciation in the alga *Chara braunii*, where distinct genotypes were found among 73 plants sampled from shallow (< 15 cm depth) and from deep (> 1 m depth) aquatic habitats at 45 locations in Japan. Genetic sequences from chloroplast DNA showed almost perfect segregation between habitat types, while genes from nuclear chromosomes indicated some recent intermixing. Results suggested that *Chara braunii* from these distinct habitat types appeared to have recently diverged from each other genetically despite a wide variety of potential mechanisms for intermixing between them. Lakes and ponds frequently occur near rice paddies, often as sources of irrigation water, and migratory birds are well known to disperse *Chara* spores among aquatic habitats. Despite these abundant opportunities for gene flow, it appears that adaptation to some characteristic(s) associated with water depth has facilitated genetic divergence between the shallow and deep ecotypes of this *Chara* species.

The final example we will examine here represents a later timepoint in species divergence. In this study, Edwards and Sharitz (2000) looked at genetic structure within and among populations of two *Sagittaria* species (*Sagittaria isoetiformis* and *Sagittaria teres*) in coastal plain wetlands in the eastern United States (Figure 9.19). Previous investigators had debated the relationships of these two species

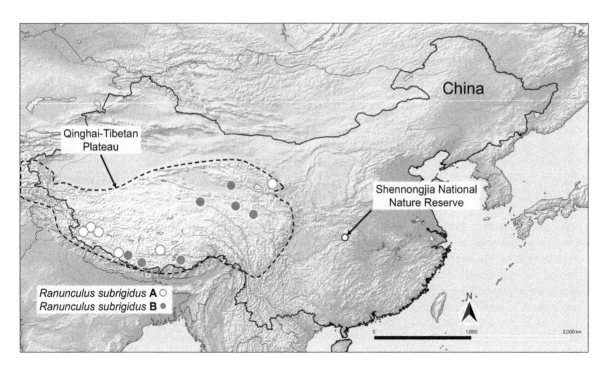

FIGURE 9.18 Study Locations for Work on *Euptelea Pleiospermum and Ranunculus Subrigidus*. Wei, Meng, and Jiang (2013) Studied 36 Populations of *Euptelea Pleiospermum* from Four River Valleys in the Shennongjia National Nature Reserve in Central China. Wu et al. (2019) Studied 13 Populations of *Ranunculus Subrigidus* from Across the Qinghai-Tibetan Plateau, in Western China. Details of Each Study are Given in the Text.

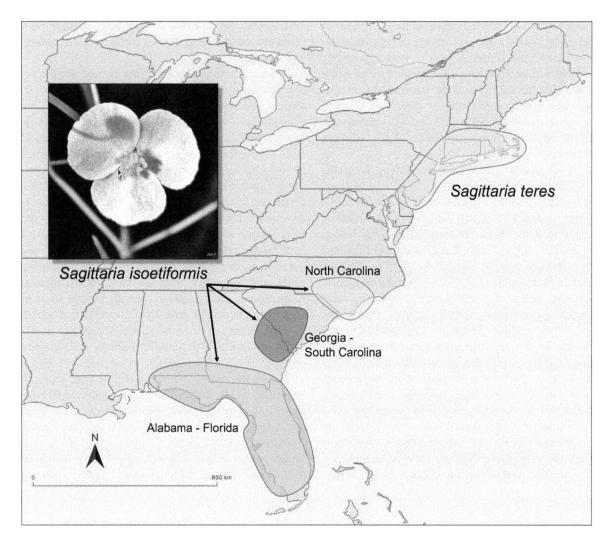

FIGURE 9.19   Population Genetic Analyses Revealed Three Distinct Populations of *Sagittaria Isoetiformis* and a Clear Separation of these from the Related *S. Teres* Among Coastal Plain Wetlands of the Eastern United States. Based on Data from Edwards and Sharitz (2000). Photo of *S. Isoetiformis* © Bob Peterson; CC BY-SA 2.0; https://creativecommons.org/licenses/by-sa/2.0/deed.en.

and another, *Sagittaria graminea*. Although it was reported that *S. isoetiformis* was reproductively compatible with some varieties of *S. graminea*, *S. teres* was not compatible with either *S. isoetiformis* or *S. graminea*. Edwards and Sharitz (2000) found strong genetic segregation between *S. isoetiformis* and *S. teres*, with *S. teres* exhibiting lower overall genetic diversity and occurring largely in wetlands of the US northeast that were created during the most recent retreat of glaciers from this area. Furthermore, they confirmed reproductive incompatibility between *S. teres* and *S. isoetiformis* in 30 attempts at crossing these species via manual pollination. Their conclusion was that *S. teres* was a more recently derived species within this group, potentially having arisen through expansion of refuge populations of *S. isoetiformis* during glacial retreat and subsequent spatial and reproductive isolation of those populations as sea levels continued to rise. Although Edwards and Sharitz (2000) did not include *S. graminea* in their analyses (thus precluding the ability to rule out this species as the one that gave rise

to *S. teres*), *S. isoetiformis* and *S. teres* share several physiological and ecological traits that *S. graminea* does not exhibit. For example, *Sagittaria isoetiformis* and *S. teres* both occur in acidic wetlands with sandy soils, whereas *S. graminea* typically lives in wetlands with more organic-dominated soils. *Sagittaria isoetiformis* and *S. teres* also both are capable of switching from C_3 to CAM photosynthesis when their leaves become submerged (see Chapter 6); *Sagittaria graminea* does not have this capability, according to Edwards and Sharitz (2000). *Sagittaria isoetiformis* and *S. teres* are both considered rare within their respective ranges, and work such as this potentially contributes to conservation efforts for these and similarly threatened species in the face of ongoing wetland loss around the globe.

9.5   EVOLUTIONARY ECOLOGY

We have seen that there are many systems employed for sexual reproduction among aquatic plants, in addition to their diversity of approaches to clonal propagation. All these methods employed for population increase in aquatic and wetland plants play a central role in responses to, and as drivers of, evolutionary change. Because of this, and because of the intimate connections between a species' evolution and its ecology, aquatic plants represent countless opportunities to better understand the role of ecology in shaping population and species dynamics (Barrett, Eckert, and Husband 1993; Eckert, Dorken, and Barrett 2016). Barrett, Eckert, and Husband (1993) pointed out that, in the early 1990s, population genetics had been poorly studied in aquatic plants, in contrast to terrestrial plants and other groups of organisms, and that this had contributed to a deficiency in understanding evolutionary processes in aquatic plants. Although there is still much to be learned in this area, as discussed by Eckert, Dorken, and Barrett (2016), there have been many studies published examining aquatic plant ecology using molecular genetics approaches in the years since Barrett, Eckert, and Husband (1993) noted the discrepancy between aquatic and terrestrial plants. In this final section of the chapter, we will look at two areas in which molecular genetics tools have yielded insight into species evolutionary ecology, within an applied context.

9.5.1 CONSERVATION GENETICS

The advent of molecular genetics has provided invaluable tools for studying responses of organisms to the environment, detecting previously cryptic species and species relationships, and revealing the true boundaries of populations, among other things. These tools allow us to investigate the genetic basis for tolerance to environmental phenomena such as drought and flooding, to understand processes that have led to current biological diversity, and to reveal mechanisms that facilitate or prevent gene flow among individual habitat patches. On one hand, the collective understanding of all these aspects of aquatic plant biology can contribute to basic scientific knowledge that may help to broaden our appreciation of botanical diversity. On the other hand, this information can provide useful insight into how populations and species respond to ongoing challenges such as climate change, habitat alteration, and species invasion (Barrett 2010).

Lobato-de Magalhães and colleagues took this latter approach in their studies of the floating-leafed *Nymphoides fallax* in wetlands of the Mexican highlands (Lobato-de Magalhães and Martínez 2020; Lobato-de Magalhães, Cabrera-Toledo, and Martínez 2019; Lobato-de Magalhães et al. 2020). Until 1969, *Nymphoides* in Mexico and Central America was thought to comprise a single species, *N. humboldtiana* (Ornduff 1969). *Nymphoides humboldtiana* is a white-flowered species, and despite this, collections of the yellow-flowered *N. fallax* (Figure 9.20), which occurs in restricted areas of high-altitude wetlands in Mexico and Guatemala at elevations from 1,500 m to 3,000 m above sea level, had been identified as *N. humboldtiana* (Ornduff 1969).

FIGURE 9.20 Locations of *Nymphoides Fallax* Populations Used to Examine Landscape Influences on Metapopulation Dynamics in this Aquatic Plant. The Number on Each Circle is the Number of Populations Sampled at Each Location; Ten Plants were Sampled in Each of the 18 Wetlands. In the Lower Photo, Graduate Student Luis Enrique Olguín Chávez is Collecting Data on Floral Morphology of *N. Fallax*. Photos Provided Courtesy of Dr. Tatiana Lobato De Magalhães, of the Universidad Autónoma De Querétaro, Mexico.

Nymphoides fallax is a distylous species, producing flower morphs having short styles and long anther filaments and other morphs with long styles and short anther filaments (Table 9.3; Lobato-de Magalhães and Martínez 2020). Strikingly, most floral traits that were measured differed in size between the short- and long-styled floral morphs, from sepal and petal length to the size of individual pollen grains (Table 9.3). Also noteworthy is that the lengths of the pistil and filaments are almost perfect complements to one another, helping to ensure consistent cross-pollination between the two morphs. Although distyly provides the ecological and evolutionary benefit of ensuring outcrossing and thus minimizing problems associated with inbreeding, distyly can be problematic in rare species or species occurring in threatened habitat types, if reductions in population size were to result in loss of one of the floral morphs (Barrett 2019). Fortunately, in their studies of *N. fallax*, Lobato-de Magalhães and collaborators found the style morphs to be almost equally represented (Lobato-de Magalhães and Martínez 2020), and they found relatively high genetic diversity within, although not among, individual wetlands

(Lobato-de Magalhães, Cabrera-Toledo, and Martínez 2019).

Lobato-de Magalhães et al. (2020) took a metapopulation approach to investigating gene flow as a measure of functional connectivity among 18 high-elevation wetlands in the central Mexican highlands (Figure 9.20). In their study, they were interested in whether wetland spatial proximity, wetland size, landscape features, or wetland connectivity played a role in genetic diversity of *Nymphoides fallax* in these high-elevation wetlands. In this context, wetland connectivity was represented as a one of three mathematical indices of the number of nearby wetlands to which a given wetland could be connected. Their results indicated that the 18 populations of *N. fallax* were highly connected by gene flow and that neither distance among wetlands nor wetland size alone was a strong predictor of genetic diversity. However, wetland connectivity and wetland area in combination provided the strongest ability to predict multiple measures of genetic diversity in this set of wetlands. They also found that forested land cover in the vicinity of these wetlands appeared to enhance gene flow among wetlands, in contrast to the expectation that forests might serve as barriers to genetic exchange. It was suggested that forests may serve as habitat for some of the more important insect pollinators of *N. fallax*; thus, forested areas could serve an important indirect role in maintaining gene flow and genetic diversity within this *N. fallax* metapopulation.

There are two important conclusions from this study, from the perspective of wetland ecosystem and species conservation. First, the importance of wetland connectivity identified by Lobato-de Magalhães et al. (2020) indicates the importance of maintaining wetlands across the landscape to serve as potential habitat islands for subpopulations of important plant species. Maintaining these habitat patches, even if they are unoccupied in some years, helps to ensure habitat availability as a metapopulation grows and shrinks over time in response to natural environmental variability. Second, the unexpected role that forest cover played in gene flow among these wetlands provides another important insight into the importance of maintaining a diversity of natural or minimally impacted land cover across the landscape. Furthermore, maintaining diverse natural land cover across the landscape benefits not only wetland species such as *Nymphoides fallax*, but also species that inhabit those other land cover types that are being conserved.

Evidence in support of this population-level vulnerability of wetland plants to habitat loss is provided in the work of (Wu et al. 2019). These researchers investigated *Ranunculus subrigidus* in 13 populations on the Qinghai-Tibetan Plateau (one of the regions we learned about in Chapter 3) to assess the relative importance of spatial isolation and environmental conditions on genetic diversity in wetlands of that region (Figure 9.18). They found these populations of *R. subrigidus* to have relatively low genetic diversity, despite the relatively low prevalence of clonal propagation in these populations, suggesting abundant self-fertilization and inbreeding. It was expected that there would be substantial genetic

TABLE 9.3

Floral measurements for *Nymphoides fallax* short-and long-styled flower morphs from a temporary Mexican highlands wetland in Guanajuato, Mexico. Asterisks indicate that the measurements differed between the two flower morphs. Pistil length is given in boldface font to highlight the distinction between the short-styled and long-styled flowers. Data from Lobato-de Magalhães and Martínez (2020).

	Short Styled Morph	Long Styled Morph
Flower part (mm)		
sepal length*	7.6 ± 1.1	6.7 ± 0.9
sepal width	1.7 ± 0.3	1.6 ± 0.3
petal length*	11.6 ± 1.4	10.6 ± 1.0
petal width	2.9 ± 0.4	2.6 ± 0.2
corolla tube length*	3.6 ± 0.8	2.4 ± 0.2
filament length*	7.7 ± 1.0	5.2 ± 0.5
anther length*	3.0 ± 0.3	2.2 ± 0.2
pistil length*	**5.3 ± 0.6**	**7.2 ± 0.6**
stigma length*	1.1 ± 0.1	1.4 ± 0.2
stigma width*	1.0 ± 0.1	1.4 ± 0.3
stigma-anther separation	2.4 ± 0.6	2.0 ± 0.5
Flower part (μm)		
pollen length*	29.8 ± 0.6	22.8 ± 0.8
stigma papilla length*	39.8 ± 5.4	83.6 ± 22.4
Count		
ovule number per ovary	21.0 ± 7.7	17.8 ± 2.6
seed number per capsule	18.9 ± 6.4	17.2 ± 4.7

diversification between the eastern and western subregions that were sampled (Figure 9.18); however, only 4% of the total observed genetic variation was attributed to divergence between these areas. The two groups that were identified (groups A and B) were intermixed spatially, and there was no evidence for genetic isolation among populations attributed to distance among wetlands (Wu et al. 2019). The authors indicated that isolation among wetlands resulting from habitat fragmentation on the Qinghai-Tibetan Plateau likely had led to shrinkage of local populations, making them vulnerable to small-population genetic issues, such as genetic drift and inbreeding.

As discussed in Chapter 3, marshes on the plateau have experienced approximately a 50% reduction in area since the 1970s, with climate change being blamed for much of the loss. Wu et al. (2019) indeed found that temperature, temperature seasonality, and annual precipitation contributed to patterns they saw in the genetic diversity of Qinghai-Tibetan Plateau *Ranunculus subrigidus*. Sensitivity of this species to climate change and the resulting reductions in population size appear to have severely impacted population structure and genetic diversity of *R. subrigidus*. Unfortunately, the forecast is not very favorable for this and similar species of the Qinghai-Tibetan Plateau, with temperatures increasing on the plateau at approximately twice the average global rate, along with increasing pressures from human land use (Zhao et al. 2015).

9.5.2 HYBRIDIZATION

Up to now, in this chapter, we have progressed gradually from discussing demographics of individual populations, to exploring spatial interactions among networks of populations in a metapopulation context, to learning about spatial and genetic isolating mechanisms that lead to the eventual divergence of new species. In Chapter 8, we learned about mechanisms that ensure accurate pollination within species as well as some of the means that plants employ to discourage inbreeding among individuals that may be too closely related. What happens, however, when isolating mechanisms such as pollinator specificity, spatial isolation, and genetic incompatibility fail to prevent pollination between individuals of different, but reproductively compatible, species?

Successful pollination between individuals of different species can result in the formation of **hybrids**, and it is thought that as many as 25% of plants species undergo hybridization in nature (Vallejo-Marín and Hiscock 2016). There are many possible outcomes of hybridization, including **hybrid inviability** (the failure of hybrids to grow and develop normally), hybrid sterility, and a process known as **hybrid breakdown**, where back-crosses between hybrids and parental genotypes or crosses among hybrids result in inviable or sterile offspring (Judd et al. 2016). Another possible outcome of hybridization is that the hybrid offspring may be both viable and reproductively fertile, that is, they may be perfectly normal. In this case, we often see that

hybrids exhibit traits that may be intermediate between those of the parental species, such as intermediate height, leaf size, or flower size or color. On the other hand, hybrids may exhibit exaggerated phenotypes that lend themselves to higher fitness than either parent, a phenomenon referred to as **hybrid vigor**.

There are various reasons for inviability in failed hybridizations, such as habitat incompatibility between parental species and hybrids, failure of hybrid embryos to mature, or molecular regulatory incompatibility issues between the two parental genomes (Judd et al. 2016). Another major problem that hybrids face is the formation of viable gametes during meiosis. Successful meiotic division during the production of gametes requires that chromosomes successfully pair prior to reduction of chromosome number in the first phase of meiosis. When the chromosome number resulting from hybrid pollination prevents successful pairing, this can prevent the production of viable gametes, which will result in failed sexual reproduction. A process that can occur in plants that solves the chromosome pairing problem is **polyploidization**.

Polyploidization results from the failure of paired chromosomes to separate during the first phase of meiosis, leading to the production of **unreduced gametes** (gametes that carry multiple sets of chromosomes), rather than the single set carried in normal haploid gametes (Figure 9.21). When two of these unreduced gametes unite, the resulting embryo will carry a full complement of chromosomes from each diploid parent; this ensures that each chromosome will have a partner with which it can pair during meiosis. Polyploidization is incredibly common among angiosperms, including aquatic taxa, and it is estimated that approximately 70% of angiosperm species have an evolutionary history that includes at least one instance of polyploidization (Lobato-de Magalhães et al. 2021; Meyers and Levin 2006).

One example of a very successful genus of wetland plants that has benefitted from hybridization and polyploidization is the saltmarsh genus *Spartina*. This genus comprises 15 to 17 species worldwide, with no extant diploid species known (Ainouche et al. 2009; Barkworth 1993). It is hypothesized that the extant *Spartina* species are derived from a tetraploid clade (i.e., a group containing species with four sets of chromosomes) that originated via hybridization between two ancient diploid ancestor species (Figure 9.22). Within the genus, we see a predominance of tetraploid species, with a subclade of three hexaploid species (having six sets of chromosomes), including the very widespread *S. alterniflora*. As with the primary tetraploid clade from which the three hexaploids derived, the hexaploid subclade is hypothesized to have originated from a hybridization event. That hybridization could have involved, for example, unreduced gametes from a tetraploid parent species (4×) and a diploid parent species (2×), as illustrated in the middle portion of Figure 9.21, unreduced gametes from one tetraploid parent (4×) and reduced gametes from another (2×), or unreduced gametes from two triploid hybrids (3×) that are not represented among known extant *Spartina* taxa.

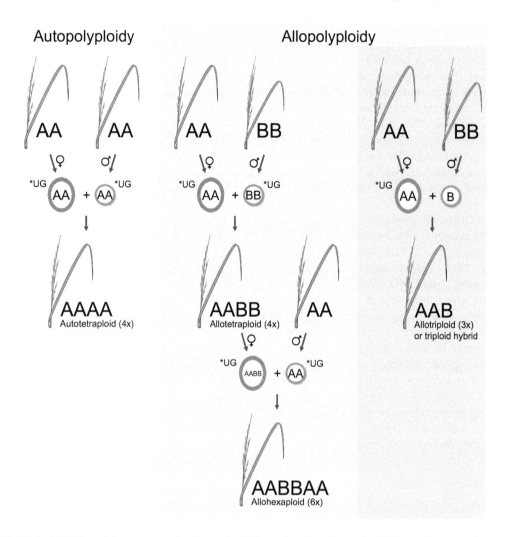

FIGURE 9.21 Polyploid Plant Lineages can be Formed within a Species (Autopolyploidy) or between Compatible Species (Allopolyploidy). The most Common Process by Which this Happens is Via the Fusion of Gametes in Which the Chromosome Numbers are not Reduced During Meiosis; "*Ug" in the Figure here Represents these Unreduced Gametes. The Examples here Show Some, but not all, of the Ways in Which Polyploids with Even Numbers or Odd Numbers of Chromosomes can be Formed. Based, in Part, on Strong and Ayres (2013).

Among the *Spartina* species in Figure 9.22, only *S. maritima* is native to Europe (Strong and Ayres 2013). The others are native to the Americas, with *S. foliosa* being native to California (USA), *S. arundinacea* native to South America, and both *S. argentinensis* (synonym of *S. spartinae*) and *S. alterniflora* being native along the Atlantic coasts of North and South America (Barkworth 1993). Of these five species, *S. alterniflora* is the most widespread, occurring from Canada to Argentina, although it is thought to be absent from Central America (Barkworth 1993). In addition to these species of *Spartina*, all of which are thought to be hybrids of now-extinct parental lineages, there are seven known hybrids that have been formed from various combinations of four of these parental species (Figure 9.23).

Spartina densiflora is thought to have formed in South America from the native *S. arundinacea* and *S. alterniflora* (Bortolus 2006; Fortune et al. 2008). *Spartina densiflora* has been very successful globally at establishing as an invasive

species and in forming additional hybrids with *S. foliosa* in California and *S. maritima* in Europe (Figure 9.23). In fact, *S. densiflora* and *S. alterniflora* are said to have formed hybrids with native *Spartina* species each time they have been introduced outside their native ranges (Strong and Ayres 2013). Of these two, *S. alterniflora* may be the more successful, also having formed hybrids with *S. foliosa* in California and two named hybrids with *S. maritima* in Europe and serving as one of the parental lineages for *S. densiflora* itself (Figure 9.23; Strong and Ayres 2013). As a result, each of the hybrids listed in Figure 9.23 is a descendant of some population of *S. alterniflora*.

The invasion of saltmarshes by *Spartina* taxa has come with ecological as well as genetic costs, including the extensive hybridization mentioned earlier (Ainouche et al. 2009; Ainouche and Gray 2016; Strong and Ayres 2013, 2016). For example, *Spartina alterniflora* invaded Willapa Bay estuary in Washington (USA), eventually covering 27,000

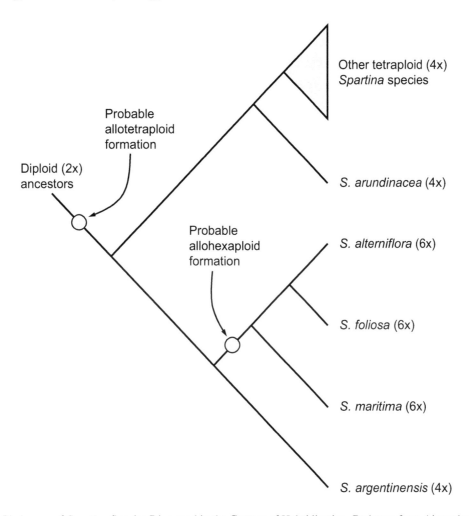

FIGURE 9.22   Phylogeny of *Spartina* Species Discussed in the Context of Hybridization. Redrawn from Ainouche et al. (2009) and Strong and Ayres (2013).

ha of habitat historically not occupied by *S. foliosa*, which occupies a narrower range within the tidal zone than does *S. alterniflora* (Strong and Ayres 2016). This invasion impacted migratory bird habitat, navigation, intertidal habitat, and a local oyster industry. After approximately 100 years of spread, an aggressive three- to four-year eradication program succeeded in reducing the invasion to a few scattered patches of invasive *Spartina*, at a cost of approximately US$30 million.

Interestingly, most of the impactful *Spartina* invasions have come not through the direct actions of introduced *Spartina* species, but through the spread of hybrids resulting from those initial introductions (Strong and Ayres 2013). For example, *S. densiflora* and the dodecaploid (12×) hybrid *S. anglica* were introduced into California in the 1970s. By the 1990s, hybrids of *S. densiflora* and the native *S. foliosa* had been detected, and by 2009, hybrids of these two taxa were known from 23 of the 32 saltmarshes where these two taxa were known to co-occur (Strong and Ayres 2013). Similarly, *S. alterniflora* was transported from the Atlantic coast of Maine and Virginia (USA) to San Francisco Bay, where it also hybridized with *S. foliosa*. The

resulting hybrids backcrossed with both parental species, and with one another, and began to spread widely through marshes around the bay naturally and by hitchhiking along with marsh restoration projects. These *S. foliosa* × *alterniflora* hybrids established in mudflat habitats where neither parent was able to establish and negatively impacted feeding habitat of migratory shorebirds and other coastal wildlife (Strong and Ayres 2016). Similar effects have been observed in Europe, Australia, and China, where hybrid *Spartina* invasions have impacted shorebird habitat, benthic food webs, and species richness of coastal ecosystems (Strong and Ayres 2013).

I began this section on hybridization by pointing out that hybridization requires the absence or removal of isolating mechanisms such as pollinator specificity, spatial isolation, and genetic incompatibility. The barrier that has been removed in most known modern hybridizations among *Spartina* taxa is that of spatial isolation. This is a result of the increased movement of people and goods from place to place around the globe during the past 500 years or so. More recently, global transport has been accompanied by unprecedented rates of change in land

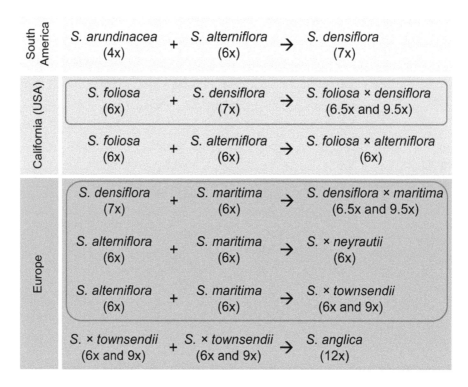

FIGURE 9.23 Noteworthy Hybrids Formed within the Genus *Spartina*. Those Taxa Bounded by a Red Border (One in the US and Three in Europe) are Considered Reproductively Sterile. Ploidy Level for Each is Given in Parentheses. Based on Information from Ainouche et al. (2009), Fortune et al. (2008), and Strong and Ayres (2013).

use, eutrophication, and other forms of environmental disturbance, which also serve to reduce some spatial and ecological barriers to gene flow among species (Vallejo-Marín and Hiscock 2016). In addition to increased potential for interaction between species, we also are seeing increased within-species gene flow among individuals from disparate regions of species' native ranges following introduction to new regions of the globe. This **intraspecific hybridization** also has the potential to result in increased vigor and invasiveness of progeny, by permitting novel combinations of genes that allow more rapid adaptation to resources and conditions experienced in the species' invasive range (e.g., Lucardi, Wallace, and Ervin 2020; Rius and Darling 2014; Thum and McNair 2018). Advancements in molecular genetics have facilitated discovery of patterns such as these, and we will see in coming years how commonly intraspecific hybridization has occurred and the degree to which it has impacted the ecology of aquatic and wetland ecosystems.

9.6   IN SUMMARY

Population biology aims to understand how factors such as reproduction, survival, and life history traits influence changes in the size of populations over time and space. Because a population is defined by the spatial co-occurrence of individuals of a species, population biology is an inherently spatial discipline. Nevertheless, the simplest

formulations of population dynamics focus on quantifying changes in the numbers of individuals within a population without consideration of spatial interactions. That is, the simplest equations (e.g., Equation 9.1) do not include spatial parameters and assume that the investigator has accounted for spatial limits of a population in acquiring data about the population. All equations describing population dynamics, however, account for the addition of individuals to (births or immigration) or removal of individuals from (deaths or emigration) a population over time.

In addition to time and space, some descriptions of population dynamics account for the effects of population density on population growth (i.e., density-dependent growth, Equations 9.7 and 9.8), while others assume growth is density independent (Equations 9.3–9.6). Still more complicated models of population growth allow us to account for the effects of individual growth stages on changes in populations over time (Figure 9.5). These stage-based population models are powerful tools for informing the value of different life history stages to current and future population change and are often used to inform decisions about conservation of rare or threatened species.

To account for the spatial relationship of individuals in a population more explicitly, we may wish to take a metapopulation approach, wherein we acknowledge that individuals, or their genes, may migrate among nearby habitat patches with varying frequencies over time. A metapopulation approach uses the same basic mathematical descriptions of population change but conceptually views interacting subpopulations as a "population of populations" and, rather than considering addition or removal of individuals, it considers the occupancy or vacancy of habitat patches in describing metapopulation dynamics over time (Figure 9.8). It is at the scale of metapopulations where we can begin to understand, mathematically and conceptually, the effects of landscape modification on population dynamics, through the effects of such modifications on connectivity among nearby habitat patches.

It is also at the scale of metapopulations where we can begin to understand the processes that lead to local adaptation within populations and to speciation, as previously interacting populations become more isolated over time. Spatial isolation, resulting from natural or anthropogenic forces, can lead to divergence in the genotypes and phenotypes of individuals in populations, which can lead, in turn, to genetic isolation among those populations. Similarly, genetic differentiation through mutation, genetic recombination, or drift can sometimes lead to genetic isolation among subpopulations, with or without spatial isolation. For example, genetic differentiation that leads to acquisition of novel pollinators may result in loss of gene flow between ancestral and derived genotypes. Persistent loss of gene flow among subpopulations can eventually lead to speciation, even when individuals live in very close proximity to one another.

The principles of population biology and metapopulation theory can be powerful tools in understanding mechanisms of species' responses to such processes as climate change, species invasion, and anthropogenic habitat modification. In combination with molecular genetics, population biology allows us to investigate gene flow among habitat patches within a metapopulation and to determine what intrinsic or extrinsic factors may facilitate or obstruct interactions among subpopulations. As with the stage-based population methods mentioned earlier, population genetics allows more insightful conservation of rare or threatened species or populations.

Finally, hybridization, whether between different species (interspecific hybridization) or between previously isolated genotypes of a single species (intraspecific hybridization), can have important implications for population biology. The examples of hybridization in *Spartina* species indicate the significant effects that hybridization, enforced through polyploidization, can have on the potential spread of a species as well as the overall ecology of habitats in which those species live.

9.7   FOR REVIEW

1. What is a population? What is the general objective of population biology?
2. What are the basic components of population change over time?
3. What is the conceptual difference between discrete and continuous population growth? Which

of these is a better descriptor of plant populations, and why?
4. How do we distinguish between density-dependent and density-independent population growth, mathematically?
5. How do stage-based population models differ from other population growth models discussed in this chapter?
6. What is a metapopulation? What are some conceptual differences between metapopulations and populations?
7. How does the theory of island biogeography apply to metapopulation theory?
8. What is an ecotype? How are common garden experiments used in studying ecotypes?
9. What are mechanisms that influence genetic divergence, or lack thereof, within and among plant populations?
10. What are some examples of conservation genetics as applied to aquatic plants? How is this related to large-scale ecological forces such as climate change and species invasion?
11. How are hybridization and polyploidization related?

9.8   REFERENCES

Ainouche, M. L., P. M. Fortune, A. Salmon, C. Parisod, M. A. Grandbastien, K. Fukunaga, M. Ricou, and M. T. Misset. 2009. "Hybridization, Polyploidy and Invasion: Lessons from *Spartina* (Poaceae)." *Biological Invasions* 11 (5): 1159–73. https://doi.org/10.1007/s10530-008-9383-2.

Ainouche, Malika, and Alan Gray. 2016. "Invasive *Spartina*: Lessons and Challenges." *Biological Invasions* 18 (8): 2119–22. https://doi.org/10.1007/s10530-016-1201-7.

Barkworth, Mary E. 1993. "*Spartina*." In *Flora of North America Editorial Committee, Eds. 1993+. Flora of North America North of Mexico [Online]*, Vol. 25. New York and Oxford. http://floranorthamerica.org/Spartina.

Barrett, Spencer C. H. 2010. "Understanding Plant Reproductive Diversity." *Philosophical Transactions of the Royal Society B* 365: 99–109.

———. 2019. "'A Most Complex Marriage Arrangement': Recent Advances on Heterostyly and Unresolved Questions." *New Phytologist* 224: 1051–67.

Barrett, Spencer C. H., Christopher G. Eckert, and Brian C. Husband. 1993. "Evolutionary Processes in Aquatic Plant Populations." *Aquatic Botany* 44 (2–3): 105–45. https://doi.org/10.1016/0304-3770(93)90068-8.

Baskin, Carol C., and Jerry M. Baskin. 2014. *Seeds: Ecology, Biogeography, and Evolution of Dormancy and Germination*. 2nd ed. San Diego, CA: Academic Press.

Bortolus, Alejandro. 2006. "The Austral Cordgrass *Spartina densiflora* Brong.: Its Taxonomy, Biogeography and Natural History." *Journal of Biogeography* 33 (1): 158–68. https://doi.org/10.1111/j.1365-2699.2005.01380.x.

Clatworthy, J. N., and John L. Harper. 1962. "The Comparative Biology of Closely Related Species Living in the Same Area: V. Inter- and Intraspecific Interference within Cultures of *Lemna* Spp. and *Salvinia Natans*." *Journal of Experimental Botany* 13 (2): 307–24. https://doi.org/10.1093/jxb/13.2.307.

Clausen, Jens, David D. Keck, and William M. Hiesey. 1948. "Experimental Studies on the Nature of Species III. Environmental Responses of Climatic Races of *Achillea*." 581. Washington, DC: Carnegie Institution of Washington.

Costa, R. H. R., C. T. Zanotelli, P. Belli Filho, C. C. Perdomo, and M. Rafikov. 2003. "Optimization of the Treatment of Piggery Wastes in Water Hyacinth Ponds." *Water Science and Technology* 48 (2): 283–89. https://doi.org/10.2166/wst.2003.0132.

Dahlgren, Johan P., and Johan Ehrlén. 2005. "Distribution Patterns of Vascular Plants in Lakes—the Role of Metapopulation Dynamics." *Ecography* 28 (1): 49–58. https://doi.org/10.1111/j.0906-7590.2005.04018.x.

Eckert, Christopher G., Marcel E. Dorken, and Spencer C. H. Barrett. 2016. "Ecological and Evolutionary Consequences of Sexual and Clonal Reproduction in Aquatic Plants." *Aquatic Botany (Forty Years of Aquatic Botany, What Have We Learned?)* 135: 46–61. https://doi.org/10.1016/j.aquabot.2016.03.006.

Edwards, Adrienne L., and Rebecca R. Sharitz. 2000. "Population Genetics of Two Rare Perennials in Isolated Wetlands: *Sagittaria isoetiformis* and *S. teres* (Alismataceae)." *American Journal of Botany* 87 (8): 1147–58. https://doi.org/10.2307/2656651.

Erwin, S., A. Huckaba, K. S. He, and M. McCarthy. 2013. "Matrix Analysis to Model the Invasion of Alligatorweed (*Alternanthera philoxeroides*) on Kentucky Lakes." *Journal of Plant Ecology* 6 (2): 150–57. https://doi.org/10.1093/jpe/rts024.

Figuerola, Jordi, Luis Santamaría, Andy J. Green, Isabel Luque, Raquel Alvarez, and Iris Charalambidou. 2005. "Endozoochorous Dispersal of Aquatic Plants: Does Seed Gut Passage Affect Plant Performance?" *American Journal of Botany* 92 (4): 696–99. https://doi.org/10.3732/ajb.92.4.696.

Fortune, P. M., K. Schierenbeck, D. Ayres, A. Bortolus, O. Catrice, S. Brown, and M. L. Ainouche. 2008. "The Enigmatic Invasive *Spartina densiflora*: A History of Hybridizations in a Polyploidy Context." *Molecular Ecology* 17 (19): 4304–16. https://doi.org/10.1111/j.1365-294X.2008.03916.x.

Fox, Gordon A., and Jessica Gurevitch. 2000. "Population Numbers Count: Tools for Near-term Demographic Analysis." *The American Naturalist* 156: 242–56.

Frighetto, Daiane Frighetto, Gustavo Maia Souza, and Alexandre Molter. 2019. "Spatio-Temporal Population Control Applied to Management of Aquatic Plants." *Ecological Modelling* 398 (April): 77–84. https://doi.org/10.1016/j.ecolmodel.2018.09.027.

Gotelli, Nicholas J., and Aaron M. Ellison. 2002. "Nitrogen Deposition and Extinction Risk in the Northern Pitcher Plant, *Sarracenia purpurea*." *Ecology* 83 (10): 2758–65. https://doi.org/10.1890/0012-9658(2002)083[2758:NDAERI]2.0.CO;2.

Gurevitch, Jessica. 1992a. "Differences in Photosynthetic Rate in Populations of *Achillea lanulosa* from Two Altitudes." *Functional Ecology* 6: 568–74.

———. 1992b. "Sources of Variation in Leaf Shape among Two Populations of *Achillea lanulosa*." *Genetics* 130: 385–94.

Gurevitch, Jessica, Samuel M. Scheiner, and Gordon A. Fox. 2006. *The Ecology of Plants*. 2nd ed. Sunderland, MA: Sinauer Associates, Inc.

Hanski, Ilkka. 1985. "Single-Species Spatial Dynamics May Contribute to Long-Term Rarity and Commonness." *Ecology* 66: 335–43.

———. 1998. "Metapopulation Dynamics." *Nature* 396: 41–49.

Harper, John L. 1977. *Population Biology of Plants*. New York: Academic Press, Inc.

Hastings, Alan. 1997. *Population Biology: Concepts and Models*. New York: Springer-Verlag.

Jordano, Pedro. 2017. "What Is Long-Distance Dispersal? And a Taxonomy of Dispersal Events." *Journal of Ecology* 105 (1): 75–84. https://doi.org/10.1111/1365-2745.12690.

Judd, Walter S., Christopher S. Campbell, Elizabeth A. Kellogg, Peter F. Stevens, and Michael J. Donoghue. 2008. *Plant Systematics: A Phylogenetic Perspective*. 3rd ed. Sunderland, MA: Sinauer Associates, Inc.

———. 2016. *Plant Systematics: A Phylogenetic Approach*. 4th ed. Sunderland, MA: Sinauer Associates, Inc.

Kato, Syou, Kazuharu Misawa, Fumio Takahashi, Hidetoshi Sakayama, Satomi Sano, Keiko Kosuge, Fumie Kasai, Makoto M. Watanabe, Jiro Tanaka, and Hisayoshi Nozaki. 2011. "Aquatic Plant Speciation Affected by Diversifying Selection of Organelle DNA Regions." *Journal of Phycology* 47 (5): 999–1008. https://doi.org/10.1111/j.1529-8817.2011.01037.x.

Lázaro-lobo, Adrián, and Gary N. Ervin. 2021. "Wetland Invasion: A Multi-Faceted Challenge during a Time of Rapid Global Change." *Wetlands* 41: 1–16.

Leeuwen, Casper H. A. van. 2018. "Internal and External Dispersal of Plants by Animals: An Aquatic Perspective on Alien Interference." *Frontiers in Plant Science* 9: 153.

Levin, Donald A. 1995. "Metapopulations: An Arena for Local Speciation." *Journal of Evolutionary Biology* 8 (5): 635–44. https://doi.org/10.1046/j.1420-9101.1995.8050635.x.

Levins, Richard. 1969. "Some Demographic and Genetic Consequences of Environmental Heterogeneity for Biological Control." *American Entomologist* 15: 237–40.

Lobato-de Magalhães, Tatiana, Dánae Cabrera-Toledo, and Mahinda Martínez. 2019. "Microsatellite Loci Transferability and Genetic Diversity of the Aquatic Plant *Nymphoides fallax* Ornduff (Menyanthaceae), Endemic to the Mexican and Guatemalan Highlands." *Limnology* 20 (2): 233–41. https://doi.org/10.1007/s10201-019-00571-5.

Lobato-de Magalhães, Tatiana, and Mahinda Martínez. 2020. "Insights into Distyly and Seed Morphology of the Aquatic Plant *Nymphoides fallax* Ornduff (Menyanthaceae)." *Flora* 262: 151526. https://doi.org/10.1016/j.flora.2019.151526.

Lobato-de Magalhães, Tatiana, Kevin Murphy, Andrey Efremov, Victor Chepinoga, Thomas A. Davidson, and Eugenio Molina-Navarro. 2021. "Ploidy State of Aquatic Macrophytes: Global Distribution and Drivers." *Aquatic Botany* 173: 103417. https://doi.org/10.1016/j.aquabot.2021.103417.

Lobato-de Magalhães, Tatiana, Yessica Rico, Dánae Cabrera-Toledo, and Mahinda Martínez. 2020. "Plant Functional Connectivity of *Nymphoides fallax* in Geographically Isolated Temporary Wetlands in Mexican Highlands." *Aquatic Botany* 164 (February). https://doi.org/10.1016/j.aquabot.2020.103215.

Lomolino, Mark V., Brett R. Riddle, and James H. Brown. 2006. *Biogeography*. 3rd ed. Sunderland, MA: Sinauer Associates, Inc.

Lucardi, Rima D., Lisa E. Wallace, and Gary N. Ervin. 2020. "Patterns of Genetic Diversity in Highly Invasive Species: Cogongrass (*Imperata cylindrica*) Expansion in the Invaded Range of the Southern United States (US)." *Plants* 9 (4): 423. https://doi.org/10.3390/plants9040423.

MacArthur, Robert H., and Edward O. Wilson. 1963. "An Equilibrium Theory of Insular Zoogeography." *Evolution* 17: 373–87.

———. 1967. *The Theory of Island Biogeography*. Princeton, NJ: Princeton University Press.

Malthus, Thomas. 1798. *An Essay on the Principle of Population, as It Affects the Future Improvement of Society, with Remarks on the Speculations of Mr. Godwin, M. Condorcet, and Other Writers*. London: Johnson.

Meyers, Lauren Ancel, and Donald A. Levin. 2006. "On the Abundance of Polyploids in Flowering Plants." *Evolution* 60 (6): 1198–206. https://doi.org/10.1111/j.0014-3820.2006.tb01198.x.

Nies, G., and T. B. H. Reusch. 2005. "Evolutionary Divergence and Possible Incipient Speciation in Post-Glacial Populations of a Cosmopolitan Aquatic Plant." *Journal of Evolutionary Biology* 18 (1): 19–26. https://doi.org/10.1111/j.1420-9101.2004.00818.x.

Ornduff, Robert. 1969. "Neotropical *Nymphoides* (Menyanthaceae): Meso-American and West Indian Species." *Brittonia* 21 (4): 346–52.

Ramsey, Justin, Alexander Robertson, and Brian Husband. 2008. "Rapid Adaptive Divergence in New World *Achillea*, an Autopolyploid Complex of Ecological Races." *Evolution* 62 (3): 639–53. https://doi.org/10.1111/j.1558-5646.2007.00264.x.

Reusch, T. B. H. 2000. "Pollination in the Marine Realm: Microsatellites Reveal High Outcrossing Rates and Multiple Paternity in Eelgrass *Zostera marina*." *Heredity* 85 (5): 459–64. https://doi.org/10.1046/j.1365-2540.2000.00783.x.

Richardson, Frank David. 1983. "Variation, Adaptation and Reproductive Biology in *Ruppia maritima* L. Populations from New Hampshire Coastal and Estuarine Tidal Marshes." Durham, NH: University of New Hampshire.

Rius, Marc, and John A. Darling. 2014. "How Important Is Intraspecific Genetic Admixture to the Success of Colonising Populations?" *Trends in Ecology & Evolution* 29 (4): 233–42. https://doi.org/10.1016/j.tree.2014.02.003.

Ruckelshaus, M. H. 1995. "Estimates of Outcrossing Rates and of Inbreeding Depression in a Population of the Marine Angiosperm *Zostera marina*." *Marine Biology* 123: 583–93.

Santamaría, Luis. 2002. "Why Are Most Aquatic Plants Widely Distributed? Dispersal, Clonal Growth and Small-Scale Heterogeneity in a Stressful Environment." *Acta Oecologica* 23 (3): 137–54. https://doi.org/10.1016/S1146-609X(02)01146-3.

Simberloff, Daniel. 1974. "Equilibrium Theory of Island Biogeography and Ecology." *Annual Review of Ecology and Systematics* 5: 161–82.

Smith, Marian, Hal Caswell, and Paige Mettler-Cherry. 2005. "Stochastic Flood and Precipitation Regimes and the Population Dynamics of a Threatened Floodplain Plant." *Ecological Applications* 15 (3): 1036–52. https://doi.org/10.1890/04-0434.

Strong, Donald R., and Debra R. Ayres. 2013. "Ecological and Evolutionary Misadventures of *Spartina*." *Annual Review of Ecology, Evolution, and Systematics* 44: 389–410. https://doi.org/10.1146/annurev-ecolsys-110512-135803.

———. 2016. "Control and Consequences of *Spartina* spp. Invasions with Focus upon San Francisco Bay." *Biological Invasions* 18 (8): 2237–46. https://doi.org/10.1007/s10530-015-0980-6.

Thum, Ryan A., and James N. McNair. 2018. "Inter- and Intraspecific Hybridization Affects Germination and Vegetative Growth in Eurasian Watermilfoil." *Journal of Aquatic Plant Management* 56: 24–30.

Tolvanen, A., J. Schroderus, and G. H. R. Henry. 2002. "Age- and Stage-Based Bud Demography of *Salix arctica* under Contrasting Muskox Grazing Pressure in the High Arctic." *Evolutionary Ecology* 15: 443–62.

Vallejo-Marín, Mario, and Simon J. Hiscock. 2016. "Hybridization and Hybrid Speciation under Global Change." *New Phytologist* 211 (4): 1170–87. https://doi.org/10.1111/nph.14004.

Van Tussenbroek, Brigitta I., Tania Valdivia-Carrillo, Irene Teresa Rodriguez-Virgen, Sylvia Nashieli Marisela Sanabria-Alcarez, Karina Jimenez-Duran, Kor Jent Van Dijk, and Guadalupe Judith Marquez-Guzman. 2016. "Coping with Potential Bi-Parental Inbreeding: Limited Pollen and Seed Dispersal and Large Genets in the Dioecious Marine Angiosperm *Thalassia testudinum*." *Ecology and Evolution*, 15–15.

Wei, Xinzeng, Hongjie Meng, and Mingxi Jiang. 2013. "Landscape Genetic Structure of a Streamside Tree Species *Euptelea pleiospermum* (Eupteleaceae): Contrasting Roles of River Valley and Mountain Ridge." *PLoS ONE* 8 (6): 2–9. https://doi.org/10.1371/journal.pone.0066928.

Werner, Patricia A., and Stephanie J. Peacock. 2019. "Savanna Canopy Trees under Fire: Long-Term Persistence and Transient Dynamics from a Stage-Based Matrix Population Model." *Ecosphere* 10 (5): 1–41. https://doi.org/10.1002/ecs2.2706.

Wetzel, Robert G., and Michael J. Howe. 1999. "High Production in a Herbaceous Perennial Plant Achieved by Continuous Growth and Synchronized Population Dynamics." *Aquatic Botany* 64 (2): 111–29. https://doi.org/10.1016/S0304-3770(99)00013-3.

Wu, Zhigang, Xinwei Xu, Juan Zhang, Gerhard Wiegleb, and Hongwei Hou. 2019. "Influence of Environmental Factors on the Genetic Variation of the Aquatic Macrophyte *Ranunculus subrigidus* on the Qinghai-Tibetan Plateau." *BMC Evolutionary Biology* 19 (1): 1–11. https://doi.org/10.1186/s12862-019-1559-0.

Zhao, Zhilong, Yili Zhang, Linshan Liu, Fenggui Liu, and Haifeng Zhang. 2015. "Recent Changes in Wetlands on the Tibetan Plateau: A Review." *Journal of Geographical Sciences* 25 (7): 879–96. https://doi.org/10.1007/s11442-015-1208-5.

10 Species Interactions

10.1 INTERACTIONS IN THE CONTEXT OF LIFE HISTORY

We began our discussion of plant reproduction in Chapter 8 by examining the relevance of life history for reproduction. Life history in that context was viewed from the perspective of longevity of the individual plant and the frequency with which the species typically undergoes sexual reproduction. The three primary life histories we considered were annual, monocarpic perennial (biennial), and polycarpic perennial life histories. We also can view life histories in the context of the way in which selective pressures in the environment shape the differences among species within each of the aforementioned three primary life histories. In this context, ecologists often use the term "strategy" to refer to the combination of traits that characterize the way attributes of a species' life history have been shaped by selection. Thus, we may consider the **life history strategy** of a particular species as being the sum of the selective adaptations by which the species has fit its niche within a community. Furthermore, the relative importance of those individual adaptations and the way they interact to the advantage, or detriment, of the species may vary from one community to another, based on the other species present and the abiotic conditions encountered in different parts of the species' range.

A critically important note regarding this term "strategy," however, is that we must keep in mind that there was no forethought on the part of the species in deciding how to allocate its resources among different traits to best fit the communities in which it lives. Quite the contrary, as was discussed in previous chapters, the species itself was very much at the mercy of the environment in determining which combinations of traits were least appropriate, in terms of fitness, and selecting against those combinations. Individuals who possessed genotypes that yielded the most fit combinations of traits, or the best strategies for allocating resources, left behind the greatest numbers of viable offspring. Trait combinations that tend to yield the highest fitness for individuals under specific sets of environmental conditions are thought of as characteristic life history strategies for habitats typically having those conditions. Thus, we can think of strategy in this context as the characteristic pattern of resource allocation trade-offs found under certain environmental conditions.

In the field of plant ecology, we often consider two general constructs for organizing life history strategies. One of these parallels our treatment of life history in Chapter 8, dealing with the longevity and reproductive patterns of the species. In this construct, two general life history strategies are considered: the r-selected life history strategy and the K-selected life history strategy (Table 10.1; MacArthur and Wilson 1967; Pianka 1970). We saw the terms r and K in Chapter 9, when we talked about the intrinsic rate of population increase and carrying capacity in population growth models. The intrinsic rate of population increase, r, was a key component of models describing unconstrained population growth, where high reproductive rates led to rapid, exponential population increase. In constrained population growth, on the other hand, carrying capacity, K, represented the maximum population density that could be sustained in a habitat. Population growth near K tended to be very slow, approaching zero growth as population density approached carrying capacity.

The r-selected life history strategy is characterized by species that invest resources heavily in rapid, early sexual reproduction (Table 10.1). To accomplish this, those species tend to have short developmental times, reach sexual maturity quickly, and produce many small offspring. Because of this proportionally large investment in sexual reproduction, r-selected individuals tend to be relatively small and short-lived, and we often consider them to exhibit a weedy behavior. One consequence of their small size and low investment in structures for acquisition and storage of resources is that these species tend to be at a disadvantage when growing in the presence of established plants that employ a K-selected strategy (Table 10.1). Thus, r-selected species tend to be more prevalent in habitats that have variable or unpredictable environmental conditions that lead to low overall plant density, or in disturbed areas, such as mudflats of a drained beaver pond, disturbed soil, or possibly on sand or gravel bars in streams. They also tend to be represented by annual species. Species that exhibit the opposite, K-selected, strategy tend to invest more heavily in traits that lend themselves to resource capture and storage and allow those species to maintain relatively consistent population densities in crowded habitats; that is, their populations will tend to approximate the carrying capacity of habitats they occupy. As a result, K-selected species tend to be polycarpic perennials, and long-lived trees often are given as examples of species exhibiting a K-selected strategy.

We saw similar comparisons in Chapter 8 regarding asexual propagation versus sexual reproduction, where sexual offspring tended to exhibit characteristics in line with r-selected species, and clonal propagules tended to better fit the K-selected strategy (Table 8.8). In fact, by definition, annual species generally rely only on sexual reproduction for population maintenance, while perennial species display a wide range in relative investment into sexual reproduction versus asexual propagation. Finally, it is important to keep in mind that these characteristics are general tendencies and not strict rules. As we have seen in the two previous

DOI: 10.1201/9781315156835-10

TABLE 10.1

Generalized characteristics of species exhibiting r-selected and K-selected life histories, or of habitats they typically occupy, from MacArthur and Wilson (1967) and Pianka (1970).

	r-selected	K-selected
Developmental time	Short	Long
Size of individuals	Small	Larger
Juvenile period	Short	Long
Reproductive frequency	Monocarpic (annuals)	Polycarpic
Reproductive rate	High	Low
Number of offspring	Large	Small
Size of offspring	Small	Larger
Mortality	Density independent	Density dependent
Population size	Variable but typically well below carrying capacity	Generally constant through time and approximating carrying capacity
Environmental conditions	Variable or unpredictable	Invariable or predictable
Density of neighbors	Low	High

chapters, considerable variation in all these traits can exist among populations within a single species. Furthermore, because species hypothetically had almost infinite flexibility in how they might have portioned out resources to different life functions (biomass allocation among parts, resource use vs. resource storage, numbers vs. size of flowers, number vs. size of mature seeds and fruit, etc.), we should think in terms of a continuum of phenotypes along an r-to-K gradient, rather than a strict r-or-K dichotomy.

The second construct of categorizing plant life history strategies that we often encounter comprises three life history strategies. These three strategies are determined by the joint effects of two environmental factors that have been hypothesized to limit the production of plant biomass: stress and disturbance (Figure 10.1; Grime 1977). **Stress**, in this context, was defined by Grime as any factor external to the plant that can be shown to restrict the rate of production of biomass by individual plants or the collective vegetation of an area (Grime 1977). Examples of stressors include toxic substances, heavy metals, temperature extremes, and deficiencies or excesses of light, water, or nutrients, including deficiencies and excesses in these materials brought about by neighboring plants. The other factor, **disturbance**, was defined by Grime as mechanisms that destroy plant biomass, such as floods, fires, wind, erosion, herbivory, pathogens, and human activities that remove biomass (e.g., soil tillage, deforestation, mowing). Grime further considered that environments that experience high levels of environmental stress *along with* high frequency or intensity of disturbance were unlikely to provide sufficient opportunities for plants to evolve resource allocation strategies that would permit them to persist in such habitats. Hence that quadrant is described as having "No Viable Strategy" in Figure 10.1.

Central to understanding Grime's C-S-R life history strategies is the hypothesis that, under relatively productive environmental conditions (i.e., habitats with ample

light, water, and nutrients) and in the absence of frequent disturbance, species that will dominate the flora in a community will be those that are best equipped to compete for available resources. We can see in Table 10.2 combinations of characteristics that enable these species to effectively exploit available resources in habitats they occupy. Key among those characteristics are production of a large, spreading root and shoot system, rapid growth, and, consequently, a relatively small allocation of resources to sexual

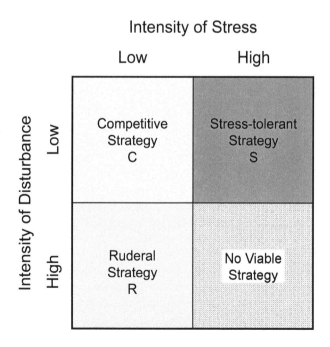

FIGURE 10.1 Evolutionary Basis for the Three Primary Life History Strategies Introduced by Grime (1977). Grime Hypothesized that Stress and Disturbance were Major Factors Influencing Plant Productivity and, thus, their Interaction would Form the Basis for Selection of Major Strategies Encountered Among Plants in Nature.

TABLE 10.2

Generalized characteristics of species exhibiting C-, S-, and R-selected life history strategies, or of habitats they typically occupy, from Grime (1977).

	Competitive	Stress Tolerant	Ruderal
General morphology	Tall, dense leaf canopy with large above- and belowground spread	Highly variable	Small plants with limited spread
Litter or detritus	Abundant and persistent	Sparse and sometimes persistent	Sparse and rapidly decomposed
Potential growth rate	Rapid	Slow	Rapid
Phenology of leaf production	Peaks of leaf production coinciding with periods of maximum potential productivity	Tend to be evergreen with variable periods of leaf production	Short period of leaf production during period of maximum potential productivity
Phenology of flowering	Flowering after (or, rarely, before) period of maximum potential productivity	Variable	Flowering at end of growth period
Proportion of resources allocated to seed production	Small	Small	Large

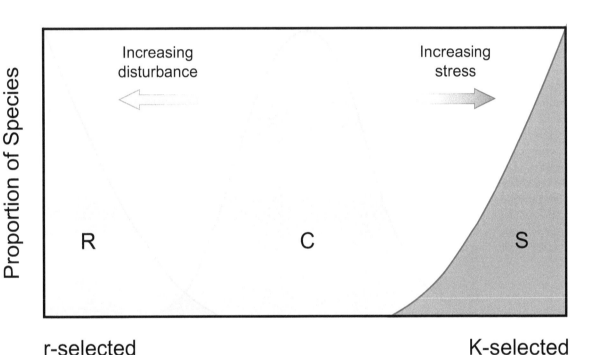

FIGURE 10.2 Qualitative Comparison of Grime's Three Primary Life History Strategies with the R- to K-Selection Continuum. as Illustrated in Figure 10.1, Competitive Species (C) are Expected to Dominate Productive, Undisturbed Habitats. As Frequency or Intensity of Disturbance Increases, the Importance of Competitive Species Will Decline, in Favor of Ruderal Species (R), Whose Characteristics Closely Resemble the Idealized R-Selected Strategy. At the other Extreme, When Resources Become Less Readily Available, Competitive Species may be Replaced by Stress-Tolerant Species (S) that Exhibit Traits Similar to K-Selected Strategists. Redrawn from Grime (1977).

reproduction. Within wetlands, herbaceous perennials, especially taxa such as *Typha* and *Phragmites* species that often dominate the wetland vegetation, might be considered to demonstrate a competitor life history strategy.

In habitats that are more frequently or more intensely disturbed, we begin to see a shift in dominance from competitive species to those that Grime refers to as ruderal (Figure 10.2). Ruderal species tend to exhibit characteristics very similar to the r-selected strategists discussed earlier (Table 10.2), and in

fact, Grime points out this similarity by directly comparing his C-S-R construct to the r- versus K-selected life history continuum (Figure 10.2). Rapid growth, large proportional investment in seed production, and the annual life history are shared characteristics of r-selected species and the ruderals of Grime's construct, and these are traits that allow these species to dominate highly disturbed habitats.

At the other end of the r-to-K life history continuum, we find Grime's stress-tolerant life history strategy. As mentioned earlier, Grime defined stress as any factor external to the plant that restricts production of biomass, which he described as taking many forms, including the conditions frequently experienced in the understory of mature forests, where the canopy trees tend to monopolize light and soil resources. In other situations, using Grime's definition of stress, we would consider very sandy habitats such as dunes along a lakeshore or beach, or perhaps sand deposits in a river floodplain to be stressful, owing to the generally low water and nutrient availability within sand-dominated soils. Under conditions such as those, species that Grime described as competitor strategists find difficulty acquiring sufficient resources to maintain their high growth rates and large amounts of biomass, and, instead, we might expect to find species employing a more conservative growth strategy (Table 10.2).

As was mentioned earlier regarding the conceptual view of r- and K-selected life histories as a continuum of allocation strategies, Grime also emphasized that the competitor, stress-tolerator, and ruderal life history strategies exist as extremes on a continuum of plant phenotypes (Figure 10.3). He discussed specific combinations of these three primary strategies as secondary strategies such as stress-tolerant competitors (C-S) or competitive ruderals (C-R) and provided examples of habitats where each might typically be encountered. He further related these primary and secondary strategies to plant growth forms and taxonomic groups, in the context of **succession** (the change in plant species composition of an area over time), which we will discuss in the following chapter.

It is worth noting, before moving along, that there has been considerable disagreement about the terms stress and disturbance, as used by Grime. The concept of disturbance is the simpler of the two to discuss, so we will discuss it

first. Grime (1977) defined disturbance as any mechanism that limits plant biomass by "causing its destruction." However, one concern about this definition was that it was unclear how or whether disturbance could be measured (Grime 1989). Indeed, both stress and disturbance in the context of this C-S-R construct appear to have been more conceptual than quantitative and seem to have been meant more for thinking about how one might summarize forces acting to shape outcomes of natural selection or that have driven plant species diversification in natural communities. Other definitions of disturbance that have been used include defining disturbance as an "unusual event . . . that upsets normality" (Begon, Harper, and Townsend 1990, p. 740), "an event that removes organisms and opens up space which can be colonized by individuals of the same or different species" (Begon, Harper, and Townsend 1990, p. 850), and a "relatively discrete event in time that causes an abrupt change in ecosystem, community, or population structure and changes resource availability, substrate availability, or the physical environment" (Gurevitch, Scheiner, and Fox 2006, p. 285). These definitions all vary in their specificity, but they all are connected by their assertion that disturbance comes in the form of some event that changes the structure (or other manifestation of "normality") of the disturbed habitat. Included in the definition used by Gurevitch, Scheiner, and Fox (2006) is a change in one or more abiotic components of the habitat along with the biotic structure of the habitat.

We will revisit the concept of disturbance in the next chapter, but for now, the role of disturbance as a potential selective force that may have influenced plant species diversification should be somewhat clear. The frequency with which a habitat experiences events capable of altering biotic structure will influence the longevity of species that can establish sustainable populations therein. Longer intervals between disturbances will allow longer-lived species to maintain viable populations.

Stress, on the other hand, is more difficult to define, perhaps in part because forces that cause stress (stressors) are not always as visible as are forces that cause disturbance. Another aspect of stress that complicates defining it clearly is that stress itself, being an intrinsic property of the stressed organism, may not be well understood by an outside observer (e.g., by humans; Otte 2001). Otte (2001) noted that plants adapted to a particular suite of environmental conditions (e.g., oxygen deficient, flooded organic sediments of a wetland) are not likely to perceive those conditions as stressful, given that it is under those conditions where the species flourishes, but others fail to survive. In this sense, the idea of a "stress-tolerant" species is likely less relevant than the idea of environmental conditions creating stressors that select against species intolerant of those conditions (i.e., factors external to the plant that restrict their rate of biomass production). Thus, stress, from the perspective of the plant, results from conditions (the stressors) whose magnitude lies outside the range under which the individual or species evolved (Otte 2001).

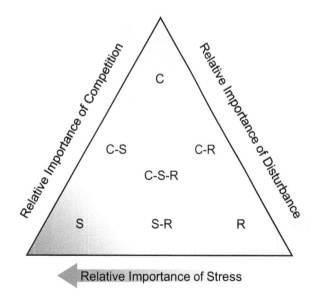

FIGURE 10.3 Depiction of Grime's Three Primary Strategies as a Three-Axis Continuum of Primary, Secondary, and Tertiary Life History Strategies. Redrawn from Grime (1977).

Think back to Chapter 1, where we discussed definitions of wetland plants. Wetland plants evolved under conditions typically encountered in wetlands, which are stressful to non-wetland-adapted species. The adaptations discussed in Chapters 6, 7, and 8 have enabled wetland plants to perform quite well under conditions that may cause significant stress to other plant species. Another complication in applying the concept of stress to plant species and species assemblages is the need to distinguish between the stress, that is the physiological perception of a stressor, and the putative stressor itself, which is an environmental factor external to the species or individual in question. This distinction, which was not specified in Grime's initial formulation of his C-S-R framework, may also have contributed to some of the confusion about the importance of stress in ecological interactions.

Despite previous disagreements about the terms that have been used and their definitions, it is useful to think about mechanisms that have structured communities and that have presented selective forces leading to the diversity of plant species we find in different habitats. We can think of these life history strategy concepts (r-to-K and C-S-R) as two examples of many efforts to disentangle the complex forces that have influenced plant and ecosystem diversification. We encountered another similar construct in Chapter 8, where Grace (1993) attempted to simplify the diversity of clonal structures in aquatic and wetland plants by organizing them along axes of plant functional traits (Figure 8.28). Moore et al. (1989), Keddy (1990), and Wisheu and Keddy (1992) similarly attempted to explain

diversification in species composition among wetlands along multiple environmental gradients, including factors of stress and disturbance, as well as biotic interactions that determine which species coexist, or not, under different combinations and levels of other environmental factors (Figure 10.4). This idea of gradient-dominated organization of communities will become more relevant in the latter portions of the present chapter.

This introduction is meant to develop a baseline for thinking about the ecological and evolutionary implications of how species interact within aquatic and wetland ecosystems. As in the discussion of ecological and evolutionary trade-offs between sexual reproduction and asexual propagation in Chapter 8, the strategies by which individuals allocate resources among functions that ensure growth, survival, and reproduction will dictate their success in interactions with other species in the community. There are potentially endless ways in which individuals can proportionally allocate resources among life functions, and there are similarly diverse ways in which they can interact with other members of the communities they occupy (Figure 10.5).

In Figure 10.5, the outcomes of interactions among species are represented by a pair of mathematical symbols, indicating whether each of the two interacting species exhibits a positive (+), neutral (0), or negative (−) response to the interaction. For example, in **mutualisms**, some of which were discussed in Chapter 7 (mycorrhizae and nitrogen-fixing symbioses) and Chapter 8 (biotic pollination), both species receive a benefit from the interaction;

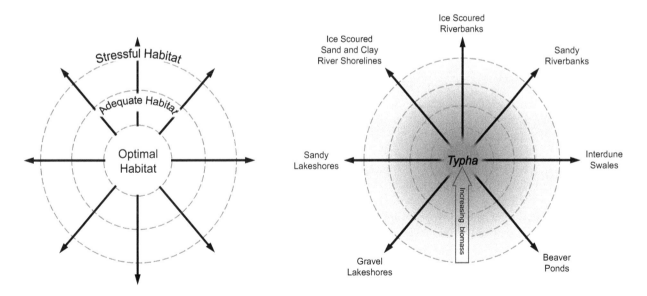

FIGURE 10.4   Gradient-Driven Community Organization is a Major Theme in the Study of Interactions in Wetlands and other Ecosystems. Left: Otte (2001) Suggested that Plants will Experience Increasing Stress as Influential Environmental Variables Approach Levels Uncommon in Habitats they Typically Occupy. Thus, Environmental Stress is Hypothesized to be a Major Variable Selecting against Persistence of Species in Habitats Where they Typically do not Occur. Right: Keddy and Colleagues have Demonstrated Gradient-Dominated Community Organization in many Freshwater Wetlands. These Patterns were Driven by Competition with Dominant Plant Species, such as *Typha Latifolia*, at One End of the Gradient and Various Forms of Stress or Disturbance at the Other. Figure at Left Modified from Otte (2001); Figure at Right Modified from Wisheu and Keddy (1992).

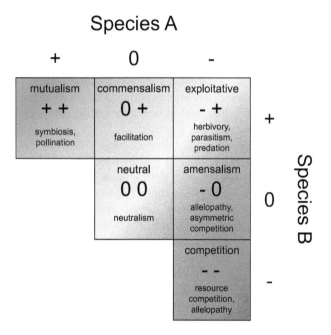

FIGURE 10.5 Summary of the Most Common Basic Ecological Interactions Among Species. The Nature of Each Interaction is Given as a Sign (+, 0, −) Representing Positive, Neutral, or Negative Effects of One Species on Another. The Placement of Some Interactions within this Scheme (E.g., Allelopathy, Epiphytes) Sometimes Depends on Magnitude of Impact of One Species on the other, the Way the Outcomes are Measured, or other Factors that can Influence Outcomes of Interactions. Modified from Barbour et al. (1999).

mutualisms, thus, are represented as (+ +) interactions. At the other extreme are interactions where both species experience negative outcomes (− − interactions), most common among those is competition, which we will discuss in more detail next.

10.2 COMPETITION (− −)

When Grime laid out his arguments for the C-S-R life history strategy concept (Grime 1977), he defined **competition** as "the tendency of neighboring plants to utilize the same quantum of light, ion of a mineral nutrient, molecule of water, or volume of space." He indicated that his intent was for this definition to represent the mechanism of competition instead of the effects of competition. A similar definition is used by Begon, Harper, and Townsend (1990, p. 197), who described competition as "an interaction between individuals, brought about by a shared requirement for a resource in limited supply, and leading to a reduction in the survivorship, growth and/or reproduction of the competing individuals." A key difference between these two definitions is that the latter incorporates a means of evaluating the outcome of the interaction, while retaining the mechanistic description of what causes that outcome. That is, because two individuals share a requirement for a resource that is in limited supply, and because they are in sufficient proximity to pull those resources from the same pool (i.e., they

are "neighboring plants"), there is a smaller quantity of the resource(s) in question available to each individual than would be the case if one or the other were not present. That reduction in per capita resource availability, then, results in decreased growth, survival, and/or reproduction of both the competing individuals, and we see a mutually negative outcome in the parameters being measured (− −).

10.2.1 A FORMAL CONSTRUCT FOR STUDYING COMPETITION

The difference between the two definitions given earlier is significant for many reasons. First, as was discussed in the chapter introduction, Grime's C-S-R construct is particularly useful as a conceptual construct for thinking about how natural selection may have driven plant species diversification in natural communities. In this respect, it is not necessary to include, in the definition, a means by which one would quantify competition between individual plants. However, the lack of an operational component (i.e., an inherent indication as to how one would measure competition in action) led to some degree of discord among plant ecologists in terms of when, where, why, and how competition is important (Grace 1991). There were several papers and a few books published in the late 1980s and early 1990s attempting to clarify various perspectives on plant competition, including a book entitled *Perspectives on Plant Competition* (Grace and Tilman 1990). The amount of time spent simply clarifying conceptual and operational definitions of competition indicates the importance of studying interactions as a component of understanding natural ecosystems. In keeping with that tradition, we will explore some quantitative underpinnings of competition before moving on to discuss details of how one measures competition and other interactions among plants.

While Grime's approach to defining competition lacked explicit operational components necessary to transition from conceptual to quantitative examination, the approach employed by another plant ecologist was quite the opposite (Tilman 1980, 1985, 1990). As described by Grace (1990), Tilman applied an operational approach to competition, defined as a situation where two or more species use shared resources that are in limited supply, where competitive success is determined by dominance of one species over the other(s). This definition is quite similar to the others given earlier, but the explicit quantitative framework within which Tilman went about defining competition distinguishes this approach from those mentioned thus far.

Because Tilman was interested in quantifying ecological processes such as competition based on how the process affects population density or abundance of the interacting species, he took a population biology approach to describing the process. Recall that, in Chapter 9, we represented the change in population density over time as dN/dt (i.e., change in N with respect to some change in time). In Chapter 9, we were talking about changes in abundance as a function of population size itself. In the context of species interactions, these population models must be modified slightly to account for the mechanism of interaction. In the case of competition, we would consider changes in population density, biomass, or other measures of abundance as a function of resource availability and use. Tilman (1980) did this in the form of

$$\frac{dN}{N\,dt} = f\left(R_k\right) - m \qquad (10.1)$$

where dN/Ndt represents the per capita rate of change in some measure of abundance of a species over time, $f\left(R_k\right)$ represents a function (f) describing the dependency of population growth on resource availability (R_k) for some k number of essential resources, and m represents mortality. The form of this equation is very similar to those we saw in Chapter 9, all of which described change in population size over time as the difference between births and deaths. Similarly, Tilman (1980) described changes in the resource supply over time as the difference between additions of the resource to a habitat, minus resource consumption by the population(s) of species occupying the habitat.

$$\frac{dR}{dt} = g\left(R\right) - \sum\nolimits_{i=1}^{s} \Sigma N_i\, f_i\left(R_k\right) h_i\left(R_k\right) \qquad (10.2)$$

In Equation 10.2, the change in some individual resource over time (dR/dt) is a function of the supply rate of that resource, $g(R)$, minus community-level consumption of the resource. Cumulative consumption of the resource by all species in the community is a function of the number of individuals present in each species (N_i) multiplied by the rate of growth of that species' population $(f_i\left(R_k\right))$ multiplied by the amount of resource required for each unit of growth in the population of the species $(h_i\left(R_k\right))$. Because consumption of a resource by one species affects availability of that resource for all species in the community, Equation 10.2 includes the summation of resource consumption by all species present in the community (summation from species 1 to s, where s is the number of species that use the resource in question).

Although seemingly complicated at first glance, these two equations simply represent changes in availability of a resource as demands for that resource change, based on changes in the population density or abundance of a focal species. These equations can be modified to account for the availability of multiple resources and the consumption of those resources by more than one species (Tilman 1980, 1985, 1990). When more than one resource and more than one species are involved, we can consider relative rates of resource supply and consumption as major determinants of ultimate outcomes of competition among the species of interest. A graphical depiction of this is given in Figures 10.6 and 10.7.

The lines labeled "Zero Net Growth Isoclines" (ZNGIs) in these figures represent the level of resource availability

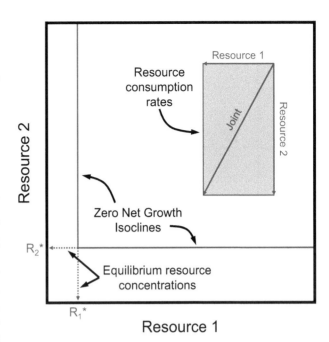

FIGURE 10.6 Graphical Representation of Components of Tilman's Resource Competition Model. For a Species of Interest, the Model Considers the Dependency of Population Growth on Resource Availability (Equation 10.1), the Rates at Which the Species Consumes Each Resource of Interest (Equation 10.2), and the Rates of Resupply into the Habitat (A Component of Equation 10.2). The Lines Labeled "Zero Net Growth Isoclines" (Zngis) Represent Levels of Each Essential Resource at Which Net Population Growth Equals Zero, or Where Population Increases Equal Losses. The Points Labeled "Equilibrium Resource Concentrations" (R_1^* and R_2^*) are Concentrations of Each Resource at Which Resource Consumption Rate Equals Rate of Resupply and at Which Net Population Growth is Zero. At Lower Levels of Resource Availability, the Population Will Decline. Modified from Tilman (1985).

required for the species to maintain net zero population growth over time, or to balance population growth with mortality (Figure 10.6). The levels of each resource at which zero net growth occurs is the carrying capacity resource concentration and is labeled as "R*" for each resource and each species. For example, in Figure 10.6, R_1^* is the concentration of Resource 1 that is required to maintain net zero population growth for the species. At levels of Resource 1 above R_1^*, the population can grow (assuming all other resources are sufficient), but at levels below R_1^*, the population will decline, even if other resources are above carrying capacity levels. Note that population decline below R_1^* assumes that Resource 1 is an essential, non-substitutable resource. An **essential resource** is one that is absolutely required for survival and growth of the species; a **substitutable resource** would be one that could be replaced by some other resource if its concentrations fell below carrying capacity concentration for the species. If the resource being modeled is anything other than essential and non-substitutable, other models would be used to describe the outcome of species interactions (Tilman 1980).

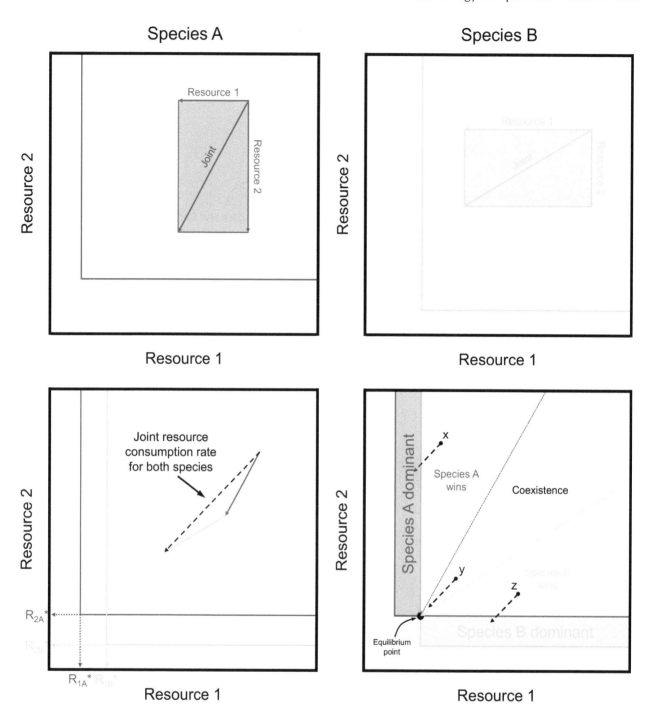

FIGURE 10.7   A Two-Species Model of Competition for Essential Resources. Panels A And B Illustrate Resource Consumption Rates and Zngis for Each of the Two Species. The Combined Rate of Resource Consumption for the Two Species, Considering Each of their Joint Resource Consumption Rates, is Shown by the Dashed Black Vectors in C and D. The Combined Resource Needs for Population Maintenance (Zngis), Resource Consumption Rates, and Availability of Resources in the Environment Will Determine the Outcome of Competitive Interactions between the Two Species (D). For Example, Given the Combined Resource Consumption Rate, if the Supply of Resources begins at Point X, Species A Will Eventually Dominate the Community because the Two Species together Will Reduce the Supply of Resource 1 below that Which is Required for Species B to Maintain its Population (I.e., below the R_1* for Species B). (Modified from Tilman (1985).

We can extend this basic, one-species model to consider two or more species (Figure 10.7). In Figure 10.7, we have two species that exhibit different rates of consumption of the two resources. Because the two species use shared essential resources that are in limited supply, we must consider the combined rate of resource consumption when anticipating the outcome of the interaction between the two species (Figure 10.7C). With this in mind, the outcome will also depend on both the rate of resupply of resources into the habitat and the concentrations of resources available in the environment when the two species encounter one another (Figure 10.7D). For example, if the beginning concentration of the two resources is at the point indicated by "z" in Figure 10.7D, then, based on the two species' consumption rates, Resource 2 will eventually fall below the level required for Species A to maintain non-negative population growth. That is, Resource 2 will be drawn below the R_2^* for Species A. At that point, Species B will dominate the community because it can still maintain non-negative population growth, while Species A is declining. Assuming sufficient resource resupply rates into the habitat, the two species can coexist within the central region indicated in Figure 10.7 because their combined rates of resource consumption pull resource concentrations toward the point at which their two ZNGIs overlap.

This model provides an explicitly quantitative approach to understanding how competition takes place as well as why a species might dominate a community in which it occurs. The specific mechanism for community dominance in this framework is the ability to maintain non-negative population growth at levels of resource availability below that at which neighboring species fail to do so. This sounds remarkably similar to what Grime referred to as stress tolerance. However, remember that Grime was discussing stress tolerance in the context of habitat characteristics that select for suites of species attributes (i.e., life history strategies), whereas Tilman's model is focused strictly on the mechanism of resource-based species interactions. For those interested in exploring these two ecologists' ideas about plant competition further, Grace (1991) discusses similarities and differences between Grime's and Tilman's approaches to plant species interactions, including ways to reconcile their two bodies of work.

10.2.2 QUANTIFYING COMPETITION

Now that we have a framework for thinking about what competition is, how it occurs, and what may determine the outcome, we will examine how one collects data to quantify competition and other interactions among plants. We will consider two components of this process. The first of these is the design of the study that will be used to measure interactions, and the second is the analysis of the data that are collected.

10.2.2.1 Design of the Experiment

As defined previously, competition is an interaction involving a shared requirement for one or more resources in limited supply that leads to a reduction in the survival, growth, or reproduction of the individuals involved in the interaction (Begon, Harper, and Townsend 1990). In measuring competition, the latter half of this definition is helpful in two ways. First, it allows one to measure something other than the actual uptake of individual units of resource, which would be a difficult undertaking, at best. Second, the explicit expectation of a reduction in survival, growth, or reproduction among competing individuals provides

specific guidance on parameters that may be measured in evaluating whether competition is taking place. That is, one should measure, in some way, the survival, growth, and/or reproductive output of the individuals we think may be competing with one another.

These measurements can be carried out in field experiments or via experiments conducted under more controlled conditions (Figure 10.8). Regardless of the system in which the study is executed, a critical component of these experiments is that plant attributes be measured in the presence and in the absence of interaction between the putatively competing individuals. Under controlled conditions, this can be accomplished by growing plants in isolation and comparing performance of those plants with performance of plants grown in the presence of a potential competitor (Figures 10.8 and 10.9). In the field, this separation can be accomplished by stepwise removal of components of competitive interaction (Figure 10.10). For example, to examine the effects of competition for light, we might manipulate canopy shading by one plant and compare growth of its neighbors versus growth of plants neighboring an unmanipulated plant of the same species (Figures 10.8 and 10.10). Removal of belowground competition for water and nutrients in a natural field setting is a bit more complicated, usually involving targeted removal of the putative competitor by precise application of herbicides, the use of physical belowground barriers, or trenching around study plots (Figure 10.10). Another important consideration under natural field conditions is that factors other than competition can influence plant performance. For example, in my own field experiment shown at the bottom of Figure 10.8, fences had to be erected around study plots to exclude herbivory by rabbits.

In controlled experimental conditions, we remove the effects of interaction with interspecific neighbors by growing plants alone or with other individuals of the same species (Figure 10.9). However, if we base our estimate of the effects of interaction with neighbors on comparison with an individual plant grown in isolation, we risk exclusion of density effects from our estimate of interaction strength. That is, if we compare growth of one plant of species A in isolation with that of one plant of species A grown with one plant of species B, we fail to include the simple effect of having twice the density of plants in our mixed-species experimental group (Jolliffe 2000). On the other hand, we may also be interested in the per capita performance of plants grown without neighbors as an index of the species' true growth potential (Gurevitch et al. 1990). This presents the investigator with a potential decision to make in whether to carry out the experiment with paired plants only, with paired plants plus individuals, or with other densities of plants to more fully explore density effects on interactions between the two species.

There is no universally agreed-upon standard for conducting such experiments, but there are a few designs that tend to be used more frequently than others (Figures 10.9 and 10.11). The simplest approach to carrying out experiments to

FIGURE 10.8   Examples of a Controlled Experiment (A–D) and a Field Experiment in a Natural Wetland ((E–F) Designed to Test Competitive Interactions Among Wetland Plants. In Controlled Experiments, Factors such as Light and Nutrient Availability and Plant Density can be Relatively easily Manipulated to Test Specific Aspects of the Interaction. The Study Shown in A–D Manipulated Density of Individuals of Two Species (*Juncus Effusus* and *Schoenoplectus Tabernaemontani*) to Examine the Effects of Plant Density on Outcomes of Competition Between these Species. In the Field Experiment Shown in E–F, the Canopy of *Juncus Effusus* Was Manipulated (Panel F) to Assess the Impact of Competition for Light Between *J. Effusus* and Neighboring Plant Species in the Talladega Wetland Ecosystem (Alabama, USA, Mentioned in Chapter 4).

Density Effects   Density Effects

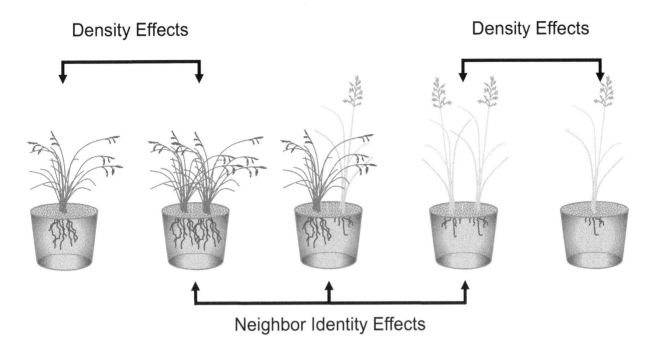

Neighbor Identity Effects

FIGURE 10.9   Controlled Pot Experiments Benefit from Providing Control Over Virtually all Conditions under which the Plants Interact. One Important Consideration in these Experiments is the Density of Plants that are being Grown together. Various Experimental Designs have been Developed to Attempt to Quantify Per Capita Effects of Interacting Individuals, while Controlling or Accounting for Plant Density. The Central Three Pots here Labeled "Neighbor Identity Effects" Represent One of the Simplest Designs, the Simple Pairwise Design, Where Intraspecific and Interspecific Pairs of Plants are Grown for Comparison of Performance in Monoculture and Mixture.

investigate potentially competitive interactions between two plant species is the simple pairwise design (Figure 10.9), in which each species is grown in monoculture and in mixed-species pairs. The performance of each species in monoculture is then compared with its performance in mixture to assess the nature of interactions between the two species. This design is the simplest form of a category of designs referred to as replacement series (Figure 10.11), in which plants are grown at a constant density (e.g., two plants per container) but individuals of one species are replaced with individuals of the other species as one progresses from a monoculture of one species to a monoculture of the other. All of these designs have been criticized for various reasons, including the potential to intermingle effects of species identities with those of plant density (Austin et al. 1988; Gurevitch, Scheiner, and Fox 2006; Jolliffe 2000). One proposed solution to these issues has been to conduct experiments using an addition series design, such as that illustrated by Figure 10.11 in its entirety, and then analyzing the results using what is referred to as a **response surface approach**. This approach considers performance of both species at all densities simultaneously to evaluate the potential performance of each species at each density, using three-dimensional mathematical modeling. Although this approach solves the issue of confounding species identity with plant density, it is mathematically complex, and some authors have suggested that the benefits from this approach may not always outweigh the complexity or the increased effort (and cost) to carry out such a large and complicated

study (Gurevitch, Scheiner, and Fox 2006). In the end, the design of the study should match the objectives of the investigators, the limitations of each candidate design should be considered, and, if necessary, multiple designs might be combined to yield the most robust estimate of the effects of competition between species of interest.

10.2.2.2   Measuring Plant Performance

As defined previously, we expect that two individuals competing with one another for resources will experience a reduction in the survival, growth, or reproduction because of the interaction. Thus, we should measure one or more aspects of survival, growth, or reproduction to support efforts at quantifying interactions among species. Survival is relatively straightforward to measure, but it has been infrequently used as an index of plant performance (He, Bertness, and Altieri 2013; Strobl, Schmidt, and Kollmann 2018; Younginger et al. 2017). Strobl, Schmidt, and Kollmann (2018) suggest this could be a result of the low variability in binary survival data (i.e., an individual either survives or it dies) in contrast with the more continuous and variable nature of growth and reproductive data. These characteristics of growth and reproduction result in those metrics providing greater statistical sensitivity to detect even subtle differences in plant performance, in contrast with the less subtle survival data.

Reproductive data can consist of measures of flower, fruit, or seed production, and they can be parsed into separate maternal and paternal metrics. In an evaluation of the

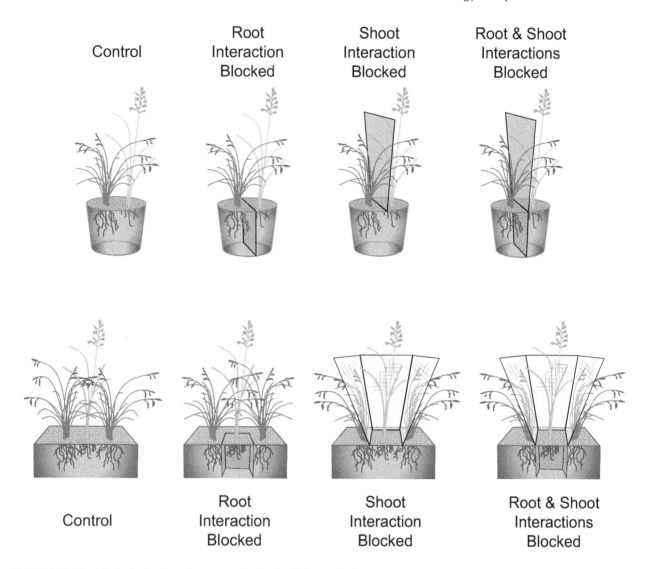

FIGURE 10.10 Manipulative Experiments can be Used to Dissect the Separate Effects of Above- Versus Belowground Interactions Between Neighboring Plants. The Upper Row of Figures Illustrate the Use of Barriers in an Experiment Using Potted Plants to Block Aboveground, Belowground, or Both Categories of Interaction. Control Treatments Allow all Interactions Between Neighbors, Blockage of Root Interactions Allows Aboveground Interaction, Blockage of Shoot Interactions Allows Belowground Interaction, and the use of Both Barriers is Meant to Remove all Interaction Between the Plants. In a Field Setting, the Same Procedures can be Used to Restrict the Type of Interactions Occurring Between a Focal, or Target Plant, and its Neighbors. This is Similar to the Approach Used in the Figure 10.8 (E–F) to Remove Shading by *Juncus Effusus*, Except in that Study, the Neighbors Collectively Formed the Target Species Assemblage. Modified from Gurevitch, Scheiner, and Fox (2006).

suitability of biomass measures as surrogates for fitness (defined in Chapter 8 as the production of viable offspring), Younginger et al. (2017) found that 58% of studies they examined used some measure of seed production as a fitness estimate. Flower production was used in 39% of the studies, and fruit production in 33%. A much smaller percentage of studies parsed reproductive effort into maternal (6%) and/or paternal (13%) fitness. Despite its broad use in studies focused specifically on plant fitness, reproductive effort can be much more difficult to quantify than biomass production, owing to the need to collect samples and data at appropriate times to accurately reflect flower, fruit, or seed maturity, as well as the potentially broad temporal variation

in each of these aspects of reproduction, in addition to the long times required to reach sexual maturity in some perennial species. As a result of these factors, most plant ecology studies use some measure of growth as their index of plant performance.

Strobl, Schmidt, and Kollmann (2018) and Younginger et al. (2017) addressed the question of whether growth is an adequate index of plant fitness, experimentally (Strobl, Schmidt, and Kollmann 2018) and by examining other published studies (Younginger et al. 2017). The experimental work of Strobl, Schmidt, and Kollmann (2018) found that in *Drosera rotundifolia* and *Eriophorum vaginatum*, reproductive effort was correlated with numerous measures of

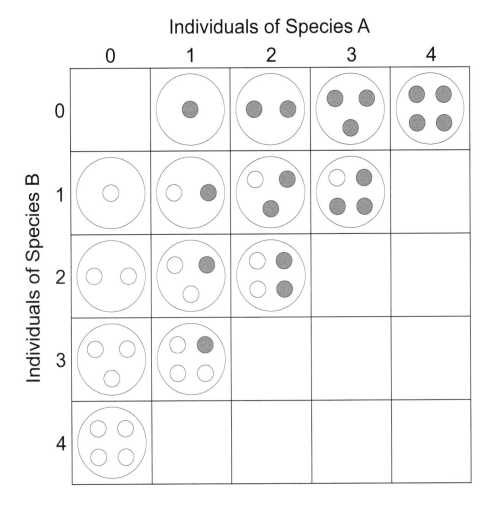

FIGURE 10.11   The Addition, or Additive, Series is a Complex Design to Investigate the Nature of Interspecific Plant Interactions. This Figure, in its Entirety, Represents an Addition Series Design. Within this Design, there are Component Designs, such as the Single-Density Replacement Series (any of the Diagonals with Two or More Plants Per Container) and the Additive Series (any of the Mixed-Species Rows or Columns). Note that the Replacement Series Utilizing a Density of Two Plants Per Container is also Referred to as the Simple Pairwise Design. Additionally, the Two Mono-Specific Series (Top Row and Leftmost Column) can be Used to Assess the Effects of Density on Monoculture Plant Performance.

vegetative growth, including overall plant diameter, number of leaves produced, height, biomass, and changes in measures of plant size over time. The literature survey conducted by Younginger et al. (2017) showed that 42% of studies assessing plant fitness included some measure of plant size as an index or surrogate for fitness. Among studies that included plant size along with an actual fitness measure, approximately two-thirds found a positive correlation between plant size and fitness; the remaining third of those studies found no statistically significant relationship between size and fitness. Those authors attributed an unspecified percentage of the studies finding no correlation to the presence of a threshold size for onset of reproduction in some species. That is, some longer-lived species must reach a minimum size before sexual reproduction is possible; beyond that size, reproductive success may be unaffected by plant size. In the end, Younginger et al. (2017) concluded that plant size may be an acceptable index

of potential fitness, but they cautioned that investigators should first consider the species in question and whether there may be limits to the applicability of size as an indicator of fitness.

This brings us to measures of plant size as indicators of performance in the presence versus absence of neighbors in efforts to determine the nature of interspecific interactions. Commonly used measures of plant size include numbers of stems or leaves, height or diameter of stems, diameter of tussocks or rosettes, leaf area, areal coverage of the plant canopy, and mass of aboveground, belowground, or total plant mass, collectively or by separate parts (Gibson 2015). All these measures except biomass can be taken at multiple points in time for a given set of experimental plantings, without destruction of the plants themselves. This is advantageous, in that it allows one to follow performance of a set of plants through time in response to changes in the environment or changes in growth, survival, or reproductive status of neighboring plants. The disadvantage to this is that none of those measures is a perfect index of the actual biomass produced by the plant during the study. However, we often are willing to accept this imperfection in exchange for the ability to determine relative performance of individual plants through time, as this may give a more complete picture of the temporal dynamics of plant interactions than would a single endpoint biomass sample (Gibson 2015). The best scenario, however, would be to collect data through time in addition to endpoint biomass data to allow greater ability to evaluate the nature of plant interactions and whether they may change over time, if one had the time and resources to do so.

Another drawback of obtaining an endpoint biomass measurement is that this approach carries the implicit assumption that all plants began with a comparable size. The problem presented by this is that initial size differences, even prior to seed germination, can have compounding effects on performance in the presence of potential competitors (Gurevitch et al. 1990; Jolliffe 2000). To avoid potential confounding effects of differences in initial plant size, a measure or estimate of initial plant size can be determined. This initial size determination can be made by weighing seeds prior to planting, by weighing seedlings or established plants prior to planting, or by weighing a representative sample of plants that can be subjected to destructive measurements. The initial and final biomass measurements then can be used to calculate the relative growth of experimental plantings.

$$Relative\,Growth\,Rate\,(RGR) = \\ \frac{\ln(Final\,Mass) - \ln(Initial\,Mass)}{Time} \quad (10.3)$$

In Equation 10.3, time can be represented in any units relevant to the study system, such as days, weeks, months, or years, depending on the species of interest and the duration of the study. The units for RGR will thus be mass produced per unit of initial mass per time, such as grams per gram per week or kg per kg per month, and these will simplify to units of "per time" because the units of mass (e.g., grams per gram) will cancel one another out. The relativization of mass production into a relative growth rate is important, again because of the impact that initial size can have on plant performance over time.

A final complication associated with measuring initial mass of experimental plants is that live plants (including embryos still within a seed or fruit) contain some amount of water, and the mass of this water can vary over time, sometimes considerably. In addition to variation over time, water content can also vary among species. These sources of variation have the potential to add unknown variation to initial size estimates based on live plant biomass. Because of this, we often measure dry plant mass to obtain a more precise and more reliably comparable estimate of plant biomass among species. The drawback to this, however, is that

once plants have been thoroughly dried for these measurements, they cannot be used as study subjects. The result is that we experience a trade-off, scientifically, between obtaining less reliable initial live biomass estimates for the actual study subjects and obtaining more accurate initial dry biomass estimates of a representative sample of plants that are not used in the experiment itself. Most plant scientists opt for the latter, more accurate, approach and use dry plant biomass of representative samples for initial biomass measurements. This is then compared with dry biomass measurement of study subjects at the termination of the experiment or taken at multiple points during and at the end of the experiment. We will next look at how one would use the data obtained from a study of plant interactions.

10.2.2.3 Quantifying Competition

At the termination of a study examining interactions between two plant species, the investigator will have data on one or more measures of plant survival, growth, or reproduction. As we saw previously, these data often will be in the form of some type of growth measurement, most commonly biomass, but any of the other measures of plant growth discussed earlier may be used. An important consideration, also discussed earlier, is that the two species very rarely will have equivalent initial biomass, unless special care is taken to ensure this. As a result, plant growth metrics should be relativized to allow more appropriate comparisons of performance between the two species in the face of potential interactions. Grace (1995) provided a detailed essay on the importance of using relative, versus absolute, growth metrics in studies of plant interactions, including examples where the use of absolute metrics had led to inaccurate conclusions (Figure 10.12).

Weigelt and Jolliffe (2003) reviewed 57 formulas for calculating competition indices for plants, giving pros and cons of each, along with recommendations for the most reliable indices among them. They divided those formulas into those representing indices of **competitive intensity**, **competitive effect**, and **competitive outcomes**. Competitive intensity was described as the degree to which competition reduces performance of an individual plant, relative to its performance in the absences of competition. Competitive effect is similar to competitive intensity, except that it accounts for density effects of neighbors, and it often is expressed on an areal basis. Finally, indices of competitive outcome represent the potential long-term dynamics of species grown in mixtures and may be thought of as indicating which species would win or lose in competition (Weigelt and Jolliffe 2003).

I have summarized the indices of competitive intensity and competitive outcome that have found the most favor in ecological studies, including one (RII) that was unavailable for the analysis conducted by Weigelt and Jolliffe (2003; Table 10.3). Because indices of competitive effects incorporate an area aspect to account for plant density, and because most ecological studies standardize the area of experimental or observational replicates (e.g., pots, plots,

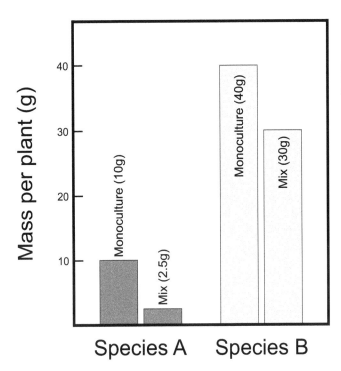

Index	Species A	Species B
ACI	7.5	10.0
RCI	0.75	0.25
LRR	-1.4	-0.3
RII	-0.6	-0.1
REI (expected)	-1.4	1.4
REI (observed)	-2.5	2.5

FIGURE 10.12 Hypothetical Example of Outcomes from a Competition Experiment Between Two Plant Species. In this Example the Two Species Started from Plantings with 1 Gram of Mass Each and were Grown Alone (Monoculture) and in Mixture with One Another. Formulas for Calculating Indices of Competition in the Table here are Given in Table 10.3. of Note here is that Only the Absolute Competition Index (Aci) Suggests That Species B Experienced Higher Competition Intensity when the Species were Grown Together; all other Indices Suggest that Species A Experienced Higher Competition Intensity. See Text for Further Discussion of These Results. Example Taken from Grace (1995).

or quadrats), I have not included any of those indices here. I have applied all the indices in Table 10.3 to the data used in Grace's (1995) hypothetical experiment to illustrate their use (Figure 10.12).

The most important note to make in the example shown in Figure 10.12 is that the absolute intensity of competition (ACI; the absolute reduction in biomass resulting from competition) is the only of these indices that suggests species B experienced a greater intensity of competition than did species A in this hypothetical experiment. When grown alone (i.e., in monoculture), species B produced four times as much biomass as species A. The greater inherent growth potential of species B meant that it could exhibit considerably greater amounts of biomass reduction than species A while continuing to experience a much smaller proportional size reduction. In contrast, when the intensity of competition is represented by the relativized RCI, we see that species A experienced three times the intensity of competition as did species B. This is much more obvious visually, as well, when we compare the relative size reduction in the graph in Figure 10.12.

The next two indices used in Figure 10.12 (LRR and RII) represent response ratio formulas. Thus, these indices represent the response of one of our species to the presence of the other. The fact that these two species were competing with one another is demonstrated by the negative sign in front of both the LRR and RII values. Recall from our

definitions of competition and from Figure 10.5 that competition is a reciprocally negative interaction between two individuals. If the value for LRR or RII had been positive, this would indicate that one (or both) of the species had been facilitated by the other. This type of interaction is discussed in a later section of this chapter. The values of LRR and RII in Figure 10.12 also substantiate those of RCI in that species A is indicated to experience the greater intensity of competition in this experiment. That is, for both of those indices, the value for species A is more negative than the value for species B.

Finally, we can evaluate the long-term prospects for these two species in this interaction by examining the relative efficient index (REI). The expected REI is the value of REI that we would expect to find as a result of this experiment if the species produced as much biomass when grown together as they did when grown in isolation. The negative value of REI for species A means that we would expect species A to be at a disadvantage in the interaction, because of its lower relative growth rate. Recall that, in this hypothetical experiment, each species began with one gram of biomass. Thus, because species B ended the experiment with four times the biomass as species A in monoculture, it has a much higher inherent growth rate; this is the cause for the expected disadvantage of species A in the interaction between these two. When we examine the REI calculated from the biomass produced by these species when

TABLE 10.3

Indices used to evaluate intensity and outcomes of plant interactions, taken from Weigelt and Jolliffe (2003) and authors listed in the table. Note that absolute competition intensity (ACI) is listed for comparison only; it is not recommended for use as an index of competition intensity.

Index Category and Index Name	Formula	Notes	Source
Competition Intensity			
Absolute competition intensity (ACI)	$P_{mono} - P_{mix}$	Fails to accurately represent proportional effects of interactions	Campbell and Grime (1992)
Relative competition intensity (RCI)	$(P_{mono} - P_{mix})/P_{mono}$	Relativizes the absolute index above, reflecting proportional impacts of interactions	Grace (1995)
	$(P_{removal} - P_{reference})/P_{removal}$	Modifies the above formula for use in mixed species assemblages, such as field experiments	Wilson and Keddy (1986)
Log response ratio (LRR)	$\ln(P_{mix}/P_{mono})$	Symmetrical for competitive and facilitative effects and has desirable statistical properties	Hedges, Gurevitch, and Curtis (1999)
Relative interaction intensity (RII)	$(P_{mix} - P_{mono})/(P_{mix} + P_{mono})$	Similar to LRR but its defined limits make it more applicable across all interaction types	Armas, Ordiales, and Pugnaire (2004)
Competition Outcome			
Relative efficiency index, observed (REI$_{obs}$)	RGR $_{A\,mix}$ − RGR $_{B\,mix}$	Observed performance trajectory of species A, relative to species B, in mixture	Connolly (1987)
Relative efficiency index, expected (REI$_{exp}$)	RGR $_{A\,mono}$ − RGR $_{B\,mono}$	Expected performance trajectory of species A, relative to species B, based on growth of species in monoculture	Grace (1995)

P = performance metric (biomass, RGR, etc.), per plant; mono, mix = plants grown in either monoculture (alone) or mixture (with another species); removal = plots where neighboring species have been removed (equivalent to monoculture); reference = unmanipulated plots of vegetation with neighbors intact (equivalent to mixture); ln = natural logarithm; RGR $_{A\,mix}$ = relative growth rate of species A in mixture with species B.

grown together (REI$_{obs}$), we see that species A was at an even greater disadvantage when grown with species B than we would have expected. Thus, the long-term prospects of species A in this interaction do not look good at all. We would expect species A eventually to be displaced from the experimental arena by species B if the study were allowed to continue.

One additional point to be made before moving on to the next section concerns the mathematical signs for the response ratios, LRR and RII. In the example in Figure 10.12, both indices have a negative sign for each species. This indicates that each species experiences interaction with the other negatively, as expected in competition. Thus, as suggested in Figure 10.5, we can use the sign in front of the value for indices such as these to determine the nature of the interaction being measured. This is the reason that, in Table 10.3, where these indices were summarized, I have indicated that they are indices used to evaluate intensity and outcomes of plant interactions, rather than for quantifying competition alone. In fact, some of these indices, the response ratios in particular, have been used much more broadly to examine responses of plants to all types of interaction. Hedges, Gurevitch, and Curtis (1999) proposed that response ratios, specifically LRR, were quite useful indices of the effects of many types of interaction in ecology, including such phenomena as predation in streams, invertebrate grazing

on benthic algae, and response of plants to increased atmospheric CO_2. We will see other examples of the use of these indices in subsequent sections.

10.3 AMENSALISM (− 0)

Competition was defined as an interaction resulting from the shared requirement for one or more resources that are in limited supply, resulting in reduced survivorship, growth, and/or reproduction of the competing individuals. The reduction in plant performance that we observe comes as a result of the competing individuals having fewer resources to allocate to growth, survival, and reproduction. We further stipulated that, from a practical standpoint, it must be possible to measure a reduction in performance to be able to conclude that neighboring individuals are, in fact, competing with one another. What happens, however, when we only detect a reduction in the performance of one of the presumably interacting individuals? By definition, we will have identified a case of **amensalism**, where one individual has been negatively affected, while the other exhibits no detectable effect from the interaction (Begon, Harper, and Townsend 1990).

It may well be that the two individuals in an amensalism share a requirement for a resource in limited supply and that each of them experiences at least a small reduction in performance because of reduced availability of that resource. However, the degree to which performance of one of the individuals is reduced may be so small that we cannot detect it with the usual tools available to us. For example, we are limited by the precision with which we can measure biomass, and the ability to discern statistical differences between groups of individuals grown in mixture versus monoculture is strongly influenced by the magnitude of natural variation among individuals within each group. This type of situation is often referred to as **asymmetric competition**, and it is quite common in nature (Begon, Harper, and Townsend 1990; Keddy 1990).

One widely cited example of this among wetland plants is a study by Grace and Wetzel (1981) investigating habitat partitioning by two cattail species: *Typha latifolia* and *T. angustifolia* (Figure 10.13). When these two species were grown in monoculture, each was able to establish at water depths ranging from 15 cm below the soil surface to 50 cm of standing water, at which point *T. latifolia* began to decline in productivity, while *T. angustifolia* continued growing out to more than 100 cm depth. In natural mixed populations of these two species, however, *T. angustifolia* was not observed at depths less than 15 cm, and even at that depth, it performed very poorly when in the presence of *T. latifolia*. Overall performance of *T. angustifolia* was reduced approximately 40% by the interaction. *Typha latifolia*, on the other hand, experienced no statistically detectable effect from the interaction (Figure 10.13). Additional

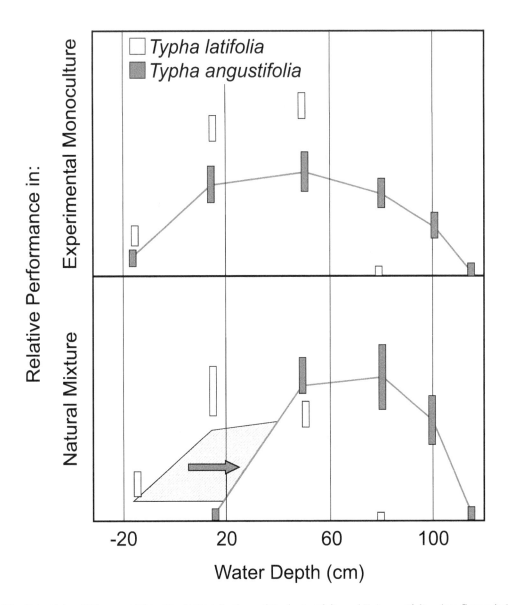

FIGURE 10.13   Potential and Observed Water Depth Distributions of *Typha Latifolia* and *T. Angustifolia* when Grown in Monoculture (Upper) and Observed in Natural Mixtures (Lower). The Hatched Area and Arrow in the Lower Panel Indicate the Area from Which *T. Angustifolia* was Displaced as a Result of Competition with *T. Latifolia* at Shallower Depths. Vertical Colored Bars Represent the 95% Confidence Intervals for Biomass of Each Species, and Lines Connect Mean Biomass Values for Each. Redrawn from Grace and Wetzel (1981).

work in this system showed that clonal propagules of *T. angustifolia* failed to establish within stands of *T. latifolia* (Grace and Wetzel 1981). The mechanism for this competitive displacement of *T. angustifolia* was concluded to be asymmetric competition for light, wherein the broader leaves of *T. latifolia* provided a significant advantage in the shallower depths to which it is adapted.

Another common example of amensalism is **allelopathy**, in which a plant produces chemical compounds that bring about a reduction in the performance of individuals that come into contact with those compounds (Gopal and Goel 1993; Gross 2003). Although the original use of the term allelopathy included stimulatory effects on the recipient plant (Gross 2003), most published examples of allelopathy deal with inhibitory effects that are presumed to provide a competitive advantage to the plant that produces the active chemical compounds (termed **allelochemicals**). Few studies investigate changes in performance of both the producer and the recipient of the allelochemicals, but in many cases investigations of allelopathy begin with the observation that performance of one species is substantially reduced when in the presence of the putatively allelopathic species (Gopal and Goel 1993; Gross 2003; Williamson 1990). Thus, many of these studies begin with the observation of a highly asymmetrical interaction between two species of interest, fitting an approximation of amensalism.

10.4 ALLELOPATHY IN AQUATIC AND WETLAND PLANTS

Gopal and Goel (1993) cited several examples of diminutive aquatic plants appearing to suppress or eliminate the growth of larger submersed or even floating-leafed species. Some of these examples involved various species of *Eleocharis* and included examples of larger plants having been inhibited by exposure to water in which *Eleocharis* species had been grown, even in the absence of the *Eleocharis* plants themselves. Ervin and Wetzel (2003) list more than 20 examples of aquatic and wetland plant species in which ecologically relevant examples of allelopathic interactions had been documented. Those examples included all growth forms of wetland angiosperms, as well as ferns (*Osmunda* species) and conifers (*Taxodium distichum*).

The missing piece in most of studies of allelopathy, however, has been identification of chemicals responsible for inducing the observed allelopathic effects. It turns out that this aspect of investigating allelopathy is quite difficult, in part the result of the diverse array of **secondary metabolites** produced by angiosperms. Secondary metabolites are chemicals that are not involved in primary plant metabolism (e.g., photosynthesis, respiration, growth, or development) and include two groups of carbon-rich compounds (terpenes and phenolic compounds) and nitrogen-containing compounds, such as alkaloids, as well as derivatives of these major groups (Taiz and Zeiger 2002; Figure 10.14, Table 10.4). These compounds often play roles in such processes as defense against or deterrence of herbivores and

pathogens, attractants of pollinators or seed and fruit dispersers, and allelopathy.

Because of the multitude of ways in which parts of these compounds can be interchanged and replaced, there are essentially an infinite number of secondary metabolites that could be present in nature, and Barbour et al. (1999) provided a count of approximately 18,000 known secondary metabolites from the three groups given earlier. As a result, chemical analysis of plant parts and extracts of those parts can be an expensive and time-consuming process. Nevertheless, there are examples of identification of biologically active allelochemicals from aquatic macrophytes. Gross (2003) reviewed many examples, including algicidal phenanthrenes (phenolic derivatives) identified from at least two species of *Juncus*, phenolics from *Typha domingensis* that inhibited growth of *Salvinia minima*, algicidal phenolics from *Nuphar lutea*, algicidal amines (nitrogen-containing compounds) from *Eichhornia crassipes*, and many algicidal phenolic compounds from *Myriophyllum spicatum*. A more recent review of anti-algal allelochemicals produced by aquatic angiosperms included production of allelochemicals by *Elodea* (phenolic derivatives), *Ceratophyllum* (small organic acids), *Pistia* (phenolic derivatives), *Potamogeton* (terpenes), and *Phragmites* (phenolic compounds; Nezbrytska et al. 2022). The predominance of studies identifying algicidal compounds likely results from a combination of the degree of sensitivity of algae to plant secondary compounds and the ease of experimental assessment of algal susceptibility, owing to their small size and short generation times, in contrast with larger, longer-lived angiosperms. However, some examples of allelopathy had their start with the observation of anti-algal activity in plant extracts. As was alluded to in Chapter 5, anti-algal allelochemicals likely are important for submersed plants because of the competition for light that occurs between the plants and planktonic or attached algae within aquatic habitats.

In my own work with *Juncus effusus*, I was interested in processes that allowed it to form almost monospecific stands in shallow wetlands in the southeastern US. Part of that work tested for aboveground competition between *J. effusus* and neighboring species (Figure 10.8), and I found significant effects of shading by *J. effusus* on it neighbors (Ervin and Wetzel 2002). Another line of investigation, however, explored the possibility of allelopathic suppression of neighbors. Those experiments were stimulated, in part, by the observation that seedlings of *J. effusus* appeared to suppress growth of algae in laboratory cultures (Figure 10.15). The unexpected discovery from ensuing experiments was that extracts of *J. effusus* leaf tissues had a much greater suppressive effect on its own seedling growth and development than on other species that were tested (Figure 10.15; Ervin and Wetzel 2000). Intraspecific allelopathic suppression is termed **autotoxicity** and has been reported in other emergent wetland plants that often form monospecific stands, such as species of *Typha* and *Phragmites* (Ervin and Wetzel 2003).

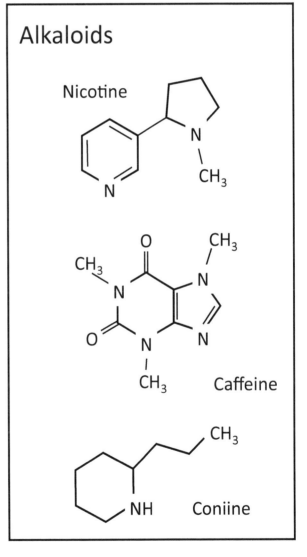

Terpenes

Monoterpene (10 carbons)

Sesquiterpene (15 carbons)

Diterpene (20 carbons)

Phenolic compounds

OH

Phenol

CH₃

O

O

OH

HO

Ferulic acid

O

OH

OH

Salicylic acid

Alkaloids

Nicotine

N

CH₃

N

O   CH₃

CH₃   N

N

O   N   N

CH₃   Caffeine

CH₃

NH   Coniine

FIGURE 10.14 Examples of the Three Major Groups of Plant Secondary Metabolites. Terpenes are Formed from 5-Carbon Subunits Called Isoprene Units and Include Insecticidal Pyrethroids, Limonoids from Citrus Fruits, and Phytoecdysones that can Function Similarly to Insect Molting Hormones. Phenol is the Starting Compound for Biosynthesis of Light-Activated Coumarin Compounds that Function in Herbivory Deterrence, Phenolic Acids (such as Ferulic Acid) that are Well-Known Allelochemicals, and Salicylic Acid, Which is Involved in Plant Pathogen Resistance and Related to Acetyl Salicylic Acid, also Known as Aspirin. Finally, the Nitrogen-Containing Alkaloids are Highly Active Animal Toxins, many of Which Act as Neurotoxins. Redrawn from Taiz and Zeiger (2002).

TABLE 10.4

Examples of plant secondary metabolites from the three major groups discussed in the text. Taken from Barbour et al. (1999), with additional information from Taiz and Zeiger (2002).

Metabolite Group and Subgroup	Examples	Activity	Distribution
Terpenes			
Mono-, Sesqui-, Diterpenoid compounds	Essential oils, latex, plant resins	Attractant odors, bitter taste, allergens, toxins	Widely distributes among angiosperms
Saponins	Saponins (soap-like compounds)	Dissociate lipid membranes, hemolyze blood cells	Found in ~70 angiosperm families
Cardenolides	Digitonin, from foxgloves (*Digitalis*)	Bitter taste, extreme toxicity	Common in Apocynaceae and Scrophulariaceae
Carotenoids	Yellow, orange, or red pigments, such as beta-carotene	Accessory pigment in photosynthesis, pollinator attractants	Common in angiosperm leaves and fruit
Phenolic Compounds			
Simple phenols	Caffeic and ferulic acids, Coumarins, Salicylic acid	Allelopathic, Photo-reactive toxins, Antimicrobial	Universal among angiosperms
Flavonoids	Colored (light-absorbing) compounds	Pollinator attractants, UV absorption, antimicrobial, animal toxins	Universal among angiosperms, gymnosperms, and ferns
Tannins	Diverse group of polymerized phenolic compounds	Bind to and deactivate proteins, including digestive enzymes	Widely distributed among angiosperms, especially woody species, often more concentrated in unripe fruit
Nitrogen-containing			
Alkaloids	Nicotine, Caffeine, Morphine, Cocaine, Coniine	Bitter taste, toxicity	Widely distributed among angiosperm roots, leaves, and fruit
Non-protein amino acids	Canavanine	Block protein synthesis, replace normal amino acids and block protein function	Common in legumes, Fabaceae
Cyanogenic glycosides	Sugar-bound cyanide molecule	Release toxic cyanide when degraded	Less common, often found in fruits and leaves
Glucosinolates	Isothiocyanates in mustards	Bitter taste, sometimes toxic	Eleven families, including the Brassicaceae

10.4.1 INTERPLAY BETWEEN ALLELOPATHY AND COMPETITION

As we have seen throughout much of this book, plants must perpetually balance responses to multiple selective pressures in their environments. While they balance need versus availability of resources with growth and allocation of acquired resources among life support functions and potential reproduction, plants also interact with neighbors that include competing plants, microbial symbionts, possible pathogens, herbivores, and, during the daytime, they also receive a continuous stream of solar radiation that can serve as a source of energy for photosynthesis but also a source of energy that can degrade proteins, cell membrane lipids, and other biochemical infrastructure. Because secondary metabolites are not required components of growth and development, their production is based on availability of some degree of excess resources.

Among the three major groups of secondary metabolites that are known to function as allelochemicals (Figure 10.14), some are carbon-rich (phenolic compounds and terpenes) and some incorporate nitrogen. We saw in Chapter 7 that nitrogen is one of the most commonly limiting nutrients for plants, whereas carbon is increasing in availability day by day, because of continuing emissions of CO_2 into the atmosphere. As a result, a potential determinant as to whether plants produce nitrogen-containing allelochemicals may be local nitrogen availability, which will be influenced not only by soil concentrations and resupply rates, but also by the local density of competing plants. At the same time, use of available nitrogen will be split between growth, development, and reproduction, with excesses potentially allocated to production of nitrogen-containing secondary metabolites, especially alkaloids (Inderjit and del Moral 1997).

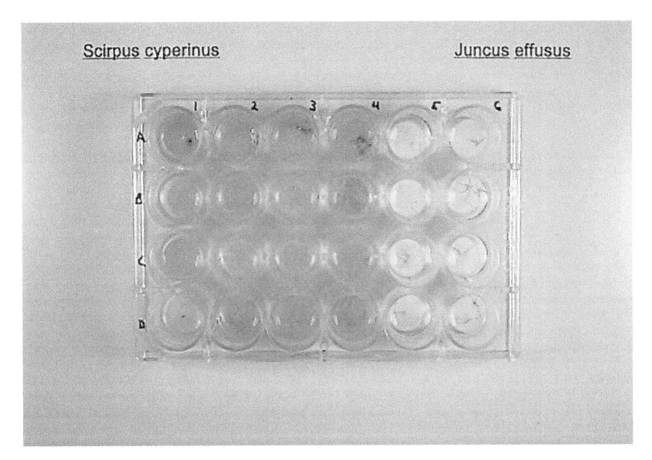

Scirpus cyperinus   Juncus effusus

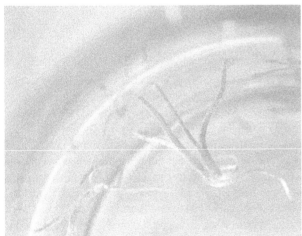
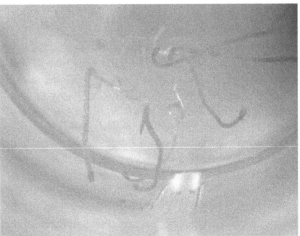

FIGURE 10.15 I Observed No Algal Growth in Water Containing Seedlings of *Juncus Effusus*, While Abundant Algae were Produced in Water with Seedlings of *Scirpus Cyperinus* and *Eleocharis Obtusa* (The Latter Not Shown Here). Investigation into Potential Allelopathic Effects of *J. Effusus* Revealed Significant Suppression in Growth and Development of its Own Seedlings (Lower Photos), whereas Seedlings of other Species Tested Showed Little or No Effect.

When nitrogen is in more limited supply, such as when neighbor density is high, plants may experience a surplus of carbon, relative to nitrogen availability for production of proteins and nucleic acids, and carbon may be shunted to other metabolites (such as terpenes and phenolic compounds) to ensure continued use of incoming solar energy. We saw a similar scenario when we discussed advantages of photorespiration in Chapter 6. Because phenolic compounds and terpenoids are involved in the production of many different subclasses of plant secondary metabolites (Table 10.4), increased production of these compounds under conditions of reduced nitrogen availability can enhance overall plant function while also helping to reduce performance of neighboring plants and algae that may be in competition for local nutrient supplies (Ervin and Wetzel 2003; Gross 2003; Inderjit and del Moral 1997). While this is clearly advantageous for the plant, the interconnectedness of these processes obscures attempts at determining whether pressures of competition have selected for production of allelochemicals under nutrient-deficient conditions or if, perhaps, shifts in metabolism under conditions of high carbon and light availability have coincidentally yielded chemicals that possess biological activity against potential competitors (Inderjit and del Moral 1997). Unfortunately, few studies have examined competition and allelopathy simultaneously, leaving many gaps in our understanding of interactions between these processes (Gross et al. 2012).

10.4.2 QUANTIFYING ALLELOPATHY IN WETLANDS

As with experiments aimed at quantifying competition between neighboring plants, the key aspect in measuring allelopathic interactions is quantifying plant performance in the presence and absence of the allelopathic compounds. In my work with *J. effusus*, for example, I measured seedling growth and development in the absence of plant extracts and compared that with the same performance measures of plants that were exposed to extracts (Figure 10.15). The use of plant extracts, rather than the growing plants themselves, is important, as a living plant neighbor could be a source of allelochemicals or a competitor for resources. Another benefit from use of plant extracts is that assays of allelopathic suppression from extracts can be conducted in a stepwise manner using progressively refined extracts to facilitate identification of the specific chemical compounds responsible for the observed suppression.

However, a disadvantage of using such an artificial system for quantifying allelopathic interactions is that we may lose the ecological context in which potential allelopathic chemicals exert their effects, as well as the ability to measure ecological outcomes of those effects. One solution to this problem has been the use of **activated carbon** in studies of allelopathy, especially field-based studies, where disentangling the multiple factors influencing plant performance can be difficult. Activated carbon, or biochar, is a material with chemical properties that allow it to adsorb and filter impurities, such as potential allelochemicals, from water, providing an effective tool for experimentally neutralizing allelochemicals (Lau et al. 2008; Zhang et al. 2021). In boreal forests, for example, Nilsson and colleagues (Nilsson 1994; Nilsson et al. 2000) used activated carbon to evaluate the mechanisms by which a dwarf shrub (*Empetrum hermaphroditum*) dominated nutrient-poor tundra ecosystems, using *Pinus sylvestris* as a test species. Using an experimental design like that in Figure 10.16, Nilsson (1994) found that the combination of root exclusion and allelochemical neutralization with activated carbon resulted in significantly larger plants of *P. sylvestris*. The use of activated carbon has been criticized because it sometimes can result in improved plant performance outside of allelochemical adsorption, so it is important to assess those effects while also assaying for potential allelopathy (Lau et al. 2008). One way of doing this would be to include monoculture assays for effects of activated carbon in the absence of potential allelochemicals, a method that also would allow one to quantify interaction effects on both the allelochemical producer and the recipient or target species (e.g., Figure 10.16). Lau et al. (2008) noted that only half of the papers they reviewed had included tests for the effects of activated carbon, and even fewer studies test for reciprocal interaction effects on both species in a putative allelopathic relationship.

Liu et al. (2020) used activated carbon to disentangle effects of competition and allelopathy between two species of *Sphagnum* in a northeastern China peatland. They included elements in their design to assess effects of activated carbon on plant performance, while also testing effects of competition, allelopathy, and hydrology, as an environmental driver of plant performance. They found no effects of activated carbon on growth of the two moss species but did find complex relationships between water table depth and interspecific interactions. Under high water table conditions, the larger of the two species experienced allelopathic inhibition from the smaller, and there were no indications of strict resource competition between the two. With a lower water table, both competition and allelopathy seemed to play roles in interactions between the two species. Again, we see that the nature of processes influencing plant performance in aquatic and wetland habitats can be complicated and involves not only interactions among the plants, but also interactions among different types of interaction, all of which may be influenced by abiotic components of the ecosystem.

10.5 COMMENSALISM (+ 0)

So far, we have examined interactions in which at least one partner experiences a negative outcome: mutually negative competitive (− −) interactions and negative interactions in which one partner is not obviously affected (− 0). Next, we will see interactions in which one partner may not be measurably affected but the other benefits. The collective term for such interactions (+ 0) is **commensalism**, and these interactions are driven largely through modification of some aspect of the environment by the presumably unaffected species. This modification may be brought about via modification of environmental factors through the unaffected species' growth form, metabolism, or some other factor of its biology that results in favorable conditions for the other species. In the more commonly recognized form of commensalism, there is often a considerable size difference between the two species, with the larger species itself serving as structural habitat for the smaller species (Figure 10.17). The smaller species is referred to as an **epiphyte**, because it grows upon (epi-) another plant (-phyte). In other, less well-known instances of commensalism, the plants sometimes are much more closely matched in size, with one species modifying habitat characteristics, rather than serving as habitat itself. A general term for this type of commensalism is **facilitation**, taking its name from the action of one species helping, or facilitating, the other.

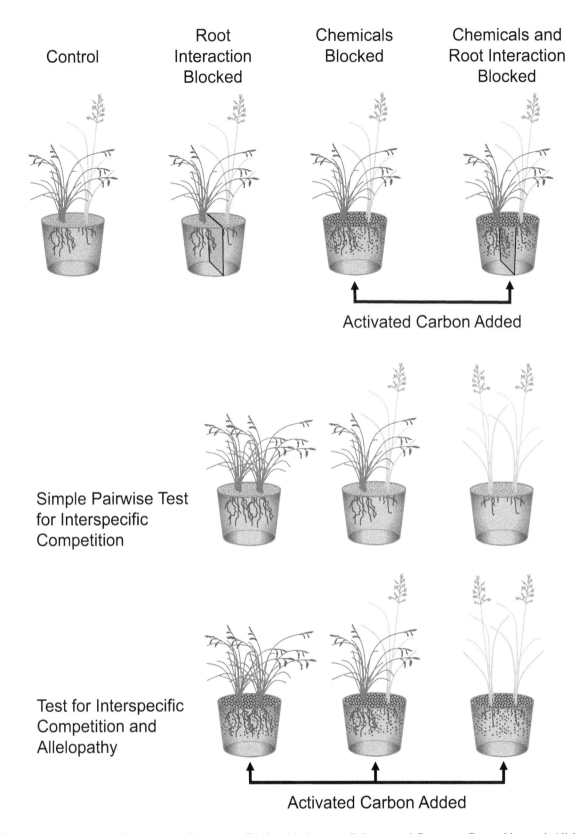

Control

Root Interaction Blocked

Chemicals Blocked

Chemicals and Root Interaction Blocked

Activated Carbon Added

Simple Pairwise Test for Interspecific Competition

Test for Interspecific Competition and Allelopathy

Activated Carbon Added

FIGURE 10.16 Potential Experimental Designs to Distinguish between Belowground Resource Competition and Allelopathy. Upper Row Shows a Simple Study Designed Solely to Distinguish between Belowground Resource Competition and Allelopathy (Modeled After Nilsson 1994). Lower Two Rows Show a More Complex Study that would not Only Help to Distinguish between the Type of Belowground Interaction (Competition Vs. Allelopathy) but also would Assess the Magnitude and Direction of Both Species' Responses to One another. That Is, the Lower Study, Via the Addition of Monocultures of Each Species, Would Inform Whether the Potentially Allelopathic Species Experiences a Reduction in Performance When in the Presence of the "Target" Species. The Lower Design also Would Provide Information on Any Potential Effects of Activated Carbon itself on Plant Performance, as Suggested by Lau et al. (2008).

FIGURE 10.17   Epiphytes, such as this Bromeliad (*Tillandsia* Species) Found Growing on a Bald Cypress in Austin, Texas (USA), Receive Considerable Benefit from the Plants Upon Which they Grow, while Causing No Readily Quantifiable Negative Effects for the Host.

In Chapters 5 and 6, we discussed obstacles to light, carbon dioxide, and oxygen availability as major stressors in aquatic and wetland ecosystems. We saw that wetland plants possess an array of adaptations that enhance their abilities to access and use these resources, but most of those adaptations serve to benefit only the individual plant that possesses them. For example, heterophylly can improve acquisition of limited light for submersed leaves and can enhance water conservation for aerial leaves, but this does not benefit the plant's neighbors. Recycling of respired carbon dioxide within aerenchyma assists individual plants in maintaining a favorable CO_2:O_2 balance for photosynthesis, but it is of little benefit to neighboring plants. Radial oxygen loss from the root system, however, modifies the sediment environment in ways that benefit all plants with roots lying within the oxygenated sediments.

This specific mechanism of wetland plant facilitation was demonstrated by Callaway and King (1996), who showed that oxygenation of wetland sediments by *Typha latifolia* at low temperatures increased the survival of neighboring willow (*Salix exigua*) and growth of forget-me-not (*Myosotis laxa*). At temperatures of 11–12 °C, willow plants grown in the absence of *T. latifolia* died within 11 days, whereas no plants died when planted in pots with *T. latifolia* neighbors. At those same temperatures, forget-me-nots grown with *T. latifolia* were approximately three times larger than plants grown alone. Oxygen concentrations in soil water within pots containing *T. latifolia* were in the range of 2.75 to 4.4 mg O_2 per liter, whereas oxygen concentrations in pots without *T. latifolia* were all 0.65 mg/L or less. Thus, it seems that amelioration of low sediment oxygen levels is a potential mechanism by which wetland plants

may facilitate one another. Performance of *Typha latifolia* was not measured in this study, so it remains unknown whether this was a true commensalism (+ 0), but facilitation of the neighbor plants was clearly present.

An experiment looking at interactions among three wetland plants in China examined the relative effects of these species on one another at three water depths (0, 20, and 40 cm depth; Luo et al. 2010). This study included experimental treatments to assess both intra- and interspecific interactions among one sedge (*Carex lasiocarpa*) and two grasses (*Glyceria spiculosa* and *Deyeuxia angustifolia*). When grown alone, all three species accumulated ten to 20 times as much biomass, or more, in unflooded conditions (0 cm water depth) than when grown at either 20 cm or 40 cm depth. Despite that substantial growth reduction under flooded conditions, being grown with a neighbor of any species resulted in greater biomass accumulation for *C. lasiocarpa* and *D. angustifolia*. *Glyceria spiculosa*, on the other hand, did not benefit from neighbors and appeared to have experienced 100% mortality when grown with either of the other two species. Under non-flooded conditions, however, growth of both *C. lasiocarpa* and *D. angustifolia* was reduced approximately tenfold in the presence of neighbors, indicating significant competition occurred in the absence of flooding. These contradictory results under flooded versus non-flooded conditions are quite common and fall under a concept known as the *stress gradient hypothesis* (Figure 10.18), wherein facilitation is expected to occur with greater prevalence in habitats experiencing high stress, while competition is expected to be more common in less stressful habitats (Bertness and Callaway 1994; He, Bertness, and Altieri 2013; Maestre et al. 2009).

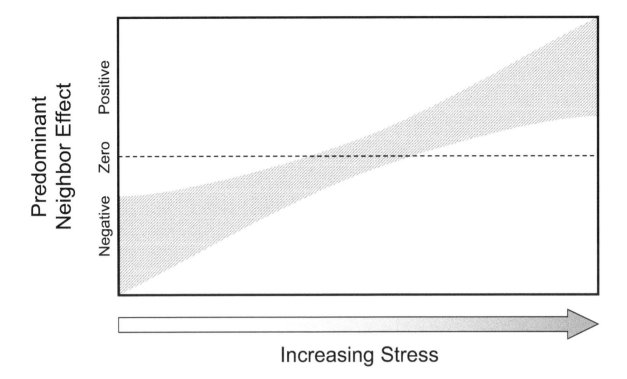

FIGURE 10.18 The Stress Gradient Hypothesis Generally States that the Nature of Interspecific Interactions Will Change from Predominantly Negative (I.e., Competitive) to Predominantly Positive, or Facilitative, Along Gradients of Increasing Stress. It should be Noted that the Nature of the Stressor may Vary (E.g., Biological or Physical Stressors) and Stressors Ought to be Relevant in the Context of the Species of Interest. Predominant Neighbor Effect in this Figure would be Represented By an Index such as Lrr or Rii, as Shown in Figure 10.12.

The stress gradient hypothesis has received a great deal of attention, as its superficial simplicity is simultaneously very attractive but seemingly improbable for natural systems. As with any general concept pertaining to natural systems, however, there are nuances that must be considered when attempting to put this hypothesis to use in either theoretical or practical settings. When we discussed stress earlier in this chapter, it was noted that stress is an intrinsic property of the stressed organism. Thus, the nature and degree of stress may vary from species to species and certainly among growth forms or across major clades of plant species, and it also may vary among ecosystems, even within a category of ecosystems such as wetlands.

For example, Fischman, Crott, and Angelini (2019) evaluated aspects of the stress gradient hypothesis along environmental gradients in coastal dune communities in the states of Georgia and Massachusetts (USA). They found that the dominant plant species in each region perceived the *a priori* defined stress gradients differently, and, consequently, restoration strategies incorporating elements of the stress gradient hypothesis provided different results in the two areas. A key conclusion from their work was that identification of species- and habitat-specific stressors and species responses are critical for successful application of the stress gradient hypothesis.

Campbell and Keddy (2022) conducted an experiment with ten species of emergent wetland plants to assess whether the stress gradient hypothesis might apply to a flooding gradient in southern Louisiana wetlands (USA). Their flooding gradient included plots that were inundated from 0 to 6.5 months of the active growing season. They found evidence of competitive displacement of species at the less frequently flooded end of the flooding gradient but no evidence of competition or facilitation under more frequently flooded conditions. Survival data suggested that flooding imposed a considerable stress, with half or fewer of the species surviving at 55% flooding frequency or greater, in either monoculture or mixture plots, at one year into the three-year experiment.

It is possible that the duration and depth of flooding in the Campbell and Keddy (2022) study were too stressful to allow for facilitative effects to be observed, given that fewer than half the species were capable of tolerating those conditions for more than one year. It is also possible that the duration of the study was insufficient to allow facilitator species to establish conditions that would allow other species to make use of them. For example, Egerova, Proffitt, and Travis (2003) found facilitation of *Baccharis halimifolia* shrubs by *Spartina alterniflora* in Louisiana salt marshes, but the *S. alterniflora* plants had established four years prior to the surveys of these investigators. They found that larger *S. alterniflora* clones were much more likely to harbor young *B. halimifolia* plants, and the larger clones hosted a greater number of *B. halimifolia* plants as well. Similarly, I found that the number of species making use of *Juncus effusus* tussocks was greater on older, larger tussocks in wetlands in Mississippi (USA; Ervin 2005). Tussocks in a six-year-old beaver wetland had no plant species living atop them, whereas those in an eight-year-old and 30-year-old wetland hosted from one to six plant species.

A broad assessment of the stress gradient hypothesis was performed in a meta-analysis by He, Bertness, and Altieri (2013). Those authors used data from 727 studies evaluating plant competition and facilitation from terrestrial, wetland, and aquatic habitats across the globe. They considered three different categories of stressors (herbivory, physical stress, and resource stress) and evaluated plant survival, growth, and reproduction as performance metrics. Support for the stress gradient hypothesis was present in the case of herbivory, representing biotic stress, where survival data indicated that interacting species tended to compete with one another at low levels of herbivory but facilitate one another's survival at high herbivory levels. In studies examining the effects of physical stressors, however, survival was always facilitated by neighbors, regardless of the stress level.

When considering physical stressors alone, effects on growth of target plants seemed to switch from competition at low physical stress (e.g., salinity or temperature extremes) to facilitation at higher stress levels, again supporting the stress gradient hypothesis. Across all stressors, however, the predominant influence of plant interactions on growth and reproduction (i.e., excluding survival data) was competitive suppression of target plants, regardless of the level of stress. In the few studies that focused specifically on wetlands (28 of the 727 examples), survival seemed to hold to expected stress gradient hypothesis patterns, but effects on plant growth seemed to always be competitive (He, Bertness, and Altieri 2013). This contrast with the general facilitation of neighbor plant growth found across the full dataset may have resulted from patterns exhibited by grasses and perennial plants, whose patterns matched that of wetlands.

Despite the variation in frequency of observations that match predictions of the stress gradient hypothesis, these studies all suggest that facilitation is a common, even if not universal, phenomenon in wetland ecosystems. Indeed, facilitation has been demonstrated many times in salt marshes, where the stresses of salinity, tidal fluctuations, and wave action are added to other stresses encountered in freshwater wetlands (Zhang and Shao 2013). Restoration methodologies in coastal ecosystems have attempted to incorporate the concept of facilitation in planting designs, as a means of enhancing the efficiency and effectiveness of restoration efforts (Fischman, Crott, and Angelini 2019; Silliman et al. 2015). However, as noted by Fischman, Crott, and Angelini (2019), considerations must be made for local species-specific responses to the most relevant stressors in the ecosystem under consideration to ensure successful application of these ideas.

The specific mechanisms by which facilitation takes place in wetlands will depend on the most important stressor for the species involved in the interaction. As discussed earlier, radial oxygen loss from the root systems of wetland plants can facilitate species that may have less well developed aerenchyma systems. In wetlands where grazing pressure is high, less palatable species can serve as refuge for species that are more susceptible to consumption by herbivores (Boughton, Quintana-Ascencio, and Bohlen 2011). In saltmarshes, shading by taller species can reduce evaporation from soils or highly salt-adapted species may preferentially take up or accumulate salts, both of which can reduce soil salinity for beneficiary species (Zhang and Shao 2013). Accumulation of soil, sediments, and organic matter in tussock-forming plants can reduce relative water levels for tussock-colonizing species, while also contributing to oxygen leakage into the soil environment atop the tussocks. In coastal marshes where wave action may make colonization of mobile sediment difficult, some species, such as *Spartina alterniflora*, can stabilize those sediments and facilitate colonization by other species (Bruno 2000). In any of these circumstances, a reduction in intensity of the stressor would be expected to result in a reduction in the benefit provided to the facilitated species, and this is essentially the basis for functioning of the stress gradient hypothesis.

10.6 EXPLOITATIVE INTERACTIONS (− +)

The final category of interactions we will examine, (− +) interactions, are referred to here as **exploitative interactions**. I have borrowed this term from Molles (2013), who describes an exploitative interaction as one in which "one organism makes its living at the expense of another." What this means is that one organism derives its energy and nutrients from the consumption of all or part of one or more other organisms. We will examine here two major categories of exploitative interaction involving plants: herbivory and parasitism/pathogenesis.

10.6.1 HERBIVORY

Herbivory is typically defined as the consumption of living plant tissue (Barbour et al. 1999; Gurevitch, Scheiner, and Fox 2006), and it usually has a net negative effect on the plant, while benefitting the herbivore (Figure 10.5). However, this definition is missing some subtle details that distinguish the act of herbivory, as we usually recognize it, from other forms of plant consumption. As Barbour et al. (1999) noted, organisms that consume plant tissue include **phytophagous** insects and nematodes (i.e., they feed on plants) and vertebrates such as browsers and grazers that feed on nonreproductive plant tissues. Other organisms that feed on living plant tissue include insects and vertebrates that feed on fruits and seeds, as well as parasitic and pathogenic microbes and even parasitic plants. The latter group of organisms usually are referred to as pathogens or parasites, and they are discussed in a later section. Among organisms that most closely fit the concept of herbivores,

grazers are those that feed on low-stature primary producers (e.g., graminoids for vertebrate grazers or tightly attached algae for aquatic invertebrate grazers), while **browsers** are those herbivores that feed on plant tissues that sit higher above the soil or substrate, such as leaves on shrubs or trees.

Organisms that feed on fruits are referred to as **frugivores**, and those that feed on seeds sometimes are considered to be **seed predators**, although seed predation technically would include only seed consumption that results in death of the enclosed embryo. Seed consumption that does not kill the embryo results in dispersal of the species, and thus is considered as dispersal, rather than herbivory. Likewise, frugivores usually are not considered to be herbivores because they often play a role in seed dispersal, and they thus are assumed to have a net positive effect on the plant (Barbour et al. 1999). It should be noted that even strict herbivory can sometimes have a quantifiably positive effect on the consumed species, through a process known as **overcompensation**, wherein consumption of vegetative tissues is followed by the plant regrowing more tissue than was consumed. Thus, if we were to consider measures of plant performance in the presence versus absence of herbivores, we might find a positive impact of herbivory. This process appears to be uncommon but is poorly understood and, as a result, poorly integrated into the ecological literature (Gurevitch, Scheiner, and Fox 2006).

10.6.1.1 Herbivores in Wetlands

Because wetlands lie at the interface of aquatic and terrestrial ecosystems and occur in freshwater, estuarine, and marine environments, they harbor a broad diversity of herbivores (Bakker et al. 2016; Batzer and Sharitz 2006; Cox and Smart 1994; Keddy 2000; Sculthorpe 1967; Wood et al. 2017). Among vertebrate herbivores, we see many species of birds, including waterfowl, such as ducks and geese. Some bird species (e.g., snow geese, *Anser* species) can have significant impacts on wetland vegetation and are the subject of substantial conservation efforts in North America. Vertebrate herbivores also are represented by fish and mammals, the latter including small herbivores such as muskrats, beaver, rabbits or hares, and nutria, as well as much larger organisms such as manatees, dugongs, moose, water buffalo, and even hippopotamus.

Invertebrate herbivores of aquatic and wetland plants include crustaceans (crabs, crayfish), molluscs (snails, slugs), nematodes, and many groups of insects. Within the insects, more than a third of insect orders contain herbivorous species, including the six most species-diverse orders: beetles (Coleoptera), wasps/bees (Hymenoptera), moths/butterflies (Lepidoptera), flies (Diptera), true bugs (Hemiptera), and grasshoppers/locusts/crickets (Orthoptera; Barbour et al. 1999; Newman 1991). Just as we saw in Chapter 8 regarding the diversity of plants and their insect pollinators, there also is a great diversity of insect herbivores, exhibiting a broad array of feeding strategies and feeding on essentially all plant parts (Bernays 1998; Figure 10.19).

10.6.1.2 Importance of Herbivory

It has been known for quite some time that aquatic and wetland plants can be important sources of both food and shelter to herbivores (Sculthorpe 1967); however, the importance of herbivory in the context of aquatic and wetland plant ecology was largely neglected until the last decade of the 20th century (Cyr and Pace 1993; Lodge 1991; Newman 1991). Cyr and Pace (1993), for example, showed that herbivores could remove roughly half the annual productivity of submersed and emergent plants, and that this held true across a range of plant productivity, from 10 to 1,000 grams carbon per m^2 per year. They further found that the rates of biomass removal for aquatic and wetland plants was three times higher than in terrestrial ecosystems, on average.

Newman (1991) noted that there had long been a predominant view of aquatic plants as a food source only for aquatic **detritivores** (animals that feed on dead organic matter) during much of the 19th and 20th centuries. This was likely to have resulted from a focus of aquatic ecologists on select aquatic insects in orders that contain few, if any, herbivorous taxa (e.g., Ephemeroptera [mayflies], Odonata [dragonflies and damselflies], and Plecoptera [stoneflies]). However, Newman went on to note that several studies had been published in the mid-20th century pointing to herbivory on aquatic plants by taxa from among the Lepidoptera and Coleoptera. He also pointed out that there were many examples of highly specialized insect herbivores, many of which had been the focus of searches for biological control agents of introduced aquatic weeds (discussed in more detail later). In fact, it seemed that specialization on one or a few plant species was far more common among insect herbivores than in other groups of invertebrate herbivores, such as gastropods or crustaceans. Insect herbivores tended to feed on an average of only two plant species, while crustaceans fed on an average of six or seven species and gastropods on nine or ten species. The high level of specialization among insect herbivores was hypothesized to be related, at least in part, to chemical defenses present in some species and evolutionary specialization to avoid or neutralize them, and this will also be discussed in the following section.

Both Newman (1991) and Lodge (1991) addressed the potential for low nutritional quality to influence avoidance of aquatic plants by herbivores and found this to be unlikely. Newman examined nitrogen content of 43 aquatic plant taxa, from among submersed, floating, and emergent species, and found an average nitrogen content of approximately 2.4%. Lodge found ranges of nitrogen content from approximately 1% to as much as 6% across emergent, floating, and submersed taxa, and those ranges all overlapped the range of values for terrestrial grasses, forbs, trees, and shrubs. Thus, these authors concluded that there was no nutritional basis for herbivores to avoid aquatic plants as a food source. Just as Newman did, Lodge (1991) noted that there exists considerable feeding selectivity among insect herbivores, in terms of aquatic and wetland plant taxa they will consume, and that this selectivity may be driven, at least in part, by chemistry of the plant species. Collectively,

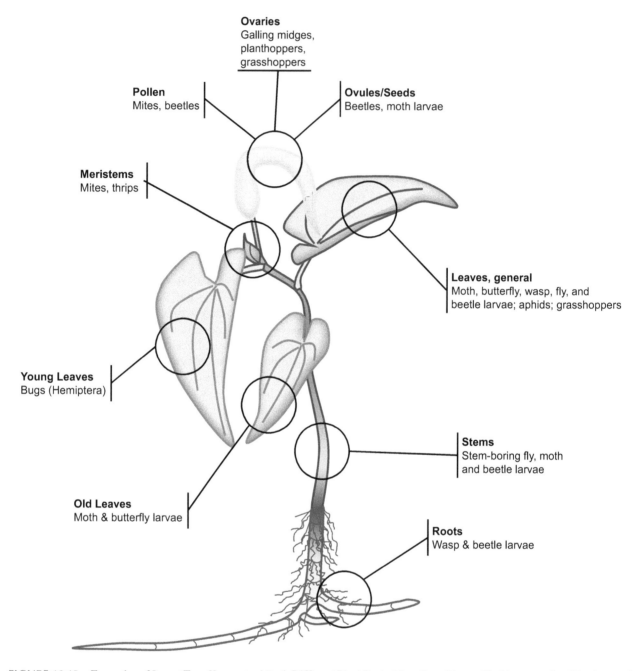

FIGURE 10.19 Examples of Insect Taxa Known to Attack Different Plant Parts. Not all are Known Herbivores or Seed Predators of Aquatic and Wetland Plants but are Known to Attack Vascular Plant Taxa in General. Adapted from Begon, Harper, and Townsend (1990), with Additional Information from Bertness, Wise, and Ellison (1987), Buckingham (1996), and Freedman et al. (2007).

the work of these two investigators provided evidence for widespread presence of herbivory on aquatic plants, not only among the plant taxa consumed but also the insect taxa that are prone to feed upon them.

A more recent study by Wood et al. (2017) aimed to more broadly assess the effect that herbivores have on aquatic plants by analyzing data from 163 studies published between the years 1961 and 2014, describing the effects of herbivory on aquatic plants on every continent and at latitudes from 3 to 73 degrees from the equator. This analysis showed that, on average, herbivores reduced aquatic plant abundance by approximately 47%, similar to the average values found by Cyr and Pace (1993), and this was regardless of the herbivore taxon or the ecosystem in which herbivory was measured (Figure 10.20). Their analysis included herbivorous birds, mammals, crustaceans, molluscs, fish, insects, and echinoderms and included studies carried out in freshwater lakes, freshwater rivers, freshwater wetlands, estuaries, salt marshes, and marine habitats. Fish and molluscan herbivores tended to cause the highest reductions in plant abundance (averaging 29% to 85% across ecosystem types), while insects and birds tended to

Habitat Type

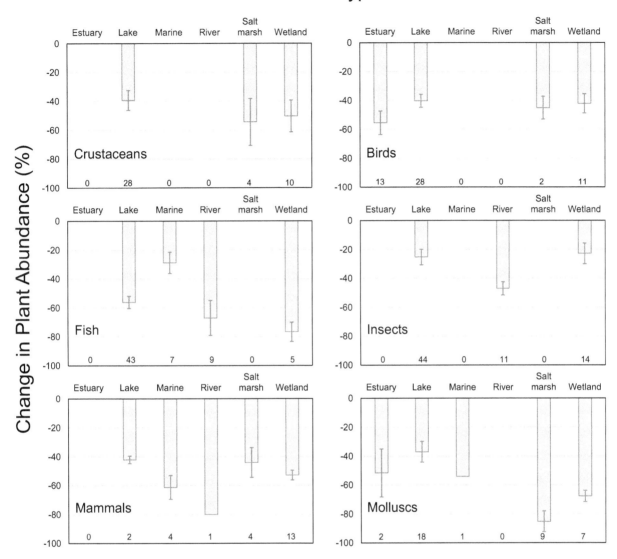

FIGURE 10.20   Mean (+/– Standard Error) Reduction in Aquatic Plant Abundance by Herbivores Across Multiple Ecosystem Types. Overall Mean Reduction was Approximately 47% (+/– 3%) Across all Studies Examined. Numbers of Studies Represent by Each Herbivore × Ecosystem Combination are Given across the Bottom of Each Individual Chart. Data Taken from Wood et al. (2017).

cause the least (averaging 23% to 55% across ecosystem types). Fish and molluscs also comprised all but one of the examples of complete removal of plant material (i.e., 100% reduction in abundance) across all the studies examined by Wood et al. (2017), with those complete removals being observed in lakes, salt marshes, and freshwater wetland ecosystems. These results strongly support widespread importance of herbivores in aquatic and wetland ecosystems, as was argued by Cyr and Pace (1993), Lodge (1991), and Newman (1991) more than 20 years earlier.

Around the same time as the Wood et al. (2017) study was published, a similar **meta-analysis** (an analysis of previously published analyses) was conducted to evaluate the importance of herbivores in coastal wetlands, in the context of other factors known to affect plant productivity,

abundance, and diversity. Authors He and Silliman (2016) evaluated the strength of herbivore control of plant performance in coastal marshes and mangrove swamps based on a collection of 178 published studies conducted across the globe (163 coastal marsh studies and 15 mangrove swamp studies). Herbivores in their study included insects, snails, crabs, waterfowl, small mammals, and livestock. They found that different herbivore taxa had consistently negative effects on aboveground biomass of salt marsh plants, as found by Wood et al. (2017), but herbivore effects on other performance metrics were much less consistent. For example, no herbivores had statistically detectable effects on areal cover of plants, and their effects on belowground biomass, plant density, and plant reproduction were variable, with some herbivores having negative effects, some

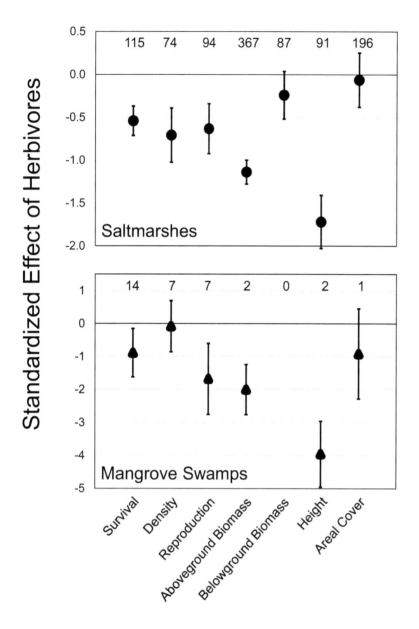

FIGURE 10.21 Standardized Effect Size of Herbivores on Performance of Saltmarsh and Mangrove Swamp Plant Species. Data are Mean +/- 95% Confidence Intervals, Taken from He and Silliman (2016). Where Error Bars Overlap with Zero, the Effect Size is not Statistically Different from Zero (E.g., Belowground Biomass in Saltmarshes). Numbers at the Top of Each Panel Represent the Sample Size for Each Metric within Each Ecosystem Type.

having no effects, and livestock having positive effects on plant density and belowground biomass. Collectively, however, He and Silliman (2016) found that herbivores negatively impacted most plant performance metrics in coastal wetlands (Figure 10.21).

In addition to the direct effects of herbivores on plant performance, He and Silliman (2016) assessed the indirect effects of herbivores on plant interactions and the interaction of herbivory with abiotic factors that are also known to influence plant performance. In terms of plant interactions, plants tended to negatively affect one another's biomass accumulation and survival in the absences of herbivores (interpreted as competition), whereas when herbivores were present the plants had no effect on one another's survival, and their effects on growth were reduced. In other words, the actions of herbivores tended to buffer the negative effects that plant species had on one another. Regarding environmental factors, eutrophication, disturbance, and

flooding of coastal wetlands all tended to exacerbate the negative effects of herbivory; the effect of herbivores on plant performance were approximately doubled by all these environmental factors.

It is clear then, in consideration of the aforementioned studies of herbivory in aquatic and wetland ecosystems, that herbivory is not only an important ecological process in these systems, but it has important interactions with other processes (Bakker et al. 2016). Herbivores have the potential to moderate interactions among the plants, while being affected by abiotic factors that are also known to impact the plants (flooding, nutrient enrichment, disturbance). In the next section, we'll examine some of the tools that plants employ to reduce the impacts of herbivory.

10.6.1.3 Chemical Defenses against Herbivory
Over evolutionary time, the relationships between plants and their insect pollinators and herbivores have resulted in

an array of plant-produced chemicals that have the effect of attracting desirable animal partners or repelling, in some cases killing, animals that may prove harmful (Bernays 1998). In fact, many of the secondary metabolites mentioned in the prior discussion of allelopathy also play a role in defense against insect enemies of plants (Table 10.4). Also as discussed in earlier chapters, diversification of plants, especially the angiosperms, has been accompanied by diversification of insects, and this is particularly the case for the six insect orders mentioned previously that contain much of the known insect herbivore diversity (Bernays 1998).

Lodge (1991) and Newman (1991) suggested that plant chemistry is an important driver of specialization in herbivores. This is a result both of diverse attractants used by plants and of the pressure to avoid or neutralize plant defenses against herbivory. Unfortunately, the historic perspective that herbivory is of limited importance in aquatic and wetland plants resulted in this process receiving little attention; consequently, we know much less about the diversity of plant defenses against herbivory among hydrophytes than for terrestrial plants. Nevertheless, the few studies that have been conducted in this area have provided considerable detail about some of the defenses these plants employ.

Shortly before Lodge and Newman published their reviews, Ostrofsky and Zettler (1986) published research in which they measured the alkaloid content in 15 species of aquatic plants, including species of *Cabomba*, *Ceratophyllum*, *Elodea*, *Heteranthera*, *Myriophyllum*, *Potamogeton*, and *Vallisneria*. They found that alkaloid concentrations ranged from 0.13 to 0.56 mg alkaloid per gram plant dry mass among these species (equivalent to 0.013% to 0.056% of plant mass). They further estimated that individual plant species produced from two to nine unique alkaloid compounds each, and compounds appeared to be relatively unique among species (~80% of potential alkaloids present occurred in only one plant species). To place their results in the broader context of plant defensive chemistry, they gave data from other studies indicating that alkaloid concentrations of 0.017% to 0.059% of plant mass have been shown to be lethal to some vertebrate animals, and concentrations of 0.02% to 0.04% of plant mass have been shown to reduce feeding and survival of some insect herbivores.

Newman, Hanscom, and Kerfoot (1992) investigated the effect of glucosinolates, another class of nitrogen-containing compounds, on feeding by amphipods, caddisflies, and snails. Watercress (*Nasturtium officinale*), in the Brassicaceae, produces a defensive system comprising glucosinolates, which consist of a glucose molecule bound to an isothiocyanate compound (Table 10.4), and an enzyme called myrosinase that can remove the glucose to release the biologically active isothiocyanate. The enzyme typically is segregated from the glucosinolate compounds in live plant tissue, but tissue damage by chewing herbivores can bring the two in contact with one another, resulting in liberation of the isothiocyanate. Newman and colleagues found that invertebrate herbivores would preferentially feed on less nutritious, yellow leaves that had glucosinolate concentrations approximately one-tenth that of fresh green tissues, despite green leaves having twice the overall nitrogen concentration of yellow leaves. Further investigations of this system showed that if the green tissues were heated to degrade the myrosinase enzyme, herbivores would readily consume those tissues, at ten times the level of unheated green leaves. Newman and colleagues conducted additional follow-up analyses that seemed to confirm that the intact myrosinase-glucosinolate system was responsible for this selectivity among aquatic invertebrate herbivores.

Phenolic compounds with feeding deterrence have been isolated from the emergent wetland plants *Saururus cernuus* and *Habenaria repens* (Bolser et al. 1998; Kubanek et al. 2000). Bolser and colleagues used a stepwise approach to identify plant species that were avoided by generalist herbivore crayfish (*Procambarus clarkii*) and then to assay progressively refined extracts of those plant species to identify the deterrent compounds. Once the least-preferred plant species were identified, the investigators freeze-dried plant tissues, ground them, and added the ground material to an artificial, agar-based plant diet, to remove effects of the plant structure on feeding selectivity. Plant tissues were subsequently extracted and chemically separated into different fractions, each of which was assayed for feeding preference and then separated further, if necessary. In the end, they identified a phenolic compound called habenariol from *H. repens*, and later work identified multiple antifeedant phenolic compounds from *S. cernuus* (Kubanek et al. 2000).

An interesting sesquiterpene compound has been isolated from the sedge *Cyperus iria*. This compound, insect juvenile hormone III, was isolated from *C. iria*, along with a precursor compound in 1988 (Toong, Schooley, and Baker 1988). Reports of grasshoppers feeding on *C. iria* indicated a high frequency of wing deformities and infertility (Bede, Goodman, and Tobe 1999), and assays of this compound against rice seedlings revealed reduced growth of rice seedlings that were exposed to insect juvenile hormone III (Bede and Tobe 2000). Bede and Tobe (2000) also found evidence of autotoxicity in *C. iria* from insect juvenile hormone, but levels of suppression of *C. iria* seedlings were five to ten times lower than in rice seedlings. Concentrations of this compound in roots of *C. iria* were as much as 40 times the amounts found in some insect species, but concentrations in other plant parts were considerably lower (Bede, Goodman, and Tobe 1999). Bede and colleagues hypothesized that this compound might be responsible, in part, for the success of *C. iria* as an important weed of rice agriculture, via suppression of neighboring seedlings, in combination with its effects on potential herbivore populations. This example highlights the potential for plant secondary metabolites to play multiple roles in plant biology, ecology, and population dynamics, enhancing the efficiency in use of resources that are allocated to their production.

10.6.1.4 Herbivory and Biological Control

Biological control, or biocontrol, is the use of an undesirable species' natural enemy, such as an herbivore, a predator, or a pathogen, to control growth or reproduction of that species. We typically see this approach used in the control of introduced nuisance species, including introduced aquatic weeds. The strategy in this method of population control is to take advantage of specialized associations between the target plant and control agent to maximize control effectiveness while minimizing human labor and cost. In the case of plants, biological control often takes advantage of the widespread occurrence of specialist insect herbivores. For example, a global survey of diet breadth among more than 7,500 insect herbivores found that three-fourths of those insect species fed on plants within a single family (Forister et al. 2015). Newman (1991) reported that insect herbivores of aquatic plants, on average, fed on fewer than three plant species each, based on data from a survey of 461 herbivorous insect taxa. Furthermore, there was very little variation among orders of insects, with average numbers of plant host species per order ranging from one to only two or three host species per herbivore. This degree of specialization among insect herbivores provides many opportunities to take advantage of these organisms as prospective biological control agents.

The potential downside of biological control of introduced aquatic weeds is that it requires the identification and subsequent introduction of an insect herbivore from the native range of the plant. Thus, biological control of the non-native plant relies on introduction of a non-native insect, and this has proven problematic with some biocontrol agents. Current efforts to develop biocontrol agents begin with extensive surveys in the native range of the potential target weed, followed by rigorous screening for potential non-target plant species effects in an effort to minimize potential non-target effects.

Efforts at developing biocontrol agents for introduced aquatic weeds in the United States began in the mid-twentieth century, with the search for potential agents to control invasive alligatorweed (*Alternanthera philoxeroides*), a South American plant species first discovered in the US in the 1890s (Buckingham 1996). Surveys in the early 1960s yielded three candidate control agents: a flea beetle (*Agasicles hygrophila*), a stem boring moth (*Vogtia malloi*), and a thrips (*Amynothrips andersoni*). Before these insects were exported to the US from their native ranges in South America, the US Department of Agriculture established a research lab in Argentina, near Buenos Aires, which was home to studies aimed at screening the biology of the insects, including dietary preferences of each. That laboratory later was moved to the city of Hurlingham, in the province of Buenos Aires, and operates today as the Fundación para el Estudio de Especies Invasivas (FuEDEI; Foundation for the Study of Invasive Species). Scientists from all over the world collaborate with FuEDEI scientists on biocontrol studies for aquatic, as well as terrestrial, invasive species.

The first three candidate biocontrol agents for alligatorweed were screened against 14 (flea beetle), 30 (moth), and 21 (thrips) plant species. Among those, the flea beetle was found to use only alligatorweed as a host, the moth fed on several species but only completed its development on alligatorweed, and the thrips was found to feed on alligatorweed and one other *Alternanthera* species not known to occur in the US. The flea beetle found greatest use out of these three biocontrol agents, but the thrips is seeing renewed interest, because it appears to have greater cold tolerance than the flea beetle, and alligatorweed is able to persist in areas beyond the cold tolerance of the beetle (Harms et al. 2021; Knight and Harms 2022).

Interest in developing biocontrol agents for introduced aquatic weeds continues today, and, within the US, the US Army Corps of Engineers' Engineering Research and Development Center is heavily involved in these efforts (Figure 10.22). Some of the current plant species under investigation there are water hyacinth, hydrilla, Brazilian peppertree, and alligatorweed. Previous efforts have involved research on biocontrol of milfoil (*Myriophyllum* species; research on control by a milfoil moth and milfoil weevil), water primrose (*Ludwigia* species; water primrose flea beetle, primrose leaf weevil), *Salvinia* species (salvinia weevil), American lotus and water lilies (*Nelumbo*, *Nuphar*, *Nymphaea* species; long-horned leaf beetles), *Hydrilla* (leaf-mining flies, Asian hydrilla moth), American lotus (American lotus borer), and cattail (cattail caterpillar; Freedman et al. 2007). Many of these insects have specialized adaptations for dealing with the aquatic environment, such as using the plants' internal aerenchyma systems for their own air supplies during the aquatic larval stages or producing air-filled cases from silk or plant material. In fact, the milfoil moth (*Acentria ephemerella*) completes almost its entire life cycle beneath the water, with adults emerging only to mate. Furthermore, the females of this moth species have no wings, and thus are restricted to swimming atop the water surface to attract mates (Freedman et al. 2007).

10.6.2 Parasites and Pathogens of Aquatic and Wetland Plants

The final interaction type that we will examine is that of **parasites** and **pathogens**. Parasites are organisms that obtain nutrients and/or energy from one or a few individuals of another species, their **host(s)**, usually without causing death to the host as a direct consequence of obtaining resources from it. A pathogen, on the other hand, is an organism with the same strategy of obtaining resources from a host but which usually causes disease and potentially death as a result. Pathogens usually are microorganisms or viruses, whereas parasites can be of essentially any taxon, including plants themselves.

It is estimated that there are approximately 4,500 species of parasitic plants, among 16 families, worldwide (Pennings and Callaway 2002; Těšitel 2016). They are

FIGURE 10.22   The US Army Corps of Engineers Conducts Aquatic Weed Biocontrol Research at their Engineering Research and Development Center in Vicksburg, Mississippi (USA). Some of their Work has Focused on (A) Brazilian Peppertree Thrips, (B) Alligatorweed Thrips, (C) Alligatorweed Flea Beetle, and (D) Water Hyacinth Planthopper. This Work is Carried Out under Strict Quarantine Conditions in Growth Chambers, such as that in (E), and Scientists at the Facility Often Maintain Multiple Genotypes of the Plant Species of Interest, to Account for Intraspecific Variation in Susceptibility to Control Agents. For Example, they have Collected Multiple Genotypes of Alligatorweed from across the US to Assess Intraspecific Variation in Plant Phenotypes and Susceptibility to the Flea Beetle and Thrips (F). I Owe Special Thanks to Nathan Harms and Ian Knight for Providing a Tour of these Facilities.

broadly grouped into two types of parasites: **holoparasites** and **hemiparasites**. Hemiparasites are capable of photosynthesis and may be found living freely, without a host. In contrast, holoparasites are (mostly) non-photosynthetic and require a host to grow, survive, and reproduce. In addition to plants that directly parasitize other plants, there are more than 400 non-chlorophyllous species that utilize mycorrhizal networks to parasitize other plants (Pennings and Callaway 2002). Parasitic plants are often selective in host choice and may use chemicals produced by potential hosts, as well as host nutritional quality, in choosing which potential host to parasitize (Pennings and Callaway 2002; Těšitel 2016). These plants can be found in most types of communities, including wetlands, where they can have important ecological impacts. One very common genus of holoparasitic plants is *Cuscuta* (Figure 10.23), a genus of approximately 145 species that infect many species of angiosperms, including important crop plants (Těšitel 2016). It is also a relatively common genus in wetlands.

Parasitic plants generally have received little attention in wetlands, but a few experimental studies in California saltmarshes have revealed complex effects of plant parasitism on saltmarsh plant assemblages. *Cuscuta salina*, a parasitic plant in salt marshes along the western coast of the US, was found to preferentially parasitize the dominant salt marsh plant *Salicornia virginica* in higher elevation zones of southern California marshes (Pennings and Callaway 1996). This selective parasitism of the dominant species indirectly facilitated rarer plant species, enhancing overall plant species diversity in those marshes. Similar studies in northern California saltmarshes again revealed strong preference of *C. salina* for dominant plant species and indirect facilitation of rarer plant species, including a hemiparasite, *Cordylanthus maritimus* (Grewell 2008). However, because *C. salina* is dependent upon hosts for growth and reproduction, decline of its preferred host led to declines in the parasite itself, leading to cycles of increasing and decreasing plant species abundance in the marshes over time (Pennings and Callaway 1996). Evidence for similar

vegetation cycles was discovered in studies of a freshwater marsh along the Delaware River in New Jersey (US). In this case, the parasite was *Cuscuta gronovii*, and clear cycles of abundance of it and its preferred host, *Impatiens capensis*, were seen in seed banks over a ten-year period (Leck and Simpson 1995). Furthermore, a codominant species, *Polygonum arifolium*, seemed to increase in abundance in seed banks during years when *I. capensis* declined. Thus, plant parasitism appears to be an important ecological phenomenon and clearly deserves more study in wetlands and other natural ecosystems.

As with plant parasites, microbial parasites and pathogens of plants have been poorly studied in wetland ecosystems (Bertness, Gaines, and Hay 2001; Crocker et al. 2016), despite an abundance of taxa and modes of pathogenesis that potentially could be found in aquatic and wetland plants (Figure 10.24). Work by Crocker et al. (2016) experimentally assessed fungal colonization of overwintering seeds of ten wetland plant species in ten freshwater wetlands of central New York state (USA). The plant species

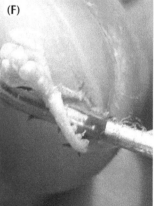

FIGURE 10.23 *Cuscuta* Species, Pictured here, are Holoparasitic Plants in the Convolvulaceae Family that Attack Diverse Host Plant Species. Pictured are (A) Infestations in a Beaver Pond in Mississippi (USA) and (B) Along the Tiber, or Tevere, River in Rome (Italy). Flowers, Floral Buds, and Trailing Stems Wrapping around Stems of Host Plants Along the Tevere River are Shown in C–D, as well as the Attachment Points (Haustoria) from an Individual in the Tevere Population (E).

tested represented a variety of both monocot and dicot emergent plant species and included both native and non-native species or genotypes. The investigators identified approximately 40 fungal species from among more than 200 cultured isolates of fungi from these seeds, the majority of which were determined to be pathogenic to plants. More than half the isolated fungi belonged to one of three pathogenic genera: *Alternaria*, *Peyronellaea*, or *Fusarium*. Despite the diversity of fungi isolated, many were found to be generalist plant pathogens, and Crocker and colleagues were unable to assess whether any of the species specialized on any one plant species. Nevertheless, one-fourth of the isolated fungi belonged to *Peyronellaea glomerata*, 85% of which was isolated from one of the genotypes of *Phragmites australis* (69% from the native genotype, 16% from an invasive European genotype). *Phragmites australis*

has been the subject of other work aimed at identification of fungal pathogens. Some of that work has been aimed at evaluating differences in susceptibility between native and non-native genotypes in North America, as with the work by Crocker et al. (2016), while other projects have been in search of causes for decline in native *P. australis* populations in the US (e.g., Galo Espinal 2021).

In the experiments conducted by Crocker et al. (2016), experimental fungal colonization did not appear to affect seed germination of the ten plant species evaluated. Seedling survival, however, was significantly reduced by experimental inoculation with isolated pathogenic fungi, and the reduction in survival was affected little by whether the fungi were isolated from the same species on which they were tested. In other words, the fungi that colonized overwintering seeds appeared to be quite general in seed

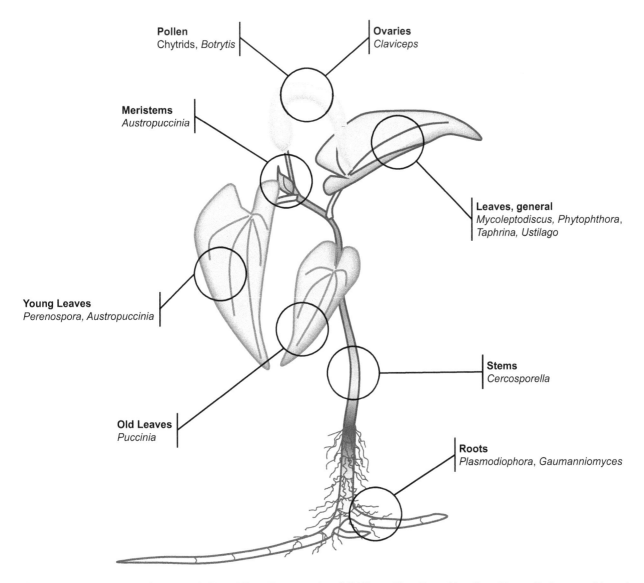

FIGURE 10.24   Examples of Pathogenic Fungal Taxa Known to Attack Different Plant Parts. Not all are Known Pathogens of Aquatic and Wetland Plants, but they are Known to Attack Vascular Plant Taxa in General. Adapted from Begon, Harper, and Townsend (1990), with Additional Information from Fernandez Winzer et al. (2018) and Freedman et al. (2007).

colonization and were capable of pathogenesis on most any species tested. Crocker et al. (2016) further suggested that it is possible some of the fungal species that were not identified to be pathogenic in their work could be pathogenic on later age classes of plants or on plant species outside the ten that were tested. For example, some of the fungal genera that were isolated but not pathogenic in their experiments are known to include pathogenic species.

In Puerto Rico, pathogenesis was investigated as a cause of dieback in the red mangrove (*Rhizophora mangle*). In that work, a single pathogenic fungal species (*Cytospora rhizophorae*) was isolated from 63% of the 164 samples collected from trees that showed symptoms of infection (Wier, Tattar, and Klekowski Jr 2000). Fungi from those samples were cultured in the laboratory and assessed for pathogenicity against greenhouse- and field-grown red mangrove plants. Mortality from inoculation ranged from 33% to 50% in those experiments, and the fungus was re-isolated from the diseased plants, which strongly suggested this pathogen to be at least partly responsible for dieback of trees in the coastal forests of Puerto Rico, which are also subject to other ongoing stresses, such as wetland drainage and climate change (WierTattar, and Klekowski Jr 2000).

Halfway around the world, in Australia, Fernandez Winzer et al. (2018) investigated a pathogenic myrtle rust fungus (*Austropuccinia psidii*) that had been accidentally introduced into Australia, first detected there in 2010. This fungus is native to Central and South America, and there was concern that it could upset the balance among tree species of diverse coastal swamps of southeastern Australia. Experiments were conducted on three Australian native tree species in the Myrtaceae (*Melaleuca quinquenervia*, *Leptospermum laevigatum*, and *Baeckea linifolia*) to assess pathogenicity of *A. psidii* and the potential impacts on dominance hierarchies among the native trees. Both *M. quinquenervia* and *L. laevigatum* developed rust symptoms after inoculation, with approximately 90% of *L. laevigatum* plants and 56% of *M. quinquenervia* plants showing signs of disease, and biomass of both species being reduced by 70%–80%. The most significant effect of *A. psidii* inoculation on *B. linifolia* was a 20% to 30% reduction in plant height. It was concluded that these results suggested potential for the fungus to alter relative species dominance in coastal swamps of Australia, where *Melaleuca quinquenervia* is regarded as an ecologically important canopy species.

Because of the potential for parasites and pathogens to reduce the performance of dominant plant species, there has been an interest in the use of pathogens of aquatic and wetland plants as potential biocontrol agents for economically important invasive species. Barreto et al. (2000) examined biocontrol studies on 13 South American plant species and indicated relatively low biocontrol success rates among some of the most important nuisance aquatic plants that have been introduced outside South America (Table 10.5). *Eichhornia* species, for example, were indicated to have a relatively species-poor assemblage of

fungal associates (Barreto et al. 2000), which reduced the potential pool of control agents from the native range. In other cases, some isolates showed limited potential but were not actively pursued as biocontrol agents, owing to the need for combination with chemical agents to obtain effective control of the target weed. Other work has shown that some pathogens identified in species' native ranges may not be effective in biocontrol, owing to low reproducibility of results in biocontrol assays. In assays of *Myriophyllum aquaticum*, it was believed that variability in plant traits, such as the waxy leaf cuticle, resulted in differential results between natural populations and those tested in biocontrol experiments (Cuda et al. 2008; Freedman et al. 2007).

Lastly, viruses are another potential group of pathogens that may play an important role in the ecology of aquatic and wetland plants but have been largely ignored. A literature review by Jackson and Jackson (2008) identified fewer than ten wetlands in which natural viral assemblages had been studied. These included three freshwater marshes, two riparian systems, and one mangrove swamp. No studies were identified of natural viral distributions in

---

TABLE 10.5

Examples of successful and unsuccessful pathogen biocontrol efforts for problematic aquatic and wetland plant species.

Plant species	Result of biocontrol studies
Alternanthera philoxeroides	weak control with fungal pathogen from native range[a]
Echinochloa polystachya	no effective fungal pathogens found in native range[a]
Egeria densa	a known generalist fungal pathogen on agronomic grasses was shown to be effective; not isolated from *E. densa*[a]
Eichhornia crassipes, *E. azurea*	no effective controls found anywhere[a]
Hydrilla verticillata	effective pathogen found on another host, outside native range[b]
Myriophyllum aquaticum	three species of fungi were found to cause severe shoot necrosis[a]; a bacterium from South Africa severely damaged aerial shoots[b]
Myriophyllum spicatum	effective pathogen found outside native range[b]
Paspalum repens	no effective fungal pathogens found in native range[a]
Pistia stratiotes	no effective controls found anywhere[a]
Salvinia auriculata	fungal pathogen from native range with some control potential but not pursued for management[a]
Typha domingensis	effective foliar pathogen was discovered in the native range[a]

[a] Barreto et al. (2000); [b] Cuda et al. (2008)

tidal freshwater marshes, non-mangrove forested wetlands, bogs, or fens. Jackson and Jackson (2008) listed four plant species or genera, aside from rice, that were known to serve as hosts to plant viruses: *Phragmites australis*, *Lythrum salicaria*, *Alisma plantago-aquatica*, and *Spartina* species. However, they indicated that it was unknown what the natural distributions are for any of these plant pathogens nor whether they cause significant impacts on plant populations or community-level species distributions.

10.7   IN SUMMARY

A common conceptual framework within which ecologists think about interactions among plants is the C-S-R construct. In this framework, it is hypothesized that the natural forces of stress and disturbance have shaped species characteristics over evolutionary time. As a result, we might expect habitats that are subject to similar and ecologically predictable levels of stress and disturbance would be occupied by species that exhibit similar suites of traits. Relatively benign habitats that are regularly disturbed might be occupied by species that have short juvenile periods and that invest a large proportion of resources in sexual reproduction. This would allow those species to quickly produce offspring, with the parents potentially dying before the next major disturbance occurred. Infrequently disturbed but stressful habitats, on the other hand, may be occupied by species that exhibit slow growth, making efficient use of limited resources. This framework can be compared to another commonly encountered r-to-K framework that contrasts population growth strategies under differing environmental conditions. Each of these is useful in some respects for thinking about why one might repeatedly encounter particular species traits in habitats with similar sets of conditions or resource availability. Each of these also has limitations in their generalizability, and those limitations ought to be considered when attempting to apply these conceptual frameworks to questions of when or why one might (or might not) expect to encounter species exhibiting particular traits in a given habitat.

In considering interactions among species, we tend to simplify by considering only two-species interactions and to think about whether each participant benefits or suffers losses from the interaction, relative to the performance of each in the absence of the interaction. For example, mutualisms, such as in nitrogen-fixing plants or mycorrhizal relationships, are considered (+ +) interactions because each participant benefits from the relationship. The opposite of this is competition (− −), wherein each participant, by definition, suffers a measurable loss as a result of the interaction.

We initially explored competition in this chapter in the context of the C-S-R life history strategy but noted that this view of competition lacked the critical ability to apply it as an operational definition for exploring interactions among plants. That is, the definition of competition that is employed in the C-S-R framework could not be used to determine whether an interaction between two individuals might result in a (− −) outcome. The definition was expanded to include an operational component, specifically defining competition as an interaction involving a shared requirement for one or more resources in limited supply that leads to a reduction in the survival, growth, or reproduction of the individuals involved. In this context, experiments can be designed to measure one or more aspects of plant performance (i.e., survival, growth, or reproduction) and assess whether performance is enhanced or reduced in the presence of a given interaction.

With data from such an experiment, we can use a relevant index of species responses to determine whether each species responded favorably to the presence of the other species. In this way, we can evaluate whether the interaction would represent mutualism (+ +), competition (− −), amensalism (− 0), commensalism (+ 0), or perhaps some other form of interaction.

Examples of amensalism among aquatic and wetland plants include asymmetric competition and allelopathy (although allelopathy is rarely confirmed to be a strict amensalism). Allelopathy relies on the production by one species of a chemical compound that suppresses a neighbor, resulting in a competitive advantage for the allelopathic individual. Because the production of potentially allelopathic chemicals is often increased in the face of nutrient limitation, neighbor density and nutrient availability may be intricately linked to the presence of allelopathic interactions among plants.

In contrast to amensalism, commensalism results in one individual benefitting from some attribute of its neighbor, with the neighbor neither benefitting measurably nor suffering losses from the interaction. Common forms of commensalism in wetlands are the epiphytic growth strategy and facilitation. Because facilitation often occurs via the amelioration of an environmental stress, it is hypothesized that there usually is a balance between facilitation and competition in natural ecosystems. This hypothesis was formalized as the stress gradient hypothesis. Because it is a relatively new hypothesis, there are some disagreements about its generality. It has been supported with data from diverse ecosystems, but studies also exist in which it has not been fully supported. As with any general rule or hypothesis about natural phenomena, the stress gradient hypothesis should be applied with consideration of factors that may lie outside its operational space or that may be exceptions to its underlying assumptions.

The final group of interactions discussed in this chapter were herbivory and parasitism, which are exploitative interactions in which some organism derives nutrients and/or energy from the consumption of plant tissues. Both broad groups of interaction have received inadequate study in wetlands, in comparison with terrestrial ecosystems, although herbivory has been much more thoroughly investigated than parasitism or pathogenesis of aquatic and wetland plants. We saw that each of these interaction types has the potential to influence species hierarchies in wetlands and thus is a potentially important driver of ecological processes in these ecosystems. We also saw that herbivory and pathogenesis both have potential application for biological control, in which an exploitative species is used to control population dynamics of undesirable plant species, such as invasive aquatic weeds.

Understanding the roles of these interactions in influencing plant species assemblages in aquatic and wetland ecosystems is critical for applying wetland ecology principles to ecosystem management. Whether one is interested in managing wetlands for conservation of natural species and processes, for mitigation of surface water pollution, for wildlife habitat, or simply for aesthetic purposes, all these interactions have the potential to influence the success of one's management efforts. We will learn more about this in the following chapter, as we look at patterns in species dynamics over time and how those patterns might influence the ecological properties of a wetland and its ability to provide important ecosystem services.

10.8   FOR REVIEW

1. What is the difference between life history and life history strategy?
2. How are Grime's C-S-R framework and the r- vs. K- continuum similar? How are they different?
3. Why is stress a difficult concept to define ecologically?
4. What is competition? Why was the question of an operational component important in defining competition?
5. What operational component was incorporated into Tilman's approach to studying competition?
6. How does one carry out an experiment to determine the nature of interaction between two species or individuals?
7. Why is absolute competition intensity a poor metric for evaluating the relative intensity of competition between two species? How is that inadequacy addressed, quantitatively?
8. What is asymmetric competition, and how does it relate to the way in which we quantify interactions among species?
9. What is allelopathy?
10. What are two broad categories of potential allelochemicals? How is the relative importance of those two groups influenced by resource competition?
11. What is facilitation? How does facilitation take place in natural wetlands?
12. What is the conceptual underpinning of the stress gradient hypothesis?
13. What are some examples of herbivorous animal taxa in aquatic and wetland habitats? What plant parts are targeted by herbivores?
14. What evidence exists to support the importance of herbivory in wetland ecosystems?
15. How are chemical defenses against herbivores related to the chemicals used in allelopathic interactions?
16. How does biological control take advantage of the relationships between herbivores or pathogens and wetland plants?
17. What are some examples of parasitic taxa in aquatic and wetland habitats? What plant parts are targeted by parasites?
18. What evidence exists to support the importance of parasites and pathogens in wetland ecosystems?

10.9   REFERENCES

Armas, Cristina, Ramon Ordiales, and Francisco I. Pugnaire. 2004. "Measuring Plant Interactions: A New Comparative Index." *Ecology* 85: 2682–86.

Austin, M. P., L. F. M. Fresco, A. O. Nicholls, R. H. Groves, and P. E. Kaye. 1988. "Competition and Relative Yield: Estimation and Interpretation at Different Densities and under Various Nutrient Concentrations Using *Silybum marianum* and *Cirsium vulgare*." *Journal of Ecology* 76 (1): 157–71. https://doi.org/10.2307/2260460.

Bakker, Elisabeth S., Kevin A. Wood, Jordi F. Pagès, G. F.(Ciska) Veen, Marjolijn J. A. Christianen, Luis Santamaría, Bart A. Nolet, and Sabine Hilt. 2016. "Herbivory on Freshwater and Marine Macrophytes: A Review and Perspective." *Aquatic Botany* 135: 18–36. https://doi.org/10.1016/j.aquabot.2016.04.008.

Barbour, Michael G., Jack H. Burk, Wanna D. Pitts, Frank S. Gilliam, and Mark W. Schwartz. 1999. *Terrestrial Plant Ecology*. 3rd ed. Menlo Park, CA: Benjamin/Cummings.

Barreto, R., R. Charudattan, A. Pomella, and R. Hanada. 2000. "Biological Control of Neotropical Aquatic Weeds with Fungi." *Crop Protection*, XIVth International Plant Protection Congress, 19 (8): 697–703. https://doi.org/10.1016/S0261-2194(00)00093-4.

Batzer, Darold P., and Rebecca R. Sharitz. 2006. *Ecology of Freshwater and Estuarine Wetlands*. Berkeley, CA: University of California Press.

Bede, Jacqueline C., Walter G. Goodman, and Stephen S. Tobe. 1999. "Developmental Distribution of Insect Juvenile Hormone III in the Sedge, *Cyperus Iria* L." *Phytochemistry* 52: 1269–74.

Bede, Jacqueline C., and Stephen S. Tobe. 2000. "Activity of Insect Juvenile Hormone III: Seed Germination and Seedling Growth Studies." *Chemoecology* 10: 89–97.

Begon, Michael, John L. Harper, and Colin R. Townsend. 1990. *Ecology: Individuals, Populations, and Communities*. 2nd ed. Cambridge, MA: Blackwell Scientific Publications.

Bernays, Elizabeth A. 1998. "Evolution of Feeding Behavior in Insect Herbivores." *BioScience* 48 (1): 35–44. https://doi.org/10.2307/1313226.

Bertness, Mark D., and Ragan Callaway. 1994. "Positive Interactions in Communities." *Trends in Ecology & Evolution* 9 (5): 191–93. https://doi.org/10.1016/0169-5347(94)90088-4.

Bertness, Mark D., Steven D. Gaines, and Mark E. Hay. 2001. *Marine Community Ecology*. Sunderland, MA: Sinauer Associates, Inc.

Bertness, Mark D., C. Wise, and A. M. Ellison. 1987. "Consumer Pressure and Seed Set in a Salt Marsh Perennial Plant Community." *Oecologia* 71 (2): 190–200. https://doi.org/10.1007/BF00377284.

Bolser, R. C., M. E. Hay, N. Lindquist, William Fenical, and Dean Wilson. 1998. "Chemical Defenses of Freshwater Macrophytes against Crayfish Herbivory." *Journal of Chemical Ecology* 24 (10): 1639–58.

Boughton, Elizabeth H., Pedro F. Quintana-Ascencio, and Patrick J. Bohlen. 2011. "Refuge Effects of *Juncus effusus* in Grazed, Subtropical Wetland Plant Communities." *Plant Ecology* 212 (3): 451–60. https://doi.org/10.1007/s11258-010-9836-4.

Bruno, John F. 2000. "Facilitation of Cobble Beach Plant Communities through Habitat Modification by Spartina Alterniflora." *Ecology* 81 (5): 1179–92.

Buckingham, Gary R. 1996. "Biological Control of Alligatorweed, *Alternanthera philoxeroides*, the World's First Aquatic Weed Success Story." *Castanea* 61 (3): 232–43.

Callaway, Ragan M., and Leah King. 1996. "Temperature-Driven Variation in Substrate Oxygenation and the Balance of Competition and Facilitation." *Ecology* 77 (4): 1189–95. https://doi.org/10.2307/2265588.

Campbell, Bruce D., and J. Philip Grime. 1992. "An Experimental Test of Plant Strategy Theory." *Ecology* 73: 15–29.

Campbell, Daniel, and Paul Keddy. 2022. "The Roles of Competition and Facilitation in Producing Zonation Along an Experimental Flooding Gradient: A Tale of Two Tails with Ten Freshwater Marsh Plants." *Wetlands* 42 (1): 5. https://doi.org/10.1007/s13157-021-01524-4.

Connolly, J. 1987. "On the Use of Response Models in Mixture Experiments." *Oecologia* 72: 95–103.

Cox, Robert John, and Grover C. Smart. 1994. "Nematodes Associated with Plants from Naturally Acidic Wetlands Soil." *Journal of Nematology* 26 (4): 535–37.

Crocker, Ellen V., Justin J. Lanzafane, Mary Ann Karp, and Eric B. Nelson. 2016. "Overwintering Seeds as Reservoirs for Seedling Pathogens of Wetland Plant Species." *Ecosphere* 7 (3): e01281. https://doi.org/10.1002/ecs2.1281.

Cuda, J. P., R Charudattan, M. J. Grodowitz, R. M. Newman, J. F. Shearer, M. L. Tamayo, and B. Villegas. 2008. "Recent Advances in Biological Control of Submersed Aquatic Weeds." *Journal of Aquatic Plant Management* 46 (1): 15–32.

Cyr, Helene, and Michael L. Pace. 1993. "Magnitude and Patterns of Herbivory in Aquatic and Terrestrial Ecosystems." *Nature* 361: 148–50.

Egerova, Jana, C. Edward Proffitt, and Steven E. Travis. 2003. "Facilitation of Survival and Growth of *Baccharis halimifolia* L. by *Spartina alterniflora* Loisel. In a Created Louisiana Salt Marsh." *Wetlands* 23 (2): 250–56. https://doi.org/10.1672/4-20.

Ervin, Gary N. 2005. "Spatio-Temporally Variable Effects of a Dominant Macrophyte on Vascular Plant Neighbors." *Wetlands* 25 (2). https://doi.org/10.1672/8.

Ervin, Gary N., and Robert G. Wetzel. 2000. "Allelochemical Autotoxicity in the Emergent Wetland Macrophyte *Juncus effusus* (Juncaceae)." *American Journal of Botany* 87 (6). https://doi.org/10.2307/2656893.

———. 2002. "Influence of a Dominant Macrophyte, *Juncus effusus*, on Wetland Plant Species Richness, Diversity, and Community Composition." *Oecologia* 130 (4). https://doi.org/10.1007/s00442-001-0844-x.

———. 2003. "An Ecological Perspective of Allelochemical Interference in Land-Water Interface Communities." *Plant and Soil* 256 (1): 13–28. https://doi.org/10.1023/A:1026253128812.

Fernandez Winzer, Laura, Angus J. Carnegie, Geoff S. Pegg, and Michelle R. Leishman. 2018. "Impacts of the Invasive Fungus *Austropuccinia psidii* (Myrtle Rust) on Three Australian Myrtaceae Species of Coastal Swamp Woodland." *Austral Ecology* 43 (1): 56–68. https://doi.org/10.1111/aec.12534.

Fischman, Hallie S., Sinead M. Crott, and Christine Angelini. 2019. "Optimizing Coastal Restoration with the Stress Gradient Hypothesis." *Proceedings of the Royal Society B: Biological Sciences* 286: 20191978.

Forister, Matthew L., Vojtech Novotny, Anna K. Panorska, Leontine Baje, Yves Basset, Philip T. Butterill, Lukas Cizek, et al. 2015. "The Global Distribution of Diet Breadth in Insect Herbivores." *Proceedings of the National Academy of Sciences* 112 (2): 442–47. https://doi.org/10.1073/pnas.1423042112.

Freedman, Jan E., Michael J. Grodowitz, Robin Swindle, and Julie G. Nachtrieb. 2007. "Potential Use of Native and Naturalized Insect Herbivores and Fungal Pathogens of Aquatic and Wetland Plants:" Fort Belvoir, VA: Defense Technical Information Center. https://doi.org/10.21236/ADA471715.

Galo Espinal, David. 2021. "Isolation of Foliar Fungi From Roseau Cane (*Phragmites australis*) in Coastal Louisiana." Baton Rouge, LA: Louisiana State University and Agricultural and Mechanical College. https://digitalcommons.lsu.edu/gradschool_theses/5385.

Gibson, David J. 2015. *Methods in Comparative Plant Population Ecology*. Oxford, UK: Oxford University Press.

Gopal, Brij, and Usha Goel. 1993. "Competition and Allelopathy in Aquatic Plant Communities." *The Botanical Review* 59 (3): 155–210.

Grace, J. B. 1991. "A Clarification of the Debate between Grime and Tilman." *Functional Ecology* 5: 583–87.

———. 1993. "The Adaptive Significance of Clonal Reproduction in Angiosperms: An Aquatic Perspective." *Aquatic Botany* 44: 159–80.

Grace, James B. 1990. "On the Relationship between Plant Traits and Competitive Ability." In *Perspectives on Plant Competition*, edited by James B. Grace and David Tilman, 51–65. San Diego: Harcourt Brace Jovanovich.

———. 1995. "On the Measurement of Plant Competition Intensity." *Ecology* 76 (1): 305–8.

Grace, James B., and David Tilman, eds. 1990. *Perspectives on Plant Competition*. San Diego, CA: Harcourt Brace Jovanovich.

Grace, James B., and Robert G. Wetzel. 1981. "Habitat Partitioning and Competitive Displacement in Cattails (*Typha*): Experimental Field Studies." *The American Naturalist* 118 (4): 463–74.

Grewell, Brenda J. 2008. "Parasite Facilitates Plant Species Coexistence in a Coastal Wetland." *Ecology* 89 (6): 1481–88. https://doi.org/10.1890/07-0896.1.

Grime, J. P. 1977. "Evidence for the Existence of Three Primary Strategies in Plants and Its Relevance to Ecological and Evolutionary Theory." *The American Naturalist* 111: 1169–94.

———. 1989. "The Stress Debate: Symptom of Impending Synthesis?" *Biological Journal of the Linnean Society* 37 (1–2): 3–17. https://doi.org/10.1111/j.1095-8312.1989.tb02002.x.

Gross, Elisabeth M. 2003. "Allelopathy of Aquatic Autotrophs." *Critical Reviews in Plant Sciences* 22: 313–39.

Gross, Elisabeth M., Catherine Legrand, Karin Rengefors, and Urban Tillmann. 2012. "Allelochemical Interactions among Aquatic Primary Producers." In *Chemical Ecology in Aquatic Systems*, 196–209. Oxford, UK: Oxford University Press.

Gurevitch, Jessica, Samuel M. Scheiner, and Gordon A. Fox. 2006. *The Ecology of Plants*. 2nd ed. Sunderland, MA: Sinauer Associates, Inc.

Gurevitch, Jessica, Paul Wilson, Judy L. Stone, Paul Teese, and Robert J. Stoutenburgh. 1990. "Competition among Old-Field Perennials at Different Levels of Soil Fertility and Available Space." *Journal of Ecology* 78 (3): 727–44. https://doi.org/10.2307/2260895.

Harms, Nathan E., Ian A. Knight, Paul D. Pratt, Angelica M. Reddy, Abhishek Mukherjee, Ping Gong, Julie Coetzee, S. Raghu, and Rodrigo Diaz. 2021. "Climate Mismatch between Introduced Biological Control Agents and Their Invasive Host Plants: Improving Biological Control of Tropical Weeds in Temperate Regions." *Insects* 12 (6): 549. https://doi.org/10.3390/insects12060549.

He, Qiang, Mark D. Bertness, and Andrew H. Altieri. 2013. "Global Shifts towards Positive Species Interactions with Increasing Environmental Stress." *Ecology Letters* 16: 695–706.

He, Qiang, and Brian R. Silliman. 2016. "Consumer Control as a Common Driver of Coastal Vegetation Worldwide." *Ecological Monographs* 86 (3): 278–94. https://doi.org/10.1002/ecm.1221.

Hedges, Larry V., Jessica Gurevitch, and Peter S. Curtis. 1999. "The Meta-Analysis of Response Ratios in Experimental Ecology." *Ecology* 80 (4): 1150–56.

Inderjit, and Roger del Moral. 1997. "Is Separating Resource Competition from Allelopathy Realistic?" *The Botanical Review* 63: 221–30.

Jackson, Evelyn F., and Colin R. Jackson. 2008. "Viruses in Wetland Ecosystems." *Freshwater Biology* 53 (6): 1214–27. https://doi.org/10.1111/j.1365-2427.2007.01929.x.

Jolliffe, Peter A. 2000. "The Replacement Series." *Journal of Ecology* 88 (3): 371–85. https://doi.org/10.1046/j.1365-2745.2000.00470.x.

Keddy, Paul A. 1990. "Competitive Hierarchies and Centrifugal Organization in Plant Communities." In *Perspectives on Plant Competition*, 265–90. Caldwell, NJ: The Blackburn Press.

———. 2000. *Wetland Ecology: Principles and Conservation.* Cambridge, UK: Cambridge University Press.

Knight, Ian A., and Nathan E. Harms. 2022. "Improving Biological Control of the Invasive Aquatic Weed, Alternanthera Philoxeroides: Cold Tolerance of Amynothrips Andersoni (Thysanoptera: Phlaeothripidae) and the Short-Term Feeding Impact on Different Host Haplotypes." *BioControl*, 1–12. https://doi.org/10.1007/s10526-022-10143-9.

Kubanek, Julia, William Fenical, Mark E. Hay, Pam J. Brown, and Niels Lindquist. 2000. "Two Antofeedant Lignans from the Freshwater Macrophyte *Saururus cernuus*." *Phytochemistry* 54: 281–87.

Lau, Jennifer A., Kenneth P. Puliafico, Joseph A. Kopshever, Heidi Steltzer, Edward P. Jarvis, Mark Schwarzländer, Sharon Y. Strauss, and Ruth A. Hufbauer. 2008. "Inference of Allelopathy Is Complicated by Effects of Activated Carbon on Plant Growth." *New Phytologist* 178 (2): 412–23. https://doi.org/10.1111/j.1469-8137.2007.02360.x.

Leck, Mary Allessio, and Robert L. Simpson. 1995. "Ten-Year Seed Bank and Vegetation Dynamics of a Tidal Freshwater Marsh." *American Journal of Botany* 82 (12): 1547–57.

Liu, Chao, Zhao-Jun Bu, Azim Mallik, Line Rochefort, Xue-Feng Hu, and Zicheng Yu. 2020. "Resource Competition and Allelopathy in Two Peat Mosses: Implication for Niche Differentiation." *Plant and Soil* 446 (1): 229–42. https://doi.org/10.1007/s11104-019-04350-0.

Lodge, David M. 1991. "Herbivory on Freshwater Macrophytes." *Aquatic Botany* 41: 195–224.

Luo, Wenbo, Yonghong Xie, Xinsheng Chen, Feng Li, and Xianyan Qin. 2010. "Competition and Facilitation in Three Marsh Plants in Response to a Water-Level Gradient." *Wetlands* 30 (3): 525–30. https://doi.org/10.1007/s13157-010-0064-4.

MacArthur, Robert H., and Edward O. Wilson. 1967. *The Theory of Island Biogeography.* Princeton, NJ: Princeton University Press.

Maestre, Fernando T., Ragan M. Callaway, Fernando Valladares, and Christopher J. Lortie. 2009. "Refining the Stress-Gradient Hypothesis for Competition and Facilitation in Plant Communities." *Journal of Ecology* 97 (2): 199–205. https://doi.org/10.1111/j.1365-2745.2008.01476.x.

Molles, Manuel C. 2013. *Ecology: Concepts and Applications.* 6th ed. New York: McGraw-Hill. www.textbooks.com/Ecology-Concepts-and-Applications—Text-Only-6th-Edition/9780073532493/Manuel-Molles.php.

Moore, Dwayne R. J., Paul A. Keddy, Connie L. Gaudet, and Irene C. Wisheu. 1989. "Conservation of Wetlands: Do Infertile Wetlands Deserve a Higher Priority?" *Biological Conservation* 47 (3): 203–17. https://doi.org/10.1016/0006-3207(89)90065-7.

Newman, Raymond M. 1991. "Herbivory and Detritivory on Freshwater Macrophytes by Invertebrates: A Review." *Journal of the North American Benthological Society* 10 (2): 89–114.

Newman, Raymond M., Zac Hanscom, and W. Charles Kerfoot. 1992. "The Watercress Glucosinolate-Myrosinase System: A Feeding Deterrent to Caddisflies, Snails, and Amphibods." *Oecologia* 92: 1–7.

Nezbrytska, Inna, Oleg Usenko, Igor Konovets, Tetiana Leontieva, Igor Abramiuk, Mariia Goncharova, and Olena Bilous. 2022. "Potential Use of Aquatic Vascular Plants to Control Cyanobacterial Blooms: A Review." *Water* 14 (11): 1727. https://doi.org/10.3390/w14111727.

Nilsson, Marie-Charlotte. 1994. "Separation of Allelopathy and Resource Competition by the Boreal Dwarf Shrub *Empetrum hermaphroditum* Hagerup." *Oecologia* 98: 1–7.

Nilsson, Marie-Charlotte, O. Zackrisson, O. Sterner, and A. Wallstedt. 2000. "Characterisation of the Differential Interference Effects of Two Boreal Dwarf Shrub Species." *Oecologia* 123: 122–28.

Ostrofsky, M. L., and E. R. Zettler. 1986. "Chemical Defenses in Aquatic Plants." *Journal of Ecology* 74: 279–87.

Otte, Marinus L. 2001. "What Is Stress to a Wetland Plant?" *Environmental and Experimental Botany*, Plants and Organisms in Wetland Environments, 46 (3): 195–202. https://doi.org/10.1016/S0098-8472(01)00105-8.

Pennings, Steven C., and Ragan M. Callaway. 1996. "Impact of a Parasitic Plant on the Structure and Dynamics of Salt Marsh Vegetation." *Ecology* 77 (5): 1410–19.

———. 2002. "Parasitic Plants: Parallels and Contrasts with Herbivores." *Oecologia* 131 (4): 479–89. https://doi.org/10.1007/s00442-002-0923-7.

Pianka, Eric R. 1970. "On R- and K-Selection." *The American Naturalist* 104 (940): 592–97. https://doi.org/10.1086/282697.

Sculthorpe, C. D. 1967. *Biology of Aquatic Vascular Plants.* New York: St. Martin's Press.

Silliman, Brian R., Elizabeth Schrack, Qiang He, Rebecca Cope, Amanda Santoni, Tjisse van der Heide, Ralph Jacobi, Mike Jacobi, and Johan van de Koppel. 2015. "Facilitation Shifts Paradigms and Can Amplify Coastal Restoration Efforts." *Proceedings of the National Academy of Sciences* 112 (46): 14295–300. https://doi.org/10.1073/pnas.1515297112.

Strobl, Katharina, Claudia Schmidt, and Johannes Kollmann. 2018. "Selecting Plant Species and Traits for Phytometer Experiments. The Case of Peatland Restoration." *Ecological Indicators* 88: 263–73. https://doi.org/10.1016/j.ecolind.2017.12.018.

Taiz, Lincoln, and Eduardo Zeiger. 2002. *Plant Physiology*. 4th ed. Sunderland, MA: Sinauer Associates, Inc.

Těšitel, Jakub. 2016. "Functional Biology of Parasitic Plants: A Review." *Plant Ecology and Evolution* 149 (1): 5–20. https://doi.org/10.5091/plecevo.2016.1097.

Tilman, David. 1980. "Resources: A Graphical-Mechanistic Approach to Competition and Predation." *The American Naturalist* 116 (3): 362–93.

———. 1985. "The Resource-Ratio Hypothesis of Plant Succession." *The American Naturalist* 125 (6): 827–52.

———. 1990. "Mechanisms of Plant Competition for Nutrients: The Elements of a Predictive Theory of Competition." In *Perspectives on Plant Competition*, 484. Caldwell, NJ: The Blackburn Press.

Toong, Yock C., David A. Schooley, and Fred C. Baker. 1988. "Isolation of Insect Juvenile Hormone III from a Plant." *Nature* 333 (12): 170–71.

Weigelt, Alexandra, and Peter Jolliffe. 2003. "Indices of Plant Competition." *Journal of Ecology* 91 (5): 707–20.

Wier, Andrew M., Terry A. Tattar, and Edward J. Klekowski Jr. 2000. "Disease of Red Mangrove (*Rhizophora mangle*) in Southwest Puerto Rico Caused by *Cytospora rhizophorae*." *Biotropica* 32 (2): 299–306. https://doi.org/10.1111/j.1744-7429.2000.tb00473.x.

Williamson, G Bruce. 1990. "Allelopathy, Koch's Postulates, and the Neck Riddle." In *Perspectives on Plant Competition*, edited by James B. Grace and David Tilman, 143–62. San Diego: Harcourt Brace Jovanovich.

Wilson, Scott D., and Paul A. Keddy. 1986. "Measuring Diffuse Competition Along an Environmental Gradient: Results from a Shoreline Plant Community." *The American Naturalist* 127 (6): 862–69. https://doi.org/10.1086/284530.

Wisheu, Irene C., and Paul A. Keddy. 1992. "Competition and Centrifugal Organization of Plant Communities: Theory and Tests." *Journal of Vegetation Science* 3 (2): 147–56. https://doi.org/10.2307/3235675.

Wood, Kevin A., Matthew T. O'Hare, Claire McDonald, Kate R. Searle, Francis Daunt, and Richard A. Stillman. 2017. "Herbivore Regulation of Plant Abundance in Aquatic Ecosystems." *Biological Reviews* 92: 1128–41.

Younginger, Brett S., Dagmara Sirová, Mitchell B. Cruzan, and Daniel J. Ballhorn. 2017. "Is Biomass a Reliable Estimate of Plant Fitness?" *Applications in Plant Sciences* 5 (2): 1600094. https://doi.org/10.3732/apps.1600094.

Zhang, Liwen, and Hongbo Shao. 2013. "Direct Plant–Plant Facilitation in Coastal Wetlands: A Review." *Estuarine, Coastal and Shelf Science* 119: 1–6. https://doi.org/10.1016/j.ecss.2013.01.002.

Zhang, Zhijie, Yanjie Liu, Ling Yuan, Ewald Weber, and Mark van Kleunen. 2021. "Effect of Allelopathy on Plant Performance: A Meta-Analysis." *Ecology Letters* 24 (2): 348–62. https://doi.org/10.1111/ele.13627.

11 Plants in the Context of Wetland Ecosystems

11.1 PUTTING IT ALL TOGETHER

In Chapter 1, I introduced the idea of asking why a particular plant is found in the place(s) where it is found: "Why does that plant grow there?" In the chapters since, we have learned about some of the challenges of wetland ecosystems that highlight slightly different questions, such as: How does that plant grow there? How does the plant deal with difficulties of accessing light, carbon dioxide, or oxygen? How does the plant cope with fluctuating hydrology of the wetland habitat? How does the plant access nutrients in the soil when faced with excess water and deficits of oxygen in those same soils?

Once the plant has acquired sufficient light, carbon, oxygen, and other nutrients to allow it to survive and grow, how does it go about ensuring that it or its offspring persist over time? We saw that questions about persistence over time are questions that have many levels. We can think about a lineage of plants (i.e., a plant and its descendants) persisting year to year in a specific part of a single wetland or across that entire wetland or across greater periods of time. We can expand our spatiotemporal window further and think about that lineage as part of a metapopulation that exists across a network of wetlands that are connected by gene flow and whose connections wax and wane over decades, along with changes in climate or in the availability of dispersal vectors.

Other factors that may influence persistence within a local population or across a metapopulation include interactions with neighboring plants, whose identities will shift over time, just as connections within the network of our focal plant species change temporally. Shifts in plant neighbors may be influenced by local extinctions or by the introduction of new species to the region. Similarly, populations and metapopulations of other species with which our plant interacts will change over time. This includes microbial symbionts, herbivores, parasites, pathogens, and even changes in the populations of humans, who may serve as dispersal vectors or herbivores, in addition to other direct and indirect effects we have on wetland plant populations.

Now, in this final chapter, we will weave together these aspects of plant biology to look at aquatic and wetland plants as central players in the functioning of wetland ecosystems. We will first examine how plant species assemblages are expected to change over time in wetlands, after which we will see some of the ways in which expected changes in dominant species may be altered by external biotic and abiotic forces, including elements of ongoing climate change. We then will examine wetlands as tools to deliver ecosystem services such as mitigation of surface water contamination and sequestration of anthropogenic greenhouse gases to combat climate change.

11.2 SUCCESSION

In the discussion of Grime's C-S-R framework in Chapter 10 (Figure 11.1), we considered that assemblages of plant species that persist under a given set of environmental conditions and resource availability might be expected to share certain traits that enhance their success under those conditions. Otherwise, we would very rarely expect to see those species under those conditions. We thought of this in terms of selective pressures that favor life history strategies that commonly appear under a given combination of conditions and resources. In Grime's framework (Grime 1977), there were two major forces thought to drive selection of plant life history strategies: stress and disturbance. He defined stress as any factor external to the plant that restricts plant performance, and disturbance was defined as any mechanism that destroys plant biomass. Examples of disturbances are floods, fires, wind, erosion, herbivory, pathogens, and human activities that remove biomass such as soil tillage, deforestation, and mowing. Examples of stress included such factors as environmental toxins, heavy metals, temperature extremes, and resource deficiencies. Resource deficiencies, as Grime indicated, included instances where the deficiency is driven by neighboring species, such as in the understory of mature forests, where the canopy trees may outcompete understory species for light and soil resources.

It was in this context of disturbance and neighbor-induced resource limitation where Grime discussed potential changes in dominant life history strategies over time, that is, during plant succession. As a reminder, succession refers to changes in plant species composition over time, usually following a disturbance that removes all or most of the dominant plant species over some area. However, there are nuances to the concept of succession that make it a bit more complicated than simple changes in species identities. As Grime indicated, there typically are characteristic suites of plant traits, including characteristic plant life history strategies, that one expects to see as time passes following a disturbance. This is in part a factor of the time required for species to disperse to an area that has been disturbed, but also driven, in part, by the time required for a species to establish and become part of the dominant vegetation of an area.

Our consideration of plant life histories in Chapter 8 dealt with plant longevity, time to sexual reproduction, and

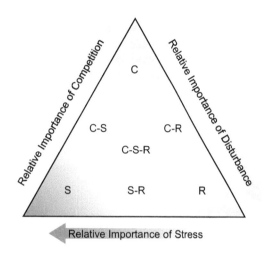

FIGURE 11.1 Depiction of Grime's Three Primary Strategies (Competitors, C; Stress Tolerators, S; and Ruderals, R) as a Three-Axis Continuum of Primary, Secondary, and Tertiary Life History Strategies. of Key Importance in the Context of Succession is the Natural Progression Over Time from Recently Disturbed Conditions Favoring Ruderal Species (Bottom-Right Corner) to Conditions that Favor the Other Two Life History Strategies, Alone or in Combination. Redrawn from Grime (1977).

frequency of reproduction, including the general categories of annuals, biennials (or monocarpic perennials), and perennials (polycarpic perennials). Grime (1977) conceptually linked his C-S-R life history strategies with longevity- and reproduction-based plant life histories, as part of placing his framework within the context of succession (Figures 11.2 and 11.3). Annuals, which are short-lived and produce copious small, readily dispersed seeds, commonly are the dominant species shortly following a disturbance of moderate to severe intensity. Annuals correspond with what Grime termed "ruderals" (Table 10.2) and what Pianka (1970) called r-selected species (Table 10.1). Biennials, because they require at least two growing seasons to complete their life cycles, typically will dominate habitats experiencing a lower frequency or intensity of disturbance, compared with annuals, or they will dominate later following disturbance. Thus, we see segregation of annuals and biennials in Grime's C-S-R triangle (Figure 11.2). Grime's reconciliation of herbaceous perennials with his C-S-R framework is reflective of the diversity represented within that group. Herbaceous perennials are a highly heterogeneous group of plants, and they can occupy a similarly heterogeneous range of habitat conditions. At the side of Grime's triangle farthest removed from disturbance, we see trees and shrubs (Figure 11.2). These species generally require long periods of time to establish themselves as the dominant vegetation and, as a result, will rarely dominate highly disturbed habitats or those that have recently experienced moderate to severe disturbances. There are exceptions to this generality,

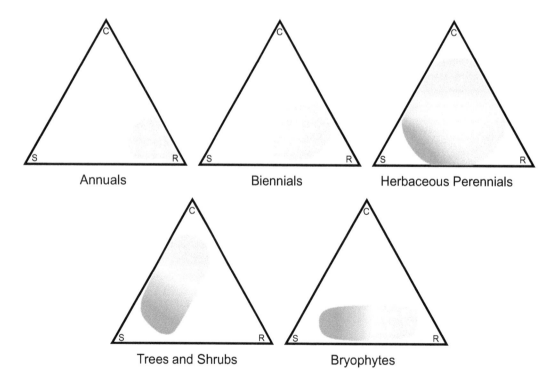

FIGURE 11.2 Regions of Grime's C-S-R Continuum That Characterize Select Groups of Plant Life Histories and Growth Forms. Recall that the Lower-Right Corner of this Triangle Represents Habitats that Have Experienced a Recent or Severe Disturbance. Redrawn from Grime (1977).

however, as some fast-growing woody species are particularly well adapted to highly disturbed conditions.

If we were to consider changes in species composition over time through the lens of Grime's C-S-R framework, we might trace paths such as those depicted in Figure 11.3. In these examples provided by Grime (1977), he was attempting to illustrate how the trajectory in dominance among different life history strategies might be influenced by the potential productivity of a site. Because his framework defines competitors as those species that tend to dominate under conditions of high productivity, the successional trajectory in a wetland with relatively high productivity will include a period during which competitors dominate the vegetation. Noteworthy in both trajectories in Figure 11.3 is that both terminate with a late-successional community characterized by relatively high levels of stress. Stress, in this case, comes in the form of low resource availability. In the more productive habitat, resource depletion is the result of later arriving and long-lived canopy trees intercepting sunlight and locking away soil nutrients in woody tissues, creating a resource "stress" for their neighbors. In the less productive habitat, resource depletion stress results from longer-lived, but typically shorter-stature, plants having accumulated biomass that allows them to continuously draw from the already limited pool of soil resources that were available.

If we compare Grime's successional C-S-R timeline with expectations in the context of the r- versus K-selected species framework, we see an early emphasis on traits associated with the r strategy, progressing towards a greater importance of K strategists later in succession

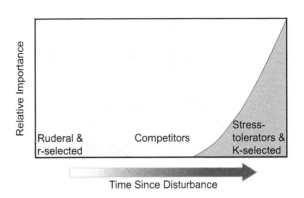

FIGURE 11.4 Shifting Importance of Different Life History Strategies across a Successional Timeline. The Initial Years Following a Disturbance Favor Species that are Readily Dispersed and Capable of Rapid Colonization of Available Habitat. Annual Species are Gradually Replaced in Dominance by Longer-Lived and Larger Species, which also will often be Replaced by even Larger and Longer-Lived Species. Late Succession is often Dominated By Species that are Very Long-Lived and Invest Relatively Small Amounts of Captured Resources in Sexual Reproduction. Modified from an Illustration in Grime (1977).

(Figure 11.4). The middle phase of succession would be dominated by species whose traits are transitional between r and K selection, moving progressively towards more and more K-strategist type species. Grime would consider that the identities of those mid-successional species will be determined to a large degree by the potential productivity of the site (Figure 11.3).

It is important to keep in mind, however, that shifts in vegetation or dominant species over time are not abruptly punctuated changes from exclusive dominance by one or two species to exclusive dominance by a second species or set of species. For example, Figure 11.2 depicts assemblages of species with similar characteristics occupying distinct triangles. However, at any given point in time, we might see species from multiple of those five triangles present at the same time in a wetland, but the dominant species might represent only one or two of the triangles. Similarly, the trajectories of change in life history strategies over time depicted in Figure 11.3 would not occur via distinct, stepwise change from one point along the trajectory to the next. Rather, we would see a gradual progression over time (and over space) from one region of the triangle to another. These concepts of spatiotemporal gradients in successional processes were expressed more than a century ago by Henry Chandler Cowles, who asserted, "there are few if any sharp lines in nature" (Cowles 1899). You may remember Cowles as the namesake of Cowles Bog Wetland, one of our examples from Chapter 4. He used that wetland and surrounding dune complexes in his studies of plant species succession in dune communities.

In the mid-1900s, a principle termed "initial floristic composition" was proposed to account for the process by which plant species shift in dominance over time (Figure 11.5; Table 11.1). Egler (1954) suggested that

FIGURE 11.3 Hypothetical Shifts in Dominant Plant Life History Strategies Following Disturbance in Habitats with Differing Levels of Potential Productivity. The Upper Path, among the Darker Circles, Represents Successional Changes in Plant Species in a Higher Productivity Wetland. The Starting Point, the Open Circle in the Lower-Right Corner, Represents Dominant Plant Species Shortly After a Severe Disturbance. Comparison of the Trajectories here with the Life History Strategies in Figure 11.1 is Informative about the Influence that Productivity Might Have on Successful Strategies at Different Times after Disturbance. Redrawn from Grime (1977).

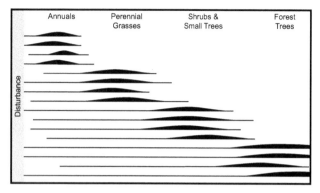

FIGURE 11.5 Egler's Description of the Initial Floristic Composition Principle Supported the View that Many, if not Most, Species that May Eventually Achieve Dominance within a Community are Represented in some Way Even Before a Disturbance Takes Place (Indicated By the Horizontal Lines). They May Exist on the Margins of the Habitat, as Small Refuge Populations within the Larger Habitat Matrix, or as Dormant Individuals in the Seed or Propagule Bank. They Will Appear Among the Dominant Species within the Community at a Point in Time when Local Environmental Conditions Favor their Dominance (Indicated by the Thickened Parts of the Lines), But they Eventually may be Replaced by Species that are Better Suited to New Conditions Resulting in Part from Modification by the Dominant Species Themselves. Redrawn from Egler (1954).

TABLE 11.1

General descriptions of models of succession discussed in this chapter.

Succession Model	Description	Reference
Environmental sieve model	Dominant plant species arise out of the potential pool of species whose seeds or vegetative propagules are present in the wetland at a given time, either in the soil seed bank or in the arrival of dispersed propagules to the wetland. The ability of species to establish within the dominant species assemblage at a given point in time is determined by their life history characteristics, their ability to tolerate then-current hydrology, and their ability to persist in the soil seed/propagule bank.	Van der Valk (1981)
Facilitation model	Early successional species colonize a newly disturbed site, and their modifications to the environment make it more suitable for establishment of later successional species. This process continues until there are no subordinate species nor potential new colonizers whose dominance is facilitated by the then-current dominant species or life history strategy.	Connell and Slatyer (1977)
Hydrarch succession	Succession of lakes or ponds towards mesophytic forest, via infilling of the water body with detritus. Although this process was hypothesized to culminate in mesophytic forest, such has not been demonstrated outside of ecosystems where hydrology is significantly altered. Many examples exist, however, of succession to coniferous forested wetlands in higher latitudes.	Cooper (1913)
Inhibition model	Dominant species at some stage of succession modify the environment such that it becomes unfavorable for both earlier and later successional species. In this model, the dominant species retain dominance until disturbance removes sufficient biomass for them to be replaced by another species or group of species.	Connell and Slatyer (1977)
Initial floristic composition	An area is hypothesized to contain a full complement of species from any or all points along a typical succession timeline immediately following disturbance. The dominant species are expected to shift during succession, based on shifts in prevailing environmental conditions and resource availability, drawing from both the pre-existing seed or propagule bank and the migration of seeds and propagules into the area during succession.	Egler (1954)
Relay floristics	One set of dominant species relatively abruptly gives way to another set as environmental conditions change over time, largely because of modification of the environment by the groups of dominant species that are being replaced.	Egler (1954)
Tolerance model	Early successional species modify the environment such that it is less suitable for early successional species, with no measurable effect on later successional species. Later successional species simply tolerate conditions within the site until such time as they can replace or displace earlier species.	Connell and Slatyer (1977)
Wetland continuum concept	Spatiotemporally dynamic model of succession that links changes in dominant plant species with hydrologic cycles in wetlands experiencing interannual variation in inputs of water from precipitation and groundwater sources.	Euliss et al. (2004)

succession had previously been viewed predominantly as a series of distinct groups of dominant species characterizing successive stages in vegetation change over time (despite Cowles' denial of the existence of sharp lines in nature). Egler termed the abruptly punctuated view of succession "relay floristics," in which one set of dominant species would give way to another as environmental conditions changed within an area over time, largely as a result of modification of the environment by the groups of dominant species

that were being replaced. He argued, however, that even before an area is disturbed, that area will receive sexual and asexual propagules of species that may represent any stage of succession. As a result, succession may begin with an initial floristic composition that includes a full complement of species from any or all points along a typical succession timeline, but the dominant species will be expected to shift during succession, based on shifts in prevailing environmental conditions and resource availability. This aspect of succession was also recognized by Clements (1928) but was not emphasized nearly as strongly as were the distinct stages of succession and the processes that bring about shifts from one stage to another.

One of the major drivers of succession on which Clements (1928) focused was competition, which is similar to the scenario in Figure 11.4, where Grime's competitors often might be the dominant life history strategy during intermediate stages of succession. A notable difference between the two, however, is that Clements felt that competition was "a universal characteristic of all plant communities," except in recently colonized bare areas where spacing among individuals prevents their interaction. In addition to competition, Clements (1928) emphasized something he called "reaction" as a key element of successional change. Reaction was the sum of the ways in which a species or an individual affects the habitat in which it grows. This included processes such as shading, production and accumulation of leaf litter, modification of the water table, moderation of local humidity, and stabilization of coarse substrates or sediments. These are some of the same processes that were discussed in Chapter 10 with respect to facilitation. In fact, all the interactions that were discussed in the previous chapter are involved to varying degrees in determining the trajectory and speed of succession following disturbance.

A similar, but somewhat more complex, view of succession was presented by Connell and Slatyer (1977). In their conceptual model of mechanisms driving succession, they proposed three routes by which succession may take place (Table 11.1). The first was a facilitation model, in which early successional species colonize a newly disturbed site, and their modifications to the environment make it more suitable for establishment of later successional species. This process would then continue until there are no subordinate species nor potential new colonizers whose dominance is facilitated by the then-current dominant species or life history strategy. At that point, further changes in the species composition would be expected only when disturbance removes a sufficient amount of biomass to allow dominance by an earlier successional species (Connell and Slatyer 1977).

Another of the three models presented by Connell and Slatyer (1977) was a tolerance model, in which early successional species modify the environment such that it is less suitable for early successional species, with no measurable effect on later successional species. Later successional species simply tolerate conditions within the site until such time as they can replace or displace earlier species. As with the facilitation model, it was suggested that this process would repeat itself until there were no species remaining to which dominance can be relinquished. The third possible route for successional change proposed by Connell and Slatyer was an inhibition model, in which dominant species at some stage of succession modify the environment such that it becomes unfavorable for both earlier and later successional species. In that model, the dominant species

would retain dominance until disturbance removes sufficient biomass for them to be replaced by another species or group of species.

Either the facilitation or the tolerance model proposed by Connell and Slatyer could be applied to Egler's initial floristic composition hypothesis (Figure 11.5). For example, under the facilitation model, the dominant annual species would modify the environment such that perennial grasses could replace them as the dominant species. This could be via accumulation of dead organic matter in the soil, which would enhance soil moisture and nutrient content, or by nutrient enrichment of the soil by annual legume species, via their nitrogen-fixing symbionts. The perennial grasses then would further increase soil organic matter content, through their greater productivity, and would increase soil stability with their extensive root and rhizome networks. The taller stature of those species might also help maintain soil moisture and could shade later successional shrub and tree seedlings that might be sensitive to excessive light levels. Over time, shrubs, and then trees, would increase in height, forming canopies that would overtop the previous species, while taking advantage of the modified environmental conditions that facilitated establishment of their seedling stages.

Viewing the initial floristic composition hypothesis through the lens of the tolerance model, we might see some later successional species present in the community immediately after disturbance. However, those species cannot achieve dominance within the community until the earlier successional species decline in abundance. At that time, the later successional species increase in abundance until their influence on the environment no longer favors their own dominance and, upon their decline, the next group of dominant species increase. These comparisons across different conceptual representations of succession illustrate the similarities that exist in this process across the many ecosystems that have been studied, as well as some of the subtle differences in the ways that one might interpret the outcomes of long-term interactions among species.

It is important to note that, although we should expect to see some later successional species present in the community at the time of disturbance, it is likely that other species will be recruited into the community over time from other locations. Perhaps it is no surprise that this also has been incorporated into conceptual models of succession, and our next example is specific to wetland ecosystems. In this model, often referred to as the environmental sieve model, species' life history traits are considered as potential predictors of how the dominant species in a wetland will change in response to changes in the wetland environment (Van der Valk 1981). The major environmental driver of change in this model, and the major factor that distinguishes wetlands from upland or terrestrial ecosystems, is hydrology (Figure 11.6). We saw this at length in Chapters 4–6, when we examined interactions between vegetation and hydrology and the many ways in which aquatic and wetland plants are adapted to life in the aquatic environment.

Flooded

Recruited from Seedbank
Annual and Perennial species establishing in standing water

Recruited via Dispersal
Annual and Perennial species establishing in standing water

Temporarily Extirpated
Annual and Perennial species intolerant of flooding,
except Perennials persisting via vegetative structures

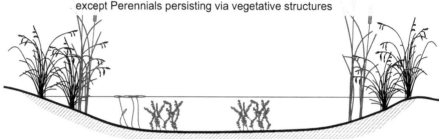

Drawdown

Recruited from Seedbank
Annual and Perennial species establishing when there is no standing water

Recruited via Dispersal
Annual and Perennial species establishing when there is no standing water

Temporarily Extirpated
Annual and Perennial species adapted to standing water, except Perennials
persisting via vegetative structures

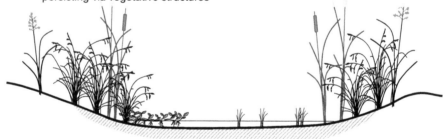

FIGURE 11.6 Van Der Valk's Environmental Sieve Model States that the Dominant Plant Species in a Wetland will be Selected from a Local Pool of Species Represented as Seeds and Vegetative Propagules Present in the Wetland Seed Bank and those Being Dispersed into the Wetland. Conditions in the Wetland at the Time Propagules Arrive Will Determine Which Species Will Establish, Based on their Individual Life History Traits. Traits of the Dominant Species also Will Determine Which Species Remain, as the Wetland Changes Over Time. Based on Van Der Valk (1981).

In the environmental sieve model, dominant plant species arise out of the potential pool of species whose seeds or vegetative propagules are present in the wetland at a given time, either in the soil seed bank or in the arrival of dispersed propagules to the wetland. Van der Valk (1981) considered three primary factors that might be expected to determine the ability of a plant species to establish within the dominant vegetation of a wetland. These included (1) the species' life history, combined with its capacity to persist vegetatively for an indeterminate period, (2) its ability to establish under flooded versus non-flooded conditions, and (3) whether it can establish from a long-lived seed (or propagule) bank or whether it must be recruited from propagules that are dispersed into the wetland.

An example of this process was given by the succession of plant species in a glacially derived shallow-lake marsh, Eagle Lake, Iowa (USA; Van der Valk 1981). In this lake, **drawdown**, which is a significant decrease in water level that exposes some or all of the wetland sediments, resulted in the establishment of annual species as well as emergent

FIGURE 11.7 Succession of Dominant Plant Species of Eagle Lake, Iowa (USA) Following Major Changes in Wetland Hydrology. Superscripts Legend: A = Annual; P = Non-Persistent Perennial; V = Perennial, Persisting Vegetatively; S = Species that Establish from a Persistent Soil Seed Bank or Propagule Bank; 1 = Establish During Drawdown; 2 = Establish with Standing Water. Redrawn from Van Der Valk (1981).

perennials that were unable to establish under flooded conditions (Figure 11.7). When water levels rose again, the drawdown-adapted annuals disappeared from the dominant vegetation and were replaced by flood-adapted annuals and perennials. If such flooded conditions were to persist, shorter-lived perennials that cannot maintain their populations vegetatively would disappear, relinquishing their place in the dominant vegetation to highly clonal species such as *Typha* species. Submersed flood-adapted annuals may persist under flooded conditions if sufficient open water is available to provide refuge from competition with the clonal emergent species. If water levels fall again, whether the result of drought or human intervention, the vegetation would return to dominance by drawdown-adapted species.

The role that hydrology plays in these cycles of dominant vegetation provides a stark contrast to expected successional trajectories in terrestrial ecosystems. In terrestrial systems, there usually is an anticipated **climax community**, or presumed final stage of dominant vegetation, that is characteristic of a particular region, or geological subdivision of a region. Much of the succession theory that was developed in the early to mid-20th century was based on successional trajectories in what are known as "old fields," which were abandoned agricultural fields. In old field ecosystems, the climax community often is a hardwood forest dominated by deciduous tree species. Water, however, presents an additional challenge that filters from among the available species, as Van der Valk described it. That is not to say that there are no climax communities that result from wetland succession, but the fluctuations in water levels that are characteristic of natural wetlands can serve to frequently "reset" successional trajectories, preventing or prolonging arrival at a climax species assemblage. Other disturbances, such as fire and herbivore outbreaks, can have similar effects, but those are not unique to wetlands and have been studied quite extensively in terrestrial ecosystems. This aspect of succession is discussed further in the following section.

With respect to climax communities in wetlands, it was suggested that succession in wetlands in a region dominated by forest biomes would eventually climax with an upland forest. This process was referred to as **hydrarch succession**, defined as succession of lakes or ponds towards **mesophytic** forest vegetation, or forest vegetation dominated by species that are characteristic of neither wet nor

exceedingly dry habitats (Cooper 1913). Cooper (1913) suggested, for example, that succession in all communities in northeastern Canada and along the Appalachian Mountains of eastern North America should progress towards mesophytic forest; however, he went on to state that sufficient evidence to support that hypothesis was lacking at the time of his studies. Nevertheless, the term hydrarch succession has since been broadly applied to the succession of lakes or ponds into wetlands.

A study of oxbow lakes in Alberta, Canada, for example, demonstrated evidence for hydrarch succession from open water to forested wetland (Van der Valk and and Bliss 1971). This study made use of an approach called a **chronosequence**, in which multiple wetlands at different successional stages were used to infer successional processes and the sequence of dominant plant species over time. Across the 15 oxbow lake wetlands in that study, there was clear evidence of succession from submersed species to dominance by emergent plants, followed by progression to a wet forest dominated by birch (*Betula* species) and willows (*Salix* species), with some of the oldest wetlands dominated by *Populus balsamifera* (balsam poplar) forests. In accordance with the environmental sieve model mentioned earlier, there also was evidence for occasional reversion of some plant assemblages following periods of flooding.

Another major factor in the rate of succession in those oxbow wetlands was the rate of biomass production by the dominant plant species (Van der Valk and Bliss 1971). The progression in dominant species over time was accompanied not only by a change in dominant growth form and life history, but also in the rate of primary productivity (up to a point). Floating-leafed species generally had higher productivity than did submersed species, and emergent plants had higher productivity than floating-leafed species, as we saw in Chapter 1. Van der Valk and Bliss (1971) suggested the increased productivity also was accompanied by a greater temporal stability in the populations of dominant species. Emergent species, owing to their stature, size, and persistence via vegetative growth, provided greater structural stability to the community during changes in hydrology, which provided the emergent-dominated communities greater resilience to changes in water levels than was exhibited by communities dominated by submersed and floating-leafed species, or by early successional annual species. This stability in the dominance of long-lived clonal perennials also was an aspect of the environmental sieve model discussed earlier.

An older study by Matthews (1914), conducted in White Moss Loch, in Scotland, illustrated the typical, concentrically ringed zonation formed during hydrarch succession (Figure 11.8). As emergent vegetation encroaches on the open water zone of the wetland, detritus accumulates annually, reducing the area of open water and allowing the emergent species to migrate farther towards the wetland center. Nearer the periphery, wetland shrubs and trees will begin to establish and migrate towards the wetland center as well, as sufficient peat accumulates to allow their establishment. We saw evidence of this type of peat accumulation in Clara

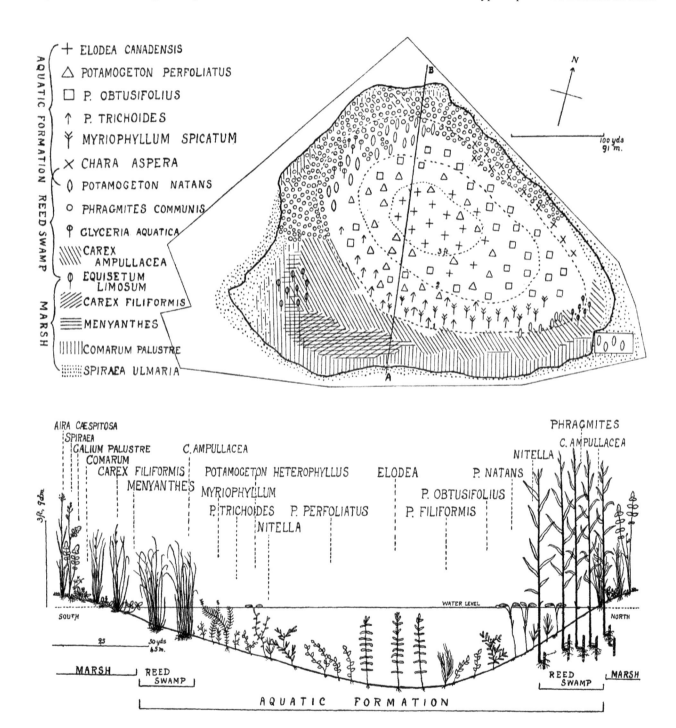

FIGURE 11.8   Distribution of Dominant Plant Species in White Moss Loch, Perthshire, Scotland, During August 1913. The Concentric Zonation of Vegetation in the Upper Diagram is Typical of Succession in Depressional Wetlands, Referred to as Hydrarch Succession. The Cross-Section in the Lower Panel Illustrates Species Along the South-North Transect Crossing the Center of the Wetland in the Upper Panel, Labeled "A-B." Not Included in the Map was a Fringing Tree Assemblage around the Wetland Dominated by *Alnus Rotundifolia* (Alder) and Several Species of *Salix* (Willows). Taken from Matthews (1914).

Bog and Cowles Bog in Chapter 4, where peat had accumulated to depths of 4 m to 10 m during succession in those wetlands.

In another study of hydrarch succession, in Cedar Bog Lake, Minnesota (USA), Lindeman (1941) found that the mat of emergent wetland vegetation fringing the lake's open water zone migrated towards the lake center at a rate of approximately one meter per five years. The mat was dominated by the creeping emergent species *Decodon verticillatus* (swamp loosestrife) on the water's edge and graded into *Typha latifolia* towards the periphery. During drier years, *T. latifolia* would migrate into the *D. verticillatus* zone, and in wetter years, *D. verticillatus* would resume dominance. Nearer the wetland periphery, scattered individuals of willow and alder were found, and there was a "distinct forest girdle" of *Larix laricina* (larch) approximately 40 m wide at the outermost edge of the wetland (Lindeman 1941).

These examples of hydrarch succession suggest that this process is a better fit to the relay floristics model of succession discussion by Egler (1954) than to his initial floristic composition model (Figure 11.9). This is an important contrast between succession in terrestrial versus aquatic and wetland communities and illustrates a fundamental aspect of succession. The initial floristic composition model is an example of **secondary succession**, in which the successional processes occur following the disturbance of a previously vegetated area. Because the area was previously vegetated, there has been an opportunity for propagules of species representative of multiple stages of succession to arrive at the site before a disturbance resets the community

to an earlier successional stage (Figure 11.5). In contrast, hydrarch succession describes succession from an aquatic ecosystem (a lake or pond) to a wetland, and it is more akin to, but still perhaps distinct from, **primary succession**. Primary succession takes place on a site that was previously unvegetated, at least during the typical lifetimes of long-lived plant species. Because the site was not vegetated previously, there are no propagules present at the time succession begins, although propagules of later successional species may arrive early during succession. Primary succession typically occurs on newly exposed rock surfaces, lava flows, sand deposits, or other sites with predominantly mineral substrate that must undergo soil formation along with successional changes in vegetation.

Depending on the type of lake undergoing succession, there may or may not be propagules of aquatic or wetland species present when succession begins. In kettle lakes formed on the tracks of receding glaciers, such as Cedar Bog Lake mentioned earlier, the lake forms on almost exclusively mineral material, and we can consider succession within that lake to represent primary succession. In oxbow lakes that are formed when a portion of a river or stream channel is cut off from the main channel by erosional processes, it is likely that the new lake hosts at least some aquatic and wetland plant life. Thus, succession in an oxbow may be better representative of secondary succession, although this could be debated, as rivers scour away previous vegetation as they meander across the landscape and floods often leave significant, unvegetated sand deposits adjacent to river channels.

Although hydrarch successional patterns in wetlands are quite common, especially at higher latitudes in the northern hemisphere, it is important to note that few, if any, examples exist of wetlands being converted to upland forest through natural successional processes. Cronk and Fennessy (2001) provided an excellent discussion of reasons for this. Central to the explanation is the dependency on oxygen-deficient conditions for peat to accumulate. As we have discussed in earlier chapters, the absence of oxygen results in very slow decomposition of detritus, allowing peat to accumulate. If or when peat reaches the water's surface or higher, either through accumulation or decline in water levels, decomposition will accelerate, reducing the depth of peat. Thus, hydrology again becomes a critical driver of succession in wetland ecosystems, even if indirectly.

One final model of wetland succession, developed from studies of wetlands in the prairie pothole region of North America, is the wetland continuum framework (Euliss et al. 2004). This framework addresses the separate roles of precipitation patterns and groundwater hydrology on wetland vegetation dynamics (Figure 11.10). This model is closely connected with the environmental sieve model, in that it places substantial emphasis on the role of wetland hydrology in determining changes in plant species composition over time. Where this model extends beyond the environmental sieve is in its explicit effort to link changes in wetland vegetation to changes in other biotic components of

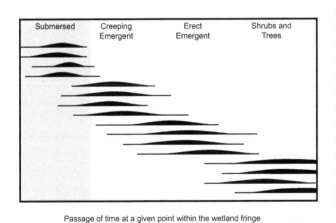

FIGURE 11.9 Hydrarch Succession of Lakes or Ponds to Wetlands is Essentially a Form of Relay Floristics, Strongly Influenced by Site Hydrology. As Emergent Species, Including Mat-Forming Creeping Emergents, Establish around the Lake Perimeter, their Detritus Serves to Decrease Relative Water Depth, Allowing them to Encroach Further into the Wetland Interior. Reduced Water Depth around the Wetland Periphery then Facilitates Establishment of Species Adapted to Still Lower Frequency and Duration of Flooding. In the Absence of Modifications to the Site's Hydrology, this Process May Continue, Ultimately Terminating in the Establishment of Wetland Tree Species. Modified from Egler (1954).

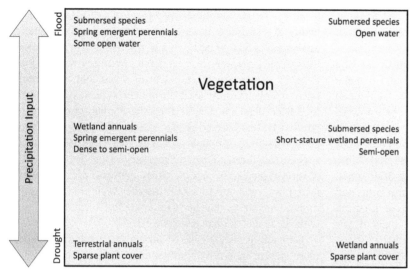

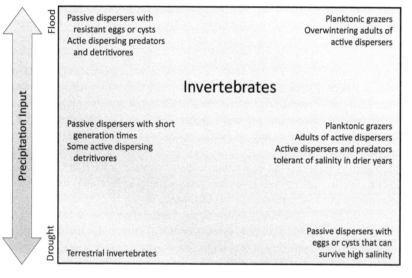

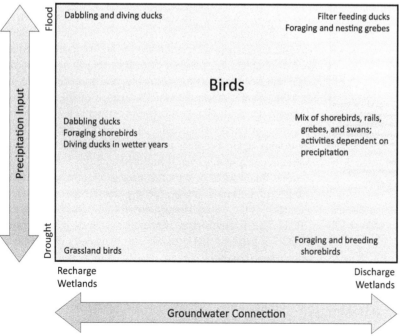

FIGURE 11.10 The Wetland Continuum Framework Considers Interrelationships of Precipitation and Wetland Connections to Groundwater in Predicting Changes in Vegetation Over Time. It also Considers that Other Biotic Components of Wetlands, such as Invertebrates and Birds, Will Respond to Changes in Hydrology and the Ensuing Changes in Plant Species Composition. Thus, it Considers the Wetland as an Integrated Biological System. The Lower-Left Corner of these Figures Represents the Driest Conditions, while the Upper Right is the Wettest. High Salinity is Found in the Drier Groundwater Discharge Wetlands because of Dissolved Ions in the Groundwater Entering the Wetland from Subterranean Sources. Redrawn from Euliss et al. (2004).

the wetland, such as invertebrates, amphibians, and birds. The extensions of plant succession to their impacts on other biota in the wetland continuum are excellent examples of the importance of wetland plants for ecosystem function, which is the foundation for services that we humans derive from aquatic and wetland ecosystems. We will consider some of these services further in the final sections of this chapter.

11.3 DISTURBANCE

We saw several definitions of disturbance in the previous chapter, one of which was used by Grime (1977) in developing his C-S-R framework, wherein he described disturbance as "any mechanism that destroys plant biomass." A more complete definition was that used by Gurevitch, Scheiner, and Fox (2006, p. 285), which elaborated on the process as any "relatively discrete event in time that causes an abrupt change in ecosystem, community, or population structure and changes resource availability, substrate availability, or the physical environment."

A similar but somewhat more concise definition was used by Keddy (2000, p. 281), who stated that disturbance includes any "short-lived event that causes a measurable change in the properties of an ecological community." Keddy additionally discussed what was meant in this context by "short-lived," specifying that disturbances are events that typically have a "duration much shorter than the life span of the dominant species in the community." Wetland disturbances, then, would include windstorms such as hurricanes, cyclones, and tornadoes; fires; floods; flood- and storm-associated erosion and deposition; wave action; ice scour; drought; herbivore outbreaks; construction of beaver ponds; and human activities such as mowing, timber harvest, dam construction, and wetland drainage. Keddy made a point to note that, because of the specification that disturbances are short-lived events, climate change would not be considered a disturbance. However, climate change does alter the frequency and intensity of disturbances such as floods, droughts, fires, and storms, along with its demonstrated effects on herbivore populations and animal migration.

In the context of succession, disturbance is a critically important force. As we have seen in many of the prior examples of succession, changes in hydrology can have marked effects on the dominant plant species of a wetland and can alter the trajectory of succession, often resetting the dominant species to an earlier successional assemblage. In the oxbows of Alberta, for example, Van der Valk and Bliss (1971) found evidence for what they referred to as rhythmic cycling between early and late successional stages, driven by cycles of flooding in the developing wetlands. In some years, flooding in those oxbow lakes deposited enough sediment (3 to 4 cm depth) to bury all established submersed vegetation, and further flooding prevented any recolonization for the remainder of the year. In the Eagle Lake wetland, alternating wet and dry periods brought about shifts

between species of differing growth forms and life histories, with submersed annuals forming much of the dominant vegetation during flooded periods, and perennial emergent species dominating during drier periods (Van der Valk 1981). The Eagle Lake example also illustrated the potential change that can result from interacting disturbances; herbivory by muskrats (*Ondatra zibethicus*) decimated populations of emergent plant species in that wetland, causing a shift to dominance by submersed and free-floating species.

Another important aspect of Keddy's treatment of disturbance was a discussion of properties of disturbance that help determine its impact on ecosystems. These properties are the frequency, intensity, duration, and spatial extent of the disturbance (Keddy 2000). All these aspects of disturbance are quantifiable, and thus can be directly correlated with effects on plant performance, as specified in Keddy's definition. That is, if we have quantifiable parameters of the plant species assemblage (e.g., biomass, species richness, annual productivity), we can determine the per-unit effects of any of these aspects of disturbance on the wetland vegetation. Examples of these attributes of disturbance for flooding would include how often flooding occurs (frequency), how deep the standing water is during a flood (intensity), how long the flood lasts (duration), and what percentage of a wetland is under standing water (spatial extent). Each of these attributes might exert its effect on the plant assemblage as a whole or differentially among plant species. Likewise, these four properties may interact with one another in their joint effects on the vegetation, and different types of disturbance may co-occur and have compounding effects on the vegetation (Buma 2015).

Drought, for example, reduces plant health, which can lead to increased susceptibility to herbivores, parasites, and pathogens. Plant mortality caused by these attacks increases fuel for fires, especially in drought-impacted forests, and this can increase the frequency and intensity of fires (Dale et al. 2001). Damage to forests resulting from windstorms can have similar effects by increasing the amount and distribution of dead biomass on the forest floor, resulting in increased intensity and spatial extent of future fires (Buma 2015). Furthermore, it is estimated that more than 75% of earth's land surface is directly impacted by human activities, such as urbanization and agriculture (Ellis, Beusen, and Goldewijk 2020; Sanderson et al. 2002). This globally widespread anthropogenic encroachment on natural ecosystems has the unfortunate consequence of magnifying effects of natural disturbances. As a result, it is important to keep in mind that the things we have learned in the past about how disturbances affect wetlands may not have exactly the same application for management of wetlands now and in the future.

Agricultural activities undergo frequent intensification in an effort to feed our growing human population, and essentially all farmable land is currently under cultivation or otherwise impacted by human activities (Sanderson et al. 2002). Because of asynchronicities between land preparation, fertilization, and biological cycles of crop species, agricultural lands are subject to considerable runoff of eroded soil and excess fertilizers, much of which finds its way into adjacent streams and wetlands (see again Chapter 7), and this can negatively impact wetland plant assemblages. Shoemaker, Ervin, and DiOrio (2017) found that high loading of sediment and nutrients, characteristic of agricultural lands in the lower Mississippi River Alluvial Valley (USA), significantly reduced the species richness, diversity, and floristic quality of wetland vegetation. Although each of these factors alone was found to negatively affect species assemblages, the combination of high sediment and high nutrient loading had a much stronger effect than either alone. A similar study was conducted by Kercher and Zedler (2004), with the addition of flooding and introduction of an invasive grass species, *Phalaris arundinacea*. They found that as little as four weeks of increased sediment loading and flooding significantly reduced plant species richness in their experimental wetlands. Furthermore, the loss of dominant species from those communities facilitated the establishment of *P. arundinacea* via increased light availability, and increased nutrient loading was correlated with a tripling in the productivity of invasive *P. arundinacea*.

This latter example of the interplay between disturbance and plant invasion is a common theme in successful species invasion (Lázaro-lobo and Ervin 2021; Zedler and Kercher 2004). Because wetlands are naturally prone to broad fluctuations in environmental conditions, often fluctuating between drought and flood and receiving inputs of excess water, nutrients, sediment, and other contaminants from adjacent uplands, plants that are adapted to aquatic and wetland ecosystems may flourish under disturbed conditions in human-influenced landscapes. As a consequence, a significant number of the world's most problematic invasive plant species are wetland plants (Zedler and Kercher 2004). Many invasive plants also exhibit multiple characteristics of r-selected species (easily dispersed, high reproductive rates, fast growth rates), which makes them highly adapted to environmental conditions of recently disturbed, early successional communities. Furthermore, many of the most problematic aquatic and wetland invasive plants are highly successful via asexual propagation (e.g., alligatorweed, hydrilla, water hyacinth, milfoils, water lettuce, *Salvinia*, *Phragmites*, cattails, *Spartina*), eliminating the need for introduction of multiple mating types or genotypes for population establishment.

Once invasive species are established, many of them have the capacity to modify ecosystem attributes such as nutrient cycling, litter accumulation, fire regimes, light availability, hydrology, and sedimentation rates, oftentimes facilitating their own dominance and inhibiting successful establishment of other species (Gordon 1998; Lázaro-lobo and Ervin 2021; Suding, Gross, and Houseman 2004; Yelenik and D'Antonio 2013; Zedler and Kercher 2004). These effects are reminiscent of the inhibition model of succession introduced by Connell and Slatyer (1977) and not only can result in what are termed **alternative stable states** (different combinations of resilient, self-perpetuating species assemblages and environmental conditions), but

also can complicate restoration efforts because the magnitude of ecosystem change sometimes makes restoration to native communities much more difficult (Suding, Gross, and Houseman 2004).

Although disturbance can strongly influence invasion of a wetland or aquatic habitat by outside species, it is not an inherently negative process. Floods, droughts, fires, and herbivory are natural processes in wetlands and are important for maintaining historic/natural dynamics of these ecosystems. Indeed, efforts at restoring and managing degraded wetlands to enhance ecosystem services they provide often rely on integration of natural disturbance regimes for successful restoration outcomes (Zedler 2000, 2003). Understanding the roles of specific elements of disturbance in maintaining the functional integrity of wetlands will continue to grow in importance as human activities continue to alter these ecosystems and their interactions with adjacent communities.

11.4 MITIGATION OF EUTROPHICATION BY WETLAND PLANTS

A major impact of the urbanization and inefficient agricultural practices mentioned earlier is an increase in the occurrence of coastal hypoxia and harmful algal blooms in both inland and coastal waters (Anderson et al. 2021; Diaz and Rosenberg 2008). It has long been recognized that wetlands can be used in the remediation of waters contaminated with runoff from human-impacted landscapes, as a result of the high capacity of wetland plants for nutrient uptake (Boyd 1970; Steward 1970). As a result of their potential for nutrient uptake and their influence on biogeochemical processes in wetland ecosystems (see, for example, Chapter 7), aquatic and wetland plants are an important, but perhaps underutilized, component of efforts at reducing the negative effects of eutrophication on inland and coastal waters (Mitsch et al. 2001; Zedler 2003).

Early work suggesting the use of wetland plants for nutrient mitigation relied on annual productivity estimates and plant tissue nutrient content as hints towards species that might function best in this capacity (Boyd 1970; Steward 1970). Steward (1970) suggested that perennial emergent and floating species, such as *Typha latifolia*, *Cyperus papyrus*, *Phragmites australis*, and *Eichhornia crassipes* ought to perform well, based on their annual productivity. Boyd (1970) similarly suggested that emergent and floating or mat-forming species were likely to provide the greatest nutrient removal benefits, listing *E. crassipes*, *T. latifolia*, *Alternanthera philoxeroides*, and *Justicia americana* as candidate species. Both authors suggested that additional experimental work was needed to better determine the efficacy of these or other species as tools for water quality improvement.

A few years later, Spangler, Sloey, and Fetter (1976) described an experimental study wherein they considered *T. latifolia*, *P. australis*, *Sparganium eurycarpum*, *Iris versicolor*, *Scirpus acutus*, and *Scirpus validus* as potential wastewater treatment species. Selection of these species was based on their local availability, ease of vegetative propagation, rapid regrowth after harvest, and tolerance of repeated harvests. After several experiments using these six candidate species in multiple growth media and water depths, and with growth from different types of vegetative propagules, Spangler, Sloey, and Fetter (1976) concluded that the two *Scirpus* species were the best candidates for continued evaluation because they met the aforementioned four conditions of availability, propagation, and growth. The other species either exhibited undesirable regrowth after harvest or were too difficult to obtain in the desired quantities. *Scirpus validus* (for which the currently accepted scientific name is *Schoenoplectus tabernaemontani*) was determined to be the best of these species for use in artificial marshes for wastewater treatment, based largely on its "favorable response to harvesting." However, because a more important conclusion of that work was that harvesting the plants did not remove an appreciable amount of phosphorus from the wetland, Spangler, Sloey, and Fetter (1976) recommended that other species may need to be considered.

The results of the experiments conducted by Spangler, Sloey, and Fetter (1976) highlight important considerations for the use of wetland plants in water quality mitigation. One of these is that, to fully remediate eutrophication of surface waters, the nutrients must be removed from the ecosystem, not simply removed from the water. What I mean here is that the biomass produced via nutrient uptake must be harvested and disposed of or used in some other manner (e.g., as an agricultural soil amendment) to ensure that the nutrients are not released back into the water column upon plant decomposition. A second important factor is that, to facilitate harvest and ensure optimal nutrient removal, clonal herbaceous perennial plants are likely the best option for nutrient mitigation. We have seen many examples throughout this text indicating that plants in this category tend to have the highest rate of biomass production, and their clonal nature facilitates cultivation, harvest, and perennation.

Following the early studies mentioned earlier, many tests were published of *T. latifolia*, *P. communis*, and *S. tabernaemontani* (alone or with other similar species) in constructed wetlands for wastewater treatment (Gersberg et al. 1986; Tanner 1996; Wu et al. 2011). Kadlec et al. (2000) reported that there was no clear evidence that performance varied among common emergent plant species and suggested that species of *Typha*, *Phragmites*, or *Schoenoplectus* (Figure 11.11) should successfully meet treatment needs. This conclusion was based on those species' growth potential, survivability, and costs of planting and maintaining populations of these plants. They further suggested that maintenance of dense populations of these plants and year-round persistence of aboveground plant structure are other important considerations in optimizing nutrient removal from surface waters.

Brisson and Chazarenc (2009) conducted a review of 35 published experimental studies on the role of wetland plants for wastewater treatment. Among the 51 species included in the studies they examined, 20 came from the Cyperaceae family (including nine *Scirpus* or *Schoenoplectus* species), and 17 from the Poaceae (four *Phragmites* taxa), and there were approximately seven *Typha* taxa represented. These three groups thus represented more than 85% of the plant species evaluated. Furthermore, approximately 90% of the plant taxa used in those studies would be considered to exhibit an emergent graminoid, or very generally grass-like, growth form. Thus, the majority of experimental studies of wetland plants as tools for wastewater treatment represented a relatively narrow subset of plant taxonomy and structural types available. Finally, their review of this literature gave little indication that true differences existed among the species tested. Individual studies sometimes showed substantial differences among species tested, but the best performing species from one study weren't always the best in other studies. Occasionally, the relative rankings of species would even reverse from one study to another. Brisson and Chazarenc (2009) indicated that few studies included measurements of nutrient removal along with plant biomass, despite the early work by Tanner (1996) demonstrating a correlation between biomass and nutrient removal. They suggested that more studies should evaluate plant biomass and plant growth over time as correlates of nutrient removal, in addition to searching more broadly for interspecific differences in removal potential. I would add that studies should not only consider differences among species but also differences among growth forms, given that 90% or more of studies to date have focused solely on emergent graminoid species from the Poales.

Another approach to address differences in species and growth forms would be to search for differences among nutrient mitigation wetlands that provide contrasting habitat or microhabitat conditions. Recall from Chapter 7 that sediment microorganisms are responsible for most of the transformations of nitrogen, a key contaminant of wetlands, especially in agricultural landscapes. Wetland plants enhance microbial activity by creating heterogeneity in soil or sediment microhabitats, which increases the diversity of microbial processes that can occur in the wetland. Similarly, different types of treatment wetlands can provide conditions that differentially facilitate microbial metabolism.

Vymazal (2013) reviewed published studies of 60 wastewater treatment wetland systems that combined two or more of the treatment wetland designs illustrated in Figure 11.12. Among the major types of treatment wetlands, there are distinct differences in the mechanisms by which they may mitigate nitrogen contamination (Vymazal 2013). Free-water surface wetlands allow nitrification, an aerobic process, in the water column as a result of photosynthetic oxygenation of the water by algae, cyanobacteria, and submersed plants, if present. If there are deposits of detritus present at the bottom of the wetland, anaerobic conditions there can facilitate denitrification. In horizontal subsurface flow wetlands, where water levels are often maintained at or near the surface of the rooting medium, anaerobic conditions may prevail, allowing denitrification to occur. In contrast, the rooting medium in vertical subsurface flow wetlands may experience fluctuating

FIGURE 11.11 Experimental Study Examining Nutrient uptake by and Competition Among *Typha Latifolia*, *Phragmites Australis*, *Schoenoplectus Tabernaemontani*, and *Juncus Effusus*. *Juncus Effusus* is not Visible in the Photo; *P. Australis* is the Tall Vegetation Behind *S. Tabernaemontani*.

aerobic and anaerobic conditions, potentially allowing cycles of nitrification and denitrification.

As a result of these differences, removal rates of different wastewater contaminants vary among treatment system designs (Vymazal 2013). However, when different designs were combined into hybrid systems, overall treatment effectiveness generally increased, as a result of combining different types of plant and microbial physiological processes. For example, total nitrogen removal was highest in hybrid systems that combined a free-water surface treatment wetland with at least one horizontal or vertical subsurface flow wetland. Despite differences among different configurations in hybrid systems, hybrid systems removed, on average, approximately twice the total nitrogen as treatment systems using only one wetland design. Similarly, hybrid systems removed significantly more ammonium nitrogen than single systems, with no difference detected among hybrid system configurations for removal of ammonium.

Unfortunately, there again was little to be said about differences among plant species, as the majority of treatment systems included in the studies reviewed by Vymazal (2013)

used either *Phragmites australis* (38 studies) or a species of *Typha* (13 studies). Among the 52 studies for which species plant taxa were named, there were only six taxa from orders other than the Poales, and only one non-monocot taxon, an unnamed species of *Hydrocotyle*. So, although wetland plants and wetland ecosystems hold great potential as an efficient and environmentally friendly approach to mitigate nutrient runoff from human-impacted landscapes, there is still much to learn about the relative capacity to provide this service among the available species and growth forms of aquatic and wetland plants.

11.5 CLIMATE CHANGE AND WETLAND CARBON STORAGE

In the prior discussion of disturbance, it was noted that climate change, in and of itself, is not considered a disturbance because it is such a lengthy process. However, we saw several examples of ways that climate change can alter the frequency and intensity of disturbance, as well as ways that it can modify interactions among various types of biotic

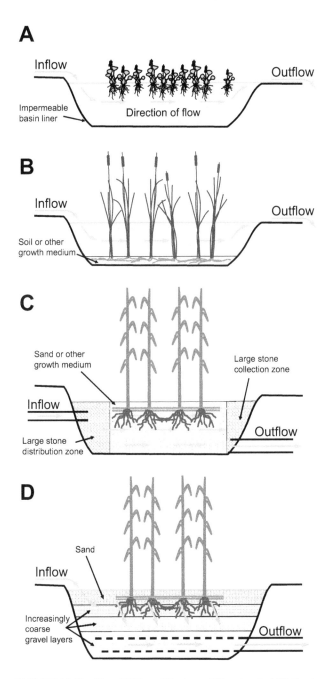

FIGURE 11.12 Four Different Designs of Constructed Wetlands for Nutrient Removal. (A) Free Water, Surface Treatment Wetland Populated with Floating Plants. (B) Free Water, Surface Treatment Wetland Populated with Emergent Plants. (C) Horizontal, Subsurface Flow Wetland Populated with Emergent Plants. (D) Vertical, Subsurface Flow Wetland Populated with Emergent Plants. Redrawn from Kadlec et al. (2000) and Vymazal (2007).

and abiotic disturbances. The influence that climate change has on interactions among disturbances is an important factor to be cognizant of, in part because climate change is expected to affect different components of climate differently in different parts of the world. For example, a global mean temperature increase of 2 °C is predicted to result in more frequent warm periods and more frequent drought in the Amazon basin, higher annual maximum temperatures

with greater maximum rainfall amounts in the southeastern US, and greater maximum rainfall and longer flooding events, with only moderate temperature increases in southeastern Asia and India (Betts et al. 2018).

If the need to consider interactions among disturbances along with differential changes in components of climate change weren't complicated enough, different plant species respond differently to those different climatic variables. For example, among mangrove tree species, those in the northern hemisphere are reportedly limited in poleward migration by winter temperature extremes, whereas southern hemisphere mangrove species are limited more strongly by consistently cold mean winter low temperatures (Osland et al. 2019). Differential establishment of tree species also has been observed in bog and fen communities of high-latitude peatlands in the Adirondack region of New York State (USA). Open-canopied bog communities were colonized by black spruce (*Picea mariana*), as might be expected in the case of hydrarch succession. However, forested bog and fen communities were found to have been colonized by broadleaf deciduous trees more characteristic of lower elevations and latitudes, such as red maple (*Acer rubrum*) and yellow birch (*Betula alleghaniensis*). Successful establishment of maple and birch in these communities was hypothesized to have resulted from the combination of climate-facilitated range expansion of these species, along with eutrophication of the landscape via anthropogenic nitrogen deposition (Langdon, Dovciak, and Leopold 2020). This latter example illustrates the potential for localized differences in the ways that plant species respond to the joint effects of climate change, eutrophication, and inherent ecosystem characteristics. The human-driven increases in nutrient availability of the typically nutrient-poor bogs and fens allowed maple and birch species to colonize wetlands that would otherwise have been inaccessible, despite their climate-driven range expansion.

The ability of migrating species to respond to such nuanced effects of global change illustrates a phenomenon discussed by Osland et al. (2016) in which subtle shifts in environmental conditions can lead to major shifts in dominant species composition. Their discussion of these processes concerned the potential for small changes in macroclimatic factors, such as aridity and winter temperature, to result in large-scale replacement of dominant coastal wetland vegetation. For example, they estimated that, under the right conditions, small increases in mean minimum winter temperature could result in the conversion of thousands of square kilometers of saltmarsh to mangrove-dominated coastal ecosystem in parts of the southeastern United States (Osland et al. 2013). The potential for such climate-driven changes is compounded by the rate of ongoing sea level rise, which threatens half or more of the world's coastal ecosystems (Mitsch and Gosselink 2015), along with the expected increases in frequency and intensity of storms in coastal areas (Cahoon 2006). These threats to natural coastal ecosystems are further exacerbated by the fact that more than 600 million people live in coastal communities around the world, with that number expected to surpass one billion by the year 2100 (Hauer et al. 2020).

Looking again to boreal and tundra ecosystems, it is expected that increasing temperatures will result in further melting of permafrost underlying tundra soils, which will lead not only to loss and alteration of tundra wetlands (Haynes, Connon, and Quinton 2018; Mitsch and Gosselink 2015; Overland et al. 2019), but also to exposure of organic matter within those soils and an increase in the susceptibility of those ecosystems to disturbance by fire (Alexander and Mack 2016). Between 1989 and 2014, fires in black spruce–dominated boreal forests of North America not only resulted in the release of carbon from accumulated detritus on the forest floor via combustion, but also, in many cases, a shift in dominant tree species that was accompanied by complex changes in other ecological characteristics of these ecosystems (Baltzer et al. 2021). In almost 20% of boreal forest sites experiencing a fire during the 25 years from 1989 to 2014, the historically dominant black spruce (*P. mariana*) was replaced by other tree species. Some of those species possessed characteristics such as more slowly decomposed wood and more quickly decomposed leaf litter that substantially shifted nutrient and energy dynamics in the novel post-fire forests (Figure 11.13; Alexander and

Prior to severe fire

Dominant species:
Black spruce (*Picea mariana*)

Lower productivity
Lower production of leaf litter
Leaf litter lower in nitrogen

Lower C:N ratios in stemwood, likely with faster decomposition rate
(8% of carbon in stemwood)

Slower leaf litter decomposition
Slower nitrogen cycling

Cooler, moister soils

Increased soil organic matter accumulation
(>80% of carbon in soil)

Thicker organic soil layer with substantial moss contribution

After severe fire

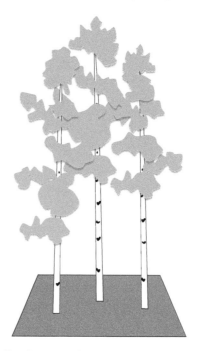

Dominant species:
Trembling aspen (*Populus tremuloides*)
or
Alaska paper birch (*Betula neoalaskana*)

Higher productivity
Higher production of leaf litter
Leaf litter higher in nitrogen

Higher C:N ratios in stemwood, with slower decomposition rate
(43% to 65% of carbon in stemwood)

Faster leaf litter decomposition
Faster nitrogen cycling

Warmer, drier soils

Reduced soil organic matter accumulation
(15% to 35% of carbon in soil)

Thinner organic soil layer with negligible moss contribution

FIGURE 11.13 Increased Frequency and Intensity of Fire in Boreal Peatland Ecosystems Can Result in Wholescale Replacement of Dominant Black Spruce by Deciduous Trees, such as Trembling Aspen and Paper Birch. This Shift in Dominant Tree Species is Accompanied by Major Changes in Other Ecosystem Structural Characteristics and Ecological Processes. Based on Research by Alexander and Mack (2016).

Mack 2016). The greater rapidity of litter decomposition reduced soil insulation, leading to warmer soils that were more conducive to rapid organic matter decomposition and greater rates of productivity for the new dominant tree species. Thus, the climate-induced increase in fire disturbance led to a new assemblage of dominant plant species that were able to maintain their dominance through significant shifts in ecosystem properties that are enforced by aspects of their own biology (Figure 11.13).

Aside from the replacement of the historically dominant black spruce in these boreal peatland systems, the loss of significant amounts of carbon from the underlying soils is an important consequence of both the thawing of permafrost and the increased frequency of fires in boreal systems. The loss of this carbon, whether via combustion or increased decomposition, transforms soils, releasing long-stored nutrients and decreasing the soil's moisture-holding capacity. It also returns that carbon to the atmosphere, and this has much broader implications for the global ecosystem. The changes in temperature that drive climate change are, themselves, driven by the ever-increasing concentration of greenhouse gases such as carbon dioxide (CO_2) and methane (CH_4) in the atmosphere (Figure 11.14). Carbon lost from wetlands or other ecosystems via combustion or decomposition is lost primarily as CO_2, although, as we saw in Chapters 5 and 7, wetlands also can be a significant source of CH_4. As a result of its high concentration and long residence time in the atmosphere, CO_2 is a more important long-term driver of temperature increase, while CH_4, which has a significantly shorter residence time, absorbs 20 to 25 times more heat than does CO_2 on a per-molecule basis (Goudie 2019). Thus, in the context of the role that wetlands might play in influencing climate change, both CO_2 and CH_4 typically are considered.

Mitsch et al. (2013) and Mitsch and Gosselink (2015) provided detailed data on the net uptake and storage of carbon, termed **carbon sequestration** (Figure 11.15), and the release of methane by wetlands (Table 11.2). Mitsch et al. (2013) further developed a quantitative simulation to determine the length of time needed for different wetland types to become net **carbon sinks** (i.e., net storage reservoirs for carbon), taking into account fluxes of both CO_2 and CH_4 over time. Their work showed that most types of wetlands have the potential to become net carbon sinks in less than 100 years after disturbance, assuming no perturbation of their natural wetland disturbance cycles. However, there were some clear latitudinal patterns in the time required to become net carbon sinks, with tropical wetlands typically requiring the greatest time for CO_2 storage to compensate for CH_4 emissions (Table 11.2, Figure 11.16). This is likely driven by the fluctuating water levels in tropical wetlands that experience distinct wet and dry seasons. During the dry season, those wetlands will experience more rapid decomposition of organic detritus than would occur under constant flooded conditions (Mitsch and Gosselink 2015). Indeed, tropical wetlands that remain continuously flooded, such as the floodplains of the Congo River in Africa, accumulate considerable peat deposits, as we saw in Chapter 3. It also was estimated that fires occurring in tropical peatlands of Indonesia during 1997 resulted in the combustion of 0.65 to 2.0 gigatons of carbon (1 GT = 10^{12} kg). That was an amount equivalent to 10% to 30% of annual global emissions of carbon via fossil fuel combustion at the time. Thus, even tropical wetlands, which tend to require more than 200 years to become net carbon sinks (Table 11.2), can store enormous quantities of carbon if they remain undisturbed.

In related research, Berkowitz (2019) found that restored bottomland forest wetlands in the southernmost region of the Mississippi River floodplain (USA) showed significant accumulation of woody debris and accumulation of soil organic matter within 25 years after initiation of restoration efforts. The 12 wetland restoration sites in their study all maintained some connection with the Mississippi River via periodic flooding during high water periods, which facilitated the development of hydric soils characterized, in part, by accumulation of partially decomposed organic material in the upper soil horizons. Statistical modeling efforts associated with that study further indicated that accumulation of soil organic carbon and woody debris were expected to plateau within 40 to 60 years after restoration (Berkowitz 2019), supporting the predictions by Mitsch et al. (2013) that temperate wetlands could become net carbon sinks within 40 years following restoration of wetland conditions (Table 11.2).

Just as we see feedbacks in the carbon-climate system that have resulted in rising temperatures and climate change (Figure 11.14), successful carbon sequestration efforts have the potential to set in motion a feedback cycle in the opposite direction. Sequestration of carbon in wetlands and other natural land cover, such as undisturbed forests and

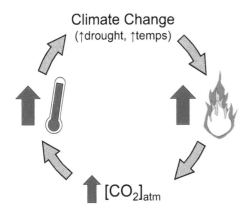

FIGURE 11.14 Changes in Fire Regimes in Boreal Forests and Other Ecosystems are Part of a Positive Feedback System Involving Major Components of Climate Change. Altered Spatiotemporal Patterns of Temperature and Precipitation Have Set Up Conditions in Which Fire has become a More Common Phenomenon. These Fires Release Carbon Dioxide into the Atmosphere, Which Contributes to Further Increases in Global Mean Temperature, Which, in Turn, Feed Continued Changes in Climatic Patterns.

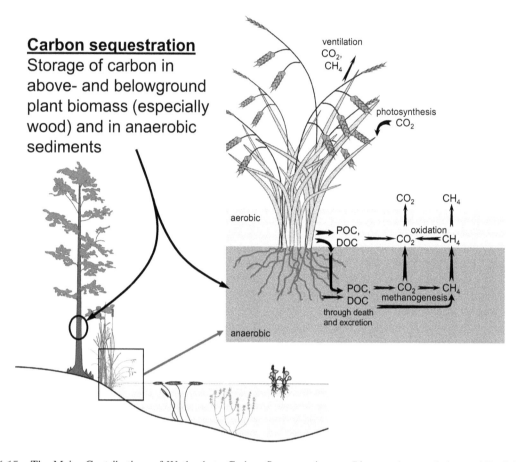

Carbon sequestration
Storage of carbon in above- and belowground plant biomass (especially wood) and in anaerobic sediments

FIGURE 11.15 The Major Contributions of Wetlands to Carbon Sequestration are Biomass Accumulation and Burial of Organic Material in Anaerobic Sediments. Trees Contribute Significantly to this Process By Locking Carbon away in Woody Tissues, While Herbaceous Species, Especially Clonal Emergent Perennials, Contribute through their High Rates of Primary Productivity. Because Herbaceous Detritus is more Readily Decomposed than Woody Detritus, Burial of Litter from Herbaceous Plants in Aerobic Sediments is a Critical Component of Preventing that Carbon from Re-Entering the Atmosphere. Wet-Dry Cycles of Wetlands Will Influence Both the Potential Oxidation of Buried Organic Matter and Rates of Methanogenesis. Methane Emission must be Balanced by Carbon Uptake and Burial to Ensure Net Positive Carbon Sequestration.

TABLE 11.2

Rates of carbon sequestration and methane emission from a global sample of wetlands. Carbon sequestration and methane emission data are from Mitsch and Gosselink (2015). Data on years to become a net carbon sink are based on simulation models of Mitsch et al. (2013).

Wetland Category	Carbon Sequestration g C m^{-2} y^{-1}	Methane Emission g C m^{-2} y^{-1}	Years to Become Net Carbon Sink
Boreal peatlands	10 to 30	7 to 20	0 to 95 median = 27.5
Tropical or subtropical freshwater wetlands	20 to 480	1 to 263	0 to 255 median = 76
Temperate freshwater wetlands	140 to 504	3 to 225	0 to 36 median = 25
Created wetlands	180 to 267	20 to 100	0 to 8 median = 4

grasslands, is an important tool in efforts to reverse the impacts of climate change (Moomaw et al. 2018). Mitsch and Gosselink (2015) further pointed out that wetland restoration and creation of new wetland areas can contribute significantly to this effort, as well. Adequate attention to hydrology in restoration and wetland creation activities allows rapid accumulation of carbon in typically carbon-starved soils, resulting in some of the highest carbon

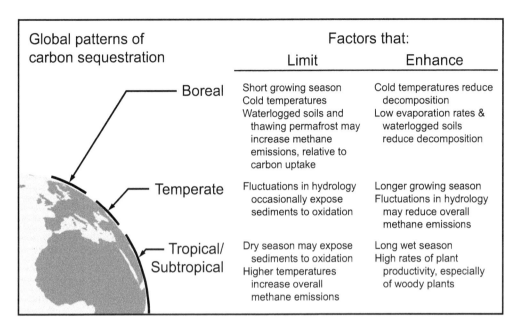

FIGURE 11.16 Latitudinal Patterns of Temperature and Precipitation Strongly Influence Potential Rates of Wetland Carbon Sequestration at the Global Scale, although Local Patterns of Temperature, Precipitation, and Flooding are also Important Drivers of Sequestration in Individual Wetlands. Based, in Part, on Information in Mitsch et al. (2013) and Mitsch and Gosselink (2015).

accumulation rates measured with approximately median levels of CH_4 emission (Table 11.2).

11.6 CLOSING THOUGHTS

Wetlands are incredibly important ecosystems. They are transitions between terrestrial and aquatic habitats, serving as buffers between the two but also serving as productive, diverse habitat for organisms uniquely adapted to life there. Many of these organisms are capable of an amphibious existence, contributing to the ecology of both terrestrial and wetland ecosystems, to wetland and aquatic ecosystems, or sometimes spanning all three ecosystem types. Because plants, as primary producers, form the energetic base of food webs in wetland ecosystems, they are essential to all organisms that utilize these habitats, including humans. An understanding of these plants thus allows one to gain a deeper appreciation for the delicate nature of balancing a life transitional between land and water, as well as an appreciation for the origins of services we derive from wetlands and aquatic ecosystems.

It is my hope that this text inspires a greater appreciation for the importance of aquatic and wetland plants, as well as a greater admiration of the many ways that these species have become adapted to life both in and out of the water. Societal needs in coming years and decades will rely more and more on the services these plants can provide, making it even more important to understand the diverse ways we can answer the question "Why does that plant grow there?" and what the answers mean for sustainable use of our natural resources.

11.7 IN SUMMARY

The biology of aquatic and wetland plants determines how these species interact with one another over successional time. Early colonizers of a new or newly disturbed wetland set in motion changes in the abiotic environment that determine which species will, or will not, succeed them over time. These changes in plant species and their environment may also influence populations of other organisms, such as herbivores or soil microbiota, affecting processes such as nutrient cycling and disturbance and potentially modifying the rate of succession.

Numerous frameworks have been developed to describe the process of succession, and examples of most of these can readily be found in the literature on wetland ecosystems. This is, perhaps, a result of the diversity of wetland types and their distribution across the globe at all latitudes and elevations. We have seen in this chapter examples that fit the facilitation, tolerance, and inhibition models of Connell and Slatyer (1977), as well as the hydrarch succession model first described by Cooper (1913). The environmental sieve model developed by Van der Valk (1981) is frequently a factor in successional change in wetlands and was expanded upon in the wetland continuum framework by attempting to disentangle the separate contributions of precipitation and groundwater to wetland hydrology (Euliss et al. 2004).

Because of the dynamic nature of wetland hydrology, the proximity of wetlands to both aquatic and terrestrial ecosystems, and their frequent occurrence in coastal areas, wetlands are subject to various forms of disturbance that may alter the trajectory of succession, sometimes substantially. These disturbances sometimes will simply reset the wetland vegetation to an earlier successional assemblage of dominant species. However, in other instances, disturbances may be accompanied by the introduction of species that completely transform the abiotic conditions in ways that ensure their continued dominance of the wetland. This latter outcome is often encountered in the case of introduction of invasive plant species to the wetland. An important consideration regarding disturbance in wetlands is that some forms and frequencies of disturbance are a natural part of wetland ecology. Effective management of wetlands or wetland restoration efforts will sometimes hinge on recognizing which disturbances are natural to a given wetland and incorporating those into management actions. Implementation of unnatural disturbance regimes can sometimes result in even native species becoming problematic, from a management perspective.

Finally, two of the most common threats to wetlands that we saw in earlier chapters involve climate change and nutrient and sediment loading from surrounding landscapes. Although wetlands are well equipped to handle low to moderate loading of nutrients and sediment, excessive loads of these contaminants can significantly alter the dominant species, sometimes permanently. Nevertheless, we often take advantage of this nutrient mitigation ability of wetlands in the treatment of wastewater, storm runoff, and runoff from agricultural lands. While considerable work has gone into optimizing the process of water treatment by wetlands, much remains unknown regarding differences in nutrient removal capabilities among species, or even growth forms, of wetland plants.

Regarding climate change and wetlands, the inherent sensitivity of these ecosystems to environmental perturbation makes them highly vulnerable to changes in temperature and precipitation patterns. There is considerable heterogeneity among plant species in their abilities to tolerate, or to take advantage of, a changing climate and the effects that has on local environmental conditions. Factors such as increased temperatures and decreased precipitation not only have direct effects on wetland hydrology but also may alter dynamics of other elements of disturbance. For example, melting of permafrost can result in wetland drainage, transforming wetlands into more mesophytic habitats, while also exposing previously protected soil organic matter to factors such as decomposition and fire. Interactions among elements of climate change and human activities are creating novel habitat conditions around the globe, further exacerbating threats to aquatic and wetland ecosystems.

11.8 FOR REVIEW

1. What is succession?
2. How does Grime's C-S-R framework relate to succession in wetlands?
3. What are some key differences between early and late successional plant species?
4. What is the difference between the relay floristics and initial floristic composition models of succession? Which of these better reflects succession in wetlands and why?
5. What is the environmental sieve model of succession? How is this related to the wetland continuum model?
6. What is hydrarch succession?
7. What is one potential flaw in the initial hypotheses about how hydrarch succession operates?
8. What are the roles of plants in nutrient mitigation for surface waters?
9. Give an example of ways in which climate change may bring about complete replacement of dominant plant species in wetland ecosystems. How do these changes relate to one or more of the major models of succession in wetlands?
10. What is the potential role of wetlands in carbon sequestration?
11. What factors determine the effectiveness of a wetland in carrying out carbon sequestration?

11.9 REFERENCES

Alexander, Heather D., and Michelle C. Mack. 2016. "A Canopy Shift in Interior Alaskan Boreal Forests: Consequences for Above- and Belowground Carbon and Nitrogen Pools during Post-Fire Succession." *Ecosystems* 19 (1): 98–114. https://doi.org/10.1007/s10021-015-9920-7.

Anderson, Donald M., Elizabeth Fensin, Christopher J. Gobler, Alicia E. Hoeglund, Katherine A. Hubbard, David M. Kulis, Jan H. Landsberg, et al. 2021. "Marine Harmful Algal Blooms (HABs) in the United States: History, Current Status and Future Trends." *Harmful Algae* 102: 101975. https://doi.org/10.1016/j.hal.2021.101975.

Baltzer, Jennifer L., Nicola J. Day, Xanthe J. Walker, David Greene, Michelle C. Mack, Heather D. Alexander, Dominique Arseneault, et al. 2021. "Increasing Fire and the Decline of Fire Adapted Black Spruce in the Boreal Forest." *Proceedings of the National Academy of Sciences* 118 (45): e2024872118. https://doi.org/10.1073/pnas.2024872118.

Berkowitz, Jacob F. 2019. "Quantifying Functional Increases across a Large-scale Wetland Restoration Chronosequence." *Wetlands* 39: 559–73.

Betts, Richard A., Lorenzo Alfieri, Catherine Bradshaw, John Caesar, Luc Feyen, Pierre Friedlingstein, Laila Gohar, et al. 2018. "Changes in Climate Extremes, Fresh Water Availability and Vulnerability to Food Insecurity Projected at 1.5°C and 2°C Global Warming with a Higher-Resolution Global Climate Model." *Philosophical Transactions of the Royal Society A: Mathematical, Physical and Engineering Sciences* 376 (2119): 20160452. https://doi.org/10.1098/rsta.2016.0452.

Boyd, Claude E. 1970. "Vascular Aquatic Plants for Mineral Nutrient Removal from Polluted Waters." *Economic Botany* 24 (1): 95–103. https://doi.org/10.1007/BF02860642.

Brisson, J., and F. Chazarenc. 2009. "Maximizing Pollutant Removal in Constructed Wetlands: Should We Pay More Attention to Macrophyte Species Selection?" *Science of the Total Environment* 407 (13): 3923–30. https://doi.org/10.1016/j.scitotenv.2008.05.047.

Buma, B. 2015. "Disturbance Interactions: Characterization, Prediction, and the Potential for Cascading Effects." *Ecosphere* 6 (4): 1–15. https://doi.org/10.1890/ES15-00058.1.

Cahoon, Donald R. 2006. "A Review of Major Storm Impacts on Coastal Wetland Elevations." *Estuaries and Coasts* 29: 889–98.

Clements, Frederic E. 1928. *Plant Succession and Indicators.* Toronto, Ontario, Canada: Hafner Press.

Connell, Joseph H., and Ralph O. Slatyer. 1977. "Mechanisms of Succession in Natural Communities and Their Role in Community Stability and Organization." *The American Naturalist* 111: 1119–44.

Cooper, William S. 1913. "The Climax Forest of Isle Royale, Lake Superior, and Its Development." *The Botanical Gazette* 55: 1–44.

Cowles, H. C. 1899. "The Ecological Relations of the Vegetation on the Sand Dunes of Lake Michigan. Part I.-Geographical Relations of the Dune Floras." *The Botanical Gazette* 27: 95–117.

Cronk, Julie K., and M. S. Fennessy. 2001. *Wetland Plants: Biology and Ecology.* Boca Raton, FL: CRC Press.

Dale, Virginia H., Linda A. Joyce, Steve McNulty, Ronald P. Neilson, Matthew P. Ayres, Michael D. Flannigan, Paul J. Hanson, et al. 2001. "Climate Change and Forest Disturbances." *BioScience* 51 (9): 723. https://doi.org/10.1641/0006-3568 (2001)051[0723:CCAFD]2.0.CO;2.

Diaz, Robert J., and Rutger Rosenberg. 2008. "Spreading Dead Zones and Consequences for Marine Ecosystems." *Science* 321 (5891): 926–29. https://doi.org/10.1126/science.1156401.

Egler, Frank E. 1954. "Vegetation Science Concepts I. Initial Floristic Composition, a Factor in Old-Field Vegetation Development with 2 Figs." *Vegetatio* 4 (6): 412–17. https://doi.org/10.1007/BF00275587.

Ellis, Erle C., Arthur H. W. Beusen, and Kees Klein Goldewijk. 2020. "Anthropogenic Biomes: 10,000 BCE to 2015 CE." *Land* 9 (5): 129. https://doi.org/10.3390/land9050129.

Euliss, Ned H., James W. LaBaugh, Leigh H. Fredrickson, David M. Mushet, Murray K. Laubhan, George A. Swanson, Thomas C. Winter, Donald O. Rosenberry, and Richard D. Nelson. 2004. "The Wetland Continuum: A Conceptual Framework for Interpreting Biological Studies." *Wetlands* 24: 448–58.

Gersberg, R. M., B. V. Elkins, S. R. Lyon, and C. R. Goldman. 1986. "Role of Aquatic Plants in Wastewater Treatment by Artificial Wetlands." *Water Research* 20 (3): 363–68. https://doi.org/10.1016/0043-1354(86)90085-0.

Gordon, Doria R. 1998. "Effects of Invasive, Non-Indigenous Plant Species on Ecosystem Processes: Lessons from Florida." *Ecological Applications* 8 (4): 975–89. https://doi.org/10.1890/1051-0761(1998)008[0975:EOINIP]2.0.CO;2.

Goudie, Andrew S. 2019. *Human Impact on the Natural Environment.* 8th ed. Oxford, UK: John Wiley and Sons, Ltd.

Grime, J. P. 1977. "Evidence for the Existence of Three Primary Strategies in Plants and Its Relevance to Ecological and Evolutionary Theory." *The American Naturalist* 111: 1169–94.

Gurevitch, Jessica, Samuel M. Scheiner, and Gordon A. Fox. 2006. *The Ecology of Plants.* 2nd ed. Sunderland, MA: Sinauer Associates, Inc.

Hauer, Mathew E., Elizabeth Fussell, Valerie Mueller, Maxine Burkett, Maia Call, Kali Abel, Robert McLeman, and David Wrathall. 2020. "Sea-Level Rise and Human Migration." *Nature Reviews Earth and Environment* 1 (1): 28–39. https://doi.org/10.1038/s43017-019-0002-9.

Haynes, K. M., R. F. Connon, and W. L. Quinton. 2018. "Permafrost Thaw Induced Drying of Wetlands at Scotty Creek, NWT, Canada." *Environmental Research Letters* 13 (11): 114001. https://doi.org/10.1088/1748-9326/aae46c.

Kadlec, Robert H., Robert L. Knight, Jan Vymazal, Hans Brix, Paul Cooper, and Raimund Haberl. 2000. *Constructed Wetlands for Pollution Control: Processes, Performance, Design and Operation.* London: IWA Publishing.

Keddy, Paul A. 2000. *Wetland Ecology: Principles and Conservation.* Cambridge, UK: Cambridge University Press.

Kercher, Suzanne M., and Joy B. Zedler. 2004. "Multiple Disturbances Accelerate Invasion of Reed Canary Grass (*Phalaris arundinacea* L.) in a Mesocosm Study." *Oecologia* 138 (3): 455–64. https://doi.org/10.1007/s00442-003-1453-7.

Langdon, Stephen F., Martin Dovciak, and Donald J. Leopold. 2020. "Tree Encroachment Varies by Plant Community in a Large Boreal Peatland Complex in the Boreal-Temperate Ecotone of Northeastern USA." *Wetlands* 40 (6): 2499–511. https://doi.org/10.1007/s13157-020-01319-z.

Lázaro-Lobo, Adrián, and Gary N. Ervin. 2021. "Wetland Invasion: A Multi-Faceted Challenge during a Time of Rapid Global Change." *Wetlands* 41: 1–16.

Lindeman, Raymond L. 1941. "The Development of Cedar Creek Bog, Minnesota." *The American Midland Naturalist* 25: 101–12.

Matthews, J. R. 1914. "The White Moss Loch: A Study in Biotic Succession." *The New Phytologist* 13: 134–48.

Mitsch, William J., Blanca Bernal, Amanda M. Nahlik, Ülo Mander, Li Zhang, Christopher J. Anderson, Sven E. Jørgensen, and Hans Brix. 2013. "Wetlands, Carbon, and Climate Change." *Landscape Ecology* 28 (4): 583–97. https://doi.org/10.1007/s10980-012-9758-8.

Mitsch, William J., J. W. Day, J. W. Gilliam, P. M. Groffman, D. L. Hey, G. W. Randall, and N. Wang. 2001. "Reducing Nitrogen Loading to the Gulf of Mexico from the Mississippi River Basin: Strategies to Counter a Persistent Ecological Problem." *BioScience* 51 (5): 373–88. https://doi.org/10.1641/0006-3568(2001)051[0373:RNLTTG]2.0.CO;2.

Mitsch, William J., and James G. Gosselink. 2015. *Wetlands.* 5th ed. Hoboken, NJ: John Wiley & Sons, Inc.

Moomaw, William R., G. L. Chmura, Gillian T. Davies, C. M. Finlayson, B. A. Middleton, Susan M. Natali, J. E. Perry, N. Roulet, and Ariana E. Sutton-Grier. 2018. "Wetlands in a Changing Climate: Science, Policy and Management." *Wetlands* 38 (2): 183–205. https://doi.org/10.1007/s13157-018-1023-8.

Osland, Michael J., Richard H. Day, Courtney T. Hall, Laura C. Feher, Anna R. Armitage, Just Cebrian, Kenneth H. Dunton, et al. 2019. "Temperature Thresholds for Black Mangrove (*Avicennia germinans*) Freeze Damage, Mortality and Recovery in North America: Refining Tipping Points for Range Expansion in a Warming Climate." *Journal of Ecology* (September): 1–12. https://doi.org/10.1111/1365-2745.13285.

Osland, Michael J., Nicholas Enwright, Richard H. Day, and Thomas W. Doyle. 2013. "Winter Climate Change and Coastal Wetland Foundation Species: Salt Marshes vs. Mangrove Forests in the Southeastern United States." *Global Change Biology* 19 (5): 1482–94. https://doi.org/10.1111/gcb.12126.

Osland, Michael J., Nicholas M. Enwright, Richard H. Day, Christopher A. Gabler, Camille L. Stagg, and James B. Grace. 2016. "Beyond Just Sea-Level Rise: Considering Macroclimatic Drivers within Coastal Wetland Vulnerability

Assessments to Climate Change." *Global Change Biology* 22 (1): 1–11. https://doi.org/10.1111/gcb.13084.

Overland, James, Edward Dunlea, Jason E. Box, Robert Corell, Martin Forsius, Vladimir Kattsov, Morten Skovgård Olsen, Janet Pawlak, Lars-Otto Reiersen, and Muyin Wang. 2019. "The Urgency of Arctic Change." *Polar Science* 21: 6–13. https://doi.org/10.1016/j.polar.2018.11.008.

Pianka, Eric R. 1970. "On R- and K-Selection." *The American Naturalist* 104 (940): 592–97. https://doi.org/10.1086/282697.

Sanderson, Eric W., Malanding Jaiteh, Marc A. Levy, Kent H. Redford, Antoinette V. Wannebo, and Gillian Woolmer. 2002. "The Human Footprint and the Last of the Wild." *BioScience* 52 (10): 891. https://doi.org/10.1641/0006-3568(2002) 052[0891:THFATL]2.0.CO;2.

Shoemaker, C. M., G. N. Ervin, and E. W. DiOrio. 2017. "Interplay of Water Quality and Vegetation in Restored Wetland Plant Assemblages from an Agricultural Landscape." *Ecological Engineering* 108. https://doi.org/10.1016/j.ecoleng.2017. 08.034.

Spangler, Frederic L., William E. Sloey, and C. W. Fetter. 1976. "Wastewater Treatment by Natural and Artificial Marshes." Ada, OK: Environmental Protection Agency, Robert S. Kerr Environmental Research Laboratory. www.google.com/ books/edition/Wastewater_Treatment_by_Natural_and_Art i/8SoYAQAAIAAJ?hl=en&gbpv=1.

Steward, K. K. 1970. "Nutrient Removal Potentials of Various Aquatic Plants." *Hyacinth Control Journal* 8: 34–35.

Suding, Katharine N., Katherine L. Gross, and Gregory R. Houseman. 2004. "Alternative States and Positive Feedbacks in Restoration Ecology." *Trends in Ecology & Evolution* 19 (1): 46–53. https://doi.org/10.1016/j.tree.2003.10.005.

Tanner, C. C. 1996. "Plants for Constructed Wetland Treatment Systems—A Comparison of the Growth and Nutrient Uptake of Eight Emergent Species." *Ecological Engineering* 7 (1): 59–83. https://doi.org/10.1016/0925-8574(95)00066-6.

Van der Valk, A. G. 1981. "Succession in Wetlands: A Gleasonian Approach." *Ecology* 62 (3): 688–96.

Van der Valk, A. G., and L. C. Bliss. 1971. "Hydrarch Succession and Net Primary Production of Oxbow Lakes in Central Alberta." *Canadian Journal of Botany* 49 (7): 1177–99. https://doi.org/10.1139/b71-167.

Vymazal, Jan. 2007. "Removal of Nutrients in Various Types of Constructed Wetlands." *Science of the Total Environment* 380 (1–3): 48–65. https://doi.org/10.1016/j.scitotenv.2006. 09.014.

———. 2013. "The Use of Hybrid Constructed Wetlands for Wastewater Treatment with Special Attention to Nitrogen Removal: A Review of a Recent Development." *Water Research* 47 (14): 4795–811. https://doi.org/10.1016/j. watres.2013.05.029.

Wu, Haiming, Jian Zhang, Peizhi Li, Jinyong Zhang, Huijun Xie, and Bo Zhang. 2011. "Nutrient Removal in Constructed Microcosm Wetlands for Treating Polluted River Water in Northern China." *Ecological Engineering* 37 (4): 560–68. https://doi.org/10.1016/j.ecoleng.2010.11.020.

Yelenik, Stephanie G., and Carla M. D'Antonio. 2013. "Self-Reinforcing Impacts of Plant Invasions Change over Time." *Nature* 503 (7477): 517–20. https://doi.org/10.1038/ nature12798.

Zedler, Joy B. 2000. "Progress in Wetland Restoration Ecology." *Trends in Ecology and Evolution* 15 (10): 402–7.

———. 2003. "Wetlands at Your Service: Reducing Impacts of Agriculture at the Watershed Scale." *Frontiers in Ecology and the Environment* 1 (2): 65–65. https://doi.org/10.2307/ 3868032.

Zedler, Joy B., and Suzanne Kercher. 2004. "Causes and Consequences of Invasive Plants in Wetlands: Opportunities, Opportunists, and Outcomes." *Critical Reviews in Plant Sciences* 23 (5): 431–52. https://doi.org/10.1080/07352680 490514673.

Glossary

A

abaxial: the lower surface of leaves, or the surface that is farthest from or facing away from the stem or main axis of the plant

achene: small, hard, one-seeded fruit formed from a single ovary, with thin outer wall (pericarp) surrounding the seed; includes the fruit of many species in the Asteraceae (where the fruit is sometimes called a cypsela), in which the sepals are modified to form a hairy pappus that assists in wind dispersal

activated carbon: biochar formed from the partial combustion of organic material; a material with chemical properties that allow it to adsorb and filter impurities, such as potential allelochemicals, from water, providing an effective tool for experimentally neutralizing allelochemicals

adaptive traits: phenotypic differences in characters that can be associated with a fitness gain in the context of aspects of the environment that influence growth, survival, or reproduction

adaxial: the upper surface of leaves or other organs, or the surface that is nearest the stem or main axis of the plant

adventitious roots: roots that form at (or near) the interface of the air and either the water column or inundated soil, typically at some point along the stem (often at nodes) where roots would not typically be encountered

aerenchyma: internal plant tissues that form air passageways throughout the plant, allowing diffusion or active transport of oxygen and other gases

aerobic metabolism: cellular metabolism that uses oxygen as a terminal electron acceptor

aggregate fruit: fruit formed from many individual, unfused carpels of a single flower

alleles: versions of a given gene or other specific location on the DNA (e.g., individual base pairs)

allelochemical: biologically active chemical compounds involved in allelopathy

allelopathy: interaction in which a plant produces chemical compounds that bring about a reduction in the performance of individuals that come into contact with those compounds

alluvial deposits: deposits of sand, silt, and clay along a stream and in its floodplain

alternative stable states: different combinations of resilient, self-perpetuating species assemblages and environmental conditions

amensalism: interaction in which one individual is negatively affected, while the other exhibits no detectable effect from the interaction

anaerobic metabolism: cellular metabolism that uses a molecule other than oxygen as a terminal electron acceptor

anammox: anaerobic ammonia oxidation, a bacterial-mediated process that results in the formation of dinitrogen gas (N_2) from ammonium and nitrite under anaerobic conditions

androdioecy: within a population, some individual plants produce only staminate flowers, while others produce bisexual or both staminate and pistillate flowers

andromonoecy: individual plants produce both staminate flowers and bisexual flowers; more common than gynomonoecy

anemochory: dispersal of fruit or seed by wind

anemophily: pollination by wind; in anemophily, copious dry pollen is released by dehiscing anthers and transported by air to stigmas of pistillate flowers

annual: species that are born from (or born as) a seed, live until they complete their only season of sexual reproduction, and then die

annual life history: life history in which an individual arises from a seed, grows, reproduces, and dies, leaving behind only seeds (if reproduction is successful)

anoxia: the absence of oxygen

anthropogenic: generated by or influenced by human activities

apomorphy: a derived, or novel, state in a character used in cladistic analyses

apoplast: the area within the cell walls of a plant but outside the cell membranes, in contrast with the symplast, which is the region of the plant within the interconnected cellular cytoplasm

aquatic acid metabolism (AAM): a carbon-concentrating pathway in which carbon capture takes place at night, using the enzyme PEP carboxylase, and that carbon is used in the Calvin cycle during the day

aquatic plants: plants that complete their life cycles with all vegetative (that is nonflowering) parts submersed in or supported by the water

aquifer: layer of rock or unconsolidated mineral material (e.g., sand) that is sufficiently porous to store water belowground

aril: energy-rich outgrowth on outer coat of a seed that encourages or rewards seed dispersers

asymmetric competition: competitive interaction in which the effect on one of the interacting individuals is sufficiently small to be undetectable, giving the appearance of an amensalism

ATP: adenosine triphosphate; an energy-carrying molecule in biological systems formed from the combination of an ADP molecule (which also carries energy) and an inorganic phosphate molecule

autapomorphies: apomorphies that are unique for a portion of the taxa within a clade in an evolutionary tree

autecology: the ecology of an individual species

autogamy: pollination within individual open flowers; requires that flowers be hermaphroditic (i.e., perfect)

autotoxicity: intraspecific allelopathic suppression

autotroph: an organism that does not rely on external sources of organic compounds for its own metabolism, as it is capable of reducing inorganic carbon from the environment, and that generates ATP with the aid of light energy or the oxidation of inorganic compounds from the environment

axil: the interior, acute angle within a branching point along a plant stem; axillary stems or branches are those produced along the length of a stem, in contrast growth at the tip or apex

B

basal: towards the "base" of the plant, or towards the older portions of the plant

berry: fleshy fruit formed from a compound ovary that may produce one or a few, but usually many, seeds; the seeds may have a fleshy covering or a fleshy appendage referred to as an aril

biennials: species that live longer than annuals, many living for decades, but they characteristically undergo only a single season of sexual reproduction before producing seed and dying

bifurcating: a branching pattern in which each split results in two branches

bioavailability: accessibility of resources via means that are available to an organism as part of its normal mode(s) of uptake and/or absorption

biological control: the use of an undesirable species' natural enemy, such as an herbivore or a predator or pathogen, to control growth or reproduction of that species; biocontrol

biomass: the amount, by mass, of living material produced by a plant; sometimes referred to by mass per unit area, which may also be called "standing stock"

bisexual flower: flowers that produce both stamens and pistils

bog: peat-accumulating wetland that is hydrologically isolated from groundwater or surface flows (both inputs and outputs), with pH usually below 6 because of this isolation and the vegetation that establishes in these wetlands; vegetation usually dominated by *Sphagnum* mosses, sometimes intermixed with conifers or ericaceous shrubs (i.e., from the Ericaceae)

boreal ecosystem: northern latitude cold forests lying just south of the tundra biome, dominated by conifers, poplar, birch, and aspen and with significant peat accumulation owing to the cold temperatures and moist soil conditions, resulting from low evaporation rates

boundary layer: layer of still air or water that surrounds objects (e.g., leaves)

browser: herbivore that feeds on plant tissues that sit higher above the soil or substrate, such as leaves on shrubs or trees

C

C4 photosynthetic pathway: a carbon-concentrating pathway in which carbon capture takes place in cells adjacent to leaf air spaces, using the enzyme PEP carboxylase, and that carbon is shuttled to cells near the vascular bundles for use in the Calvin cycle

calcareous: geological material that is rich in calcium

calyx: the whorl of outermost parts (sepals) of a complete flower

capsule: dehiscent fruit formed from an ovary comprising multiple carpels that splits open at maturity (often along sutures between carpels) to release seeds

carbon-concentrating mechanism: physiological mechanism for concentrating inorganic carbon in the vicinity of the Calvin cycle to enhance photosynthesis and minimize photorespiration

carbon cycle: network of physical and chemical transformations of carbon-containing compounds throughout an ecosystem or globally

carbon fixation: process of incorporating inorganic carbon into organic molecules; CO_2 reduction during photosynthesis

carbon sequestration: uptake and storage of carbon, as in wood or sequestered soil organic carbon

carbon sink: a net storage reservoir for carbon

carnivorous plant: plant that derives at least a portion of its nutrients or energy from animal prey that are (usually) captured via some form of modified leaf that acts as a trap

carpel: the ovule-producing chamber within a pistil

carrying capacity: the number of individuals that can be supported by available resources at the location in which the population occurs

caryopsis: single-seeded fruit with pericarp fused to the seed coat, in contrast to achenes and nuts, where the pericarp simply surrounds the seed

catabolic: metabolic processes that allow plants to break down and extract energy from carbohydrates

cellular respiration: cellular process in which organic molecules, such as sugars, are broken down to convert chemical bond energy in those compounds into ATP

chasmogamy: sexual reproduction in angiosperms wherein the flowers open for pollination

chemolithotrophs: autotrophs that rely solely on inorganic carbon molecules as their carbon source and derive the energy for carrying out metabolism from breakdown of hydrogen atoms into their constituent protons and electrons

chronosequence: a method used to study succession in which multiple ecosystems at different successional stages are used to infer successional processes and the sequence of dominant plant species over time

circumboreal: throughout boreal ecosystems of the cold northern latitudes

clade: a group of taxa in an evolutionary tree that share a common ancestor

cleistogamy: pollination occurring within individual closed flowers; requires hermaphroditic flowers

climax community: a presumed final stage of succession, often characteristics of a region, owing to that region's climate and geology

clonal propagation: the process of numeric increase of ramets in a clonal plant

clonal propagules: dispersal units generated through asexual propagation

clone: the collection of all ramets arising from a single genet

cohort: a group of individuals that begin some process at the same time, such as seeds that germinate during the same year or ramets of a plant that are produced during the same season

commensalism: interaction in which one partner is not measurably affected but the other benefits from the interaction

common garden experiment: an experiment in which plants are grown at one or more common locations from seeds that have been collected from multiple natural populations of interest

community: all individuals of all species that co-occur within some specified space

compatibility: the state when pollen that lands on a stigma is capable of successful fertilization, given that the stigma is receptive at the time of pollination

competition: an interaction between individuals, brought about by a shared requirement for a resource in limited supply, and leading to a reduction in the survivorship, growth and/or reproduction of the competing individuals

competitive effect: the degree to which competition reduces performance of an individual plant, relative to its performance in the absences of competition, accounting for density effects of neighbors, and usually expressed on an area basis

competitive intensity: the degree to which competition reduces performance of an individual plant, relative to its performance in the absences of competition

competitive outcomes: the potential long-term dynamics of two or more species grown in mixtures

constitutive: a process occurring as a pre-programmed aspect of development

continuous time model: model that considers some process to occur at infinitely small intervals of time, such as growth of a plant; contrasted with a discrete time model

corolla: the (often) colorful whorl of petals in a flower

cosexuality: state where both stamens and pistils can be found on individual plants; includes both hermaphroditic and monoecious plants

cotyledons: embryonic seed leaves

crassulacean acid metabolism (CAM): a carbon-concentrating pathway in which carbon capture takes place at night, using the enzyme PEP carboxylase, and that carbon is used in the Calvin cycle during the day

cryptic dioecy: situation in a monoecious or hermaphroditic population where some flowers produce nonviable pollen and others produce nonviable ovules

cryptovivipary: process in which seeds germinate while attached to the parent plant, but they disperse prior to any significant growth of the seedling

culm: unbranching stemlike structure, usually emerging from a node that branches at ground level; this term is often used for grasses or graminoid plant species

D

decarboxylation: removal of one carbon, in the form of CO_2

dehiscent: describes a fruit that dries and opens along one or two suture lines at maturity

denitrification: a series of microbial metabolic processes that result in conversion of nitrogen-containing molecules to N_2 gas, which can diffuse from the wetland into the atmosphere

density-dependent growth: population growth that is constrained by the per capita use of available resources

density-independent growth: population growth that is unconstrained by the per capita use of available resources

detritivore: organism that feeds on dead organic matter

detritus: dead organic matter, such as the accumulated dead plant material in wetland soils and sediments

dichogamy: separation of sexes in time, as in sequential hermaphroditism

dicliny: separation of stamens and pistils into separate flowers; standard monoecy and dioecy are examples of dicliny

dicot: as historically treated (now known to be a non-monophyletic group), plants characterized morphologically by flower parts in multiples of four or five, netted leaf venation, embryos having two seed leaves, and vascular bundles arranged in a ring around the stem, which allows for secondary (radial) growth; non-monocot taxa recognized now as basal angiosperms (e.g., Nymphaeales) and the Magnoliids, however, exhibit flower parts in multiples of three

dioecy/dioecious: staminate and pistillate flowers produced on separate individual plants

diploid: individual that has two complementary sets of chromosomes

discharge: the amount (volume) of water flowing along a stream or river per unit time

discharge wetland: a wetland in which water moves into the wetland from the groundwater; the groundwater discharges into the wetland

discrete time model: model that considers some process from the perspective of specific intervals of time, such as annual reproduction; contrasted with a continuous time model

dispersal shadow: the area across which an individual disperses its pollen or seeds

dissolved inorganic carbon: dissolved carbon compounds such as CO_2, HCO_3^-, and CO_3^{2-}

dissolved organic carbon: molecules in water smaller than approximately 0.22 microns that contain carbon that was derived, at some point, from living matter

distal: towards the tip or apex of the plant, or towards the younger portions of the plant

disturbance: short-lived event that causes a measurable change in the properties of an ecological community; the degree of change experienced will be influenced by the frequency and magnitude of the disturbance

distyly: production of two different floral morphs, e.g., long and short styles or filaments

dormant: in seeds, those that are incapable of immediate germination and require some ecophysiological mechanism to remove their dormancy before germination can take place

drawdown: a significant decrease in water level that exposes some or all the wetland sediments

drupe: fleshy fruit containing a hard pit that encloses, usually, one seed

E

ecosystem: integrated biotic community and abiotic components of an area, typically having some unifying characteristics, as in the case of wetlands

ecotone: a habitat or ecosystem that represents a gradient between two other, usually more distinct, ecosystems

ecotypes: discrete populations or subpopulations of species that have genetically controlled, usually adaptive, differences from other populations

efflux: an outflow of gases or other substance

electron carriers: molecules referred to as coenzymes that serve to move electrons during a variety of biochemical processes, such as photosynthesis and cellular respiration (NADP+: nicotinamide-adenine dinucleotide phosphate; NAD+: nicotinamide-adenine dinucleotide; FAD: flavin-adenine dinucleotide)

electronegativity: tendency of an atom to strongly attract electrons when bonding with other atoms, resulting in uneven sharing of those electrons, as between hydrogen and oxygen in water

emergent plant: plant that is rooted in the soil or sediment but whose vegetative and (usually) reproductive parts are exposed above the water

enantiostyly: production of mirror-image floral morphs, such as flowers with styles that bend either left or right of the flower's centerline

endosperm: tissue formed from fertilization of the central cell in the ovule of an angiosperm

energy crisis metabolism: a suite of anaerobic physiological processes initiated in the face of an energy deficit caused by the inability to use oxygen as the terminal electron acceptor for the electron transport chain within the mitochondrion

entomophily: pollination by insects; in entomophilous plants, sticky pollen attaches to insects and is carried to stigmas of pistillate flowers during insect movement from flower to flower

epihydrophily: hydrophily in which pollen is transported on the water surface

epinasty: the downward curvature of leaf petioles resulting from elongation of the petiole's upper surface

epiphyte: plant or other organism (e.g., alga, bacterium) that grows upon another plant

equipotential lines: lines of equivalent hydraulic head in a map of subsurface water

essential resource: a resource that is absolutely required for survival and growth of the species

eudicots: the true dicots; evolutionarily, those taxa that diverged after the monocots, Magnoliids, and Ceratophyllales

eutrophication: addition of excess nutrients to an ecosystem, usually leading to an increase in primary productivity; often used in reference to anthropogenic nutrient inputs, referred to as "cultural eutrophication"

evapotranspiration: the combination of evaporation of water from the wetland and transpiration of water through the plants, abbreviated as ET

exine: outer covering of a pollen grain

expansigeny: a process of aerenchyma formation wherein division and growth of cells surrounding a pre-existing intercellular space result in expansion of that space, leading to formation of the aerenchyma tissue

expansins: proteins involved in loosening the chemical bonds among cell wall fibers, allowing expansion of the cells

exploitative interaction: interaction in which one organism derives its energy and nutrients from the consumption of all or part of one or more other organisms

extant: currently living, as opposed to extinct

F

facilitation: interaction in which one plant, as a result of its normal growth or physiology, modifies habitat characteristics in such a way that another species benefits

facultative plant species: plant species expected to occur under wetland conditions at 34%–66% frequency; equally likely to occur under wetland conditions as non-wetland conditions

facultative upland plant species: plant species expected to occur under wetland conditions at 1%–33% frequency

facultative wetland plant species: plant species expected to occur under wetland conditions at 67%–99% frequency

fen: peat-accumulating wetlands that are connected to groundwater and/or surface inflows and outflows and thus have less acidic soils (usually with pH in the range of 6 to 8); vegetation usually dominated by graminoids (Poaceae, Cyperaceae) and a broad variety of shrub and tree species

fermentation: breakdown of glucose or other energy-containing molecules to yield energy without using oxygen- or nitrate-associated electron transport

fitness: the ability of individuals of a species to produce viable offspring

fixation: the exclusivity of a given alleles or genotypes within a population

floating-leafed plant: plant that is rooted in the soil or sediment and whose leaves and reproductive parts usually are supported at the surface of the water

florets: individual small flowers that form part of a larger cluster of flowers

follicle: dehiscent fruit formed from a single carpel that opens along one longitudinal suture

free-floating plant: plant whose vegetative parts, including roots, are supported at the surface of the water, with roots, if present, hanging within the water column

frugivore: organism that feeds on fruit

G

geitonogamy: pollination between different flowers on the same plant; usually found in monoecious species but can occur among separate flowers in species with bisexual flowers

gene flow: movement of genes from one area to another, via migration of individuals carrying that gene

genera: plural of genus

genet: the complete body (thallus) of a plant that has arisen from a single seed, including all modules of that individual that may have arisen through asexual propagation

genetic drift: a process in which the frequencies of alleles (different versions of a gene) change randomly over time due to random chance

genome: an individual's entire complement of genes, or DNA

genome duplication: generation of an additional copy of an organism's complete genome (i.e., all of its chromosomes), resulting from failure of homologous chromosomes to segregate during meiosis. Offspring arising from reproduction involving gametes that have undergone genome duplication will be polyploid, meaning that they will possess more than the normal diploid (2n) set of chromosomes.

genotype: the specific combination of genetic information carried by an individual

geomorphology: three-dimensional shape of the land, as influenced by geological processes

glandular trichomes: hairs on surfaces of a plant that are associated with a gland of some type; some are involved in prey capture by carnivorous plants, others are involved in defense against plant enemies

graminoids: plants with a grasslike morphology, especially in the grass (Poaceae), rush (Juncaceae), or sedge (Cyperaceae) families

gravitational potential: gravitational pressure potential, which usually is negligibly small

grazer: herbivore that feeds on low-stature primary producers

groundwater: water within the saturated layers of soil

gynodioecy: condition in which, within a population, some individual plants produce only pistillate flowers, while others produce bisexual or both staminate and pistillate flowers; more common than androdioecy

gynomonoecy: sexual system in which individual plants produce both pistillate flowers and bisexual flowers

H

haploid: individual that has only one set of chromosomes

haplotypes: groups of individuals exhibiting the same DNA sequences

hemiparasites: plant hemiparasites obtain some of their resources from other plants but are capable of photosynthesis and may be found living freely, without a host

herbivory: the consumption of living plant tissue

herkogamy: separation of sexes in space; may be found in conjunction with dichogamy, where staminate flowers, produced in one part of an inflorescence, mature prior to pistillate flowers, which are produced in another part of the inflorescence

hermaphroditic flower: flowers that produce both stamens and pistils

hermaphroditic: plants that produce flowers having both stamens and pistils (bisexual flowers) or produce both staminate and pistillate flowers on the same individual plant (monoecious plants)

hermaphroditism: plants produce flowers having both stamens and pistils (bisexual flowers) or produce both staminate and pistillate flowers on the same individual plant

heteroblasty: the condition wherein young, or juvenile, structures differ morphologically from those of older, mature, or adult structures or individuals

heterocysts: cells in cyanobacteria that provide anaerobic conditions necessary for nitrogen fixation to occur

heterophylly: the tendency for individual plants to produce markedly differing leaf or stem morphologies in response to environmental cues, such as submersed versus aerial leaves on aquatic plants

heterostyly: production of flowers differing in lengths of styles or filaments, leading to "morphs" that encourage outcrossing between, versus within, morphs

heterotrophic: describes an organism that relies on organic compounds from external sources to provide carbon and energy

holoparasites: parasitic plants that are (mostly) non-photosynthetic and require a host to grow, survive, and reproduce

homozygosity: presence of two identical alleles at a sampled chromosome location, in contrast with heterozygosity, wherein the two alleles differ

host: organism from which a parasite obtains its nutrients and/or energy

hybrid: an individual produced via successful mating between two genetically dissimilar individuals, often referring to successful mating between individuals of different species

hybrid breakdown: process in which back-crosses between hybrids and parental genotypes or crosses among hybrids result in inviable or sterile offspring

hybrid inviability: the failure of hybrids to grow and develop normally

hybrid vigor: condition wherein hybrids exhibit exaggerated phenotypes that lend themselves to higher fitness than either parent

hydrarch succession: succession of lakes or ponds towards mesophytic forest vegetation

hydrated: the state of a molecule being fully dissolved into an aqueous form

hydraulic conductivity: the ability of porous material (e.g., soil) to transmit water; usually given in units of distance per time, such as meters per second

hydraulic gradient: the change in hydraulic head over the distance that water travels through a layer of interest

hydraulic head: the total pressure on the water at a given point, measured as a distance above or below a known reference point

hydric soil: soil that exhibits characteristics of submersed or saturated conditions that regularly lead to hypoxia or anoxia

hydrochory: dispersal of fruit or seed by water

hydrogen bond: bond between atoms or molecules caused by the electrical attraction of positive ions (cations) and negative ions (anions), as in water

hydrogeomorphic setting: the three-dimensional shape of the wetland and its connections with underlying geology and with inflows and outflows of water

hydrograph: a record of changes in the amount of water in a site over time

hydrology: the depth, duration, frequency, and chemical composition of water on a site, such as a wetland; includes atmospheric water as well as above- and belowground water at a location

hydroperiod: in wetlands, this refers to the temporal pattern of water level and is usually represented with a hydrograph

hydrophily: pollination mode that uses water as the means of conveying pollen from anther to stigma

hydrophytic vegetation: plant species possessing adaptations that allow them to survive and reproduce in habitats regularly experiencing standing water or saturated soils during the growing season.

hypohydrophily: hydrophily in which pollen is transported below the water surface

hyponasty: upward movement of a leaf resulting from elongation of the petiole's lower surface

hypoxia: oxygen concentrations below 2.5 mg per liter

I

imbibe: to absorb water from the environment, as in a seed prior to germination

imperfect flower: flowers that produce only stamens or pistils

inbreeding: breeding among genetically closely related individuals

inbreeding depression: condition in which one or more phenotypic traits resulting from inbreeding leads to a decline in fitness for inbred individuals

incompatibility: physiological or biochemical condition in which pollination cannot occur between two individuals

indehiscent: fruits that do not open to release seeds

inducible: a process initiated in response to some environmental signal

inflorescence: a cluster of flowers on an individual plant

influx: an inflow of gases or other substance

infrared: wavelengths of light greater than 700 nm; heat

infructescence: a cluster of fruits on an individual plant

integument: an outer protective layer, such as the outer covering of a seed

interception: in hydrology, this refers to precipitation that is intercepted by surfaces, such as the plant canopy; intercepted water may be evaporated or may enter a wetland via stemflow or throughfall

internodes: the sections of the stem between the nodes

intraspecific hybridization: within-species gene flow, among individuals from disparate regions of species' native ranges, usually following introduction of one or both of those individuals to new geographic regions

ionic gradients: electrochemical gradients across cellular or organellar membranes, resulting from an imbalance of dissolved ions; ionic gradients function in cellular communication and trans-membrane transport

iteroparous: plants (or other organisms) that have the potential to undergo multiple seasons of sexual reproduction

L

lattice: spatially distributed, patterned arrangement of water molecules that results from a combination of hydrogen bonding among molecules and the arrangement of atoms within the molecules

leaching: dissolution of soil nutrients and their transport out of the root zone

legume: dehiscent fruit formed from a single carpel that opens along two longitudinal sutures

lenticels: areas, usually on stems, where the outer water-proof bark is interrupted, allowing exchange of gases with underlying tissues; typically, tissues just inside the lenticels are characterized by numerous intercellular air spaces, further facilitating gas exchange

life history: the typical patterns of birth, growth, reproduction, and death in a species

life history strategy: the sum of the selective adaptations by which a species has fit its niche within a community

littoral zone wetlands: wetlands that occupy the margins of lake ecosystems

loment: fruit formed from a single, multi-seeded carpel that splits transversely into multiple, one-seeded segments at maturity

lumen: internal space of the thylakoid membrane within a chloroplast

lysigeny: a process of aerenchyma formation wherein cells die, and their collapse leads to the formation of voids within the tissue

lysimeter: cylindrical device used to collect water from belowground sources

M

macromolecules: large biological molecules such as proteins, lipids, carbohydrates, and nucleic acids

macronutrients: essential nutrients that are required in relatively large amounts

mangrove swamp: marine coastal wetlands (saline or brackish) dominated by species of mangrove trees or shrubs; occur in tropical and subtropical regions

marsh: broad category of shallow wetlands dominated by herbaceous vegetation

matric potential: negative pressure potential resulting from matric attractive forces between water and surfaces, such as soil particles or cell wall fibers

megagametophytes: female gametophytes; the embryo sacs produced within the ovules

meristem: a growing point within a plant

mesophytic: vegetation or an ecosystem that is neither wet nor exceedingly dry

meta-analysis: a statistical analysis of previously published analyses

metapopulation: a collection of local populations of a given species, among which individuals may interact via dispersal, colonization, and interbreeding; often summarized as being a population of populations.

methanogenesis: anaerobic metabolism that results in the production of methane

microgametophytes: male gametophytes; the pollen grains produced within pollen sacs of the anthers

micronutrients: essential nutrients that are required in relatively small amounts

micropyle: entry point of pollen tube to the ovule and exit for the growing seedling at germination

monocarpic: plants that undergo only one season of sexual reproduction

monocot: monophyletic group of plants characterized morphologically by flower parts in multiples of three, parallel vascular venation in leaves, scattered vascular bundles in stems, and embryos having a single seed leaf; note that taxa recognized now as basal angiosperms (e.g., Nymphaeales) and the Magnoliids, however, also exhibit flower parts in multiples of three

monoecy/monoecious: plants that produce both staminate and pistillate flowers on the same individual

monophyletic: a monophyletic group is one that contains the most recent common ancestor of members in the group, all descendants of that ancestor, and no other taxa

morphological dormancy: dormancy in which the embryo is undifferentiated or underdeveloped; following dispersal and imbibition, the embryo will complete development, and seeds will germinate shortly thereafter

morphophysiological dormancy: dormancy in which the embryo is undifferentiated or underdeveloped; following dispersal and imbibition, the embryo completes development, but seeds fail to germinate within the next month or so

multiple fruit: fruit formed from the carpels of many clustered flowers

mutation: a change in one or more nucleotides within an individual's DNA

mutualism: a type of symbiosis in which each partner receives some benefit from the interaction

mycorrhizae: mutualism between plant roots and a fungal symbiont; may involve intracellular arbuscular fungi or intercellular ectomycorrhizal fungi

N

natural selection: a process driving evolutionary diversification among populations, requiring heritable variation in adaptive traits that leads to differential success in producing viable offspring. Characteristics of the environment thus select for (or against) individuals that carry beneficial (or deleterious) versions of the traits.

niche: the sum of the physical and biological properties of the environment to which the species is physiologically and behaviorally adapted and under which the species can maintain non-negative population growth over time

nitrification: stepwise transformation of ammonia-nitrogen into nitrite and then nitrate; carried out by bacteria in the genera Nitrosomonas and Nitrobacter

nitrogen fixation: bacterial-mediated capture of nitrogen gas from the atmosphere into organic form

nodes: points along the stem where branching may occur and where leaves and/or flowers may be produced

nondormant: a seed, with differentiated and fully developed embryo, imbibes water, after which root and shoot emergence usually occur within a few days

nut: relatively large, one-seeded fruit formed from a compound ovary in which a single carpel is functional; thick pericarp surrounds the seed

nutlet: one-seeded fruit formed from separate (i.e., unfused) carpels

O

obligate upland plant species: plant species expected to occur under wetland conditions less than 1% of the time

obligate wetland plant species: plant species expected to occur naturally at a frequency of > 99% under local conditions indicative of wetlands

ombrogenous wetland: wetland that receives water inputs solely from precipitation, a characteristic of bogs; also referred to as ombrotrophic wetlands

outcrossing: successful interbreeding between genetically distinct individuals

outgroup: in cladistic analyses, a group that is not among the focal taxa; used to help determine the orientation of evolutionary processes with the focal group (the ingroup)

overcompensation: outcome of herbivory in which consumption of vegetative tissues is followed by the plant regrowing more tissue than was consumed

overland flow: water moving across the surface of the land, as opposed to subsurface or groundwater flows; in wetland hydrology, this may include runoff and stream discharge, but in stream hydrology, it will only refer to surface runoff prior to entering the stream

oxidation: process wherein a molecule loses one or more electrons, increasing (numerically) its electrical charge

P

palisade parenchyma: column-like layer of mesophyll cells just inside the epidermis of a leaf

palustrine wetlands: inland, nontidal wetlands that are generally isolated from streams or lakes, or whose hydrology is not dominated by stream hydrology, if they receive stream inflows

parasites: organisms that obtain nutrients and/or energy from one or a few individuals of another species, their host

passive diffusion: passive movement of gases or dissolved substances from an area of higher concentration to an area of lower concentration

pathogens: organisms that obtain resources from a host but which usually cause disease and potentially death as a result

peatland: peat-accumulating wetland, typically in cold high-latitude regions or high-elevation alpine ecosystems

peduncle: the stalk of an individual flower or stalk of an inflorescence

PEP carboxylase: phosphoenolpyruvate (PEP) carboxylase; an enzyme functioning in the C4, CAM, and AAM carbon-concentrating pathways

perennial: plants that live multiple years and undergo multiple seasons of reproduction

perennial life history: life history exhibited by a plant that is capable of living for multiple growing seasons and undergoing multiple rounds of sexual reproduction

perfect flower: flowers that produce both stamens and pistils

perianth: collective term for the calyx and corolla combined; floral parts surrounding the stamens and/or pistil

permeability: the degree to which dissolved materials or water are capable of crossing cell walls or membranes

peroxisome: a cellular organelle responsible for highly oxidative reactions via the enzyme catalase

petiole: the stem that attaches a leaf to the branch or axil from which it was produced

phenotype: the physical form or other observable characteristics of an organism

phenotypic characters: visible or measurable attributes of an organism

photoperiod: the length of daylight in the diurnal cycle

photorespiration: a process in plants, resulting from unfavorable $CO_2:O_2$ ratios, that results in release of some of the carbon previously captured via photosynthesis

phylogenetic grouping: a grouping based on evolutionary relationships

physical dormancy: dormancy in which the seed contains a differentiated and fully developed embryo, but the seed does not imbibe water; scarified seeds imbibe water and usually germinate within a few days

physiological dormancy: dormancy in which the seed, with a differentiated and fully developed embryo, imbibes water, but emergence of root or shoot is usually delayed by at least a month

phytochromes: protein-based pigments that are regulated primarily by absorption of red (650–680 nm wavelength) or far-red light (710–740 nm wavelength)

phytoglobins: shortened form of phyto-hemoglobins; plant proteins capable of binding oxidative molecules within the cell

phytophagous: describing an organism that feeds on living plants

piezometer: cylindrical device with a porous tip used to measure water pressure at a specific depth belowground

pistil: the central whorl of a perfect flower that includes the carpel(s) and ovules, along with the stigma(s) and style(s)

pistillate flower: flower that produces only pistils

plesiomorphy: the ancestral state of characters used in a cladistic analysis

pneumatophores: porous, woody structures associated with root systems of trees or shrubs that allow gas exchange with the atmosphere

polar molecule: molecule that possesses a slight positive charge in one or more regions and a slight negative charge in others, as a result of uneven sharing of electrons among its component atoms

pollination: the process of transferring pollen from anther to stigma

polycarpic: plants that have the potential to undergo multiple seasons of sexual reproduction

polygamous: plants that produce staminate, pistillate, and perfect flowers all on the same plant

polyploidization: process of producing gametes (and then zygotes) with more than the normal number of chromosome sets, resulting from the failure of paired chromosomes to separate during the first phase of meiosis

pome: fleshy fruit with soft outer tissue surrounding papery to cartilaginous structures that enclose the seeds

population: all individuals of a given species that co-occur within some specified space

population biology: the process of analyzing the roles of reproduction, survival, and life history on changes in populations over time

population density: the number of individuals in a population

porosity: a measure of the percentage of pore space within a volume of soil

potential evaporation: the amount of evaporation expected to occur in an area, as determined by local climatological conditions

pressure potential: a positive water potential exerted by hydrostatic pressure within plant tissues; an effect of the rigid plant cell walls

primary production: the mass of plant material produced in a given area or in a given area over a specified amount of time

primary succession: succession that takes place on a site that was previously unvegetated, at least during the typical lifetimes of long-lived plant species

productivity: the mass of plant material produced per unit of time; often given as mass per area per year to compare among ecosystem types or vegetation types

programmed cell death: a process whereby some environmentally or biotically induced cell signaling pathway becomes activated, resulting in the death of a region of cells within plant tissues

propagules: a dispersal unit of a plant that gives rise to a new plant or a new clone of a pre-existing plant

protandry: mechanism of dichogamy in which anthers mature prior to pistils

protogyny: mechanism of dichogamy in which pistils mature prior to anthers

proton gradient: an electrochemical gradient created when protons accumulate in the internal spaces of the chloroplast or mitochondrion; used to generate ATP

R

radial oxygen loss: leakage of oxygen from roots into the surrounding soil

radicle: embryonic root

ramets: interconnected modules of a clonally propagating plant

rapid shoot elongation: a mechanism in which cells within the petiole rapidly elongate, providing relatively rapid access of the shoot to atmospheric oxygen

reactive oxygen species: biochemically oxidizing atoms or molecules, with the potential to degrade or damage biological molecules

receptacle: the tip of the modified branch bearing the flower, or cluster of flowers

receptivity: a biological state of the stigma in which the processes of fertilization will occur once the pollen reaches a compatible stigma

recharge wetland: a wetland in which water moves from the wetland into the groundwater; the wetland effectively "recharges" the groundwater storage

reciprocal transplant experiment: a common garden experiment in which plants from all locations are grown in common gardens at all locations to further explore environmental effects on variation

recombined genomes: genomes comprising different mixtures of genetic information than was found in either parent

reduction (chemical): process wherein a molecule gains one or more electrons, numerically reducing its electrical charge

regeneration niche: conditions specifically associated with successful replacement of individuals within the population, especially factors that influence seed production, dispersal, germination, and seedling establishment

response surface approach: in studies of competition, an analytical approach that considers performance of both competing species at all densities simultaneously to evaluate the potential performance of each species at each density, using three-dimensional mathematical modeling

rhizome: a belowground stem that grows horizontally below the soil surface

rhizosphere: volume of soil or other rooting medium immediately surrounding and metabolically affected by the roots

riparian wetlands: wetlands associated with a stream or river, such as along the banks or in the immediately adjacent floodplain

RuBisCO: the enzyme ribulose bisphosphate carboxylase-oxygenase, used in carbon capture and/or carbohydrate production during photosynthesis

runoff: movement of water across the surface of land prior to entering wetland water storage; runoff sometimes includes both above- and belowground water movement

S

salt marsh: wetlands dominated by salt-tolerant grass species and typically occurring in tidal zones of coastal ecosystems

samara: one- or, rarely, two-seeded fruit in which an outgrowth of the ovary wall forms a wing that functions in wind dispersal of the fruit

saturated zone: zone of soil that is at its maximum water-holding capacity

scape: a leafless flowering stem

scarification: physical or chemical abrasion of the outer coat of a seed prior to germination

schizocarp: fruit derived from an ovary with two or more carpels that splits into multiple, usually one-seeded, segments (mericarps); the mericarps themselves often structurally resemble achenes, samaras, or other single-seeded fruits

schizogeny: a process of aerenchyma formation wherein cells split apart from one another to form openings that later fill with gases

seagrass beds: near-shore aquatic habitats in coastal marine systems dominated by herbaceous aquatic angiosperms, all of which belong to families in the order Alismatales (Cymodoceaceae, Hydrocharitaceae, Posidoniaceae, Ruppiaceae, and Zosteraceae)

secondary metabolites: chemicals that are not involved in primary plant metabolism (e.g., photosynthesis, respiration, growth, or development) and include two groups of carbon-rich compounds (terpenes and phenolic compounds) and nitrogen-containing compounds, such as alkaloids, as well as derivatives of these major groups

secondary succession: successional processes occur following the disturbance of a previously vegetated area

seed bank: dormant seeds in or on the soil, from which future plant species assemblages may emerge

seed predator: organism that feeds on seeds, resulting in death of the embryo within

selection: process in which the environment, in the case of natural selection, favors individuals exhibiting a particular trait or acts against individuals carrying some alternative version of the same trait

self-pollination: fertilization of ovules of one plant by pollen from the same plant

semelparous: plants (or other organisms) that undergo only one season of sexual reproduction

senescence: systematic, controlled death of plant tissues; often discussed in the context of perennial plants where one set of leaves or shoots die back as a newer set is produced

sequential hermaphroditism: individual plants initially produce either staminate or pistillate flowers only (usually staminate), then switch to the other flower type; the switch sometimes is gradual, with both staminate and pistillate co-occurring briefly during the transition

sexual selection: process in which potential mates favor a particular trait during mate choice; in plants, this often will act through pollination vectors selecting in favor of or against certain floral traits

silique: dehiscent fruit formed from an ovary having two carpels; the two halves of the fruit open outwards, away from a central partition between the carpels

sister groups or sister taxa: two clades within an evolutionary tree that share a common ancestor

soil-plant-air continuum: a continuum of declining water potential from the soil or water, through the plant, and into the atmosphere

soil water: water lying within the unsaturated, usually upper, layers of the soil

solute potential: negative pressure potential induced by dissolved substances in water

speciation: the formation of new species

specific heat: amount of heat required to increase one gram of water, or other substance, by 1 °C

sporangia: spore-producing organs

stage: the depth of water within a wetland; will consider subsurface water level, when possible, as a negative stage height

stage-based model: population growth model that considers the contributions of each growth stage of the individuals within the population, rather than treating the population as a homogeneous entity

stamens: the pollen-producing whorl of structures in a flower; each stamen comprises a filament and an anther

staminate flower: flower that produces only stamens

stele: the vascular tissues (xylem and phloem) and surrounding pericycle layer or other supporting tissues

stemflow: precipitation that is intercepted by vegetation and enters the wetland water storage by flowing down plant stems

stolon: an aboveground stem that grows horizontally along the soil surface

storage: in hydrology, refers to the net amount of water remaining within a wetland after considering the dynamic inflow and outflow processes

stratification: process of storing seeds in or on a moist substrate at temperatures similar to those encountered in nature after dispersal but prior to the typical time of germination

streamflow: water flowing within a stream channel; stream discharge

stress: physiological condition in which an individual's growth or survival are negatively impacted, as a result of experiencing conditions (stressors) in the environment outside of the range to which the individual is adapted

stressor: factor external to the plant that restricts the rate of biomass production

stroma: the fluid inside the chloroplast but outside the thylakoid membrane

suberin: a complex of fatty acids that is deposited in cell walls of roots (and other tissues) that aids in protection against water loss and pathogen entry

suberized: the state of tissues having been impregnated with suberin

submersed plant: plant that is rooted in the soil or sediment and whose leaves and, sometimes, reproductive parts usually lie below the surface of the water

subsidence: loss in elevation of the wetland soil or sediment surface, resulting from settling of sediments over time.

substitutable resource: a resource that could be replaced by some other resource if its concentrations fell below carrying capacity concentration for the species

substrates: molecules on which an enzyme exerts its activity

succession: the change in plant species composition of an area over time

swamp: (usually) wetlands dominated by trees, or sometimes shrubs; may occur in or around oxbow wetlands where sufficient time has passed for development of a forest canopy; note that the term "swamp" may refer to herbaceous-dominated wetlands in some regions (e.g., European reedswamps)

symbiosis: an intimate biological relationship between individuals of two species

symplast: the region of the plant within the interconnected cellular cytoplasm

synapomorphies: apomorphies that are shared among taxa within a clade in an evolutionary tree

syntrophy: a type of mutually beneficial metabolism in which one organism utilizes products of another organism's metabolism as substrates for its own metabolism and vice versa

T

tepals: outermost whorls of indistinguishable sepals and petals in some flowers

terminal electron acceptor: final molecule at the end of a chain of electron transfers during cellular respiration

thermo-osmotic convective airflows: movement of gases within a plant driven by a combination of heat-induced pressurization and osmotic gradients across membranes

throughfall: precipitation that falls through the plant canopy and enters the wetland water storage

thylakoid: structures inside the chloroplasts that are formed from stacks of membrane, each enclosing a volume of cytoplasm in a hollow internal compartment referred to as a lumen

topology: in cladistic analyses, the general branching pattern of a clade or tree

totipotent: the ability of plant cells, under appropriate conditions, to dedifferentiate and reinitiate differentiation into other cell and tissue types

transition probabilities: values, in a stage-based model, of the probability that an individual in one stage will move into another stage during the next time interval of the model

transmit: to allow the penetration of something such as light or water through a material

transpiration: movement of water through a plant into the atmosphere, via xylem, intercellular spaces, and stomata

tristyly: production of three different floral morphs, such as having short, medium, and long styles and anther filaments

tundra: high-latitude or high-elevation ecosystem characterized by cold temperatures, low water availability, short growing season, and a dominance by mosses, lichen, herbaceous vegetation, or dwarf tree species

turgor: the water pressure exerted within cells that maintains cell shape and prevents wilting in the plant

U

unisexual flower: flowers that produce only stamens or pistils

unreduced gametes: gametes that carry multiple sets of chromosomes; successful union of unreduced gametes results in polyploid offspring

utricle: single-seeded fruit with a thin pericarp that sits loosely around the seed

UV-induced photodegradation: physical degradation of organic matter resulting from exposure to high-energy UV radiation

V

vacuoles: large, central organelles that make up most of the interior volume of plant cells; used to store chemical intermediates for critical metabolic processes, as well as some cellular toxins

W

water budget: a summary of the addition of water to and loss of water from a site, such as a wetland; these may be very detailed or may focus on only major inflows and outflows

water potential: a measure of the free energy of a sample of water, relative to pure water; gives an indication as to the tendency for that water to move along a gradient, from high to low water potential

water storage: the amount of water within a wetland, or other site; should include both above- and below-ground water

water table: the interface between soil that is fully saturated with water and soil that remains unsaturated

well: a device that allows access to subsurface water for collection of water samples or measurement of characteristics of water at a given point

wetland indicator status: an indicator rating assigned to a plant species, based on the estimated likelihood that a species would occur in a wetland versus a non-wetland habitat

wetland plants: plants adapted to live in areas where the soil is saturated or inundated for a biologically significant portion of the year

wet meadow: palustrine wetlands occurring where a shallower depth or shorter duration of flooding allows establishment of a greater diversity of plant species, including many species of grasses, sedges, and rushes

whorls: concentric rings of plant parts, such as sepals, petals, and stamens

X–Z

xenogamy: pollination that results in genetic outcrossing

zoochory: dispersal of fruit or seed by animals

Index

Note: Page numbers in *italics* indicate a figure and page numbers in **bold** indicate a table on the corresponding page.

Common names

A

American water lotus, *see Nelumbo*
ash, *see Fraxinus*

B

balsam poplar, *see Populus balsamifera*
beech, *see Fagus*
birch, *see Betula*
bladderwort, *see* Lentibulariaceae; *Utricularia*
bulrush, *see Schoenoplectus; Scirpus*
buttercups, *see* Ranunculaceae; *Ranunculus*
buttonwood, *see Conocarpus*

C

cattail, *see Typha*
common reed, *see Phragmites australis*
cordgrass, *see Spartina alterniflora*
cypress, *see Taxodium*

D

dogwood, *see Cornus*
duckweeds, *see* Araceae; *Lemna; Spirodela;*
    Wolffia; Wolffiella

E

elm, *see Ulmus*

F

fir, *see Abies*

G

grass, *see* Poaceae

H

hornwort, *see* Ceratophyllaceae;
    Ceratophyllum

L

larch, *see Larix*
loosestrife, *see* Lythraceae

M

maple, *see Acer*
milfoil, *see Myriophyllum*
muskgrass, *see Chara*

P

parrot feather, *see Myriophyllum aquaticum*
pine, *see Pinus*

pitcher plant, *see* Sarraceniaceae; *Sarracenia*
pondweed, *see* Potamogetonaceae;
    Potamogeton

R

reed canary grass, *see Phalaris arundinacea*
rush, *see* Juncaceae; *Juncus*

S

sedge, *see* Cyperaceae; *Carex*
spruce, *see Picea*
sundew, *see Drosera*

T

tamarack, *see Larix laricina*
tupelo, *see Nyssa*
turtle grass, *see Thalassia*

V

Venus' flytrap, *see Dionaea*

W

water celery, *see Vallisneria americana*
water chestnut, *see Trapa*
water cress, *see Rorippa*
water hyacinth, *see Eichhornia; Pontederia*
water lettuce, *see Pistia stratiotes*
water lily, *see* Nymphaeaceae; *Nymphaea*
water lotus, *see Nelumbo*
water meal, *see Wolffia*
water pennywort, *see Hydrocotyle*
water plantains, *see* Alismataceae
water primrose, *see Ludwigia*
water shield, *see Brasenia*
water smartweed, *see Persicaria;*
    Polygonum
water starwort, *see* Callitrichaceae
willow, *see Salix*

Scientific names

A

Abies balsamea, 35
Abies fraseri, 126
Abies lasiocarpa, 35
Acanthus ebracteatus, **38**
Acanthus ilicifolius, **38**
Acentria ephemerella, 328
Acer, 92, **241**
Acer rubrum, 108, 242, 353
Achillea lanulosa, 279, 280, 281, 282
Achillea millefolium, 279, 280
Aconitum, **24**
Acrostichum aureum, **38**
Acrostichum danaeifolium, **38**
Acrostichum speciosum, **38**

Aegialitis annulata, **38**
Aegialitis rotundifolia, **38**
Aegiceras corniculatum, **38**
Aegiceras floridum, **38**
Aeschynomene, 198, **241**
Agasicles hygrophila, 328
Aglaia cucullata, 37
Aglaophyton major, 201, 202
Alcaligenes, 190
Aldrovanda, 203, **203**, 208, 252, 24
Aldrovanda vesiculosa, 203, 208
Alisma, 24, 55
Alisma plantago-aquatica, 333
Alismataceae, 20, 24, 41, 55, 67, 91, 151, 155,
    173, 174, 241
Alismatales, xiii, 12, 19, 20, 24, 41, 45, 47,
    51, 55, 60, 91, 93, 151, 152, 160, 175, 203,
    224, 370
Alnus, 61, 74, 112, 156, 200, 202, 203, 210,
    239, 346
Alnus incana, 200
Alstonia, 71
Alternanthera, 254, 328
Alternanthera philoxeroides, 15, 253, 254, 328,
    332, 350
Alternaria, 331
Althenia, 41, **225**, 234
Althenia bilocularis, 234
Amphibolis, 41, **225**, 234
Amynothrips andersoni, 328
Anabaena, 191, 197, 198
Anabaena azollae, 30
Andropogon, **24**, 173
Anemone, **24**
Anemopsis, **24**
Anser, 323
Anthoceros, 198
Anthoceros punctatus, 198
Anthoxanthum odoratum, 261
Aphanizomenon, 191
Apios, 52, 198
Aponogeton, **24**
Aponogetonaceae, **24**, 45, 241
Appertiella, **225**, 231
Arabidopsis thaliana, 178
Araceae, 24, 41, 49, 67, 71, 91, **241**
Armoracia, 253
Aster, **24**, 44, 91
Asteraceae, 24, 44, 52, 67, **68**, **69**, 91, 93, 146,
    151, **241**, 274, 279, 361
Asterales, **24**, **151**, 152
Austropuccinia psidii, 332, 335
Avicennia, 92, 158, 246
Avicennia alba, **37**
Avicennia bicolor, **37**
Avicennia germinans, 46, 56, 247, 359
Avicennia integra, 37
Avicennia marina, **37**, 157, 158, 178, 180
Avicennia officinalis, **37**
Avicennia rumphiana, **37**
Avicennia schaueriana, **37**

Subject Index